Some Strangeness in the Proportion

A Centennial Symposium to Celebrate the Achievements of

Albert Einstein

Donors

American Can Company
American Hoechst Corporation
Aspen Institute for Humanistic Studies
Bell Laboratories
Claire and Albert Arenberg Fund
Geraldine R. Dodge Foundation, Inc.
Educational Testing Service
EMR Photoelectric
Exxon Corporation
Tibor and Dorothy Fabian
Federal Ministry of Education and Science, Federal Republic of Germany
The Ford Foundation
Benjamin Gittlin Foundation
Sibyl and William T. Golden Foundation
Goodall Rubber Company Charitable and Welfare Foundation
Frederick A. Hauck
Joseph H. Hazen Foundation, Inc.
The William and Flora Hewlett Foundation
The Lillia Babbitt Hyde Foundation
IBM
Brunson S. McCutchen
McDonnell Foundation, Inc.
The Andrew W. Mellon Foundation
The Ambrose Monell Foundation
National Endowment for the Humanities
National Science Foundation
Max Planck Institut
Princeton Applied Research Corporation
The Scherman Foundation, Inc.
Alfred P. Sloan Foundation
Supermarkets General Corporation
Fritz Thyssen Stiftung
The Raymond John Wean Foundation

Honorary Sponsors

American Academy of Arts and Sciences
American Council of Learned Societies
American Institute of Physics
American Philosophical Society
American Physical Society
International Council of Scientific Unions
International Union for the History and Philosophy of Science
International Union of Pure and Applied Physics
Israel Academy of Science
Princeton University
The Smithsonian Institution

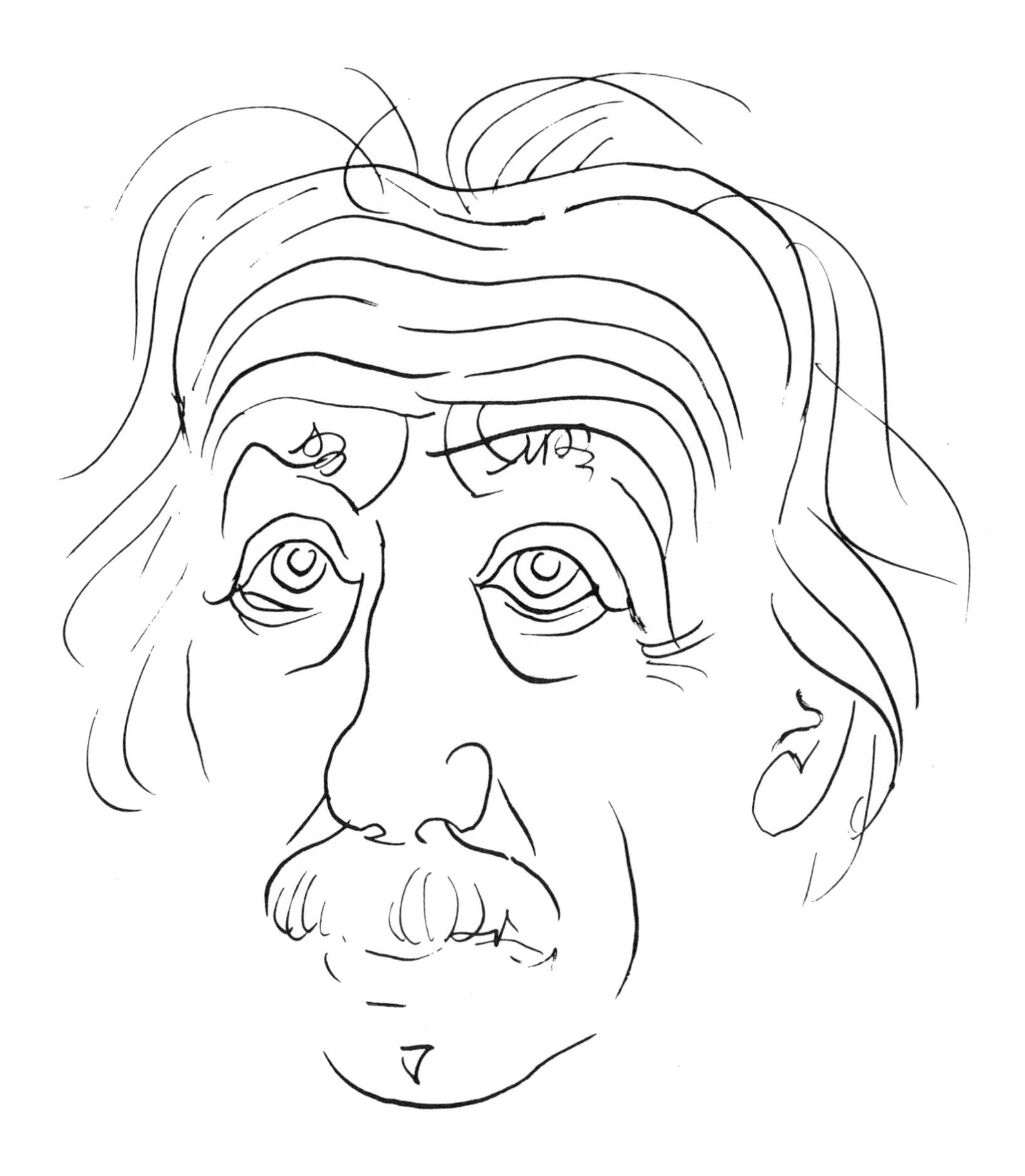

Drawing by Josef Scharl
(Courtesy of The Institute for Advanced Study, Princeton, New Jersey)

There is no excellent beauty that hath not
some strangeness in the proportion.

—Francis Bacon

F. Bacon, "On Beauty," *The Essays of Francis Bacon,* (edited by M. A. Scott, New York, Charles Scribners' Sons, 1908), p. 198.

Some Strangeness in the Proportion

A Centennial Symposium to Celebrate the Achievements of

Albert Einstein

Edited by

Harry Woolf

The Institute for Advanced Study, Princeton, New Jersey

1980

Addison-Wesley Publishing Company, Inc.
Advanced Book Program
Reading, Massachusetts

London • Amsterdam • Don Mills, Ontario • Sydney • Tokyo

Cover page illustration

Einstein Memorial Sculpture, *Arrival*, by Jacques Lipschitz
Gift of Joseph H. Hazen to The Institute for Advanced Study
to commemorate the centennial of the birth of Albert Einstein

Library of Congress Cataloging in Publication Data

Main entry under title:

Some strangeness in the proportion.

1. Einstein, Albert, 1879–1955—Congresses.
2. Relativity (Physics)—Congresses. 3. Quantum theory—Congresses. 4. Astrophysics—Congresses.
I. Einstein, Albert, 1879–1955. II. Woolf, Harry.
QC16.E5S63 530.1 80-20111

ISBN 0-201-09924-1

Published simultaneously in Canada.

Manufactured in the United States of America

Contents

Preface

To celebrate the achievements of Albert Einstein by observing, in his American home, the Centennial of his birth, March 14, 1879, was a responsibility most willingly assumed by The Institute for Advanced Study. Here he lived and worked from 1933 to 1955, the first and foremost member of the Faculty. In his spirit, like Chaucer's clerk, we continue to "gladly . . . learn and gladly teach," so that in celebrating Einstein's accomplishments we also celebrate our shared values: the continuing quest for fundamental knowledge, the search for peace and justice, and the rejection of those very conceits and distempers which such pursuits sometimes engender.

So many helped to bring this superb symposium into being, to manage its daily evolution so well, to facilitate its transmission to a wider world so effectively as to render a conventional expression of thanks insufficient for gratitude deeply felt and forever unforgotten. We cannot name, here, each and every one whose work and spirit brought it all about and infused the entire enterprise with enthusiasm and harmony. But there is one whose unseen hand kept our ship on course and brought it safely in, and I should like to acknowledge here our enormous indebtedness to my friend and associate, John Hunt, not only for service beyond duty, but for taste and judgment commensurate with the subject of our celebration.

Of the Institute staff special praise is due Mary Wisnovsky (ably assisted by Gary Farmer), peerless coordinator of the conference, whose energy and enthusiasm matched her multiple talents. Aida La Brutte and Judith Grisham kept our communications system functioning in an efficient and orderly way, and Valerie Nowak turned tapes into typescript with speed and clarity. We are especially grateful for the high professional quality which Michael Blackwood and Mead Hunt of Blackwood Productions brought to the task of filming the symposium, and the same respect and gratitude are due Emmet Gowin and Allen Hess for superb photographs, some of which appear in this volume. Mr. Herbert Bailey and Bruce Campbell, along with Jim Wageman, cooperated freely in matters of printing and design for the symposium. William Turnbull and the Educational Testing Service helped us with housing and other services. The impeccable musical taste of Edward Cone complemented the extraordinary performance of the Juilliard and Emerson Quartets. In this our intuition infers that Einstein would have approved!

Freeman Dyson, Herman Feshbach, Marvin Goldberger, Gerald Holton, Martin Klein, Abraham Pais, John Wheeler, and occasionally Heinz Pagels and Joseph Hazen joined me in planning the substance of the symposium. Without their concern and cooperation its achievements would not have been possible. My research assistant, David Cahan, proved to be much more than that; I wish to express my thanks here for his help.

My younger children, Sara and Aaron, were privileged to have shared in the occasion, to touch and be touched by a tradition that, through the life of the mind, brings all mankind together. The human and intellectual framework of these events was enhanced by the labor and love which my wife gave to the entire endeavor. Thanking Pat at the end of this prefatory statement brings it to a close in a familiar way, but it does not diminish the force of the idea that the last shall not be least.

HARRY WOOLF

The Institute for Advanced Study

Contributors/Discussants	Contribution pp.	Discussion pp.
Stephen L. Adler *The Institute for Advanced Study*		494-495, 503
E. Amaldi *Università degli Studi, Rome*		284-287
Valentine Bargmann *Princeton University*		284, 480-481, 486, 487, 488
Peter Bergmann *Syracuse University*		156, 286, 456, 478-480, 481
Hans Bethe *Cornell University*		475-489, 493-506
Brendan Byrne *Governor, State of New Jersey*	509-510	
Jimmy Carter *President, U.S.A.*	510	
Shiing-shen Chern *University of California, Berkeley*	271-280	284, 285, 286
L. Cooper *Brown University*		385
Gerard H. de Vaucouleurs *University of Texas at Austin*		337
R. Dicke *Princeton University*		142-144, 156-158
P. A. M. Dirac *Florida State University*		92, 110, 381-386, 423, 471
Freeman J. Dyson *The Institute for Advanced Study*	376-380	498-500, 502
J. L. Ehlers *Max Planck Institut, Munich*		193, 306, 455

Contributors/Discussants	Contribution pp.	Discussion pp.
George B. Field *Harvard University*	308-330	306, 336, 384
Felix Gilbert *The Institute for Advanced Study*	13-27	
Marvin L. Goldberger *California Institute of Technology*		493-506, 509
Maurice Goldhaber *Brookhaven National Laboratory*		109-110
F. Gürsey *Yale University*		457
Stephen W. Hawking *University of Cambridge*	145-152	157-158, 336
A. Hermann *University of Stuttgart*		110, 195, 267, 488
Banesh Hoffmann *Queens College*		111, 266, 475-489
Gerald Holton *Harvard University*	49-65	110
Res Jost *Eidgenössische Technische Hochschule*	252-265	193, 195
Martin J. Klein *Yale University*	161-185	192, 193, 195
Thomas S. Kuhn *Massachusetts Institute of Technology*	186-191	194
B. Kursunoglu *University of Miami*		469
D. Laughton *Princeton University*		385

Contributors/Discussants	Contribution pp.	Discussion pp.
Robert Mann *The Juilliard School of Music*	526-527	
Arthur I. Miller *Harvard University*	66-91	110
Charles W. Misner *University of Maryland, College Park*		143
Ernest Nagel *Columbia University*	38-45	
Yuval Ne'eman *Tel-Aviv University*	429-443	156-157, 192, 267, 284, 306, 336, 383, 385, 453-455, 457, 473, 505, 506
J. Ostriker *Princeton University*		384
H. R. Pagels *The Rockefeller University*		424
Abraham Pais *The Rockefeller University*	175-251	266, 267, 268, 505-506
Wolfgang K. H. Panofsky *Stanford Linear Accelerator Center*	94-105	
Phillip J. E. Peebles *Princeton University*	302-305	
R. Penrose *Mathematics Institute, Oxford*		284, 453
Frank Press *Science Advisor to the President of the U.S.A.*	511-517	
Edward M. Purcell *Harvard University*	106-108	

Contributors/Discussants	Contribution pp.	Discussion pp.
I. I. Rabi *Columbia University*		471-472, 485-486, 487, 504
N. Razinsky *Fusion Energy Foundation*		194, 195
Martin J. Rees *University of Cambridge*	291-301	306-307, 336
Tullio Regge *The Institute for Advanced Study*	281-283	
W. Rindler *University of Texas, Dallas*		109
Ilse Rosenthal-Schneider *Sydney, Australia*	521-523	
Wallace L. W. Sargent *California Institute of Technology*	331-335	
P. A. Schilpp *Southern Illinois University, Carbondale*		486
Julian Schwinger *University of California, Los Angeles*		192-196, 266-268
Dennis W. Sciama *Universities of Oxford* and *Texas, Austin*		142, 158, 306-307, 336-337, 423-425
E. Segrè *University of California, Berkeley*		336-381, 504-505
R. Sexl *University of Vienna*		109
Irwin I. Shapiro *Massachusetts Institute of Technology*	115-136	157
A. G. Shenstone *Princeton University*		488

Contributors/Discussants	Contribution pp.	Discussion pp.
H. D. Smythe *Princeton University*		193
John Stachel *The Institute for Advanced Study* and *Boston University*		157, 196, 266, 287, 423-424, 486-487
Ernst G. Straus *University of California, Los Angeles*		481-485, 486, 487, 488
J. H. Taylor, Jr. *University of Massachusetts, Amherst*		142
S. Treiman *The Institute for Advanced Study*		336-337
George E. Uhlenbeck *The Rockefeller University*	524-525	
W. G. Unruh *University of British Columbia*	153-155	158, 423
L. van Hove *CERN*		109
Peter van Nieuwenhuizen *State University of New York at Stony Brook*	444-452	
Steven Weinberg *Harvard University*		495-497, 503, 506
V. F. Weisskopf *Massachusetts Institute of Technology*		453-457, 469-472
John Archibald Wheeler *University of Texas at Austin*	341-375	381, 382, 383, 384, 385-386
E. Wigner *Princeton University*	461-468	194, 195-196, 285-286, 381, 382, 383, 457, 470, 471, 472, 488-489

ALBUM

Opening Ceremony
left to right: Philip Handler; Tullio Regge; Jürgen Schmude, Federal Minister of Education and Science, Federal Republic of Gemany; Howard C. Petersen; Joseph Hazen; J. Richardson Dilworth; Harry Woolf

Presentation of the gift of the Federal Republic of Germany to The Institute for Advanced Study
left to right: Harry Woolf; Jürgen Schmude, Federal Minister of Education and Science, Federal Republic of Germany

Plenary Session

Tullio Regge

Harry Woolf; Edward M. Purcell

Herman Feshbach; Alan S. Lapedes; Stephen W. Hawking; Mrs. Selma Lapedes; Eugene Wigner

Panel: Einstein and the Physics of the Future
left to right: Stephen Adler; Steven Weinberg; Marvin L. Goldberger; Freeman J. Dyson; C. N. Yang

Yuval Ne'eman; Peter van Nieuwenhuizen; Eugene Wigner

left to right: John A. Wheeler; Claudio Teitelboim, William O. Baker

Stephen Adler; Steven Weinberg

Enrico Bombieri; Shiing-shen Chern

P. A. M. Dirac; Dennis W. Sciama

Panel: Working with Einstein
left to right: Peter G. Bergmann; Ernst G. Straus; Banesh Hoffmann; Valentine Bargmann

Marvin L. Goldberger; Freeman J. Dyson

Abraham Pais; Martin J. Klein

left to right:
foreground: Scott McVay; Ambassador Berndt von Staden; Reimar Lust; I. I. Rabi
back to camera: Jürgen Schmude
2nd row: James Wolfensohn and Mrs. Wolfensohn; Philip Handler
background: Enrico Bombieri; Trude Goldhaber

I. Prologue

PROLOGUE

In transforming days of discussion and sounds of celebration into pages of text, I have tried to edit out the partial phrases, the redundancies, and the excessive or irrelevant soliloquies. Nevertheless, an international gathering of distinguished scientists, scholars, and friends of learning, young and old, heroes and antiheroes, with educational backgrounds and cultural origins almost as diverse as the human race, created special conditions for communication. The English common to our exchange was shaped by its users: Italianated, Indianized, sieved through the Low Countries, stressed Germanically, pressed Britannically, haunted by Hungarian, antiphonally Oriental, and loosened by American. My aim has been to preserve those elements of personal style, of language at its growing tip; and so this flexible, fluid aspect of our exchange remains, at times at the expense of correct grammar and the rules of syntax.

As a proper measure of the moment and with respect for the historical record, I have set as a standard for the occasion, the ancient admonition of German historicism to present the story *wie es eigentlich gewesen ist.* In this respect, the reader, perhaps, should also be reminded that the eye is not the ear, the past not the present, that tonalities of discourse, pregnant pauses, empathetic moods, and ironic inflections do not transmit fully via the printed page. In brief, as at the battle of Agincourt, those who were not there may find meaning in the words that Shakespeare gave King Henry V on the eve of that event:

> We few, we happy few, we band of brothers;
> For he today that sheds his blood with me
> Shall be my brother; be he ne'er so vile,
> This day shall gentle his condition;
> And gentlemen in England, now a-bed,
> Shall think themselves accurs'd they were not here . . .

and take comfort as well as learning from this volume, which aspires to share this extraordinary experience with the republic of letters.

The happy few who gathered at the Institute for Advanced Study for four days in March 1979 to memorialize a great man were indeed, like the archers and knights at Agincourt, a band of brothers, their conditions gentled by their cause. The papers presented and the discussions that

Harry Woolf (ed.), Some Strangeness in the Proportion: A Centennial Symposium to Celebrate the Achievements of Albert Einstein
ISBN 0-201-09924-1

followed upon them, as I believe the contents of this volume demonstrate, brought out the best in all of us, as in homage to Einstein everyone reached beyond his grasp. Thus the entire symposium was infused with a spiritual beauty, a richness of mind, and a warmth in social exchanges that reflected the man it honored. In that fusion of science and human values it raised its participants to levels of life not normal to the daily routine and gave our ceremony of celebration an unexpected, redemptive role, a renewal of purpose for ourselves and our institutions.

As chairman of the board of trustees of the Institute for Advanced Study, Mr. Howard Petersen formally initiated the celebration with the following welcoming statement:

The one-hundredth anniversary of Albert Einstein's birth is a special occasion for all of us and particularly for those associated with the Institute. His is a permanent presence here; our history is mingled with his, and we are indebted to him in countless ways.

It is most fitting, it seems to me, that the Einstein Centennial should open a period of commemoration at the Institute. Next year, 1980, will mark our fiftieth anniversary, and as we honor Einstein tonight, we will in the ensuing months recall Kurt Gödel, Marston Morse, Hermann Weyl, John von Neumann, Erwin Panofsky, Millard Meiss, Oswald Veblen, J. Robert Oppenheimer — these and all the others who were Einstein's colleagues and who did so much to create this unique institution.

Some of you no doubt have been here as members in past years. In talking to some people before dinner, I learned that there was a quality of reunion to this meeting here tonight which lends a unique and important warmth to it. Many of you have undoubtedly had colleagues who have been in residence at the Institute at some time in their careers. You know what the Institute has meant, and continues to mean, to the world of scholarship and research. Our present concern here is to do everything possible to see that the next half century will be as flourishing a period for the Institute as that which lies behind us.

I am sure that a scholarly history of the Institute will one day be written, and I trust and I would be confident that among the very special moments there recorded will be the Einstein Centennial program. It certainly is a very proud moment for me, and I trust will be a memorable occasion for all of you. Thank you, indeed, for joining us this week. We have done everything possible we can see to make this an appropriate celebration, and now the most important part of it is in your hands.

One whose role over the years at the Institute has been of the greatest significance is J. Richardson Dilworth. His wisdom, his patience, and unfailing support have been exemplary. As senior financial adviser to the Rockefeller family, trustee of Yale and Rockefeller universities and the Metropolitan Museum of Art, in addition to numerous directorships, his responsibilities — not to say his burdens — have been immense. Yet no one could possibly have been more generous with his time and counsel. Called upon to acknowledge the support and generosity of Mr. Joseph H. Hazen on behalf of the Einstein Centennial, he said:

Miss Einstein, Miss Dukas, distinguished guests, ladies and gentlemen:

It was almost exactly two years ago to the day that the donor of the Einstein memorial sculpture with generosity matched only by his foresight and his sense of history pledged the funds required to launch our present symposium. He referred to this initial gift as seed money, and I imagine that tonight, as he contemplates this remarkable commemorative gathering, even he — whatever his initial expectations — must be somewhat surprised and, I hope, pleased.

Having provided the initial support for the symposium, his interest in Albert Einstein and the Institute's centennial celebration took still another form, and some months after his initial gift he expressed interest in a possible Einstein memorial sculpture for the Institute. His interest became a reality when he donated from his personal collection a work of art created in 1941 by the sculptor Jacques Lipchitz, entitled *Arrival.*

Last January, he wrote these lines to the director: "*Arrival* departed this morning and with it went my emotional attachment to it. Since Lipchitz was a personal friend of long standing, I shared the terror of his *Flight* from the Nazis and his jubiliation of his *Arrival* in America."

The letter was signed by the donor, Mr. Joseph H. Hazen, who in the course of a long legal career with the film industry became a prominent and discriminating collector of late nineteenth- and twentieth-century art.

As of this evening, *Arrival* can be seen in the Institute's reading room at the rear of this dining hall. It will serve as a perpetual reminder of the larger meaning of the Einstein Centennial.

On behalf of the trustees, the members, the staff, and friends which comprise our Institute community, it is my privilege to express our deepest gratitude to Mr. Hazen for his meaningful, splendid gift.

Among the special guests at the opening was Dr. Jürgen Schmude, Federal Minister of Education and Science of the Federal Republic of Germany. Born in 1936, Dr. Schmude was educated at the universities of Göttingen, Berlin, Cologne, and Bonn. He was awarded a doctorate in jurisprudence by Bonn University.

Dr. Schmude has participated in the political life of the Federal Republic on the local and national levels, as a member of the Social Democratic party. A member of the Bundestag since 1969, he has been Secretary of State to the Federal Minister of the Interior, Chairman of the Foreign and Security Policy Working Group of the Social Democratic Council in the Bundestag, and since early 1978, Federal Minister of Education and Science. Dr. Schmude's presence bore a special significance for the Institute best revealed in his own words:

Mr. Chairman of the Board, ladies and gentlemen: In a few days from now, Albert Einstein would have been one hundred years old. As always on such occasions, many Einstein orations will be made, many Einstein essays written, and many Einstein books will be published. For you assembled today, this Einstein Centennial is an occasion for scientific work. By holding a symposium dealing with the specific aspects of his scientific work, you have chosen to pay homage to Albert Einstein in a way that is singularly apt, spreading the influence of his work even further afield so that it may bring increasing benefit to mankind. This promises to be a very worthwhile undertaking, and I wish you every success.

Although I cannot contribute to the scientific side of your discussion, I am grateful for the opportunity to speak here as the representative of the Federal Republic of Germany at this symposium and to announce officially to the Institute for Advanced Study the endowment of a chair by the federal government. This endowment will finance a chair and two fellowships for a period of five years to be devoted to research in the scientific work and career of Albert Einstein.

The idea for this endowment was put to the German federal government by the principal German scientific institutions and in particular by the president of the Max Planck Society, Professor Reimar Lüst. Our wish is to provide a new generation of scientists with the opportunity to do research work in Einstein's field and interest at Princeton as well as in his manner. It is an expression of our respect for Albert Einstein. Albert Einstein was forced to leave his native country, Germany, because under the regime of the National Socialists it could no longer be home to him and because Germany itself did not wish to count him among its sons any longer. Here in this country and in this city he found a new home. Albert Einstein did not return to Germany after the war. He had no desire to do so. During the last years of his life he stated frankly that he was not willing to reconcile himself with the country of his past. That is why he repeatedly declined honors which German institutions wished to confer on him. His old friends in Germany could not persuade him to change his mind. We do not know whether the consolidation of democracy in the Federal Republic of Germany and later the election of another emigrant, Willy Brandt, to the chancellorship would have made any difference. Einstein did not live to see that.

The endowment is not an attempt to claim Einstein for Germany posthumously. It manifests our respect for him, our respect also for his negative attitude towards Germany. It is of course a great pity that it was not possible to reconcile Einstein with present-day Germany. With this endowment we wish to make a contribution to the international appreciation of Albert Einstein, the acclaimed international scientist, and Albert Einstein, the man. As from 1933 vast numbers of people from Germany and other European countries in the grip of National Socialism left their country and came to the United States. They were admitted not as guests but as refugees and expellees in search for a new home. In these hard and bitter years Einstein untiringly helped other immigrants. For that we remember him with particular gratitude on the occasion of his centennial. Many of those immigrants and expellees were scientists. As citizens of their new homeland they have served science and recorded great achievements. Many of them have become catalizers of scientific cooperation with the new and democratic Germany.

I have come to Princeton to thank the United States for having given those people a chance to find a new home. The endowment I am announcing today is meant as a gesture of gratitude to the scientific community of this country, which enabled them to continue their work here, and I thank in particular the Institute for Advanced Study, where Einstein worked until his death.

I am not a scientist but a policymaker with a legal background. Therefore, you will not be surprised to learn that I am fascinated most of all by Einstein the philosopher, pacifist, and nonconformist. When Einstein died in 1955, Thomas Mann, who was a kind of spiritual leader of the immigrants, the representative of the other, nobler Germany said, "May his scientific genius, of which the nonscientist can have no more than a vague notion, be praised by those qualified to do so. What I love, admire, and will always cherish is the marvelous strength with which he stood up for his convictions with great courage, always devoted to the concept of humanity and superior to all conformism. With Albert Einstein died a vindicator of the honor of mankind whose name will never be forgotten."

Einstein never evaded the social and ethical issues of his time. Instead of praising scientific progress, Einstein asked why it had brought so little to mankind. In times of war, he said, scientific progress has enabled men to mutilate each other even more effectively, and in times of peace it had made them slaves of the machine. Scientific and technological development has meanwhile advanced further. The question Einstein posed in 1931 is even more difficult to answer today. Many people are losing hope that science will mean progress towards a brighter future. They have recognized that there are serious problems involved in the application of research findings, and they are at the same time bewildered by the impenetrable complexity of these problems. Above all, the unpredictability of the consequences of the scientific findings makes more and more people skeptical about science and technology. It is possible that Europeans who live in such a densely populated continent feel science and technology to be an even greater threat than people living in other parts of the world. The increasing concern and distrust in the Federal Republic of Germany finds expression in a growing number of citizens' pressure groups. They protest against the peaceful utilization of nuclear energy and against new industrial settlements, for instance. Many members of political parties, including members of Parliament, share the concern about the consequences of technological progress.

The gap between scientific knowledge and political responsibility threatens to grow wider, above all, when the substance of scientific findings is no longer intelligible for the layman, and in this sense policymakers are laymen, too. The problem of the responsibility of scientists for the consequences of science itself cannot be solved merely by ensuring a better rapport between scientists and policymakers. They have deliberated behind closed doors too long. The exclusiveness has, if anything, increased rather than allayed public mistrust. In the Federal Republic of Germany, we are seeking new platforms for discussion and dialogue with the public in order to achieve consensus on which the necessary decisions can be based. The credibility of our democratic system will have been proved once we have succeeded in solving the problems inherent in science and technology by democratic means. If we fail to translate these problems into language which a nonscientist can understand, we will have failed altogether. This is a challenge which the scientific community too must face up to if it is to live up to its responsibility towards society.

Policymakers in a democracy are accustomed to having to explain and defend their actions and often to having to work hard to win public confidence. Scientists are increasingly faced with the same necessity. They have not only to account for the use of the ever-growing funds raised by society for research and for new developments, but moreover to assume responsibility for the future which is shaped also by the consequences of new scientific findings. It will not be sufficient for the scientist to devote all his efforts to scientific work simply to achieve significant results. It is equally important for him to possess the strength and patience to explain these findings to the public, including the consequences for society. For the scientist this is something entirely new, something that contrasts sharply with the traditional view of his own role. But only this dialogue will enable him to obtain a realistic picture of the needs of the community at large and to derive from that picture new ideas for his scientific work while trying to win public confidence.

The social commitment and responsibility of the scientists features even more prominently in public debate on science policy in Germany; both within and outside the universities we need researchers who conceive their task as an opportunity to take up of their own accord the concerns and expectations of the citizens. This also includes the need to ponder, to investigate, and to explain the social consequences of science.

Through his work and actions, Albert Einstein showed that he perceived his specific social responsibility as a scientist. His endeavors with other scientists and the pursuit of peace and human living conditions are a lasting commitment for us. Dr. Woolf, I have the honor and pleasure to present to you the deed of endowment for an Albert Einstein Research Chair. This endowment provides for a guest professorship and two fellowships of your choice for a period of five years at the Institute for Advanced Study. I hope that in this way, even without a commitment as to nationality, another page will be turned in the book of relations between our two countries for the benefit of science, but also for the benefit of the people all over the globe. Thank you.

In expressing our gratitude to Dr. Schmude for this most appropriate and generous gift to the Institute in memory of Albert Einstein, I should like to lay special emphasis on the unique tribute paid to the ideal of academic freedom that the gift and its conditions embody. The Institute owes its greatness in very large part to the freedom and flexibility that have characterized it from the beginning, and this superb Einstein Centennial contribution not only reinforces that tradition, but sets an example for governmental support from other countries as well. Let me say too a word of special thanks to our academic colleagues in Germany, in particular, Reimar Lüst, director of the Max Planck Institut and many others who were so helpful in determining the character of this memorial gift. For our part, we are sure that it will do much to strengthen the intellectual and institutional ties between the Institute and the world of higher learning in the Federal Republic of Germany, ties which reach across the entire history of the Institute.

The final item of the opening program was the announcement of the Einstein Award. The presentation was made by Dr. Philip Handler, a man whose service to his profession and to his country are a matter of record and who, for the past decade, has been president of the National Academy of Sciences. The Einstein Award was authorized by Einstein himself in 1951 and is given from time to time by the Lewis and Rosa Strauss Foundation of Washington, D. C. Lewis Strauss was for many years a trustee of the Institute for Advanced Study and we were particularly happy to welcome the participation of his widow and his son in the Institute's celebration. Additionally, we were and are grateful for the honor accorded us by choosing this Einstein Centennial Symposium for the presentation of the Einstein Award. Of the previous winners of the Einstein Award, Professors Stephen W. Hawking, Yuval Ne'eman, Marshall Rosenbluth, Julian Schwinger, John Wheeler, and Eugene Wigner were with us for this symposium. Dr. Handler spoke as follows:

Mr. Woolf, Mr. Petersen, Mr. Dilworth, Mr. Minister, Your Excellencies, Mrs. Strauss, Mr. Strauss, ladies and gentlemen: As president of an organization which is dedicated to the fostering of scientific excellence and its application to human welfare, I am especially pleased to have been asked to participate in this evening's ceremonies, ceremonies opening a momentous symposium, the purposes of which have been so eloquently put both by Harry Woolf and by the minister.

The field of elementary particle physics continues to be characterized by new, exciting results, new ideas, and new methods. Progress in the field has provided remarkable insights and still baffling questions. Will the next new penetration into the structure of matter yield more questions or more answers? Will there ever be an end to the search, and, if so, how will we know it? Will we arrive at a closed form of the laws of physics, or will the subject remain endlessly open-ended? These and other similar questions will, I am certain, continue to stimulate and challenge those of you who will participate in this symposium and your successors for many years to come.

Now I cannot claim any expertise in the subject matter of this symposium, but I have talked with many of my colleagues and I do know that Professor Tullio Regge has contributed brilliantly to understanding of elementary particle phenomena. For his role in introducing the idea of complex angular momenta into elementary particle physics, he received the Danny Heinemann Prize in 1964. That idea led to new insights into the scattering of particles at very high energies and became a cornerstone of scattering theory. This suggested a speculative approach to the relativistic treatment of the scattering of elementary particles, from which emerged an entire class of models and theories, and these, in turn, stimulated a new approach to analyzing experiments resulting in many new investigations.

The Regge theory in its various modifications is still proving to be an important basis for the analysis of various elementary particle phenomena and further experimental ideas. Equally significant has been Professor Regge's contributions to the understanding of general relativity and the symmetries hidden in the dynamics of fields and particles. I am informed that his current interests include the analytical properties of Feynmann relativistic amplitudes and the investigation of the semiclassical limit of group representations.

Candidly, much of Professor Regge's work is a mystery to me. I am still a card-carrying biochemist. However, my colleagues in theoretical physics are as one regarding the genuine, thoroughgoing fundamental importance of Professor Regge's contributions. In trying to understand them, I was led to read from Albert Einstein, and the statement which seemed to come closest I would like to read to you. He said, "I am convinced that we can discover by means of purely mathematical constructions the concepts and the laws connecting them with each other which furnish the key to the understanding of natural phenomena. Experience may suggest the appropriate mathematical concepts, but they most certainly cannot be deduced from it. In a certain sense, therefore, I hold it true that pure thought can grasp reality as the ancients dreamed." Well, several of the experimentalists out in front of me may not entirely share that view, but I am sure that all of you will agree that it is a most provocative statement and exactly fits the contributions for which this evening we honor Professor Regge. As you heard, he joins Kurt Gödel, Stephen Hawking, Willard Libby, Edward Teller, Johnny Wheeler, Yuval Ne'eman, Julian Schwinger, Richard Feynmann, Leo Szilard, and Eugene Wigner in this pantheon of heroes, recipients of the prize which the *New York Times* called the highest of its kind in the United States.

It is my pleasure, therefore, to be able to make this presentation on this evening. To understand it best, let me read to you from the citation which is here before you. It says, "Mathematician and physicist, he explored the deep mathematical structures that underlie nature's landscape in many areas of science. He achieved a new understanding of particle-scattering processes of the stability of collapsed objects in general relativity and of the hidden symmetries in the dynamics of fields and particles." And for all of that, I am pleased to be able on behalf of the Lewis and Rosa Strauss Foundation to be able to make this presentation; and it comes in several pieces. One you see before you. The second is this piece — the handsome gold medal which I am sure Tullio will be glad to show you later — a medal whose design was approved by Albert Einstein. And finally, there is in my pocket one more little piece which perhaps is most meaningful of all, a check for $15,000.

The symposium that followed upon these presentations, and which is now documented in this volume, examined closely the nature and the continuing consequences of Albert Einstein's achievement. Ceremony and circumstances aside, I think he would not really have objected to our doing so, for the assessment, critical reevaluation, and extension of science were tasks normal to his being and primary to his purpose. That we have not included for elaborate discussion in our program the details of his private life, or the history of his commitment to the Jewish cause, or the rights of man, or world peace will disappoint some. Our decision is not meant to diminish the importance of those issues in his life — and in ours; that we have enlarged the scientific scope of our meeting to deal with the cultural setting in which he developed, the impact of his ideas on twentieth-century intellectual life, and some controlled speculations as to where in science we go from here, seemed appropriate to our setting and our style. Finally, the addition of music to our program was no mere entertainment. It was essential to Einstein's life; it is to ours. It is the voice of love in these deliberations.

II. The Cultural Cosmos

1. EINSTEIN'S EUROPE

Felix Gilbert

The knowledge that I bring to the presentation of this paper is that of a professional historian working in the field of European history. I had no close personal contacts with Albert Einstein, nor do I know more — perhaps even less — than the average layman about science and the history of science. I am not unduly perturbed by my lack of special knowledge. Historians are inclined to regard the perspective given by distance from their subject as an advantage rather than a handicap.

Yet there is a troublesome problem inherent in the title of this paper. "Einstein's Europe" appears to assume the existence of a close connection between Einstein and Europe; and it is certainly true that Einstein was a public figure in Europe. At the same time, however, Einstein's intellectual world and concerns were remote from those of the Europe of his time, from those of the Europe in which he grew up and lived for more than half of his adult life. It seems to me that the ideas which he advocated and for which he strove as a public figure, the influence he exerted but also some of the impediments and disappointments which he encountered, were shaped by the fact that Einstein lived at a distance from the prevailing concerns of his European contemporaries, that he was an outsider on the European scene. To understand Einstein's role in Europe, one must clarify the experiences and the attitudes that isolated him from the mainstreams of European life.

When you meet a European, you usually try to find out — either openly, or somewhat indirectly — what is his nationality? Is he English or Italian, Spanish, German, or French? This question was even more prominent in the years before World War I — years of heated nationalism. What was Einstein's nationality? The answer to this question is by no means simple. He was born in Germany at Ulm,[1] but when he left Germany at the age of sixteen, without having completed the gymnasium, he renounced his German citizenship, and some years later he acquired Swiss citizenship. He insisted on remaining a Swiss citizen when in 1914 he returned to Germany as a member of the Prussian Academy and professor at the University of Berlin. It was unusual for a man of this rank in the Prussian civil service not to be a German. It led to a diplomatic dispute between the German and Swiss governments when in 1922 it was announced that Einstein's work had received the Nobel Prize.[2] Einstein was then in Japan. If the winner of a Nobel Prize is unable to attend the ceremony at which the King of Sweden hands out the prizes, the diplomatic representative of the country to which the winner belongs accepts the prize on his behalf. But who was en-

Harry Woolf (ed.), Some Strangeness in the Proportion: A Centennial Symposium to Celebrate the Achievements of Albert Einstein
ISBN 0-201-09924-1

titled to act on Einstein's behalf, the German or the Swiss minister? In the end, the German minister represented Einstein. This dispute had an amusing sequel. The German government was eager to establish without question that Einstein was a German citizen. Since before 1918 German citizenship was dependent on being a citizen of one of the German *Laender,* the Prussian Minister of Education tried to find out whether Einstein still possessed the citizenship of the *Land* in which he had been born. Ulm, Einstein's birthplace, lies on the frontier between Bavaria and Wuertemberg. So the Prussian officials turned first to the Bavarian government, but when they failed to get a positive response, they turned to the Wuertemberg government, which also had to admit that Einstein's name could not be found in their records. Then the Prussian government cut the Gordian knot; since Einstein was a member of the Prussian Academy, it was decreed that this membership implied Prussian citizenship, and Einstein agreed. Certainly, this story has its comic aspect, but it also throws light on Einstein's true feelings and convictions. In the course of the discussions about his citizenship Einstein was asked by the Prussian Minister of Education to explain his views about the question of his citizenship; he declared that he was a Swiss citizen but that he felt himself to be a German scientist.[3] The official to whom this statement was made could hardly have regarded it as very helpful, and Einstein was probably well aware that this vague view of the importance of nationality and citizenship would not please the bureaucratic mind. Actually, it is a striking example of Einstein's dislike, almost contempt, for a basic feature of modern political society: for the value of national individuality and the importance of national distinctiveness. It was a conviction that he held throughout his life and that he proclaimed and emphasized frequently. He rejected not only chauvinism but nationalism in any form. Nationalism, in his words, was a hereditary malady.[4]

But if Einstein had little liking for being tagged with the label of a particular nationality, as a scientist he was a member of a profession, and this placed him in formal organizations of European society. However, Einstein's position in the institutional structure of the scientific community of his time was unusual. This was the result of intellectual traditions and institutional arrangements that had shaped the role of the natural sciences at the German universities as well as the result of personal factors. Intellectual life at the German universities — when they were reorganized, almost newly founded in the early nineteenth century, in the times of rising nationalism and romanticism — was dominated by a historical approach to learning. The consequences were twofold. One was that the natural sciences did not fit in the overall plan of the purposes of a university education; instruction and research in the natural sciences were relegated to specialized technical schools. Another consequence of the emphasis on a historical approach to learning was that in the various European countries scholarly disciplines developed along distinctive national lines. Later, in the middle years of the nineteenth century, the natural sciences were given a place of greater weight in the universities, but even when they had been incorporated in the universities, they were less oriented toward the aims of political and social education that had inspired the organization of the universities. The natural sciences continued in the traditions of the medieval universities, or of the Enlightenment academies, searching for generally valid laws. They formed a select group distant from and above the concerns of society. Natural scientists were more international in their outlook than scholars in the humanistic disciplines and, of course, they have remained so.

But soon after the natural sciences had found a niche in the universities, an opposite trend began to develop: they were drawn to the pursuit of practical interests that were contradictory to an apolitical, cosmopolitan attitude and had political implications. It was realized that the benefits of research in the natural sciences could be of enormous usefulness in political and social life. It is characteristic of this development that Alexander von Humboldt, a scientist of international fame and cosmopolitan outlook, wrote in his last work, the *Kosmos,* "only serious cultivation of chemical, mathematical and natural-scientific studies can prevent political decline." Those peoples "among whom the respect for such activities does not permeate all classes of society will unavoidably lose their wealth, and this will most certainly happen if neighboring states where science and industrial activities interact march forward in renewed strength."[5] Indeed the connection of science with economic progress and prosperity, and — last but not least — with *power,* was a primary motif for the astounding rise of the natural sciences in the later part of the nineteenth and the early twentieth century. University budgets were increased and a good amount of the increase went to the natural sciences, to the building of scientific institutes, and to the creation of new professorial chairs for work in the scientific fields. The task of the professor was not limited to training scholars but extended to instructing men who would make contributions in industry and business, and many professors in the natural sciences enjoyed close connections with industry and were concerned with its practical needs.[6] This development was favored by the popularity that scientists had gained among the masses, who were dazed by the changes and improvements modern scientific discoveries had brought into their lives. Scientists became the heroes of the intellectual life of the modern age.

In the development of modern science, the tension and interaction between these two opposite tendencies — of purely theoretical concern and practical interest — was important. Einstein was one of the few scientists whose work was exclusively dominated by the older tradition, the search for the laws of the universe. The personal factor, both natural inclination and the external circumstances of his early life, played a role in this.

At the beginning, Einstein's career was what, in comparison to a normal career, might be called a castastrophe; in Einstein's words, it was "a comedy."[7] He did not complete his course at the gymnasium in Munich. When he then attempted to enter the technical high school (Technical University Institute might be more correct) at Zurich, he failed the entrance examination and he was allowed to begin his studies only after he had gone to the Humanist Gymnasium in Aarau (Switzerland) for another year. The manuscript that he submitted as his dissertation in 1901 was rejected. Another manuscript that he submitted in a second attempt was again rejected as too brief, but then accepted after one sentence had been added. His first attempt to acquire the right to lecture at the University of Bern failed, but finally, in 1907, he succeeded. By then, the articles that directed attention to him had appeared and he advanced quickly in his academic career: Extraordinarius in Zurich, Professor in Prague, then back to Zurich in 1912. There was never a sustained period of university teaching, however, because in the summer of 1913, he received the offer from Berlin. This offer deserves to be described in some detail because it shows that although with this position he had reached the top of his profession, at the age of thirty-four, he remained something of an outsider.[8] He became a member of the Prussian Academy and also a professor at the University of Berlin with the right to teach but no teaching obligations; and he made only very limited use of the right to lecture at the university.[9] He received a salary as a member of the Prussian Academy: one-half of it was paid by the Prussian government and the other half from a fund of the Physical-Mathematical section of the Academy. This fund relied on payments provided by a rich industrialist, Leopold Koppel; Koppel was one of the chief benefactors of the Kaiser Wilhelm Gesellschaft. It had been founded in 1911 in order to further the needs of basic research and was to be financed chiefly from private sources.[10] There is hardly any doubt that Koppel provided half of Einstein's salary because he was told that Einstein was a suitable person to become director of the institute for physics to be founded by the Kaiser Wilhelm Gesellschaft. Einstein, too, had been told about this possibility when he was asked to come to Berlin, although as a letter to Planck shows, his enthusiasm for undertaking the directorship of such an institute was limited. Actually, the Institute for Physics, with Einstein as director, was established only in 1917, three years after Einstein's arrival in Berlin, and then the war and the financial difficulties of the postwar period made the construction of an appropriate building impossible. In 1921, Max von Laue was elected to the board of directors of the Institute, functioned as Einstein's deputy, and then became his codirector. Nevertheless — to quote from an official report of the 1930s[11] — the Institute for Physics differed from all other institutions of the Kaiser Wilhelm Gesellschaft by never having its proper scientific working space or staff. It had funds that were used for the acquisition of scientific apparatus, and for paying Einstein's assistants, and for supporting research projects all over Germany; Einstein took much trouble in evaluating these applications.[12] Nevertheless the Institute remained a skeleton rather than a working institution. Einstein's principal connection was with the Academy and that meant that he had full time to pursue his own research, and that his connection with organized scientific life remained loose and distant.

Certainly the position that Einstein held in Berlin fitted him perfectly. Some innate force resisted entanglement with the daily present. He wrote in his autobiography: "In a man of my type the turning-point of the development lies in the fact that gradually the major interest disengages itself to a far-reaching degree from the momentary and the merely personal and turns towards the striving for a mental grasp of things."[13]

Einstein's writings and letters contain revealing statements emphasizing the distance between the immutable world of the laws that rule the universe and the fickle and fugitive events and changes on the human scene. The best known is the one in which Einstein expresses his religious beliefs: "I believe in Spinoza's God who reveals himself in the orderly harmony of what exists, not in a God who concerns himself with the fate and actions of human beings."[14] A somewhat different version, which is interesting because it suggests the choice of values that is involved, can be found in a letter to Max Born: "Life in flux and clarity are antipodes; one withdraws if the other arrives."[15] The most incisive formulation of Einstein's view of the unbridgeable gap between the permanent world of nature and the ephemeral world of man is in a letter to the novelist, Herman Broch. Broch had sent him his most recent work, entitled "The Death of Virgil." Einstein wrote to him: "The book shows me clearly what I fled from when I sold myself, body and soul to science — the flight from the 'I' and 'We' to the 'It'."[16] In his very beautiful book, Broch, inspired by some analogies to his own time, dwells on the reasons for the weakness and decline of Roman culture at the moment of its high point of power. If Einstein felt alien to Broch's undertaking, the reason was his disinterest in the ephemeral, his rejection of history. To this subject, I can add something from my own experience. In the early years of the Second World War, I went once or twice a week to Einstein's house to visit Mrs. Winteler, Einstein's sister, whom I had known in Europe, and to play chess with her. On the occasion of one of these visits, Einstein said to me that he had heard much talk about the historian Gibbon, and asked whether I thought he ought to read him. In youthful missionary zeal, I brought six elegantly bound volumes of Gibbon's *Decline and Fall* the next time I came to play chess. On my next visit, two or three days later, Einstein had the six volumes in his arms and said, "Take them back. I can't read that stuff" ("Das Zeug kann ich nicht lesen"). At that time I speculated that the failure of my mission had been caused by the rather ironic and superior tone with which Gibbon narrates the cruel and terrible events of late classical and early medieval times, a tone that jars when you yourself live in times of a terrible war. It took me some time to realize that Einstein's rejection of Gibbon was part and parcel of his general attitude toward history. It was not based on a lack of knowledge of what historians are doing. As a member of the Academy, Einstein was expected to attend fortnightly meetings of the Plenum of the Academy, and, as the files of the Prussian Academy show, he did attend these meetings with praiseworthy regularity. Many of the papers were presented by historians, and a few of them are still considered to be of fundamental importance for our understanding of the past.[17] Clearly, Einstein found the study of the past a very irrelevant occupation because it did not matter to him to know whether and how the present was tied to the past: "For us, convinced physicists, the distinction between a past, a present, and a future has the character of a tenaciously held illusion."[18] Einstein was concerned with the permanent, and he disregarded the ephemeral. But, undoubtedly, this lack of interest in the past was a further reason for Einstein's distance from the contemporary European scene. Historical consciousness, awareness of the past reaching into the present and forming the external circumstances and the inner world of a person, gives the feeling of being rooted in a wider community — of being part of trends formed in the past and of companionship in your own time even though this bond might restrict the absoluteness of your judgment and make the ideals of the future more distant. Einstein did not like to acknowledge the limitations set by a historically conditioned situation. His criterion was reason, which either justified or condemned an institution. Such an attitude had its strength, but it also made him, as he wrote in a letter to Max Born, "A man nowhere really rooted — everywhere an alien."[19]

But, how did it happen that a man so remote — willingly remote — from the contemporary scene became a public figure in Europe? Two developments forced Einstein to take a stand in public life: the First World War and the arrival of fame.

Einstein emerged as a man with pronounced political views almost immediately after the outbreak of the First World War. In September 1914 there appeared the notorious manifesto of ninety-three German intellectuals that asserted Germany's innocence for the outbreak of the war, defended every aspect of German diplomatic and military conduct, and declared German militarism to be an essential element in German culture.[20] The list of signers contained the best-known names of German academic and artistic life. Einstein was not among the signers. In the mad atmosphere of war enthusiasm, the overwhelming majority of German academics approved the manifesto, but some well-known scholars refused to sign because they believed the war was a power struggle and nothing could be gained by transforming it into an ideological conflict. But Einstein went beyond mere refusal; he was willing to sign a countermanifesto, which, however, found so few supporters that it was not published.[21] Einstein's opposition to the manifesto of the ninety-three was based on principles that ruled his thinking on public affairs as long as he lived in Europe. Like others, Einstein was deeply troubled by the disruption of contacts between the scientists of the warring countries, by the collapse of the international scientific community — war was the reason for all this. As long as there was war, as long as there were armies, the progress of science and the existence of civilization was in danger. From his school days in Munich on, Einstein abhorred the militaristic discipline that he encountered in German life. He now became a convinced and radical pacifist and he maintained this attitude into the 1930s, giving his support to pacifist organizations and causes.

During the First World War, favored by the fact of his Swiss citizenship, he established relations with Romain Rolland, who had left France for Switzerland to fight against the war.[22] In Germany, Einstein became a leading member of the pacifist and internationalist Association New

Fatherland (Bund Neues Deutschland), which was founded in the early months of the First World War, and in the 1920s amalgamated with the League for Human Rights (Liga fuer Menschenrechte).[23] In the late 1920s, when political tension in Europe grew, Einstein recommended that young men refuse to do military service, and he advocated withholding tax payments for military purposes.[24] His opposition to the war did not make him popular in German officialdom. His name appears as No. 9 on a "blacklist" of pacifists and defeatists, and they could be refused visas for travel outside Germany.[25] Since this list was put together around the time when Einstein became director of the Physical Institute of the Kaiser Wilhelm Gesellschaft and member of the Board of the Federal Physical-Technical Institute, one must assume that either one German bureaucrat did not know what the other did, or that they were tolerant enough to separate science and politics.

For Einstein, among the losses that war inflicts upon humanity, the threat of a break-up of the international scientific community loomed large; his concern about this possibility comes out clearly in an action that he undertook in the spring of 1918, and which, in some respects, was intended to repair the damage that the manifesto of the ninety-three had done. Einstein suggested to a number of people the publication of a book in which distinguished intellectuals from various countries would express their belief in the need for international cooperation.[26] In justification of this enterprise he wrote: "In times of general nationalistic blindness, men active in the sciences and arts have frequently made public declarations which have done incalculable harm to the feeling of solidarity which, before the war, had hopefully grown among those devoted to higher and freer aims. The shoutings of narrow-minded defenders and advocates of the power principle is so loud, and public opinion is so misled by censorship that those who are differently minded feel hopelessly isolated and dare not raise their voices. . . . This grave situation demands of those whose intellectual achievements have won for them respect and prestige among the intellectuals of the entire civilized world a task which they must undertake."[27] The answers that Einstein received to his initiative were not encouraging. The sociologist and theologian, Ernst Troeltsch wrote: "An appeal to reason will not save us, for reason is powerless everywhere."[28] The mathematician, David Hilbert wrote from Göttingen: "The word, 'international' has on our colleagues the effect of a red rag on a bull." Hilbert believed that the enterprise that Einstein suggested would do more harm than good because it would inspire a vehement nationalistic counteraction: "The main thing is we would use up our ammunition at the wrong moment and against the wrong persons. I can well understand that you want to do something, but my advice is to wait until the hurricane of folly has passed and it is possible for reason to return, and certainly this time will come."[29] Despite sympathy with Einstein's ideas, the two answers imply that Einstein's appeal to reason and his confidence in the power of reason were remote from the contemporary scene.

When, one-half year later, the war ended with Germany's defeat and the overthrow of the imperial regime, Einstein thought that the establishment of a better world order had become a possibility, and felt that it was his duty to play his part. When, in the turmoil of the revolution, students attempted to take over control of the university, Einstein talked with Friedrich Ebert, the chairman of the People's Commissars and as such head of the government, and he pleaded with the students to preserve "academic freedom."[30] When his former colleagues from the University of Zurich hoped that the difficulties in war-torn Germany might give them a chance to persuade Einstein to return to Zurich, Einstein, in the revolutionary month of November 1918, told the Prussian Ministry of Education that he had declined the offer because he was entirely satisfied with his position in Berlin and that he did not want to be separated from his excellent colleagues at the Academy and University.[31] Even after the wave of revolutionary enthusiasm had receded, Einstein remained optimistic; he was not particularly worried about the harshness of the Treaty of Versailles because it would soon become evident that the Treaty could not be carried out, and he had high hopes for the League of Nations.[32] Later, when Einstein was reminded of his activities in November 1918, he considered his belief that he could help in transforming Germans into "honest democrats" to have been incredibly naive: "I can only laugh when I think of it."[33] Indeed, his enthusiasm for the German Republic was short-lived; already in the latter part of 1919 his political optimism was fading.

Yet Einstein could not resume the life he had led before the outbreak of the war, turn away from the problems of the wider world, and withdraw from the involvement in public affairs in which the indignation about the war had entangled him, because by the end of the world war, fame had come to him.

During the war, in 1916, Einstein had completed work on the general theory of relativity, and in 1919, observations connected with the total eclipse of the sun confirmed his theory. In 1922, he received the Nobel Prize.[34] Einstein was now world-famous and he received invitations to lecture from all over, even from countries that had fought against Germany. Of course, the person whom people wanted to hear and to see was most of all the great scientist. But an added attraction was Einstein's opposition to the war and to the German militaristic regime, and this warmed his welcome wherever he went. At the same time, this meant that in the years immediately following the First World War, Einstein's lecture tours in the United States, in England, in Japan, and in France had particular consequences for Einstein's position in the German academic world. For many years after the war, up to the middle of the 1920s, no bonds existed between German scientists and scholars and those of the allied countries. The German academic world remained boycotted. Einstein was almost the only German scholar who was able to break through the wall that had been erected around Germany. In consequence, in Germany, he was considered to be more than a scientist. He became a figure who aroused political discussion. Some regarded him with envy and hope and expected that he would use his connections in other countries to obtain an end of the intellectual boycott against Germany. Others saw in his making use of opportunities that his academic colleagues did not enjoy a disloyalty to his class; had he refused to lecture in former enemy countries, they argued, the foolishness of continuing the international boycott would be amply demonstrated. Nationalists resented that hostility to the German war effort should become an advantage. The contradictory feeling aroused by Einstein's contacts in foreign countries are reflected in an incident that happened at the Prussian Academy in 1921 after he had been in the United States and in Great Britain. In the Plenum of the Academy, he reported on his impressions of these countries, but when the minutes were presented for approval, objections were raised against placing an account of a "travel undertaken in the service of Zionist propaganda" into the minutes of the Academy. Einstein explained that "on all his trips, he had always presented himself as a German and Berlin Professor," and the minutes were then approved as originally proposed.[35]

But if Einstein's international acclaim aroused attention among German academics, it was also of great interest to the German foreign office. The German representatives abroad reported in detail about his travels and contacts;[36] the German foreign office was aware that he could be a significant asset in creating a more favorable attitude towards Germany. When in 1922 Einstein was invited to lecture at the Collège de France, he declined, but Walter Rathenau, then German foreign minister, persuaded Einstein to change his mind.[37]

Einstein's own interest in reestablishing and fortifying an international scientific community and a more peaceful international order, his delicate relationship with the German academic world and the value of his prestige for German foreign policy — all these facts came into play in the tortuous story of Einstein's membership in the League of Nations Commission for International Intellectual Cooperation.[38] This commission, which might be considered a precursor of UNESCO, was formed in September 1921, and its members were appointed by the League of Nations as individuals — not as representatives of the intellectual community of their nations. Germany was then still excluded from the League of Nations, and Einstein's participation in the work of the League accentuated his exceptional position in the German academic world. Moreover, in order to make the appointment of a German palatable to the French, his opposition to the prevailing nationalist German tendencies was publicly emphasized at the occasion of his appointment. Einstein accepted the appointment to the commission very hesitantly. One month after he accepted, the German foreign minister, Rathenau, was assassinated; Einstein, sensing this deed as pointing to a wave of nationalism and reaction in Germany, declared that as a Jew, as a radical pacifist and internationalist, he was unsuited to act as mediator between the German and the non-German intellectual world, and expressed his intention to resign his appointment.[39] Madame Curie and the British classicist Gilbert Murray implored him to remain on the commission.[40] Murray pointed to the "danger of having the Latin element overrepresented" if Einstein were to resign. Madame Curie wrote that a serious attempt at international intellectual cooperation was particularly important to counter the rising wave of nationalism. These interchanges went on until the spring of 1923, when Einstein definitely resigned. He now justified this step in a different way. He stated that he had lost confidence in the League of Nations. "In its present form it badly discredits the notion of pacifism. I can't get rid of the impression that the League of Nations, as it functions today, is nothing but a misleadingly named, obedient instrument of that group of powers which, by means of its military strength, brutally dominates Europe."[41] Einstein emphasized that his decision was not influenced by "political prejudice or chauvinism."[42] But it seems hard to believe that the change in the manner in which he justified his withdrawal — from being unrepresentative of German science to doubts about the usefulness of the League of Nations — was really unconnected with the political situation: it was the time of the occupation of the Ruhr by French troops. Einstein certainly felt the pressure rising from the tense political situation. At the end of 1923, he refused to go to a Solvay Congress in Belgium because no other Germans were invited, and he declined an invitation to go to Paris to participate in the celebration of the twenty-fifth anniversary of the discovery of radium.[43] This brought to Einstein a rather remarkable letter from Madame Curie. She told him that in justifying his resignation from the League of Nations Commission, it would have been enough if he had argued with Germany's exclusion from the League; it was unnecessary to attack the work of the League in general:

> The League is not perfect, but it has no chance of being perfect because human beings are not perfect. . . . Don't you find it sometimes exasperating to be unable to make any move that is not of world importance? You say that contacts among cultivated people ought to be easier than contacts among average men. True, but, the opposite takes place. Average men find communication easy whereas intellectuals seem to agree to erect barriers against each other which they are unable to remove. That shows that civilization contains an antihuman element — undoubtedly as consequence of the fact that teaching is not sufficiently emancipated from political influences. And at the risk of your accusing me of political prejudices and of lack of understanding of relativity, I think that this inhuman element is more noticeable in Germany than in France, because, until now, the Germans have had less political freedom.[44]

What impression Madame Curie's gentle chiding made on Einstein, we do not know, but he was glad to rejoin the Commission when Germany was on the way to enter the League of Nations. He wrote that his withdrawal had been influenced more by a passing mood of disillusionment than by clear thinking. "The League of Nations ought still to be regarded as that institution which offers best promises of effective action to those who honestly work for international reconciliation."[45] But even now, Einstein did not enjoy this activity. With Germany a member of the League of Nations, he was now really the representative of the German academic world, and he felt rather unhappy about the rights and claims the German academic community demanded and expected him to support.[46] He was equally unhappy about the domination the French exerted over the institute for international intellectual cooperation that the commission had established in Paris.[47] He participated in fewer and fewer meetings, and in 1931 he resigned definitely. His justification was that the commission could not accomplish anything useful because the League of Nations was not equal to its task. It lacked the determination "necessary to achieve real progress toward improving international relations." "It is precisely because of my profound desire to give whatever help I can toward the establishment of a supra-national arbitrating and regulatory authority that I feel compelled to resign from the committee."[48] Despite Madame Curie's admonition to recognize that nothing can be perfect, Einstein insisted on the absolute.

When Einstein resigned for the first time from the commission for international intellectual cooperation, he had stated that one of the reasons he felt unsuited to play the role of a mediator between the German and non-German world was that he was a Jew. His Jewishness and his relation to Zionism became increasingly important to Einstein. His parents were not particularly religious. His first wife was a Greek Orthodox Christian, although she later characterized herself as freethinking.[49] The first time Einstein described himself formally as a Jew was when he had become professor in Prague and the officials charged with taking his oath declared that Emperor Francis Joseph did not believe a freethinker could swear an oath. The emperor's view was that an oath is valid only to believers in God, which meant only to members of an established religious community. In Prague, Einstein was in contact with German-Jewish literary figures like Kafka and Max Brod;[50] in the German-Jewish circles of Prague, Zionism played an important role. But at that time, Einstein had an almost contemptuous attitude towards Zionism.[51] As a result of his experiences in Germany and his contacts with German academics, he became much more aware of the difficulties which Jews faced in German society. In a letter to Max Born, in 1919, he suggested that Jews and Germans should remain as far as possible separated and that the work of Jewish scholars and scientists should be financed, outside the universities, by wealthy Jews.[52] These remarks were clearly influenced by the reemergence, in Germany, of a reactionary nationalism, embodying a strong strain of anti-Semitism. Einstein and his relativity theory became a special target of German nationalists;[53] a lecture that he gave at the university early in 1920 was broken up, the relativity theory was condemned as a distortion of truth by a "Group of German natural scientists for the Preservation of the Purity of Science," and this organization arranged public meetings and hired speakers assaulting Einstein and his work and accusing him of courting publicity. These attacks reached their high point at the General Meeting of German Natural Scientists in September 1920, where Einstein and Lenard became involved in a sharp discussion, not free from anti-Semitic undertones. Of course, the attacks were countered by demonstrations of sympathy and appreciation. The Minister of Education expressed his "feelings of pain and humiliation" about the attacks to which Einstein had been subjected.[54] He was also defended in public declarations by the leading physicists of the University of Berlin and other prominent citizens.[55] Einstein was chosen to be the speaker at the annual celebration of the Foundation of the Academy,[56] and he was elected member of the Order "Pour le Mérite."[57] There was fear that in disgust, Einstein might accept one of the many offers that he had received from other countries. But in the letter in which he thanked the Minister of Education for his sympathy, Einstein wrote that he had become aware in these days "that Berlin is the place with which I am most closely connected by human and scientific relationships. I would accept an offer from another country only if I am forced to it by external circumstances."[58] It cannot be doubted, however, that the anti-Semitism that he encountered increased his concern with the situation of the Jews in Germany and his interest in Zionism.[59]

The man who converted him to Zionism was a prominent German Zionist, Kurt Blumenfeld, and according to him the decisive conversation took place in 1920 — the year when Einstein had to struggle with the anti-Semitic opponents of the relativity theory. According to Blumenfeld, Einstein said when he finally agreed to support Zionism: "I am against nationalism but for the Zionist cause. Today, I finally realized why. If a man has two arms and he declares repeatedly, I have a right arm, that man is a chauvinist, but if a human being has no right arm then he must do everything to replace this lacking limb."[60] Nevertheless, Einstein first remained cautious and somewhat hesitant in his support of Zionism. He had observed the difficulties that young Jews had in obtaining the positions in academic life that were their due, and, encouraged by Weizman, he became active in supporting the development of a university in Jerusalem. His visit to the United States in 1921, together with Weizman, had the purpose of raising funds for this project. Einstein came back to a Germany in which the political and economic situation was steadily deteriorating and nationalistic extremism, with a strong component of anti-Semitism, had become more vocal and more violent. In January 1922, Walther Rathenau became German Foreign Minister; Einstein, together with Kurt Blumenfeld, went to him and tried to persuade him to give up this position — of course, without success.[61] Einstein clearly feared that the holding of such a crucial position by a Jew would give further impetus to anti-Semitism and could have dangerous consequences for Jews in Germany.[62] When, in the summer of 1922, Rathenau was assassinated, Einstein was deeply shaken, and had some fears for his own life, since his name had been found on some lists containing names of those marked for assassination by the Nationalists. Einstein declined to appear before the annual Congress of German Natural Scientists, where he had been scheduled to give a chief address. He was happy to be able to accept an invitation to Japan, not only because he was very eager to get to know that country but also because, after the Rathenau murder, he "welcomed to have the opportunity of a lengthy absence from Germany, where, at least temporarily, some danger for my life had threatened."[63] The events since 1920, culminating in the assassination of Rathenau, gave added urgency to the creation of a Jewish homeland in Israel, and Einstein intensified his efforts on behalf of Zionism; they were further spurred by a visit to Palestine, which he made on his way back from Japan, in 1923. He was enthusiastic but he also thought that the land was not very fertile and would not be able to absorb many immigrants. This did not disturb him. It had advantages: "Because of its small size and its dependence, it will be refrained from competition for power."[64] In Israel he believed was the possibility of a genesis of a nation without those disastrous qualities that he considered to be inherent in nationalism. It is certain that Einstein felt increasingly bound to Israel and that Israel inspired in him an attachment that he had for no other nation or state; but I suspect that it is not quite accidental that this nation was not in Europe and that for most of Einstein's life it existed in the future rather than in the present.

From the First World War on, when Einstein became a public figure until the time when he left Europe in the early thirties, three issues stood in the center of Einstein's public concerns and activities: the abolition of war and the formation of a supranational government, the establishment and maintenance of an international scientific community, and efforts to establish a Jewish homeland in Palestine. He considered these issues as deserving his time and effort because the existence and happiness of human beings depended on their handling and their solution; his feelings of human decency were involved in these issues. It was entirely true when he wrote about himself in a letter to Hadamard, "When it comes to human affairs, my emotions are more decisive than my intellect."[65]

One can possibly also draw a reverse conclusion from this statement: His interventions in the public sphere were not based on systematic and consistent assumptions about the functioning of the historical process. He was a democrat, but he would also say that the masses, misled by priests and power-minded governments, could not be trusted. He said he was a socialist, but I wonder whether he ever took a serious look at Marx. He could wish that the rule over human affairs were entrusted to an elitist group of intellectuals but he could also say that intellectuals know little about the reality of life. In his first thirty-five years he lived far from the concrete texture of European life, and something of this remoteness remained even after he had become a public figure. Yet, Einstein felt bound to Europe. In a statement that has almost programmatic character, he said, "I am a man, a good European, a Jew."[66] Elsewhere he explained at some length what he understood to be the valuable part of the European legacy: "When I speak of Europe, I do not mean the geographical concept of Europe but rather a certain attitude towards life and society which has grown up in Europe and is characteristic of European civilization. I speak of that spirit which was born in ancient Greece and which, more than a thousand years later, at the time of the Renaissance, spread from Italy to the whole continent — the spirit of personal liberty and respect for the rights of the individual. To what do we owe our knowledge of the laws of nature and the heretofore unparalleled technological possibilities for improvement of our human conditions if not to the realizations of bygone days that only the individual is truly capable of creating what is new and worthwhile."[67]

Einstein regarded classical antiquity and the Italian Renaissance as the crucial periods for the origin of the ideas that were Europe's great achievements. The ideas on which Einstein placed emphasis and that contain for him the essence of the European legacy are only part of that complex of ideas, attitudes, and institutions that can be comprised under the name of European civilization. There was one century, however, in which the ideas that Einstein valued were particularly prominent and crucial: the Age of Enlightenment. It is striking how close Einstein's views frequently appear to those of the eighteenth-century *philosophes:* the condemnation of war and power politics, the belief in the realization of eternal peace, the struggle against the misdirection of education by priests and churches, the fight for equal rights for the Jews, and, most of all, the supranational community of the intellectuals reflected in the international status of the academies. One might, perhaps, add that for Einstein music reached its acme with Bach and Mozart, and he saw the coming of decline in Beethoven's romanticism. It has an irresistible attraction to think that Newton, the man who discovered "the laws of nature and of nature's God," and Einstein, who revised Newton's image of the universe, intellectually shared the same age.

NOTES

1. I wish to thank the executors of the Einstein estate, Mr. Otto Nathan and Miss Helen Dukas, for permission to quote from unpublished material in the Einstein papers. Miss Dukas, with her unequaled knowledge of Einstein material, has been of the greatest possible help to me in my study of these papers. I also wish to acknowledge helpful suggestions that I received in conversations with David Cahan, Armin Hermann, and Fritz Stern.

   Among the many books on Einstein, that by Banesh Hoffmann with the collaboration of Helen Dukas, *Albert Einstein — Creator and Rebel* (Viking, New York, 1972) is probably the best informed and most reliable; if I indicate no other source, data about Einstein's life and career are taken from this book.

2. The documentary material has been published in *Albert Einstein in Berlin 1913–1933,* Teil I: Darstellung und Dokumente, edited by Christa Kirsten and Hans-Juergen Treder, Studien zur Geschichte der Akademie der Wissenschaften der DDR, Vol. VI (Berlin, 1979), pp. 113–18; see also the official report about the ceremony in *Les Prix Nobel en 1921–1922* (Stockholm, 1923), p. 48. The German government officials themselves disagreed on the question whether Einstein was Swiss or German, and the German Minister in Stockholm, Nadolny, happily made use of the view expressed by the Prussian Academy according to which Einstein was German. This was hardly Einstein's opinion, although he later accepted that, in addition to his Swiss citizenship, as member of the Prussian Academy he was also German. For the question of Einstein's citizenship and the discussions among officials and with Einstein, see also *Albert Einstein in Berlin 1913–1933,* Teil II: Spezialinventar, Studien zur Geschichte der Akademie der Wissenschaften der DDR, Vol. VII (Berlin, 1979), pp. 29–34.

3. *Einstein in Berlin,* Teil II, op. cit. in n. 2, p. 29.

4. See *Einstein on Peace,* edited by Otto Nathan and Heinz Norden (Schocken Books, New York, 1960), p. 108, but similar remarks are frequent; "insanity of nationalism" (p. 13), "the monstrously exaggerated spirit of nationalism" (p. 152), etc. See also Einstein's remark to Miss Seeley: "My citizenship is a strange affair. Although my real citizenship is Swiss, I am a German citizen on account of my official position. However, for an internationally minded man citizenship of a specific country is not important. Humanity is more important than national citizenship." (p. 211).

5. Alexander von Humboldt, *Kosmos, Entwurf einer physischen Weltbeschreibung* (original edition), p. 36.

6. An outstanding example of the close connection between university science and industry was Walther Nernst; next to Planck, he had been the chief supporter of Einstein's appointment to Berlin, and had been chiefly responsible for getting the industrialist Koppel to underwrite half of Einstein's salary, see *Einstein in Berlin,* Teil I, op. cit. in n. 2, pp. 95–9, and in general Kurt Mendelssohn, *The World of Walther Nernst: The Rise and Fall of German Science 1864–1941* (Pittsburgh, 1973). The contrast between Einstein and Nernst is emphasized in the article by Wolfgang Yourgrau, "Einstein and the Vanity of Academia" in *Albert Einstein: His Influence on Physics, Philosophy and Politics,* edited by Peter C. Aichelburg and Roman U. Sexl (Friedrich Vieweg, Braunschweig, 1979), p. 216, but it ought to be stressed that Nernst unfailingly defended Einstein against all political attacks.

7. Hoffmann, op. cit. in n. 1, p. 55.

8. For the arrangements connected with Einstein's Berlin appointment, see the text of the most important documents in *Einstein in Berlin,* Teil I, op. cit. in n. 2, pp. 95–106, and the checklist of documents in Teil II, op. cit., pp. 17–22. I discuss the main features of the appointment in some detail, because, in the literature on Einstein, there is a remarkable confusion about the relation of Academy and Kaiser Wilhelm Gesellschaft and Einstein's position in these institutions.

9. Einstein gave from time to time rather general lectures at the University and was quite willing to see a wider public, not only students of physics, in these lectures, see *Einstein in Berlin,* Teil II, op. cit. in n. 2, pp. 60–6 and Teil I, op. cit., pp. 201–2. Einstein probably had no great interest in instruction of students by lectures on the general problems of physics, and he was, according to Yourgrau in the article mentioned in n. 6 "not an inspiring, not a stimulating academic teacher."

10. See Lothar Burchard, *Wissenschaftspolitik im Wilhelminischen Deutschland: Vorgeschichte, Gruendung und Aufbau der Kaiser Wilhelm-Gesellschaft zur Foerderung der Wissenschaften.*

11. See die "Festschrift": 25 *Jahre Kaiser Wilhelm-Gesellschaft zur Foerderung der Wissenschaften* (Berlin, 1936), p. 54.

12. For financial support given to various projects under Einstein's direction of the Institute, see *Einstein in Berlin,* Teil I, op. cit. in n. 2, pp. 151–4; I owe Professor Armin Hermann thanks for the information that Einstein evaluated these applications personally and with great care.

13. Published in *Albert Einstein: Philosopher-Scientist,* edited by Paul Arthur Schilpp (The Library of Living Philosophers, New York, 1951), p. 7.

14. See Hoffmann, op. cit. in n. 1, p. 95.

15. The German text runs: "Lebendiger Inhalt und Klarheit sind Antipoden, einer raeumt das Feld vor dem anderen." The passage can be found in *Albert Einstein, Hedwig und Max Born, Briefwechsel 1916–1955,* with comments by Max Born (Nymphenburger, Muenchen, 1969), p. 135.
16. Hoffmann, op. cit. in n. 1, p. 254.
17. *Einstein in Berlin,* Teil II, op. cit. in n. 2, the section "Regesten von Sitzungsprotokollen der Berliner Akademie der Wissenschaften," pp. 207–84.
18. "Pour nous, physiciens croyants, cette séparation entre passé, présent et avenir, ne garde que la valeur d'une illusion, si tenace soit elle" [*Albert Einstein, Michele Besso Correspondance 1903–1955* (Collection Histoire de la Pensée, Paris, 1972), p. 539].
19. "als nirgends wurzelnder Mensch. . . .Ueberall ein Fremder" [*Einstein-Born Briefwechsel,* op. cit. in n. 15, p. 48]. It deserves mentioning that appreciations of Einstein, published or republished on the occasion of the centenary of his birth, and written by outstanding physicists, emphasize this particular feature. See Robert Oppenheimer: "I had the impression that he [Einstein] was also in an important sense, alone" ["On Albert Einstein" in *Einstein. A Centenary Volume,* edited by A. P. French (Harvard University Press, Cambridge, Mass., 1979), p. 47]; Carl Friedrich von Weizsaecker: "It was just his aloofness which enabled Einstein to remain true to his naive directness in his political judgment" ["Einstein's importance to Physics, Philosophy and Politics" in *Einstein,* op. cit. in n. 6, p. 167]. Finally see A. Pais in this volume: "If I had to characterize Einstein by one single word I would choose 'apartness.' "
20. About the manifesto of the ninety-three and also about the attitude of the German academic and intellectual world in the years of the First World War, see Klaus Schwabe, *Wissenschaft und Kriegsmoral: die Deutschen Hochschullehrer und die politischen Grundfragen des Ersten Weltkrieges* (Musterschmidt, Goettingen, 1969), and the literature quoted therein. Hans Wehberg's *Wider den Aufruf der 93* (Charlottenburg, 1920) is still of value in clarifying the motives of those who signed, and of the few who did not.
21. It is now published in *Einstein on Peace,* op. cit. in n. 4, pp. 4–6.
22. On Einstein's meetings with Romain Rolland, see ibid., pp. 14–21.
23. Einstein belonged to several other organizations of this character: the "Gesellschaft der Freunde des Neuen Russland," the "Juedische Friedensbund," etc., see *Einstein in Berlin,* Teil I, op. cit. in n. 2, p. 219, and Teil II, op. cit., p. 14.
24. See *Einstein on Peace,* op. cit. in n. 4., Chap. 4, pp. 90–126.
25. See *Einstein in Berlin,* Teil I, op. cit. in n. 2, pp. 198–9.
26. This correspondence, previously unpublished, is preserved in the Estate of Albert Einstein, Princeton, N.J.
27. Zu unzaehligen Malen haben in diesen Traurigen Jahren der allgemeinen nationalistischen Verblendung Maenner der Wissenschaft und Kunst Erklaerungen in die Oeffentlichkeit gesandt, die dem vor dem Kriege so hoffnungsvell entwickelten Solidaritaetsgefuehl derer, die sich hoeheren and freieren Zielen widmen, bereits unberechenbaren Schaden zugefuegt haben. Das Geschrei enghersiger Priester und Knechte des oeden Machtprinzips erhebt sich so laut, und die oeffentliche Meinung ist durch zielbewusste Knebelung des ganzen Publikationswesens derart irregefuehrt, dass die Besser-Gesinnten im Gefuehle trostloser Vereinsamung ihre Stimme nicht zu erheben wagen. . . Diese ernste Situation stellt diejenigen welche durch glueckliche geistige Leistungen ein ueberlegenes Ansehen bei den geistigen Arbeitern der ganzen zivilizierten Welt erlangt haben, eine Aufgabe, der sie sich nicht entziehen duerfen. [Draft by Einstein, dated April 1918.]
28. "Ein Appell an die Vernunft rettet uns nicht, denn diese ist ueberall ohnmaechtig" [Ernst Troeltsch, letter to Albert Einstein, 1 May 1918].
29. "wirkt doch schon das Wort international auf unsere Kollegen, wenn sie sich in corpore fuehlen, wie das rote Tuch. Aber auch der Sache wuerden wir schaden . . . Was aber die Hauptsache ist, wir wuerden unser Pulver zu unrechter Zeit und vielleicht auch auf unrechte Personen verschiessen. Ich kann so recht das Beduerfnis verstehen, das ich ebenso wie Sie empfinde, etwas zu thun; aber ich moechte raten zu warten, bis der Wahnsinnsorkan ausgetobt hat und die Vernunft wiederzukehren die Moeglichkeit hat — und diese Zeit wird sicher kommen" [David Hilbert, letter to Albert Einstein, 5 May 1918].
30. See *Einstein-Born Briefwechsel,* op. cit. in n. 15, pp. 202–6; also Hoffmann, op. cit. in n. 1, pp. 135–9.
31. *Einstein in Berlin,* Teil I, op. cit. in n. 2, p. 108.
32. *Einstein-Born Briefwechsel,* op. cit. in n. 15, pp. 29, 32.
33. Ibid., p. 202.
34. The 1921 Nobel Prize for physics was awarded to him in 1922.
35. *Einstein in Berlin,* Teil I, op. cit. in n. 2, p. 128.

36. German diplomatic reports about Einstein's travels between 1920 and 1931 are printed in *Einstein in Berlin,* Teil I, op. cit. in n. 2, pp. 225–40. On some of his trips Einstein visited the German diplomatic representative or was his guest at a luncheon or dinner; on other occasions, for instance on his trip to Paris in 1922, he had no contact with the German embassy.

37. *Einstein in Berlin,* Teil I, op. cit. in n. 2, p. 210.

38. A careful, concise statement about political aspects in the formation and the work of the Commission can be found in the article by Brigitte Schroeder-Gudehus, "Les professeurs allemands et la politique du rapprochement," Annales d'Etudes Internationales, **I**, 23–44 (1970). See also *Einstein on Peace,* op. cit. in n. 4, Chap. 3, pp. 58–89, and the checklist of relevant documents in *Einstein in Berlin,* Teil I, op. cit. in n. 2, pp. 149–56. For my presentation of Einstein's relations to the League of Nations Commission, I have used unpublished documents preserved in the Estate of Albert Einstein in Princeton; quotations from the text of these documents were translated into English, but in such cases I have put the original German or French in the notes.

39. "Nicht nur bei Gelegenheit des tragischen Todes von Rathenau, sendern auch bei andren Gelegenheiten habe ich wahrgenommen, dass in der Schicht, die ich gewissermassen beim Voelkerbund zu vertreten habe, ein sehr starker Antisemitismus sowie ueberhaupt eine Gesinning solcher Art herrscht, dass ich mich nicht dazu eigne, vertretende, bzw. vermittelnde Person zu sein." [Albert Einstein, letter to Madame Curie, 7 July 1922, from the Estate of Albert Einstein.]

40. Gilbert Murray's letter is dated 10 July 1922, Madame Curie's 4 July 1922; both letters are in the Estate of Albert Einstein.

41. "dass der Voelkerbund, so wie er jetzt seines Amtes waltet, die schlimmste Diskreditierung des pazifistischen Gedankens bedeutet. Ich kann mich des Eindrucks nicht erwehren, dass der Voelkerbund als Ganzes, so wie er heute arbeitet, nichts anderes ist als ein unter irrefuehrendem Namen funktionierendes gefuegiges Instrument derjenigen Maechtegruppe, welche durch ihre militaerische Macht gegenwaertig in Europa hemmungslos dominiert" [A. Einstein, letter to Comert, an official in the League of Nations, 11 April 1923, from the Estate of Albert Einstein.]

42. "in keiner Weise durch politische Voreingenommenheit oder gar durch Chauvinismus" [A. Einstein, letter to Gilbert Murray, 25 May 1923, from the Estate of Albert Einstein.]

43. *Einstein on Peace,* op. cit. in n. 4, pp. 63–5.

44. Il était certainement difficile pour vous d'en faire partie tant que l'Allemagne ne fait pas partie de la S.d.N., et il me semble qu'il eût été suffisant de donner cette raison en vous retirant. Car si la S.d.N. était complètement constituée, les défauts qu'elle peut avoir seraient moins graves. — Je vous accorde bien volontiers que la S.d.N. n'est pas parfaite. Elle n'avait aucune chance de l'être puisque les hommes sont imparfaits. . . . Je regrette que vous ne puissiez venire à Bruxelles pour revoir vos amis belges et francais. Ne pensez vous pas quelquefois qu'il est exaspérant de ne pouvoir faire un geste sans que cela ait une significance mondiale? Vous dites que les relations entre les hommes cultivés devraient être plus faciles qu'entre eux du commun public. Sans doute, mais c'est le contraire qui a lieu. Les communs mortels communiquent assez librement tandis que les intellectuels ont réuni à dresser entre'eux des barrières qu'ils ne savent pas faire disparaître. Cela prouve qu'il y a dans la culture un élément antihumain qui dépend sans doute de ce que l'enseignement n'est pas suffisament dégagé de considérations politiques. Et au risque d'être par vous accusée de parti pris et d'incompréhension de relativité, je crois que cet élément est plus prononcé en Allemagne qu'en France, parce que le régime politique de l'Allemagne a été jusqu'ici beaucoup moins libre. [Madame Curie, letter to Einstein, 6 January 1924, from the Estate of Albert Einstein.] This letter was an answer to a letter by Einstein, dated 25 December 1923, the German text of which — in parts very similar to that to Comert of 11 April 1923, op. cit. in n. 41, published in Carl Seelig, *Albert Einstein und die Schweiz* (Europa-Verlag, Zürich, 1952), pp. 163–4.

45. *Einstein on Peace,* op. cit. in n. 4, p. 66.

46. See Einstein's letters to Madame Curie and Painlevé, from March and April 1930 and the correspondence with Kruess, the head of the German National Committee from April 1930 in the Estate of Albert Einstein; see also Einstein, letter to Haber, 20 July 1931: "der daemliche und eitle Kruess." See also *Einstein in Berlin,* Teil II, op. cit. in n. 2, p. 156.

47. See *Einstein on Peace,* op. cit. in n. 4, pp. 86 and 98.

48. The letter to the undersecretary of the League of Nations, by which Einstein resigned from the League of Nations Commission for International Intellectual Cooperation, is given in English in *Einstein on Peace,* op. cit. in n. 4, p. 110, but the English translation contains an error in translation. Einstein is quoted as saying that he had wanted to help toward the establishment of "an international arbitrating and regulative authority," but the German word he used is "ueberstaatlich," the correct translation of which is "supranational," not "international." His ideas were more radical than the translation "international" suggests.

49. Seelig, op. cit. in n. 44, p. 45, and Hoffmann, op, cit. in n. 1, pp. 39, 94.

50. It has been said that the Kepler in Brod's novel *Tycho Brahes Weg zu Gott* is modeled after Einstein, but I find it impossible to discover any resemblances.

51. In *Einstein-Born Briefwechsel,* op. cit. in n. 15, p. 21, Einstein mentions a "zionistisch verseuchten kleinen Kreis" in Prague.

52. Ibid., p. 36.

53. See in *Einstein in Berlin,* Teil II, op. cit. in n. 2, the materials listed in the section "Kampf um die Relativitaetstheorie," pp. 135–49, but details about the personal reactions of Einstein and of some of his physicist friends can be found in *Einstein-Born Briefwechsel,* op. cit. in n. 15, pp. 57–60, and in *Albert Einstein/Arnold Sommerfeld, Briefwechsel,* edited by Armin Hermann (Schwabe, Basel, 1968), pp. 69–72. The debate on the Nauheim Congress between Einstein and Lenard is described by Hermann Weyl, "Die Relativitaetstheorie auf der Naturforscherversammlung in Bad Nauheim," *Jahresbericht der Deutschen Mathematiker-Vereinigung* **31**, 59–63 (1922).

54. See the exchange of letters between the Minister of Education Haenisch and Einstein in *Einstein in Berlin,* Teil I, op. cit. in n. 2, pp. 203–4.

55. See *Einstein in Berlin,* Teil II, op. cit. in n. 2, pp. 138, 140.

56. See Ibid. p. 48; Einstein's "Festvortrag" was entitled "Geometrie und Erfahrung."

57. See *Einstein in Berlin,* Teil I, op. cit. in n. 2, p. 209; the peace section (Friedensklasse) of the "Pour le Merite" consisted of thirty German and thirty non-German members (scientists, scholars, writers, artists, musicians); in the case of the death of a member, the other members selected a successor.

58. "habe ich in diesen Tagen erlebt, dass Berlin die Staette ist, mit der ich durch menschliche und wissenschaftliche Beziehungen am meisten verwachsen bin. Einem Ruf ins Ausland wuerde ich nur in dem Falle Folge leisten, dass aeussere Verhaeltnisse mich dazu zwingen."

59. In general see Gerald E. Tauber, "Einstein and Zionism" in *Einstein: A Centenary Volume,* op. cit. in n. 19, pp. 199–207.

60. Kurt Blumenfeld, *Erlebte Judenfrage* (Deutsche Verlagsanstalt, Stuttgart, 1962), p. 127.

61. Ibid., p. 142.

62. There are some difficulties in explaining Einstein's demand for a withdrawal of Rathenau. Einstein himself was by no means inclined to refrain from political activities; his work at the League of Nations Commission, which extended through the 1920s, was, as he himself admitted, not without political implications, and in July 1932 he was one of the prominent signers of an appeal for forming a united front against Fascism by the German Socialdemocratic and the German Communist party (see the reproduction of the appeal in *Einstein in Berlin,* Teil I, op. cit. in n. 2, pp. 224, 225). Briefly, he did not see any reason for refraining from participation in German politics. Einstein might have drawn a line between exerting one's citizen rights and assuming a government job. But it is also possible that, in his conversation with Rathenau, he was influenced by Kurt Blumenfeld, who, as a Zionist, was certainly opposed to participation in German politics. Then the influence of Blumenfeld might also be seen in the highly strange message of condolence that Einstein sent to Rathenau's mother after the murder, in which he characterized Rathenau as "eine von den grossen Juedischen Gestalten, die dem ethischen Ideal der Menschenversoehnung mit Aufopferung des Lebens sich hingegeben haben" — an unusually inappropriate characterization, as Blumenfeld (*Erlebte Judenfrage,* op. cit. in n. 60, p. 145) himself later implied. Briefly, the most likely explanation — which I gave in the text — is that, in counseling Rathenau to retire, Einstein believed himself to be acting in the interest of German Jews, who would be less under attack.

63. See *Einstein in Berlin,* Teil I, op. cit. in n. 2, p. 231, where Einstein tells the German Ambassador in Tokyo that "nach dem Rathenau-Mord begruesste ich es allerdings sehr, dass mir die Gelegenheit einer laengeren Abwesenheit aus Deutschland gegeben war, die mich der zeitweilig gesteigerten Gefahr entzog."

64. See Albert Einstein, *Lettres à Maurice Solovine* (Gauthier-Villars, Paris, 1956), p. 30.

65. *Einstein on Peace,* op. cit. in n. 4, p. 100.

66. Ibid., p. 237.

67. Ibid., p. 240.

2. ALBERT EINSTEIN: ENCOUNTER WITH AMERICA

Harry Woolf

In one of the most moving moments in the Oratorio *Judas Maccabeaeus,* Georg Friedrich Händel has the tenor proclaim: "Sound an alarm, sound an alarm, your silver trumpets sound, and call the brave, and only brave around. Who listeth follow, to the field again. Justice with courage is a thousand men."[1]

We who celebrate the centennial of Albert Einstein's birth should sound those silver trumpets, of course — and in our collective bravery gathering together, indeed we do! But equally, we should take alarm at the fanfare, for although his achievements, his values, and his beliefs propelled him to the center of the world's arena, we remember his preference for the quieter wings off stage, for places like Zürich and Caputh, and Princeton, New Jersey.

Yet it would be irresponsible in the extreme not to take notice of this anniversary, and the world will, reflecting its own vanities and values, from the self-serving to the nobly sacrificial, from gross display to refined homage, and all that falls between. Yet whatever personal discomfort we may feel at all the sound and the fury, let us remember how rarely we are given the kind of historical opportunity to celebrate simultaneously great genius and the good man in whom it arose, one whose public concerns and private decencies were as warm and wide as his learning was deep.

In our desire to tell it all, during this celebratory season, too much light perhaps will be thrown on too little, and the small shadows that reveal a simple human shape vanish under the flattering intensive glare. Thus, in the course of these days here and elsewhere we will cast Einstein gigantically in bronze, analyze his childhood, scrutinize his vocabulary, dissect him minutely on questions of quanta, and consider a universe of supergravity and black holes that is a direct consequence of his revolutionary recasting of the fundamentals of physics. All this is perhaps as should be as the house of learning assembles to celebrate one of its own, but in adding my small voice to the great chorus, I wish only to sing one song, that of the itinerant and the migrant who found his peace in America.

Einstein's encounter with America took several forms and began well before the general intellectual migration to the United States launched by the accelerating success of Fascism and Nazism and the collapse of traditional European culture before the Second World War. Others

Harry Woolf (ed.), Some Strangeness in the Proportion: A Centennial Symposium to Celebrate the Achievements of Albert Einstein
ISBN 0-201-09924-1

have detailed elsewhere, in reminiscences, analyses and assessments, the wider aspects of the intellectual movement to America in the thirties. In concentrating on Einstein as part of that story, I do not mean to minimize the larger issues — the receptivity to science, which already spoke of a sophistication and maturity in the United States that Americans at large were unaware of, as well as some of the same sensibility, awareness, and capability that were equally there for the congeries of historians (especially art historians), architects, psychologists, and others who came. On the other hand, some artists, literary critics, and philosophers never found quite as congenial a cultural home in America, and glibly perhaps we have attributed this to the practical, the empirical, even the commercial overtones of American life. However, I think this is too convenient a self-inflicted wound, one that removes us too readily from the painful, searching campaign required to reach a greater understanding and comprehension of the matter.

Albert Einstein first came to America in 1921, two years after the dramatic announcement, at a joint session of the Royal Society and the Royal Astronomical Society, of the confirmation of the prediction of the deflection of light by the sun's gravitational field — an idea and a calculable consequence derived directly from his general theory of relativity. Certain aspects of his initial intellectual encounter with America were brought to a sharp focus by the unexpected drama of these scientific events and are perhaps best summed up for us by a sequence of headlines that appeared in *The New York Times* from November 9 to November 21, 1919:

1. "Eclipse Showed Gravity Variation. Diversion of light rays accepted as affecting Newton's Principles. Hailed as Epochmaking. British Scientist calls the discovery one of the greatest of human achievements."[2]
2. "Lights Askew in the Heavens. Men of Science more or less agog over results of eclipse observations. Einstein theory triumphs. Stars not where they seemed or were calculated to be, but nobody need worry. A book for 12 wise men. No more in all the world could comprehend it, said Einstein, when his daring publishers accepted it."[3]
3. "Don't worry over new light theory. Physicists agree that it can be disregarded for practical purposes. Newton's law is safe. . . ."[4]
4. "A new Physics, based on Einstein. Sir Oliver Lodge says it will prevail, and mathematicians will have a terrible time. . . ."[5]

Although by 1905 he had made extraordinary contributions to basic science, having published the three great papers that revolutionized physics — the statistical analysis of Brownian motion, which definitively established the atomic nature of matter, the electrodynamics of bodies in motion (the special theory of relativity), and the quantum explanation of the photoelectric effect (the work for which he was later given the Nobel Prize) — it was strangely the demonstration of that predicted measurement of less than two seconds of arc that brought the ascendent trajectory of his reputation to an extraordinary height. It carried his scientific fame beyond the boundaries of physics and brought him public attention beyond the borders of national life. But he did not come to America in 1921 for the cause of science.

Out of a somewhat ambivalent relationship with Judaism that had characterized his upbringing, out of his differences with the Jewish community of Berlin in 1920, and through both of those experiences symbolically more than substantively at first, Einstein emerged to identify himself with the Jewish struggle for justice and ultimately for survival. Although a pacifist and a socialist virtually by nature, and beyond nationalism by instinct and by conscience, he recognized in the rising virulence of anti-Semitism an assembly of malevolent forces not only capable of destroying his own people, but of infecting the larger culture with the disease to end all decencies. Thus, by 1921, when Chaim Weizmann asked Einstein to join him in an American fund-raising campaign to establish a Jewish university, he accepted gladly. In recalling the crossing to America with Einstein, Weizmann said: "Einstein explained his theory to me every day and on my arrival, I was fully convinced that he understood it."[6]

The news of his forthcoming trip brought a flood of invitations and a thunderous, enthusiastic welcome when he actually arrived in New York on April 2, 1921. But well before that had taken place, an enterprising reporter from the *New York Evening Post* had interviewed him just before his departure, and the translated comments by Einstein that came to the American public this way revealed his early sense of America — part measure, part vision:

> There can be no peace . . . nor can the wounds of war be healed until . . . internationalism is restored. . . . Internationalism, as I conceive the term, implies a rational relationship between nations, mutual cooperation for mutual advancement. . . . Here . . . is where science, scientists and especially the scientists of America, can be of great service to humanity. Scientists . . . must be pioneers in this work.
>
> In this matter of internationalism America is the most advanced among nations. It has what might be called an international 'psyche.' The extent of this internationalism was evidenced by the initial success of Wilson's ideas for a world organization as well as by the popular acclaim with which his ideas were greeted by the American people. That Wilson failed to carry out his own ideas is beside the point. What is important and indicative of the American state of mind is the enthusiastic response of the American people to those ideas. American scientists should be among the first to attempt to broaden and carry forward the idea of internationalism.[7]

Thus, he projected the unity of science and peace, and his intensive four-week tour in turn gave him a first reading of the country that, though he did not know it then of course, was to become his final home. Keys to the city, an invitation to the White House, and sitting silently for the most part besides an eloquent Weizmann on one platform after another — for he hardly spoke any English as yet — gave him the impression of an energetic and philanthropically vigorous America, for they raised enough money to begin building the medical faculty of the future Hebrew University. But interspersed between these events were visits to Columbia, Harvard, and Princeton for contacts and exchanges with scientific colleagues that brought him an awareness of the actual American academic scene and prepared the intellectual ground for his future implantation on American soil. Princeton University was wise enough to accord him an honorary degree and to capture him for four lectures eventually translated and published as the "Meaning of Relativity," the many editions of which printed since that time continue to testify to his great expository skills.

The first American visit, excessively rapid and scattered as it was, and severely limited by Einstein's modest command of English, nonetheless confirmed two significant matters for him: the sense of his own role as a Jew and his firm commitment thereafter to the Zionist cause, on the one hand, and the vigor and quality of American sicentific life on the other. Interviewed by a Berlin newspaper in July 1921 after his return, he informed the reporter that in some scientific areas the baton of leadership was about to pass into American hands. He was charmed too, and I quote from the same Berlin newspaper, by "an unprejudiced, youthful and sensitive people . . . [with their] simple camaraderie and inoffensive chummy way."[8] "Much is to be expected from American youth," he also said, "a pipe as yet unsmoked, young and fresh."[9]

The press and public stayed with him throughout the trip, with the theory of relativity making the front page and his name and his photograph omnipresent in the media. Interviews were often silly and repetitious and after the lengthy one that initiated them all, he closed the press conference with the quip, "Well gentlemen, I hope I have passed my examination."[10] The months that followed clearly proved that he had.

A decade was to pass before he returned again, years of turmoil and tensions that bore witness to the collapse of the German Republic in which he believed politically. At the same time there emerged before him a rising American scientific Cytheria towards which his scientific life was drawn. Invitations to lecture in the United States, and to remain, abounded, and in at least one case of record — the invitation from the Johns Hopkins University — he felt that he had been offered too much money for his talent. By the winter of 1930, however, he returned for the first of a series of winter visits to the California Institute of Technology, responding to Robert Millikan's invitation to share in the burgeoning scientific life of America there and elsewhere. But his travels were now that of a world figure, a personage to be followed, an oracle to be consulted. From his ship at anchor in New York, where again he was besieged by reporters, he said:

> As I am about to set foot again on United States soil, after an absence of ten years, the thought uppermost in my mind is that this country, today the most influential on earth . . . constitutes a bulwark of the democratic way of life. . . . Your country has demonstrated, by the work of its heads and its hands, that individual freedom provides a better basis for productive labor than any form of tyranny. . . . Your political and economic position is so powerful . . . that . . . you can break the tradition of war from which Europe has suffered throughout its history. . . . Destiny has placed this historic mission in your hands. . . . Inspired by these hopes, I salute you and the soil of your country. I eagerly look forward to renewing old friendships and to broadening my understanding in the light of what I shall see and learn while I am among you.[11]

Like De Tocqueville long before him and other travelers to the New World, Einstein came to express the European dream of America, the risen giant, unsullied by a dark and dismal past, clean and pure, the nation with a mission to rectify the twisted tale of yesteryear. In turn, the enthusiastic American embrace of so unorthodox a hero must surely have been based on the same vision, shared by so many — a vision of manifest destiny reenforced.

The visit to California, largely scientific, initiated new friendships, enriched old ones, and through both enhanced the scope and depth of scientific discussion in the United States. Robert Millikan, George Ellery Hale, Albert Michelson, Edwin Hubble, names to conjure with in the story of American science, were part of that early Pasadena circle. Crowds formed wherever he went, and newspaper headlines dutifully noted their presence, while also checking off visitors like Helen Keller and Norman Thomas. In Germany the drums of disaster were beating louder and Einstein would record in his shipboard diary of December 6, 1931: "I decided today that I shall essentially give up my Berlin position and shall be a bird of passage for the rest of my life. Gulls are still escorting the ship, forever on the wing. They are my new colleagues."[12]

It was also in the warm sunshine of Pasadena that the trajectories of Albert Einstein and Abraham Flexner first crossed. The author of the famous Flexner report on medical education — a document that led to the dramatic reform of the nation's medical education and research efforts and set the model for most medical education since — Abraham Flexner was then at the zenith of his fame as an educator. He was soon to publish his analysis of higher education, *Universities; American, English, German,* and he had already been approached by Louis Bamberger and his sister Mrs. Felix Fuld as to how they might best employ their fortune for the advancement of learning in America. Informal discussions between Flexner and the Bambergers took place through 1929 and 1930, with the educator and the philanthropists already coming together in the design of a new institution devoted to research and learning and, it is sometimes forgotten, to its transmission to others as well! By the spring of 1930 the formal process to establish the Institute for Advanced Study was underway with the Bambergers seeking the services of Flexner as its first director and writing to prospective trustees to say:

> It is our hope that the staff of the institution will consist exclusively of men and women of the highest standing in their respective fields of learning, attracted to this institution through its appeal as an opportunity for the serious pursuit of advanced study. . . . It is fundamental in our purpose, and our express desire, that in the appointments of the staff and faculty as well as in the admission of workers and students, no account shall be taken, directly or indirectly, of race, religion, or sex. We feel strongly that the spirit characteristic of America at its noblest, above all the pursuit of higher learning, cannot admit of any conditions as to personnel other than those designed to promote the objects for which this institution is established, and particularly with no regard whatever to accidents of race, creed or sex.[13]

Within the framework of this philosophy and with the economic wherewithal with which the Bambergers were initially to endow the Institute for Advanced Study, Flexner could not refuse the invitation to become its first director. Although he was sixty-four in 1930 and entitled to the venerable retirement that a long and successful career had generated, the excitement of new intellectual possibilities captured him. As Flexner came to form the actual character of the new Institute, partial models were to be found in All Soul's College at Oxford, the Collège de France, and the young Rockefeller Institute for Medical Research that his brother Simon then directed. But his particular paradigm was the Johns Hopkins University and the very model of a model educator, its first president, Daniel Coit Gilman. In his autobiography Flexner clearly establishes the affiliation:

> In retrospect, President Gilman's wisdom and courage appear unmatched in the history of American education. . . . Mr. Gilman set an example that few university executives have yet imitated: he travelled through America and western Europe in order to confer with outstanding scholars and scientists of the world before making an important appointment. Having chosen his key men, he let them alone. Sixty years later, in creating the Institute for Advanced Study and in selecting its original staff, I adopted Gilman's procedure.[14]

There is a double significance in this account, for just as the establishment of the Johns Hopkins University in 1876 represented the migration of the German higher academic system to America, so too did the establishment of the Institute for Advanced Study, albeit unwittingly, profit from the transfer of German educational and research achievements, and innocently guarantee its continuity in a new environment. Devoted, in Flexner's words, to "minds that are fundamental in their searching, whatever the spring that moves them — curiosity, pity, imagination, or practical sense,"[15] the Institute for Advanced Study also came to mark the visible maturity of American education and research, and within its own modest boundaries to create the locus for the interaction and the extension of a European and an American ideal.

Thus, in January 1932, when Einstein arrived at the California Institute of Technology, Flexner was out recruiting, reiterating the Gilman enterprise in seeking a faculty for the new institution. Robert Millikan had invited Einstein to that institute, where under his own leadership as well as George Ellery Hale's another transformation in American higher education and research was underway, as Einstein's second visit there, at the time, demonstrates. Obviously, we do not know exactly what was said when Flexner and Einstein met, save that Flexner intended only to consult with Einstein about the nature and future form of the Institute, and in turn, he received an invitation from Einstein to continue the conversation in Oxford later in the year. There in the following spring Flexner offered him a professorial position with the words, "You would be welcome on your own terms." Einstein expressed serious interest in the possibility and asked Flexner to join him at his summer home in Caputh outside Berlin.[16] There, in June, at the conclusion of a day's discussion about how the Institute for Advanced Study would develop, Einstein said: "Ich bin Flamme und Feuer dafür."[17] Thus the bird of passage was to take wing again, but this time to a final nest.

In their origin, as in their individual histories, the story of Albert Einstein in America and the growth of the Institute for Advanced Study are an intertwined tale too complex to tell here, with innumerable divergent chapters separate from our present purpose. His permanent arrival in 1933, its meaning for science and learning in America, is best summed up in a comment by Paul Langevin, a good friend and scientific colleague. When he heard that Einstein had accepted Flexner's offer, he said: "It's as important an event as would be the transfer of the Vatican from Rome to the New World. The Pope of physics has moved and the United States will now become the center of the Natural Sciences."[18] The gravitational effect of his presence at the Institute for Advanced Study was enormous, attracting others, even outside his field, like the distinguished mathematician Hermann Weyl, to come and stay.

Not all of America welcomed him; some superpatriotic groups protested his political and pacifist views, urging the State Department to deny him a visa. Indeed, in 1932, an extremely vocal women's group of this kind succeeded in having his visa temporarily delayed, an act that elicited from Einstein the kind of irony that often characterized his wit. "Never before," he wrote, "have I been so brusquely rejected by the fair sex; at least never by so many of its members at once!"[19]

With Einstein's appointment in hand, and the Institute for Advanced Study launched with so extraordinary a first step, Flexner turned to Oswald Veblen for help in formally establishing a School of Mathematics. As a significant contributor to mathematics itself, with a broad knowledge of the profession and an entrepreneurial skill matching Flexner's, they were able, between 1933 and 1935, to persuade James Alexander, John von Neumann, Marston Morse, and Hermann Weyl to join the permanent faculty. By 1935 the School of Economics and Politics was formed and in 1936, the School of Humanistic Studies, with such luminaries as Erwin Panofsky and Benjamin Meritt as faculty. The four schools — Mathematics, Natural Sciences, Historical Studies, and Social Science — of today's Institute for Advanced Study are thus natural derivatives of the original design. With a visitor program of about 160–170 scholars per year and a permanent faculty of twenty-three testifying to ongoing vitality, the proper place of the Institute for Advanced Study in the academic system of the free world is clearly defined. It is where it ought to be, at the intersection between the known and the unknown, between the commitment to search and to understand and the commitment to teach and to transmit. In this sense, if it is not presumptous to say so, the Institute for Advanced Study is an institutional embodiment of Einstein's impact upon America.

As the world's descent into disaster accelerated through the thirties, Einstein's Princeton peace eroded. The succession of quiet, pleasant days devoted to study and research seemed to characterize his life at the Institute, but even a cursory glance at his correspondence, the calendar of his visitors, or the record of his actions suggest otherwise. In the spring of 1932 he had shared the speaker's platform with Lord Ponsonby at a meeting of the Joint Peace Council about disarmament failures. Under the auspices of the International Institute of Intellectual Cooperation, he had joined Freud in the work that led to the publication of *Why War?* in 1933. January of that year found him addressing a student group in Pasadena; July, founding (with Lord Davies) the New Commonwealth Society, devoted to the idea of an international police force; October, speaking at Royal Albert Hall at a mass meeting for the Refugee Assistance Fund, participating there with James Jeans, Rutherford, William Beveridge, and others; and December, at a World Peaceways' dinner that included Sinclair Lewis, Frank Kellogg, and Irving Langmuir. Indeed we could outline the years 1934 to 1939 in the same way, and more, delineating his participation in innumerable meetings and conferences from the National Conference of Christians and Jews to the tercentennary of higher education in America (1936). And even though, in February 1939, in lieu of a statement he could not personally deliver on behalf of the "Defense Fund for the Humanities and Intellectual Freedom," he was to say: "I am neither an effective speaker nor sufficiently versed in the English language to express myself orally without the danger of misunderstanding,"[20] there was never any doubt about where Einstein stood, or about the meaning of his endorsement of a cause!

Primarily to help Leopold Infeld improve his financial future, but also because he believed in its importance, Einstein agreed to join him in producing "A popular book containing the principal ideas of physics in their logical development."[21] The publication of *The Evolution of Physics* in 1938 accomplished its purpose, bringing the desirable financial rewards to Infeld and undesirable publicity to Einstein, for it was reviewed everywhere. *Newsweek* referred to him as "Germany's no. 1 refugee" and *Time* spoke of "Exile in Princeton," with "Notes on Evolution of Physics by Einstein and Infeld."[22]

"Germany's no. 1 refugee" stood at the top of a list of more than 1600 scholars, of which about one-third were scientists. His assistance to Infeld, while collaborative and obviously more elaborate than the help he gave others, nonetheless typifies much of what he did in support of so many. More than 100 refugee physicists, mostly of German or Austrian origin, became part of American physics between 1933 and 1941, and Einstein helped many in the United States and elsewhere, with letters of support and recommendation. Indeed, his generosity sometimes exceeded his judgment, and during that epoch alone, the value of an Einstein letter could be momentarily and partially discounted. Nevertheless, for the American public and the world at large, he became, as Laura Fermi has remarked in her book *Illustrious Immigrants,* "A symbol of persecuted European genius and . . . a measure of the stature of the cultural migration."[23]

The troubles of the thirties, the sense of a Göttedämerung brought despair to Einstein as it did to so many others. A lonely man to begin with, his mood of separation and distance was enhanced by personal losses of family and friends, and is perhaps best captured in this paragraph from a letter of January 1939 to his good friend the Queen of Belgium:

> I have been too troubled to write in good cheer. The moral decline we are compelled to witness and the suffering it engenders are so oppressive that one cannot ignore them even for a moment. No matter how deeply one immerses oneself in work, a haunting feeling of the inescapable tragedy persists.[24]

But even in such a disheartened state, Einstein's resiliency and the inner sources of his strength are revealed in the very next passage:

> Still, there are moments when one feels free from one's own identification with human limitations and inadequacies. At such moments, one imagines that one stands on some spot of a small planet, gazing in amazement at the cold yet profoundly moving beauty of the eternal, the unfathomable: Life and death flow into one, and there is neither evolution nor destiny; only being.[25]

The American citizenship that Einstein formally acquired on October 1, 1940, finally stilled the occasional social and political hostility he had known in previous years. An editorial in the *Christian Century* epitomized the virtually universal sentiment of the nation in reaction to his citizenship: "Professor Einstein has been an exile in recent years, but he is an exile no longer. He is at home — an American in America."[26] Whether or not that appellation would ever really apply, Einstein certainly anticipated his citizenship eagerly. Commenting on its forthcoming on his sixtieth birthday in 1939, he said:

> My birthday affords me the welcome opportunity to express my feelings of deep gratitude for the ideal working and living conditions which have been placed at my disposal in the United States, and I am also very happy over the prospects of becoming an American citizen in another year. My desire to be a citizen of a free republic has always been strong and precipitated me in my younger days to emigrate from Germany to Switzerland.[27]

He spoke frequently in the years to come about the significance of America's political freedom, and about the importance of the moral character of its people in the unfolding struggles with Fascism. Thus he participated in the "I am an American" series sponsored by the United States Immigration and Naturalization Service, and with the outbreak of the war for the United States, on December 7,1941, he sent to the White House a "Message for Germany" unequivocally restating his long-established position:

> This war is a struggle between those who adhere to the principles of slavery and oppression and those who believe in the right of self-determination both for individuals and for nations. Man must ask himself: Am I no more than a tool of the state? Or is the state merely an institution which maintains law and order among human beings? I believe the answer is that . . . the only justifiable purpose of political institutions is to assure the unhindered development of the individual and his capacities. This is why I consider myself particularly fortunate to be an American. America is today the hope of all honorable men who respect the rights of their fellow men and who believe in the principles of freedom and justice.[28]

Perhaps the most powerful interaction between Albert Einstein and the United States occurred through a pair of famous letters dated August 2, 1939, and March 7, 1940, to President Franklin Roosevelt. In these, Einstein was more the courier than the originator of the contents, for they expressed the thoughts of Leo Szilard, Eugene Wigner, and Edward Teller about the urgency of developing nuclear weapons. Nuclear fission had been demonstrated in Germany by Otto Hahn and Fritz Strassmann. The German occupation of Czechoslovakia had brought a halt to the sale of uranium and evidence pointed to intensive nuclear research by German scientists. The story of the enormous atomic research effort that these letters brought into being is widely known and has been well told elsewhere. Except for some minor consulting for the U. S. Navy, these letters constitute Einstein's scientific war effort. His political statements, of course, also served, as did his name in appropriate places. For example, an unexpected contribution took place when two of his manuscripts were donated to a war bond rally. Their sale brought in $11.5 million!

Although derived from some of his own ideas and stimulated by his letters to Roosevelt, the use of nuclear weapons disturbed him greatly. In response to the growth of a postwar policy about their development with which he disagreed and the rise of a reactionary national mood, Einstein emerged again to speak publicly about his concerns. In May 1946 he became a member of the Emergency Committee of Atomic Scientists (publishers of the *Bulletin of Atomic Scientists*); he served the committee faithfully for more than two years in a variety of ways — participating in radio broadcasts, fund raising, conferences, and the like. He joined in the call for the international control of atomic energy and sought to educate the American people about its uses. If there was danger, there was also opportunity, for the existence of atomic weapons, he felt, demanded a supranational system, a world government. Towards this end, he devoted his public statements, not as naively as some would have us believe, for he was well aware of the limitations of such government, but he thought it "the only conceivable machinery which can prevent war [for] . . . with technology at its present level, even a poor world government is preferable to none."[29] The deep concern with world peace suggested a return to the pacifism of his earlier years, but this is best set right by his own statement in the *Bulletin of Atomic Scientists* of January 1951:

> I am not what you might call a religious pacifist. Besides I consider it preferable for men to fight rather than to allow themselves to be butchered without lifting a finger. That was just about the alternative in the case of Hitler Germany. Nor do I favor unilateral disarmament. What I advocate is an armed peace under supranational control.[30]

Such a position, enhanced by his sensitivity to expanding militarism, the sense that armament races succeed, that is, that someone momentarily wins and war follows, also led him to criticize strongly the foreign policy of the major powers, that of the United States above all. His personal history, as a European if nothing else, was enough to alert him to the erosion of freedom that such policies generated, and he sought again and again to sound that alarm for the American people.

The result was a new encounter with America, as the reactionary press, attacking him for his position on a specific issue or his support of a particular person under scrutiny, now sought to belittle his judgement. The right-wing *American Mercury* said: "It is time that the old professor's pretensions to political sagacity were deflated," and it referred to him as "pro-communist."[31] It is difficult to say how much of the American public shared these opinions at that time, and even though *Newsweek* and *The New York Times* occasionally editorialized against some of his statements, he was not as isolated as he felt himself to be. Others, many others, shared his values and joined him in the never-ending process that is part and parcel of the life of a free people. More than anything else perhaps, his public defense of J. Robert Oppenheimer — a man with whom he differed profoundly on many issues — epitomized Einstein's primary emphasis on political liberty and human rights.

As he had earlier, he continued to work for Israel and the Jewish cause until final illness prevented him from doing so. The presidency of Israel was offered to him in 1952. Unable to accept, he wrote to say:

> I am deeply moved by the offer from our State of Israel and at once saddened and ashamed that I cannot accept it. All my life I have dealt with objective matters, hence I lack both the natural aptitudes and the experience to deal properly with people and to exercise official functions. For these reasons alone I should be unsuited to fulfill the duties of that high office, even if advancing age was not making increasing inroads on my strength.
>
> I am more distressed over these circumstances because my relationship to the Jewish people has become my strongest human bond, ever since I became fully aware of our precarious situation among the nations of the world.[32]

That statement reveals again the essential honesty, with self as with others, that characterized his entire life. His encounter with America, from his gratitude for its hospitality, through his pride in its traditions of freedom and tolerance, to his despair when it fell from its own proclaimed standards bore that same honest imprint. In that way he served his final homeland well, for he fulfilled its own ideal of the good citizen in a free society.

NOTES

1. Georg Friedrich Händel, *Judas Maccabaeus. An Oratorio in Vocal Score,* edited and the pianoforte accompaniment arranged by Vincent Novello (Novello, Ewer, New York, n.d.), pp. 119–20.
2. The New York Times, 6 (9 November 1919).
3. Ibid., 17 (10 November 1919).
4. Ibid., sec. II, p. 1 (16 November 1919).
5. Ibid., 17 (25 November 1919).
6. Quoted in Helen Dukas and Banesh Hoffmann, *Albert Einstein, The Human Side* (Princeton University Press, Princeton, N. J., 1979), p. 64.
7. *Einstein on Peace,* edited by Otto Nathan and Heinz Norden (Schocken Books, New York, 1968), pp. 44–5.
8. "Einsteins amerikanische Eindruecke," interview in Vossische Zeitung, suppl. 1 (10 July 1921).
9. Quoted in Philipp Frank, *Einstein: His Life and Times* (Knopf, New York, 1957), p. 186.
10. Quoted ibid., p. 180.
11. *Einstein on Peace,* op. cit. in n. 7, pp. 115–16.
12. Ibid., p. 155.
13. Louis Bamberger and Mrs. Felix Fuld, "Letter Addressed By Founders To Their Trustees," Newark, N. J., 6 June 1930, in the Institute for Advanced Study, "Organization and Purpose," *Bulletin No. 1* (Oxford University Press, New York, 1930), pp. 2–3.
14. Abraham Flexner, *An Autobiography,* rev. version of Abraham Flexner, *I Remember* (Simon and Schuster, New York, 1960), p. 27.
15. Ibid., p. 244.
16. Ibid., p. 251.
17. Ibid., p. 252.
18. Langevin is quoted in Robert Jungk, *Brighter than a Thousand Suns: A Personal History of the Atomic Scientists* (Harcourt Brace and World, New York, 1956), p. 46.
19. *Einstein on Peace,* op. cit. in n. 7, p. 206.
20. Ibid., p. 282, as paraphrased by Nathan from the original letter.
21. Leopold Infeld, *Quest: The Evolution of a Scientist* (Doubleday, Doran, New York, 1941), p. 312.
22. "Einstein and Progress," Newsweek **11,** 21 (4 April 1938); Time **31,** 39–44 (4 April 1938).
23. Laura Fermi, *Illustrious Immigrants: The Intellectual Migration from Europe, 1930–41* (University of Chicago Press, Chicago, 1968), pp. 104–5.
24. *Einstein on Peace,* op. cit. in n. 7, p. 282.
25. Ibid.
26. Anon., "Einstein Is Now an American," The Christian Century **57,** 1268 (16 October 1940).
27. Einstein's statement is in *Science,* new series, **89,** 242 (1939).
28. *Einstein on Peace,* op. cit. in n. 7, p. 320.
29. Ibid., p. 469.
30. Ibid., p. 553.
31. J. B. Matthews, "Professor Einstein vs. Citizen Einstein," American Mercury **78,** 25–7 (June 1954).
32. *Einstein on Peace,* op. cit. in n. 7, pp. 572–73.

3. RELATIVITY AND TWENTIETH-CENTURY INTELLECTUAL LIFE

Ernest Nagel

Theories in the natural sciences often succeed in unifying apparently unrelated findings in diverse areas of research and are invaluable guides and rich sources of ideas in further inquiries in those disciplines. But such theories are frequently also employed to deal with questions that are radically different from those for whose resolution the theories were constructed, including questions arising in the study and conduct of human affairs. Indeed, theories that have been notably fruitful in the natural sciences have also been repeatedly used for ideological purposes, to endorse or to condemn current social institutions and practices, to support or to oppose proposed changes in public policy, or to serve as a foundation for some philosophical or theological system. The extension of a scientific theory to matters outside its initial scope may be justified by its fruits, and has in fact been highly successful in numerous instances. However, such proposed extensions are not always warranted, for they sometimes involve serious misconceptions of a theory's content or make use of vague, dubious, or irrelevant analogies.

A number of theories in the natural sciences can be cited to illustrate these general observations, but two familiar examples drawn from modern science will suffice to clothe them with some flesh. The Newtonian theory of mechanics and gravitation was for two centuries the foundation for systematic inquiry in most branches of physics. The theory was initially proposed to account for the motions of bodies, such as the planets, whose dimensions are negligible when compared with their mutual distances. But eventually the scope of the theory was enlarged to include the motions of solids and fluids, the thermal properties of gasses, and a variety of acoustic, optical, and electrical phenomena. The theory became the basis for the conceptions educated men formed of the cosmos; it inspired atomistic analyses of mental phenomena; it was used as a premise in arguments for a theistic religion and for a monarchical form of government; and it colored the imaginations of writers of *belles-lettres.* Newtonian theory was unquestionably fruitful for a long time in many branches of physical science; but it had questionable merit as a tool for exploring the subject matters of psychology and social inquiry.

Similarly, the Darwinian theory of evolution was first proposed as an explanation for the variety and the geographical distribution of biological species; but it eventually came to play a dominant role in just about every division of biological science, such as systematics, ecology, and

Harry Woolf (ed.), Some Strangeness in the Proportion: A Centennial Symposium to Celebrate the Achievements of Albert Einstein
ISBN 0-201-09924-1

embryology. Moreover, the theory was also employed to serve as the rationale for laissez faire economics, for imperialistic foreign policies, for strongly individualistic moral standards, and for various educational programs — uses of the theory that are now commonly recognized to be unwarranted.

Like other major theories of the natural sciences, Einstein's theory of relativity has not only entered into technical researches in the physical sciences and brought about a renewed interest in physical cosmology but has also made an impact on the general conceptions currently entertained of the nature of science itself, of the powers of the human intellect, and of the world men inhabit. It would be presumptuous of a nonscientist to attempt an account of the role relativity theory has played in the development of various special branches of science. I will make no such attempt, and will deal only with the import of the theory for a number of broad "philosophical" issues.

Perhaps the clearest and least controversial example of how relativity theory has affected the general climate of opinion is its influence on the philosophy of science.

The most obvious contribution relativity theory has made in that domain is the strong support it has given to the by no means novel idea that scientific theories cannot be established with demonstrative certainty. It has become a commonplace that even apparently firmly entrenched theories may have to be significantly modified, their scope limited, or abandoned in their entirety, because of new problems, new experimental findings, or new theoretical developments. Although most educated men have read about or heard of scientific revolutions, comparatively few in the present century had experienced a scientific revolution at first hand until the special theory of relativity made evident that classical mechanics could no longer be regarded as the secure foundation for all physical inquiry. To be sure, a number of professional scientists and philosophers continued for some years to maintain that Newtonian assumptions are unavoidably presupposed in all physical measurement and are therefore certainly true. But in the end many if not most students of the subject came to believe that the real lesson of the revolution brought about by relativity theory is that completely certain knowledge is unattainable in the empirical sciences, and that a search for it is therefore quixotic. In fact, an inclusive and eventually highly influential philosophy of science was constructed on the assumptions that not only is certainty unobtainable, but also that we can never ascertain whether a proposed theory or even a statement about some individual occurrence is true — so that what is called knowledge of the events and regularities of nature is no better than a "guess." It is remarkable that this essentially skeptical and pessimistic assessment of science has been endorsed by numerous practicing scientists, including some Nobel laureates. However this may be, that enervating wholesale skepticism must be distinguished from confessions of ignorance of most of the detailed mechanisms of nature — for example, from the view of a distinguished biomedical scientist that recognition of the depth and scope of our ignorance "represents the most significant contribution of twentieth-century science to the human intellect".[1] The humility manifested in such a confession is not an expression of skepticism, but of a conviction that the range of knowledge we have achieved is only a minute fraction of what goes on in nature.

Scientists have in general adjusted themselves easily to the notion that the human mind is fallible, in large measure because the belief that this is so is rarely an obstacle in conducting research. In fact, although a wholesale skepticism of the kind just described is sometimes *professed* by scientists, their behavior when engaged in inquiry is not congruous with their skeptical philosophical pronouncements: they show no hesitation in using and overtly acting upon scientific theories and experimental reports they declare to be merely guesses. Nevertheless, the profession of an unmitigated skepticism of the conclusions of inquiry may affect public opinion and become a source of strong opposition to the scientific enterprise, especially among those having little if any familiarity with the work of scientists. Moreover, as one critic of that enterprise has put it recently, the belief may become widespread that if every conclusion of scientific inquiry is forever open to revision, so that like Penelope's web the work of science is never done, science cannot provide a true conception of the world. But if the executive order of nature cannot be discovered through scientific inquiry, so many skeptics have concluded, the true nature of things must be found in some nondiscursive fashion — for example through intuition, mystic experience, or revelation. A wholesale skepticism of discursive reason often fortifies the romantic antirationalism and hostility to science that occur in every age.

A crucial step in Einstein's argument for the special theory of relativity was his reexamination of the way spatial and temporal intervals are measured in physics. As is well known, he found that although the notion of the simultaneity of two events plays a basic role in synchronizing clocks occupying positions at some distance from one another, especially when one clock is moving with a constant velocity relative to the other, physicists had assumed that the notion was sufficiently clear and required no further analysis. Einstein showed that assumption to be mistaken, and he proposed a definition of "simultaneity" in terms of procedures, or overt "operations" — operations that included among other things the transmission of light signals — used in applying the term in concrete situations. The definition, as Einstein was careful to point out, is neither true nor false, but is a convention whose adoption can be justified only by its convenience. On Einstein's definition the magnitude of the length of a body, as measured from a second body relative to which the first is moving with a constant velocity, is a function of that constant velocity and the velocity of light; and analogously, the mass of a body as well as the time interval between two events, when measured from such a moving frame of reference, are also functions of those velocities. In short, with the help of definitions that specified the procedures to be used in measuring distances, time intervals, and masses, Einstein derived the familiar Lorentz–Fitzgerald equations from a number of physical assumptions.

The idea of defining physical concepts in terms of the procedures to be employed in applying them was not original with Einstein. He was quite familiar with Ernst Mach's *The Science of Mechanics,* and responded favorably to Mach's positivistic approach in analyzing scientific concepts; and in Mach's operational definition of "mass," he had before him a model of how concepts should be explicated. Einstein's operational definition of "simultaneity" and of other concepts employed in mechanics was seen by many commentators as the methodologically fundamental feature of the special theory of relativity. An influential school of philosophic thought centered in Vienna, which eventually called itself "logical positivism" or "logical empiricism," was strongly sympathetic to Mach's phenomenalism. The school agreed fully with Mach's insistence that all concepts employed in science must be defined in terms of the procedures to be followed in applying them to concrete subject matters and that the procedures themselves are to be defined in terms of immediately apprehended sensory qualities. This strong requirement for admissible theory constructions was believed by members of the school to be completely satisfied by the special theory of relativity.

In the United States the physicist P. W. Bridgman also took that theory to be a model for satisfactory theory construction; and independently of logical positivism he developed the view that all concepts entering into scientific statements must be operationally defined. Bridgman's book expounding his views made quite a stir, and it became fashionable to profess operationalism.

Operationalism was often uncritically adopted in less well-developed disciplines, such as psychology and the social sciences, as a cure-all for their failings. The introduction of operational standards of definition and analysis undoubtedly contributed to the clarification of concepts in many domains of inquiry. On the other hand, the puritanical rejection of ideas that sometimes followed the adoption of operationalism, because no operational definitions could be given for them (for example, for the notion of photons), but which nonetheless may play important roles in inquiry, became an obstacle in some branches of science to the construction of imaginative and promising theories.

Einstein acknowledged his indebtedness to Mach's critique of Newtonian concepts, especially of the concepts of space, time, and mass. However, despite the fact that during the first decade of the century he frequently used the language of sensationalistic empiricism, it is not entirely clear whether at any time Einstein subscribed fully to Mach's phenomenalism. In any case, in later years he rejected as excessive the operationalist requirement that *every* term of a theory must be operationally defined. Thus, in his rejoinder to an essay by Bridgman, in which the latter criticized the general theory of relativity on the ground that, unlike the special theory, it employed nonoperationally defined ideas, Einstein pointed out that the full operational requirement for admissible theory construction is too restrictive. He admitted that the spatiotemporal coordinates employed in the general theory of relativity are not operationally defined or definable in terms of operations with measuring rods and clocks. He noted, however, that the accepted facts of observation and experiment do not *uniquely* determine a theory in science, and that a comprehensive theory (such as his general theory of relativity) is a "free creation" of the intellect, which may be *suggested* by such facts but is not "derivable" or "abstracted" from matters of sensory experience. He declared that therefore the general theory shows that Newton was mistaken in claiming that the concepts and postulates of mechanics can be deduced from observational data. But he also maintained that some of the *logical consequences* of a theory, though not necessarily all of them, must be statements all of whose terms are operationally defined, so that those statements, and thereby the theory itself, can be subjected to observational or experimental test. According to one report, when Einstein was once reminded that at one time he appeared to have embraced a thoroughgoing operationalism, he replied that this may be so, but that even a good joke should not be repeated too often. Although versions of phenomenalism continue to have adherents — indeed, some outstanding philosophers, among them Bertrand Russell, have contributed to the vast literature supporting that doctrine — no one has yet succeeded to the satisfaction of most students in defining any concept of either theoretical or experimental science exclusively in terms of directly apprehended sense data. Phenomenalism is at best a program of analysis, a program that in the opinion of a majority of students is most unpromising. Logical positivists eventually abandoned their earlier adherence to that doctrine in favor of a moderate operationalism like the one to which Einstein finally subscribed. It is such a moderate operationalism that has dominated most of the discussions in the philosophy of science during the second half of this century.

For the better part of two centuries, Newtonian mechanics was recognized by most physicists as thoroughly "intelligible" as well as true, and as the basic theory to which all other theories in physics must be reduced. As has been already mentioned, such a reduction was accomplished for a number of physical theories, among others for the theory of heat. To be sure, difficulties arose when attempts were made to reduce to mechanics (that is, to deduce from the basic Newtonian assumptions supplemented by a few others) a number of other physical laws, in particular laws dealing with electrical and magnetic phenomena. Nevertheless, since no phenomena had been encountered that clearly required a major modification of Newtonian assumptions, it continued to be believed that such attempts at reduction would eventually be successful. But this belief was finally abandoned when the mechanics of the special theory of relativity, which required substantial alterations in Newtonian mechanics, turned out to be in better agreement with the relevant experimental facts than is classical mechanics. On the other hand, the special theory of relativity was entirely compatible with Maxwell's electromagnetic theory, so that the latter theory could be retained substantially unchanged. In consequence, it was Maxwell's theory, rather than Newtonian mechanics, that came to be regarded as the fundamental theory to which all physical theories were to be reduced.

That status of electromagnetic theory was only temporary, and before long quantum mechanics replaced it as the basic theory in the physical sciences. It is of course impossible to say whether this state of affairs is a permanent one, even though attempts by Einstein and others to devise a theory more comprehensive than quantum mechanics, which would cover gravitational as well as electromagnetic and quantum phenomena, have thus far been unsuccessful. The successful reduction of a major scientific theory to another and hitherto apparently independent theory is an important event in the history of ideas. It is a step in the unification of different branches of science — a step illustrated by the absorption in the present century of chemistry by quantum theory — which may also produce large changes in men's conceptions of what the world is made of. The complete unification of the various special sciences by reducing their distinctive theories to a single basic one is an ideal that has been pursued since ancient times. It is still an unrealized ideal, and its pursuit has been and remains controversial.

One reason for the controversy is that the nature of reduction in the empirical sciences is often misunderstood. For the reduction of one theory to another has been construed as the *elimination* of some familiar and perhaps highly valued qualities from existence as "unreal," and in consequence as an impoverishment of the world. But such a construal is mistaken. For example, the reduction of the theory of heat to mechanics consists in the *logical derivation* of thermal laws, which contain terms referring to thermal properties such as temperature, from a set of premises that include various laws of mechanics. But the derivation would be impossible if the premises did not also contain statements with terms also referrring to those thermal properties. However, the reference to thermal properties in the premises is a tacit acknowledgment of their "reality." Accordingly, the reduction of the theory of heat to mechanics could not possibly establish the "unreality" of thermal properties.

It will be helpful for understanding the nature of reduction to consider also the currently much discussed question whether biological laws (for example, laws about the inheritance of eye color) are reducible to physicochemical ones. Distinguished biologists have argued that the reduction is impossible, and that the supposition that the reduction may be accomplished in the future makes no sense. The reason sometimes given for these claims is that color is an undeniable feature of human eyes, a fact that allegedly would have to be denied were the reduction possible. But the reduction would require no such denial. For were the biological law reducible to some physicochemical theory, it would have to be deducible from premises that must include, in addition to physicochemical assumptions, some statement to the effect that eye colors occur under specified physicochemical conditions — a statement that recognizes the existence and reality of colors. Were such a statement not contained in the premises, the supposed deduction would be spurious, and reductions in science would be self-defeating. Accordingly, the hostility that continues to be shown to efforts at unifying the sciences through the reduction of theories to some basic theory is directed at a fantasy that corresponds to nothing in scientific practice.

Unlike the advent of Newtonian theory in the seventeenth-century and of Darwin's evolutionary theory in the nineteenth, the replacement of classical mechanics and gravitation theory by the general theory of relativity has inspired little if any noteworthy poetry and music. Part of the reason for this is the undoubted fact that relativity theory is highly abstract, its ideas are remote from the common experiences of most men, and its content cannot be easily represented by literary images or patterns of sound. The point of these remarks was in effect endorsed by Marjorie Nicholson, a distinguished student of the influence of scientific ideas on literature. At a symposium on the unity of knowledge, at which a number of outstanding physicists and mathematicians discussed the current state of their disciplines, she expressed her feeling of frustration when she declared that she had little difficulty in understanding the work of Newton, but that the presentations of the symposiasts were completely unintelligible to her. Nonetheless, relativity theory has led to changes in the general philosophies men profess about the nature of things. A few of these changes must be examined, if only to illustrate some of the extrascientific uses to which relativity theory has been put.

It has been repeatedly claimed that according to relativity theory the events of nature cannot be characterized — as they had been on the strength of Newtonian theory — as the products of "blind" forces. On the contrary, the claim continues, events must be seen as the outcome of interactions between bodies, each of which moves "freely" on the shortest path (or geodesic) available in the region in which it is located, rather than being impelled to do so by "external" forces. Moreover, so it is repeatedly asserted, the "ultimate furniture" of the universe can no longer be regarded as consisting of inherently simple elements, each having a location and a motion in a space that exists independently of its contents. For relativity theory requires us to believe that the world is made up of events each of which has a structure, and whose components are "internally" (or "organically") related to one another and to the whole of which they are components. Indeed, the supposition that the world has constituents (whether these be bodies, events, or processes), each with an intrinisic nature that does not depend on the relations the constituent entity has to all the others, was baptized by the mathematician and philosopher Alfred Whitehead as a fallacy, "the fallacy of misplaced concreteness." Whitehead eventually developed an inclusive "philosophy of organism," which he believed only make explicit some of the "implications" of the general theory of relativity. According to that philosophy organically related processes are the ultimate realities, the stability of the system of processes being assured by a deity. Whitehead's philosophy was not the only one that was constructed on a foundation avowedly supplied by relativity theory. But it is perhaps the most influential metaphysical system in recent Anglo-American thought.

A quite different conclusion was reached by writers who saw in that theory a support for a "spiritual" or "mentalistic" conception of the universe. Expositions of the theory sometimes explain that according to the theory the spatiotemporal dimensions of objects are always relative to some "observer." Such formulations are then construed as saying that "observers" enter essentially into the composition of nature, that an uneliminable "subjective factor" is present in physical science, and that in consequence the age-old conflict between science and religion has become less sharp.

It is indisputable that the general theory of relativity, unlike Newtonian theory, does not invoke forces to account for changes in the velocity of bodies. Relativity theory formulates such changes (in particular those attributed in Newtonian theory to gravitational forces) in *geometrical* language; and Newtonian talk of forces is replaced by statements about the paths bodies follow, where the paths are geodesics specified by a geometry whose structure is determined by the distribution of matter. It is therefore misleading to interpret this state of affairs so as to suggest that because according to relativity theory bodies are not constrained to move by any *forces,* bodies are in some vague sense goal-directed when they move "freely" on their trajectories. It is equally misleading that according to relativity the motions of bodies conform to teleological laws, just as it is misleading to assert that because Newtonian laws can be formulated so as to illustrate the principle of least action the laws of classical mechanics are really teleological. It is of course true that according to relativity theory the motions of bodies along geodesics are their "unconstrained" or "natural" motions. But it is nonetheless also misleading to assert that relativity theory is a return to the physics of Aristotle, on the ground that according to him the "natural" motions of bodies are directed to their "natural places"; for one could say with equal warrant that Newtonian physics is also a return to Aristotle, since according to Newton the "natural" motion of a body not acted upon by any force is motion along a straight line with constant speed.

These various accounts of the import of relativity theory are all misleading, because their interpretations of the technical language of the theory depend on the use of vague and dubious analogies. This is not the occasion for asssessing Whitehead's philosophy of organism; but it is appropriate to note that his philosopy makes heavy use of such questionable analogies. It is also pertinent to point out that if Whitehead's fallacy of misplaced concreteness is indeed a fallacy, then every discursive utterence of daily life and the sciences commmits it, for every statement must of necessity deal with only a selection of the innumerable relations into which things enter.

Moreover, the claim that relativity theory requires a "mentalistic" view of the universe rests on a gross misunderstanding. The "observer" to which explanations of relativity theory refer is not a sentient being, but is a physical frame of reference equipped with some measuring device. Nothing can be concluded from this fact concerning the extent to which matter is mindlike or concerning the role of minds in the world.

A host of writers have tried to use relativity theory to support a variety of ethical, sociological, and epistemological theses. In many of these attempts it is tacitly assumed that the sense of the word "relativity" in the name of the theory is somehow similar to the meaning of the word "relativism" as the designation for doctrines in ethics, anthropology, or the theory of knowledge. For example, it is sometimes supposed that the substance of relativity theory is rendered by the statement that a body moving relative to one frame of reference may be at rest relative to another. This obvious truth, which is far from expressing the content of relativity theory, is then taken to yield such conclusions as that what is held to be true by one person or in one society may be rejected as false by another person or in another society, that what is regarded as moral in one culture may be judged as immoral in another, or that a policy favoring the interests of one social class may be inimical to the interests of another. The fallaciousness of such reasoning is too obvious to require comment, as is the inference from the theory of relativity that tolerance in human affairs is a virtue.

It has also been claimed that, since according to relativity theory the magnitude of the linear dimension of a rod is always relative to the reference frame in which the measurement is made, the theory shows that the "real" or "objective" length of the rod cannot be known by human beings, so that the objects of knowledge are our private or subjective sense data. Again, since the theory deals only with certain selected or abstracted features of the world, the theory is said to be about properties of our own "conceptual scheme," so that what is supposed to be the "objective and material universe" in fact consists of "little more than constructs of our own minds." But neither claim is sound. The "objective" linear magnitude of a rod *is* (by definition) the value obtained by measurements relative to some given reference frame; and this is something men *can* (and often do) know. As for the second claim, it confounds two quite different things: relativity *theory,* which is admittedly a creation of the scientist, and *the features of events and processes* that are prescinded in the theory. It is clearly not the *theory* itself, which expresses a "conceptual scheme" and is a linguistic entity, that is the subject matter of physics; it is those *features* that physicists study.

The moral of these critical remarks is that a proposed interpretation of a scientific theory that does not take into account how the theory actually functions in research and in the solution of concrete problems, but relies heavily on vague analogies suggested by the language of the theory, is bound to be seriously misleading if not entirely erroneous. Unfortuntely, many of the nonscientific uses to which the theory of relativity has been put illustrate the failure to observe this moral.

NOTE

1. Lewis Thomas, *The Medusa and The Snail* (Viking, New York, 1979), p. 73.

III. Developments in Relativity

4. EINSTEIN'S SCIENTIFIC PROGRAM: THE FORMATIVE YEARS

Gerald Holton

Albert Einstein's contributions will be studied as long as our civilization exists. But while scientists from their student days on will, on the whole, know of his work indirectly through the textbooks, Einstein's actual words, in his wide-ranging publications and correspondence, will be scrutinized chiefly by the historians of science. One may hope that Einstein would have approved; not only did he publish many essays on historical developments in science,[1] but he was on record, more than once, that a means of writing must be found that conveys the thought processes that lead to discoveries — showing how scientists thought and wrestled with their problems. Moreover, Einstein made the task of the historian easier than did many other scientists, because of the characteristic frankness and consistency in his writings. These traits of his will aid my task today, in a conference chiefly devoted to contemporary physics, to speak about some of the early steps on his path to relativity — steps that made that path a high road and caused the other, more fashionable ones of that day, to be seen eventually as blind alleys.

1. "Recognize the Unity of a Complex of Phenomena"

When viewed in terms of Einstein's early publications,[2] the road that brought him to the threshold of relativity began in apparently quite unimpressive territory. Einstein's first published article, entitled "Consequences of the Capillarity Phenomena," was sent to the *Annalen der Physik* in December 1900, a half year after Einstein had graduated from the E. T. H. in Zurich and a half year before his getting his first temporary job as a substitute teacher.

At the time, all the excitement in physics lay in a quite different direction. It was just a few years after the discovery of X rays, radioactivity, the electron. New experimental findings and new theories chased one another at a dizzy pace. The prolific Ernest Rutherford wrote to his mother in 1902: "I have to keep going, as there are always people on my track. I have to publish my present work as rapidly as possible in order to keep up the race."[3] Einstein was not ready for any of that. As he characterized it later, he was in the middle of years of "groping in the dark." He often remarked that his formal training had been spotty, although on his own he had worked his way through the volumes of classic lectures of Kirchhoff, Helmholtz, Hertz, and Boltzmann's *Gastheorie,* not to speak of Ernst Mach, to whose work Michael Besso had introduced Einstein

Harry Woolf (ed.), Some Strangeness in the Proportion: A Centennial Symposium to Celebrate the Achievements of Albert Einstein. Addison-Wesley Publishing Company, Inc., Advanced Book Program. ISBN 0-201-09924-1 soon after their first meeting in 1896. Einstein confessed later, "I had no technical knowledge."[4] But he added: "It turned out soon that the general overview ["allgemeine Übersicht"] over physical connections is often more valuable than specialist knowledge and routine."

But capillarity was by no means as dull as it now may seem. Even some five years later, a long paper on capillarity by G. Bakker in the famous volume 17 of the *Annalen der Physik* starts with the panegyric: "The theory of capillarity of Laplace was one of the most beautiful achievements of science"; he goes on to sing the praises of the subject as handled by Gauss, Young, Gibbs, and F. Neumann. Young Niels Bohr's first research, completed in 1906, was on the closely related problem of surface tension of water. For Einstein, the capillarity paper was the first of nine publications indicating his deep interest in thermodynamics and, later, statistical mechanics, which he published in the *Annalen der Physik* between 1901 and 1907.[5]

The problem to which Einstein is attending, in this first paper and in the next one, is "the problem of molecular forces." The mechanical work done in a cycle that involves isothermal increases of surfaces of liquids should be zero; but, he says, this is contrary to experience. Therefore, "there is nothing left to do but to assume that the change in surface area is attended by a conversion of work to heat." He promises to proceed from the simplest assumptions ["Annahmen"] about the nature of molecular forces of attraction, and check their consequences in terms of their agreement with experiment. In this work, he says, "I let myself be guided by analogy with gravitational forces." At the end of the paper, he summarizes:

> We can now say that our fundamental assumption has been validated: to each atom there corresponds a molecular force of attraction which is independent of the temperature and independent of the way the atom chemically combines with other atoms. . . . The question whether and in what respect our forces are related to gravitational forces must be left completely open.

What preoccupies him here, as he writes in a letter to his friend Marcel Grossmann (April 14, 1901), is "the question concerning the inner relationship of molecular forces with Newtonian forces at a distance."[6] Now, this is not a problem without ambition! Newton himself, who had hoped for a relationship between gravitational forces and molecular forces, would have been not uninterested to read this work.

The idea for Einstein's work seems to have been based on Wilhelm Ostwald's *Allgemeine Chemie*. This is the book first mentioned in all of Einstein's writings. Indeed, he sent a reprint of the paper to the scientist-philosopher Ostwald on April 5, 1901, together with a request for a job in Ostwald's laboratory and the remark that his book had stimulated this paper. Ostwald, on his side, was not overwhelmed; he did not even reply. But Einstein's interest in the program of the unification of the forces of nature was making its tentative first appearance right here. He felt he was working on important problems, and despite his inability to find a steady job, he was evidently in good spirits. In the letter to Grossmann he says he is as merry as a bird, and he adds: "It is a magnificent feeling to recognize the unity ["Einheitlichkeit"] of a complex of phenomena which to direct observation appear to be quite separate things." It is only April 1901, but this is already a familiar Einstein, here searching for bridges between the phenomena of microphysics and macrophysics.

Einstein's second paper, again, looks not very promising on the surface. It is entitled "Concerning the Thermodynamics of Potential Difference between Metals and Fully Dissociated Solutions of their Salts, and Concerning a New Method to Investigate Molecular Forces." It is dated April 1902, some two months before he started his job at the patent office. Over a year has elapsed since his first paper.

He begins with a section entitled "A Hypothetical Extension of the Second Law of the Mechanical Theory of Heat." His earlier method now has undergone a significant change. In the first paper he had started with the phenomena, listed reams of experimental data from the literature, and discussed consequences drawn from them, as was more or less the rule in papers of the time. Now, in the first section, he postulates a statement of the second law which he recognizes to be outside the limits set by the available phenomena. By generalizing ["verallgemeinern"] beyond experience, he proposes to adopt the following statement: "One remains in harmony with experience when one applies the second law to physical mixtures upon whose individual components are acting any arbitrary conservative forces." Moreover, he warns quickly, "We will rely on this hypothesis in what follows even if it does not appear to be absolutely necessary." He has now begun to go for discoveries of postulates frankly beyond the "present experience." The appeal of the *generalization* is taking over, and becomes a directive for research.

The third paper follows, sent in from Bern in June 1902. It is entitled "Kinetic Theory of Thermal Equilibrium and the Second Law of Thermodynamics," the first of three papers in 1902–04 extending Boltzmann's ideas in thermodynamics and statistical mechanics. Einstein introduces an interpretation of statistical probability in which systems run through the various possible states over and over in irregular fashion, unlike thermodynamic descriptions in which equilibrium states once reached are persisted in indefinitely. This is an important bridge to his future work. His power has begun to show now, although it did not attract much attention at the time. Without having read J. Willard Gibbs' *Elementary Principles of Statistical Mechanics,* which appeared in the same year, Einstein is getting some of the same results.

The introductory paragraph of the third paper lays forth the ambition to generalize:

> Despite the success of the kinetic theory of heat in the area of the theory of gases, it has so far not been possible, by the laws of mechanics alone, to provide a sufficient foundation for the general theory of heat. Maxwell's and Boltzmann's theories have come close to this goal. The purpose of the following considerations is to fill the gap. At the same time there will be given an extension of the second postulate which will be of importance for the application of thermodynamics. Moreover, a mathematical expression for entropy will be obtained, from a mechanical point of view.

The first numbered section follows immediately and is entitled "1. Mechanical Picture ["mechanisches Bild"] for a Physical System." The only footnote in this paper is to Boltzmann's *Gastheorie*, where, incidentally, the important term "mechanisches Bild" is also used in one of the headings.

As if to make sure that we take seriously what is implied by this term, Einstein comes back to it toward the end of his paper. In one sentence he gives the main conclusion: "The second law thus appears as a necessary consequence of the mechanical world picture ["notwendige Folge des mechanischen Weltbildes"].[11] With this, he has used the term in print for the first time. He was to return to it often later, not least in his two great critiques of the "mechanical world picture" in his "Autobiographical Notes" of 1946, and of the whole unsatisfactory development from the "mechanische Programm" of the nineteenth century to our "heutiges Weltbild," as spelled out in his earlier lecture on "Aether and Relativity Theory" (1920).

So as of 1902, we see the twenty-three-year-old Einstein entering publicly on the ground where the great fight among German scientists has been in progress between the chief rival world conceptions, one holding to mechanics and the other to electrodynamics as the ground of fundamental explanation. In this 1902 article, he is still concerned with investigating the mechanical world picture as one of the main options. And while he finds it there of use, he ends with a suspicion: "The results are more general than the mechanical representation used to arrive at them." Soon he will see it is in fact far too limited; it cannot handle, for example, Brownian movement. He will also find that its most prominent alternative, the so-called electromagnetic world picture, cannot handle fluctuation phenomena of light. And even the victory of one of the conceptions would, as he later expressed it, leave us with "two types of conceptual elements, on the one hand, material points with forces at a distance between them, and, on the other hand, the continuous fields . . . an intermediate state in physics without a uniform basis for the entirety."[7] He will have to try to build his own *Weltbild.*

The fourth paper, entitled "A Theory of the Foundations of Thermodynamics," is sent off from Bern in January 1903. It is his only paper published that year. In addition to constructing a *Weltbild*, he is also building a career, and of course a family. He and Mileva Marič are married on January 6, 1903, and in the first of his surviving letters to his friend Besso, Einstein writes in that month: "Well now, I am a married man and lead a nice, comfortable life with my wife. She takes excellent care . . . and is always cheerful." He goes on to describe the fourth paper, which he has just sent off,

> after much reworking and correction. But now it is completely clear and simple, so that I am quite satisfied with it. After postulating the energy principle and the atomic theory, there follow the concepts of temperature and entropy; with the additional aid of the hypothesis that the distribution of states of isolated systems never go into less probable ones, there follows also the second law in its most general form, namely the impossibility of a perpetuum mobile of the second kind.[8]

Hindsight makes it easy to see that the structure of the argument in this paper is perhaps its most interesting feature: first, the postulation of general principles, now with hardly a nod to the detailed phenomena; then the derivation of logical consequences; and at the end, the test against experience — in this case, experience of the most general kind.

More than a year later, on April 14, 1904, Einstein writes to his friend Conrad Habicht: "In a few weeks we are expecting a young one. I now have found in a very simple manner the relation between the elementary quanta of matter and the wavelengths of radiation."[9] In fact, he has just sent off his fifth paper to the *Annalen der Physik*, "On Molecular Theory of Heat." This paper announces itself modestly, in the first sentence, as merely an addition to work published the previous year. But the plot has thickened. In his usual introductory paragraph, he sets out the general plan. He will derive an expression for the entropy of a system which is

> completely analogous to that found by Boltzmann for ideal gases and postulated by Planck for his theory of radiation. . . . Then a simple derivation will be given of the second law. Thereupon the meaning of a universal constant will be investigated, which plays an important role in the general molecular theory of heat. Finally there follows an application to the theory of black body radiation, where a highly interesting relation is developed between the universal constant mentioned above [he means $R/2N$] (one which is determined by the magnitude of the elementary quanta of matter and of electricity), on the one hand, and the order of magnitude of the wavelength of radiation, on the other, without making use of any special hypotheses.

Thus Einstein introduces his long-lasting concern with fluctuation phenomena. He shows that the size of energy fluctuations ε in the system, and therefore its thermal stability, is determined by a universal constant (which we would write as $k \equiv R/N$). Having applied it to mechanical systems and thermal phenomena, he then makes the original and daring jump to an "application to radiation": In order to determine the universal constant from the size of observable energy fluctuations, he proposes to go to the only kind of physical system where, he says, experience allows one to "suppose" ("vermuten") that there *are* energy fluctuations — the case of an otherwise empty space filled with thermal radiation. Obviously, by now he has indeed learned to "scent out the path that leads to fundamentals," as he later put it.

There follows now an ingenious argument that shows how much he is at home with borderline cases between two different fields as defined in contemporary physics, not having to choose between one and the other, but rather using both — indeed, "recognizing the unity of a complex of phenomena." He considers a volume v with dimensions of the order of the wavelength λ of the radiation in it ($v \simeq \lambda^3$). In that case, the energy fluctuations will be of the same order as the magnitude of the energy itself, or $\overline{\varepsilon^2} = \overline{E}^2$. Now he uses the Stefan-Boltzmann law in the form $\overline{E} = cvT^4$ and deduces that $\lambda \simeq 0.4/T$. But "by experience" — Einstein does not mention that it is Wien's law (1893) — we know that indeed λ_{max} is $0.293/T$. So his result has the same general lawfulness with respect to T, and the same order of magnitude for the constant. Einstein concludes that in view of the "great generality of our assumptions," this agreement cannot be an "accident." And with that, the paper ends. (The coincidence of prediction within a factor of about 2 remains characteristically an indication for Einstein that things are going quite well.)

But what a great launching of work, in some eight pages! Fluctuation phenomena will remain one of Einstein's trusted tools, not only in the Brownian motion paper, which is almost ready to come over the horizon, or in the papers on quantum theory of radiation and on Bose–Einstein statistics, but even in some of the smaller papers that have received little notice so far. (There is, for example, one of Einstein's short publications in 1908 on a new electrostatic method for measuring small quantities of electricity. With Conrad and Paul Habicht, friends of the earliest days of the Olympia Academy in Bern, Einstein had been interested in building a device for measuring small voltages by multiplication techniques. In fact, Einstein wrote a patent application for the device, which claimed to measure potential differences down to 5×10^{-4} volts. He even found a "clever mechanic" who attempted to build the thing. But Einstein's real interest shows up in the last paragraph of the short 1908 paper that described the device.[10] In research, for example on radioactivity, an electrostatic method of measurement of highest sensitivity might be useful. But "I was led to this plan by thinking how one might find and measure the spontaneous charge appearing on conductors which should exist analogously with the Brownian motion that is required by the molecular theory of heat."

The next publication is Einstein's inaugural dissertation on "A New Determination of Molecular Dimensions," dated April 30, 1905.[11] We are now in the miraculous year of 1905, the year that will continue to be a challenge for historians of science for a long time. The work does not yet concern itself with Brownian movement. As Einstein later said, he had only just discovered the existence of that problem:

> Not acquainted with the investigations of Boltzmann and Gibbs which had appeared earlier and actually exhausted the subject, I developed the statistical mechanics and molecular-kinetic theory of thermodynamics which was based on the former. My major aim in this was to find facts which would guarantee as much as possible the existence of atoms of definite, finite size. In the midst of this, I discovered that, according to atomistic theory, there would have to be a movement of suspended microscopic particles open to observation, without knowing that observations concerning the Brownian motion were already long familiar.[12]

The inaugural dissertation paper is concerned with the viscosity and diffusion of liquid mixtures, and the calculation of Avogadro's constant from these. Einstein starts by saying that the size of molecules can be found from the kinetic theory of gases, but not yet from the observable physical phenomena of liquids. He proposes now to bring about a fusion between these portions of micro- and macrophysics.

Einstein regards the solute molecule to behave like a solid body suspended in a solvent, and applies hydrodynamic equations of motion to the large molecules, assuming an approximate homogeneity of the liquid. At the end of the paper, applying kinetic theory to liquid solutes with known diffusion coefficients, he gets for Avogadro's number *N* the value 2.1×10^{23}, and adds quite simply that *N* found by this method coincides in order of magnitude with *N* found by other methods "in a satisfactory way."

Einstein's letters show us that his thesis was really part of an older research interest. He had written to Besso more than two years earlier (March 17, 1903),

> Have you by now calculated the absolute magnitude of ions under the assumption that they are spheres, and so large that the equations of hydrodynamics of viscous liquids are applicable? Since we know the absolute magnitude of the electron [charge], this should be a simple matter. I would have done it myself but I don't have literature or time; you can also use diffusion in order to get information about neutral salt molecules in solution. . . . If you don't understand what I mean, I'll write you more explicitly.[13]

His thesis, though not done for ions, is very close to this proposal.

2. Exploiting the Confrontation of Theories

This brings us to the three great papers of 1905, sent off to the *Annalen der Physik* at intervals of less than eight weeks. When I first became interested in the history of physics of this century, it struck me that while these three papers — on the quantum theory of light, on Brownian movement, and on relativity theory — seemed to be in quite different fields, they could be traced in good part to the same general problem, namely fluctuation phenomena. Indeed, in the Archive at the Institute for Advanced Study is a letter Einstein wrote to Max von Laue on January 17, 1952, in which the connections are indicated. Einstein discusses von Laue's textbook on relativity theory, and registers a small objection:

> When one goes through your collection of verifications of the Special Relativity Theory, one gets the impression Maxwell's Theory may be unchallengeable. But already in 1905 I knew with certainty that it [Maxwell's Theory] leads to wrong fluctuations in radiation pressure, and hence to an incorrect Brownian movement of a mirror [suspended in] a Planckian radiation cavity. In my opinion one can't get around ascribing to radiation an objective atomistic structure, which of course does not fit into the framework of Maxwell's Theory. Naturally, it is comforting that the Special Relativity Theory in essence rests only upon the constant *c*, and not on a presupposition of the reality and fundamental character of Maxwell's fields. But unhappily the 50 years which have elapsed since then have not brought us closer to an understanding of the atomistic structure of radiation. On the contrary![14]

Here is made explicit the chief connection between Einstein's work on Brownian motion of suspended particles, the quantum structure of radiation, and his more general reconsideration of what he called later "the electromagnetic foundations of physics" itself.[15]

It also struck me forcibly that the style of the three papers was essentially the same, and fitted with the way Einstein had come to see his task in the course of his earlier work. Contrary to the sequence one finds in many of the best papers of the time, for example H. A. Lorentz's 1904 publication on electromagnetic phenomena, Einstein does not start with a review of a puzzle posed by some new and difficult-to-understand experimental facts, but rather with a statement of his dissatisfaction with what he perceived to be formal asymmetries or other incongruities that others might dismiss as being of predominantly aesthetic nature. He then proposes a principle of great generality. He shows that it helps remove, as one of the deduced consequences, his initial dissatisfaction. And at the end of each paper, he proposes a small number of predictions that should be experimentally verifiable, although the test may be difficult.

The way Einstein starts the paper on the quantum theory of light is typical. He writes: "There exists a deep-going, formal distinction between the theoretical representations which physicists have formed for themselves [note it is always the physicists' notions which are at fault] concerning gases and other ponderable bodies on the one hand, and the Maxwellian theory of electromagnetic processes in so-called empty space on the other hand." The problem Einstein sees is that the physicists consider the energy of light and other purely electromagnetic phenomena as continuous functions of space. On the other hand, the energy in ponderable bodies (he names atoms and electrons) is the sum of discrete entities, therefore not to be considered in arbitrarily small elements or in terms of continua. It was a problem of particle/field duality that Einstein did not solve in this paper or, for that matter, to the end of his life.

Indeed, while he was deeply attached to the continuum approach, he used the occasion of this paper to concentrate on the fact that the continuum approach leads to contradictions with experience when applied to phenomena involving the generation and transformation of light (that is, it leads to what later was termed the "ultraviolet catastrophe.") This meant importing the assumption of a discontinuous distribution of light energy into the realm that previously was thought to be covered by the continuist theory of Maxwell — a heuristic act that characterized theories that he considered transient and, though convenient, not fundamental.[16] Even Walter Nernst, an enthusiast for quantum theory, called it in 1911 still only a rule for calculation, "a very odd rule, one might even say a grotesque one."[17]

The heuristic character of the paper was assured also because, as Einstein said frankly, he would have to proceed "without putting at the foundation any picture about the generation and propagation of radiation." He therefore was forced to make the tested law of blackbody radiation the head of an axiom system, and deduce the consequences, finally coming to the photoelectric effect for which the paper is most remembered (announcing, in a courageous, almost off-hand way in two sentences results that should eventually be verifiable in the laboratory).

Nothing was said here explicitly about either the mechanical or the electromagnetic world view, but the paper carried the clear message that neither one nor the other can deal with the phenomena. The importation into electromagnetic theory of the mechanistic thema of discontinuity produced a mixture with which Einstein was never content, even when others had learned to accept it. Two years later, Einstein referred back to this paper to point out that it showed the "electromagnetic world picture" to be "unsuitable," although he had to add that "for the time being we do not possess a complete world picture corresponding to the relativity principle."[18]

The second great paper of 1905 is usually called the "Brownian movement" paper. In the recently finished paper on the quantum theory of light, he had put at the head his dismay over the apparent clash between two theories, based on continuity and discontinuity, respectively; in the Brownian motion work, he deals with and exploits an analogous clash. As he explained in the second part of that series: "According to classical thermodynamics which differentiates *principally* between heat and other forms of energy, spontaneous fluctuations do not take place; however, they do in the molecular theory of heat. In the following we want to investigate according to which laws those fluctuations must take place.''[19] Thus the range of classical thermodynamics should be found to be limited, even in volumes large enough to be observable under the microscope. This leads to the prediction of the mean displacement of small objects as a function of time.

And then comes the third great paper of 1905, his first one on relativity theory. It has been discussed and analyzed so many times that I now need to remind you only of some key points of relevance to the present purpose. His first section sets forth the total plan of the paper, and begins by drawing attention to a formal asymmetry in the calculations for determining the current generated during the relative motion between a magnet and a conductor — a phenomenon physicists had long regarded as adequately understood since Faraday's induction experiments of 1831. Einstein's paper does not invoke explicitly any of the several well-known experimental puzzles with which so many others were struggling, as in the interpretation of the ether-drift experiments — not even when the opportunity arises for him to show in what manner his relativity theory accounts for them.

Once more, two separate theories are confronted: classical mechanics, where Newton's laws are thought to hold and Galilean relativity to obtain, and on the other hand, electrodynamics, in which it was thought that a preferred reference system, the aether, existed. To deal with this dichotomy, Einstein proposes to raise a courageous conjecture (''Vermutung''), which he refers to as the principle of relativity, to an assumption or presupposition (''Voraussetzung''). It amounts to a generalization of Galilean–Newtonian relativity, with the result that when properly phrased, the laws of electrodynamics and optics *and* of mechanics have the same validity with respect to transformation between all inertial systems. With this, an ancient wall between mechanics and the rest of physics is breached, and the whole question whether the mechanistic or electromagnetic world view is preeminent is axiomatically asserted to be meaningless. In addition, Einstein makes another presupposition, that light in empty space is propagated with a definite velocity that is independent of the state of motion of the emitting body.

These two presuppositions suffice him to produce a theory of both electromagnetic and mechanical phenomena that is most parsimonious in its assumptions, free of contradictions, and unburdened of long-standing puzzles. It reveals unexpected connections between previously separate concepts, and is ready to permeate every field of physics; for example, while Lorentz and others applied the transformation equations only to electromagnetic phenomena, Einstein had the ambition from the start to let them apply to all of physics. What has to be given up is the support of long and widely used concepts such as the luminiferous aether, absolute space, and absolute simultaneity — and also the old crutch of constructing theories inductively from the mass of present data and puzzling phenomena.

3. The Postulational Method and Its Progress

Einstein's previous work had prepared him to make the necessary sacrifices when adopting a more ecumenical and independent ''Gesichtspunkt'' with which to escape the clash existing between two different current theories. As our study has shown, he had learned not to trust any of the existing *Weltbilder*. In the four years since his first paper on capillarity, he had formed a clear view about the structure of nature: At the top is a very small number of eternal, most general principles or laws by which nature operates. They are not easy to find — God is subtle — but they surely do not stop at the boundaries between fields that happen to be occupied by different theories. Nature, being a harmonious totality, would not allow deep qualitative differences between the laws of mechanics and of electrodynamics, for example.

Below those few, grand laws lies the plane of hardwon, well-established experimental facts; these experiences or key phenomena are the necessary consequences, in visible compliance of the general laws.

Between these two levels is the uncertain and shifting region of concepts, theories, and recent findings from the laboratories. They deserve to be regarded skeptically; they are manmade, limited, fallible, and if necessary disposable. This attitude appears throughout his writings, but is perhaps best expressed in a report which I heard from one of his colleagues in Berlin, the physical chemist Herman F. Mark: "Einstein once told me in the lab: 'You make experiments and I make theories. Do you know the difference? A theory is something nobody believes except the person who made it, while an experiment is something everybody believes except the person who made it.'"

Einstein was initially reluctant to present even his own work as a new theory; and after it had been so christened by Planck (1907) and others, he frequently referred to it in print as the "so-called relativity theory."[20] It is worth noting here that if he had chosen to give it a name, he would, in my opinion, have preferred to call it something like *Invariantentheorie*, which is of course much more true to the method and aim of the theory.[21]

In accord with the structure of nature, Einstein also had developed by this time the modern version of the method by which one can hope to reach the general principles of nature. We saw it at work: It is a bold, postulational method, appropriate for dealing with what Einstein came to call "theories of principle" that have the large ambition of handling the "totality of sense experiences." At the top of the methodological hierarchy, Einstein puts a well-established experimental rule, recast in its most general form, and raises it to a postulate — or, if none is available to serve this function, as in the case of relativity theory, his own *Vermutungen* have to be put into this place. From this postulate system he then draws the logical deductions, and these in turn point to eventual tests against experience. Old puzzles disappear, and new questions come to the fore. As Einstein saw first perhaps when, at the age of twelve, the "holy booklet" of Euclidian geometry came into his hands, with the right axiom system at the top, the right consequences develop by necessity. All else is broken crockery.[22]

Insofar as his contemporaries understood what he was doing — and at least until 1911 most did not — it must have seemed a dangerous high-wire act, the contact with experience being supplied by only a very few carefully selected support ropes anchored in solid ground.[23] It is not surprising that few others cared to take such risks, that the progress of Einstein's style among his contemporaries was initially quite slow. And for all the effort, and at the cost of giving up so many established notions and of paradoxical by-products, what did one really get from Einstein's relativity that was not available in a more congenial way from the work of Lorentz, Poincaré, Abraham, and others? If one did not share Einstein's anguish over the limited scope, asymmetries, and ad hoc hypotheses in the other systems, there was nothing inescapable about accepting Einstein's relativity.

The contemporary reception of relativity shows all this clearly. Among established scientists there was really only one who appreciated Einstein's principle of relativity from the beginning: Max Planck in Berlin. He was at hand when that very unconventional paper was received at the *Annalen der Physik*,[24] and gave it valuable exposure and defense, after its September publication, in talks at the German Physical Society on March 23, 1906 and at the Stuttgart Congress of German Scientists and Physicians on September 19, 1906. And by that time Einstein needed a friend indeed, for there had just appeared, immediately upon publication of his own paper, what seemed to be an inescapable experimental disproof of his 1905 paper in the *Annalen der Physik*, by the eminent experimentalist Walter Kaufmann.[25] Lumping together, as was widely done for some years, the work of Lorentz and Einstein since their predictions coincided nearly enough, it announced in italics: "*The measurement results are not compatible with the Lorentz-Einsteinian fundamental assumption.*"

The effect of Kaufmann's announcement was instant and devastating on Lorentz, and also shook Poincaré.[26] Planck, however, showed in a patient analysis that Kaufmann's measurements were not unequivocally either against Einstein's system or for those of his rivals, for example that of Max Abraham.[27] (The latter, characterized by a rigid, undeformable electron, had for many the advantage of fitting in fully with the electromagnetic world picture.)

The published debate shows that nobody in the audience was converted by Planck to what was being called there "die Relativtheorie" (the baby still had only a nickname). When pressed to the wall, Planck had to confess that in the absence of persuasive proof or disproof, the Einsteinian postulate that no absolute translation could be found made the difference: "I find it more appealing ["Mir ist das eigentlich sympatischer"]." As Cornelius Lanczos said later, Planck, although a conservative, was caught by the beauty of Einstein's paper.

Three years later, when Wilhelm Wien was converted to what Planck called the small band of relativists, it was not because some crucial experiments left him no other choice, but mainly on quasi-aesthetic grounds. Using words that Einstein must have appreciated, Wien said,

> What speaks for it most of all is the inner consistency which makes it possible to lay a foundation having no self-contradictions, one that applies to the totality of physical appearances, although thereby the customary conceptions experience a transformation.[28]

Other adherents to the relativistic point of view had recourse to similar predilections. Thus Hermann Minkowski, in the same September 1908 meeting at Cologne at which he delivered his famous address on "Space and Time," also attended a session that touched on the merits of Abraham's rigid-electron theory, then still being held up by many as the chief option, as against the theories of both Einstein and Lorentz. Minkowski exclaimed in the discussion period:

> The rigid electron is in my view a monster in relation to Maxwell's Equations, whose innermost harmony is the relativity principle. Going at Maxwell's Equations with the idea of the rigid electron seems to me just like going into a concert after stopping up one's own ears with cotton wool.[29]

Lesser scientists had a great deal more trouble with Einstein's work, of course. No wonder that when Einstein submitted his 1905 relativity paper to the Physics Faculty of the University of Bern in 1907 as his *Habilitationsschrift* to gain the right to teach as *Privadozent*, it was rejected with the grade of "unsatisfactory." The professor of experimental physics returned Einstein's copy with the curt remark, "What you have written here, that I don't understand at all."

I have wondered of course how Einstein himself — a previously unknown young fellow without academic connections or credentials — reacted to the tumult caused in 1906 by Kaufmann's "disproof" and the claim of the rival theories. If he had been a naive falsificationist, he should have eagerly embraced this evidence of "progress" in science, and gone on to other things. But in fact, for about two years he seems to have paid no attention to the Kaufmann publication that had stunned much more experienced scientists. I could find no reference to Kaufmann's attack in his publications or letters until, urged by Johannes Stark as editor of the *Jahrbuch der Radioaktivität und Electronik*, he responded in a survey published early in 1908.[30]

In a manner that may well have seemed rather haughty, Einstein allowed that Kaufmann's calculations seemed to be free of blatant error, although he added he would wait for a greater variety of observational material before the experiment could be declared free of systematic errors. But in any case, Einstein had no evident worry about the outcome, because, as he put it, the theories of Abraham and Bucherer, which Kaufmann had claimed to be proving, "have a rather small probability, because their fundamental assumptions concerning the mass of moving electrons are not explainable in terms of theoretical systems which embrace a greater complex of phenomena." Even though the "experimental facts" seemed clearly to favor the theory of his opponents, Einstein found the limited scope and ad hoc character of their theories more significant and objectionable than the apparent disagreement between his own theory and the new results of experimental measurements. It is really a classic demonstration of the justified confidence a rare type of scientist can have in the soundness of what Einstein once called his "scientific instinct," assigning "probabilities" to theories on the basis of judgments made even in the face of seemingly contrary evidence.[31]

4. "My Need to Generalize" and Its Dissatisfactions

Eventually, Einstein's 1905 relativity paper was hailed everywhere as one of the chief historic advances of science. But, again, Einstein himself saw it differently. In his own evaluation he stressed the limitations. Thus in his Nobel Prize speech[32] he did admit that "the special relativity theory resulted in appreciable advances," and he gave a short list of these — significantly, first of all that "it reconciled mechanics and electrodynamics," the respective bases of the opposing *Weltbilder* of that age. Next, he agreed that the relativity theory "reduced the number of logically independent hypotheses. . . . It enforced the need for a clarification of the fundamental concepts in epistemological terms. It united the momentum and energy principle, and demonstrated the like nature of mass and energy." He could have added, as he had done at the 1909 Salzburg conference, that energy and mass appeared now to be equivalent magnitudes, as were heat and mechanical energy. And he could have gone on to show other resulting unifications, resulting from the discarding of old conceptual barriers that had been simply "unbearable" for him.[33]

But then Einstein, in his Nobel Prize talk, adds "Yet, it was not entirely satisfactory." The foremost dissatisfaction, of course, was that the special theory "favors certain states of motion — namely those of the inertial frames. . . . This was actually more difficult to tolerate than the preference for a single state of motion, as in the case of the theory of light with a stationary ether," for now there was not even an imagined reason for preference, namely, the light ether. "A theory which from the outset prefers no state of motion should appear more satisfactory." In his "Autobiographical Notes" he added, "That the special theory of relativity is only the first step of a necessary development became completely clear to me only in my efforts to represent gravitation in the framework of this theory." This happened precisely in 1907. Even as the special theory of 1905 was very slowly gathering converts on its way toward the eventual public triumph, Einstein had already put it behind himself, and was hard at work on a generalized theory, driven by what he later called, in an unpublished letter to W. de Sitter (November 4, 1916), "my need to generalize" ["mein Verallgemeinerungsbedürfnis"].

As is well known, Einstein was always far more interested in what remained to be done than in what he had accomplished. There was so much more that kept escaping the task of grand unification that increasingly stood before him as the real goal. It is one of the great ironies that the very extent and depth of the advances Einstein himself helped to launch — including his contributions to quantum theory — eventually made it impossible for the physical phenomena to be gathered in one grand, relativistic *Weltbild* in his own time. As we have seen, writing as early as 1907, he recognized that the task was far from being accomplished; while critical of both the mechanical and the electromagnetic world pictures throughout his life, he continued to long for a unified world picture, the "Einheitlichkeit des Weltbildes."[34] But the goal was always eluding him, and in the last decade of his life, he looks ahead and asks wistfully "how the theoretical foundation of the future will appear. Will it be a field theory; will it be in essence a statistical theory?"[35]

The question has not been answered yet. Today we are surrounded by various competing, fragmentary, largely unarticulated scientific world views. It reminds one of the remark of Edward Gibbon in *The Decline and Fall of the Roman Empire* (Chapter II, Section I): "The various modes of worship which prevailed in the Roman world were all considered by the people as equally true, by the philosophers as equally false, and by the magistrates as equally useful."

5. Parts of the Legacy

A large portion of any physicist's approach to the study of nature today can be traced directly to the influence of Einstein's vision of an overarching relativistic world picture. No matter how far modern physical scientists regard themselves to have moved away from Einstein's own aims, they share at least three basic components with him.

The first is the hope of, and a method for, achieving progressively more unified theories that provide us with a sense of order by encompassing the immense range and variety of phenomena. As I have stressed, Einstein's postulational method for constructing the deep "theories of principle" came to the fore even in his early papers: On the basis of a few key phenomena and a sensitive interpretation of them, exert the scientific imagination to the utmost to formulate axiom systems of great generality. There follows the sober deduction of consequences, and then the tests against sense experience. If, perhaps after a long series of such cycles, a theoretical structure is attained that seems firm enough in one limited field, the process is further generalized to encompass larger and larger portions on the plane of experience. In Einstein's words,

> Thus the story goes on until we have reached a system of the greatest conceivable unity, and of the greatest paucity of concepts of the logical foundations which is still compatible with the properties of our sense observations. We do not know if this ambition will ever result in a definite system. If asked one's opinion, one is inclined to answer No. While wrestling with the problems, however, one is sustained by the hope that this highest goal can really be attainable to a very high degree.[36]

It is therefore clear that Einstein himself was destined, from the start, to go on from the special to the general theory of relativity, and then beyond it to a unified field theory. For, as he said, "the theory could not rest permanently satisfied with this success. . . . The idea that there exist two structures of space independent of each other, the metric-gravitational and the electromagnetic, was intolerable to the theoretical spirit."[37]

The second component of Einstein's way of building a picture of the world lies in the choice of, and unshakeable devotion to, a relatively few themata by which he judged whether a theory meets what he called the criterion of "inner perfection." He knew well that these are not easily defended, or even easy always to specify; their "exact formulation . . . meets with great difficulties. . . . The problem, here is not simply one of a kind of enumeration of the logically independent premises . . . but that of a kind of reciprocal weighing of incommensurable qualities,"[38] hence a judgment into which presuppositions, aesthetic consideration, and other preferences of that sort can and do enter prominently.

The chief themata which I find to have guided Einstein in theory construction from the early years, and thus to have shaped his scientific program, include these: primacy of formal rather than materialistic explanation; unity or unification; generalizability and egalitarian application of laws throughout the total realm of experience; logical parsimony and necessity; symmetry; simplicity; causality; completeness; continuum; and of course constancy and invariance. These fundamental conceptions define, as it were, the boundary conditions within which a particular sense of order operates. (Thus John A. Wheeler has recorded that Einstein gave a talk at Princeton in which he made the point that "the laws of physics should be simple." Someone in the audience asked, "But what if they are not simple?" "Then I would not be interested in them.")

Many of these themata appear of course also in the physics of his predecessors and contemporaries; that is what assures the continuity of the total scientific enterprise. But this particular set, and the tenacity with which it was held, characterize Einstein's own style. Thus, adherence to these themata helps explain, in specific cases, why Einstein would obstinately continue his work in a given direction even when testing it against experience was difficult, discouraging, or not possible. It also explains the deep personal anguish he must have felt when other physicists whom he loved — Lorentz, Planck, above all Bohr — based some of their work on key presuppositions that were antithetical to his own. Most physicists today believe that he was not right in giving such loyalty to some of these thematic presuppositions. But on this, too, we do not know what the future holds; the balance may look very different when our successors discuss it in the historical section of the Einstein bicentennial celebration.[39]

The third part of our Einsteinian legacy concerns what lies beyond the ability of mere humans to perceive regularity and necessity in nature. In later years, Einstein confessed that his progress on the road to relativity had tested not only his physical knowledge but his physical and psychological fortitude as well. It is a fitting comment on the success of his labors that today beginning physics students can enter easily into this once so fought-over field of knowledge. Moreover, they are hardly aware that they have inherited his grand goal, as well as so many conceptual and methodological tools for its pursuit — the goal of pushing toward the attainment of a synoptic and coherent overview of the vast sea of phenomena.

It is, however, again typical of Einstein that, in his role as scientist as well as humanist, he saw how the very progress already made brings us face to face with a deep puzzle. In a famous paragraph, he wrote in 1936:

> It is a fact that the totality of sense experience is so constituted as to permit putting them in order by means of thinking — a fact which can only leave us astonished, but which we shall never comprehend. One can say: the eternally incomprehensible thing about the world is its comprehensibility.[40]

After the publication of this paragraph, Einstein received a plaintive letter from one of his oldest and best friends, Maurice Solovine. They had met in Bern in 1902, when Einstein was just beginning on the road I have here described, and when Solovine was a young philosophy student. With Conrad Habicht they had banded together in their Olympia Academy to meet regularly and discuss science and philosophy.

Now, a half century later, Solovine had come upon this passage while translating it for a French edition of essays in Einstein's book *Mein Weltbild*. Solovine was worried. How could there be a puzzle about the understandability of our world? Was this not a dangerous notion to allow into science, mankind's most rational activity?

Einstein's reply to Solovine tried to set his mind at ease:

> You find it remarkable that the comprehensibility of the world (insofar as we are justified to speak of it at all) seems to me a wonder or eternal secret. Now, *a priori* one should expect a chaotic world that can in no way be grasped by thinking. One could, even *should*, expect that the world turns out to be lawful only insofar as we intervene to provide order. It would be the sort of order like the alphabetic ordering of words of a language.
>
> The kind of order which is provided, for example, by Newton's Theory of Gravitation is of a quite different character. Even if the axioms of the theory are put forward by human agents, the success of such an enterprise does suppose a high degree of order in the objective world, which one has no justification whatever to expect *a priori*. Here lies the sense of "wonder" which increases ever more with the development of our knowledge. . . .
>
> The nice thing is that we must be content with the acknowledgement of the wonder, without there being a legitimate way beyond it.[41]

At this end of Einstein's century, we can indeed be content with what he has done to increase not only our sense of rational order, but through it also our sense of wonder.

Acknowledgements

I wish to express my indebtedness to Miss Helen Dukas and the Estate of Albert Einstein for help and for permission to cite from the writings of Einstein. I benefitted from discussions with a number of colleagues, above all A. I. Miller and J. Stachel. I am happy to acknowledge research support by grants from the National Science Foundation and the National Endowment for the Humanities.

NOTES

1. See, for example, the collections *Mein Weltbild* edited by C. Seelig (Verlag Ullstein, Frankfurt, 1977) and A. Einstein, *Aus meinen späten Jahren* (Deutsche Verlags-Anstalt, Stuttgart, 1979).
2. I shall refer to the first nine papers of Einstein:
   1) "Folgerungen aus den Kapillaritätserscheinungen," Ann. d. Phys. ser. 4, **4**, 513–23 (1901).
   2) "Über die thermodynamische Theorie der Potentialdifferenz zwischen Metallen und vollständig dissoziierten Lösungen ihrer Salze, und eine elektrische Methode zur Erforschung der Molekularkräfte," Ann. d. Phys. ser. 4, **8**, 798–814 (1902).
   3) "Kinetische Theorie des Wärmegleichgewichtes und des zweiten Hauptsatzes der Thermodynamik," Ann. d. Phys. ser. 4, **9**, 417–33 (1902).
   4) "Eine Theorie der Grundlagen der Thermodynamik," Ann. d. Phys. ser. 4, **11**, 170–87 (1903).
   5) "Zur allgemeinen molekularen Theorie der Wärme," Ann. d. Phys. ser. 4, **14**, 354–62 (1904).
   6) "Eine neue Bestimmung der Moleküldimensionen," (Wyss, Bern, 1905), 21 pp.
   7) "Über einen die Erzeugung und Verwandlung des Lichtes betreffenden heuristischen Gesichtspunkt," Ann. d. Phys. ser. 4, **17**, 132–48 (1905).
   8) "Die von der molekularkinetischen Theorie der Wärme geforderte Bewegung von in ruhenden Flüssigkeiten suspendierten Teilchen," Ann. d. Phys. ser. 4, **17**, 549–60 (1905).
   9) "Zur Elektrodynamik bewegter Körper," Ann. d. Phys. ser. 4, **17**, 891–921 (1905).
3. Quoted in E. Hiebert, "The State of Physics at the Turn of the Century," in *Rutherford and Physics at the Turn of the Century,* edited by M. Bunge and W. R. Shea (Dawson and Science History Publications, New York, 1979), p. 3. The article is a useful survey of the range and intensity of new findings that held the attention of physicists at the time Einstein began his researches.
4. Quoted by M. Besso in *Albert Einstein, Michele Besso, Correspondence 1903–1955*, edited by Pierre Speziali (Hermann, Paris, 1972), p. 550.
5. Not to speak of twenty-two newly discovered abstract reviews of books and papers that Einstein published on these subjects in the *Beiblätter* of the *Annalen der Physik* between 1905 and 1907. See M. J. Klein and A. Needell, "Some Unnoticed Publications by Einstein," ISIS **68**, 601–4 (1977).

6. Quoted in Carl Seelig, *Albert Einstein, eine dokumentarische Biographie* (Europa Verlag, Zurich, 1954), pp. 61–62.

7. A. Einstein, "Autobiographical Notes," in *Albert Einstein, Philosopher-Scientist,*, edited by P. Schilpp (Library of Living Philosophers, Evanston, Il., 1949), p. 27. Where necessary I have provided corrected translations of quotations from Einstein's original German essays.

8. *Einstein, Besso Correspondence*, op. cit., p. 3.

9. Seelig, op. cit. in n. 1, p. 74.

10. A. Einstein, "Eine neue elektrostatische Methode zur Messung kleiner Elektrizitätsmengen," Phys. Z. **9**, 216–17 (1908). He had laid the theoretical base for the method in Ann. d. Phys. **22**, 569–72 (1907).

11. The sixth paper, using the sequence given in Margaret Shield's bibliography in *Einstein, Philosopher-Scientist*, op. cit. in n. 7. With a brief *Nachtrag*, this paper was published in Ann. d. Phys. **19**, 289–306 (1906).

12. In *Einstein, Philospher-Scientist*, op. cit. in n. 7, p. 47. Even in the *Annalen der Physik*, in which Einstein published all his early papers, there had been an article by F. M. Exner in 1900 that showed that microscopic particles move with greater average speed at higher temperature.

13. *Einstein, Besso Correspondence*, op. cit. in n. 4, p. 14.

14. A similar attitude underlies Einstein's address at the Salzburg meeting in 1909. See "Über die Entwicklung unserer Anschauungen über das Wesen und der Konstitution der Strahlung," Phys. Z. **10**, 817–26 (1909).

15. In *Einstein, Philosopher-Scientist*, op. cit. in n. 7, p. 47.

16. It was however, perhaps just for that reason, bound to seem "very revolutionary," as Einstein put it in his high-spirited letter to Conrad Habicht in the spring of 1905; see Seelig, op. cit. in n. 6, pp. 88–89.

17. Quoted in M. Klein, "Einstein, Specific Heats and the Early Quantum Theory," Science **148**, 177 (1965).

18. "Über die vom Relativitätsprinzip geforderte Trägheit der Energie," Ann. d. Phys. **23**, 371–2 (1907).

19. "Zur Theorie der Brownschen Bewegung," Ann. d. Phys. **19**, 372 (1906).

20. E.g., in "Entwurf einer verallgemeinerten Relativitätstheorie und einer Theorie der Gravitation. I. Physikalischer Teil," Z. f. Math. und Phys. **62**, 225 (1913) and similarly in Ann. d. Phys. **38**, 1059 (1912).

    In his first review article, written in good part for didactic purposes in 1907 (see n. 30), he presented his work on relativity as "the unification of the Lorentzian *theory* and the relativity *principle*" (italics supplied). His reference at the Salzburg 1909 meeting was similar (see op. cit. in n. 14). In the titles of his papers until 1911, Einstein used the term "Relativitätsprinzip" rather than "Relativitätstheorie."

    Incidentally, in the titles of Einstein's first papers on general relativity he repeatedly used "verallgemeinerte Relativitätstheorie," rather than the later phrase "allgemeine Relativitätstheorie." In his correspondence, Einstein continued to use the earlier phrase; cf. his letter to M. von Laue, 17 January 1952 (in the Einstein Archive at the Institute for Advanced Study).

21. He used the term informally, e.g. in correspondence with Besso (*Einstein, Besso Correspondence*, op. cit., p. 526). He showed his willingness to adhere to the basic invariant space-time element *ds* even when it became clear that it threw in doubt the "physical meaning (measurability-in-principle)" of the individual coordinates; see "Entwurf" op. cit. in n. 20, p. 230, and Ann. d. Phys. **35**, 930 ff. (1911). And he confessed that, all things considered, the term would have been preferable; in a letter of 30 September 1921 to E. Zschimmer of Jena in the Archive at the Institute for Advanced Study, he writes:

    > Now to the name relativity theory. I admit that it is unfortunate, and has given occasion to philosophical misunderstandings. The name "Invarianz-Theorie" would describe the research *method* of the theory but unfortunately not its material content (constancy of light-velocity, essential equivalence of inertia and gravity). Nevertheless, the description you proposed would perhaps be better; but I believe it would cause confusion to change the generally accepted name after all this time.

    J. L. Synge expressed himself similarly: "Much as I dislike the name [relativity theory] (I would much prefer to follow Minkowski, but it is now too late)." [In *Albert Einstein's Theory of General Relativity*, edited by G. E. Tauber (Crown Publishers, New York, 1979) p. 199.] Synge alludes to H. Minkowskis's remark in his 1908 talk "Space and Time": "that the word 'relativity-postulate' for the requirement of an invariance . . . seems to me very feeble [sehr matt]." Others wrote in the same vein; e.g. Arnold Sommerfeld, "Philosophie und Physik" [1948], in A. Sommerfeld, *Gesasmmelte Schriften,* Vol. 4 (Friedr. Vieweg, Braunschweig, 1968), pp. 640–1.

22. I have discussed this method in detail, based on Einstein's writings, in the essay "Einstein's Model for Constructing a Scientific Theory," in *Albert Einstein, His Influence on Physics, Philosophy and Politics*, edited by P. C. Aichelburg and R. Sexl (Friedrich Vieweg, Wiesbaden, pp. 109–36); reprinted in Am. Scholar **48**, no. 3, 309–40 (Summer 1979).

Needless to say, Einstein's published papers in their architectural details do not necessarily correspond point for point with the sequence of his actual thought processes in arriving at his conclusions.

23. As his work proceeded, Einstein became more aware of this feature of his method, and more daring still. See, for example, his "Autobiographical Notes": "I have learned something else from the theory of gravitation. No ever so inclusive collection of empirical facts can ever lead to the setting up of complicated equations. A theory can be tested by experience, but there is no way from the experience to the setting up of a theory. . . . Once one has those sufficiently strong formal conditions, one requires only little knowledge of facts for the setting up of a theory." [*Einstein, Philosopher-Scientist*, op. cit. in n. 7, p. 89.]

    Of course, this does *not* mean that Einstein had no interest in experimental facts as such; indeed he respected greatly some of the "artists" in the field of experimental physics or astrophysics, and often enjoyed puzzling over new experimental results or apparatus. Moreover, he insisted that the ultimate goal of theory construction must be the detailed and complete coordination of theory and experience — e.g., in the remark, "A theoretical system can claim completeness only when the relations of concepts and experienced facts are laid down firmly and unequivocally. . . . If one neglects this point of view, one can only attain unrealistic systems." [Quoted in F. Herneck, Forschungen und Fortschritte **40**, 133 (1966).]

24. Planck was listed as special editorial consultant on the masthead of the *Annalen*. Einstein's relativity paper was received at the *Annalen* on 30 June 1905. The editor, Paul Drude, famous for his writings on light and ether, had just moved from Giessen University to Berlin. Until just two and one-half months earlier, the *Annalen* requested that all manuscripts be sent to Drude in Giessen. One can only speculate what the fate of the manuscript might have been there.

25. W. Kaufmann, "Über die Konstitution des Elektrons," S. d. k. Preuss. Akad. Wiss. **45**, 949–56 (1905), and "Über die Konstitution des Elektrons," Ann. d. Phys. **19**, 487–553 (1906).

26. As discussed in G. Holton, *Thematic Origins of Scientific Thought* (Harvard University Press, Cambridge, Mass., 1973), pp. 189–90, and in detail in A. I. Miller's essay in this volume.

27. Max Planck, "Die Kaufmannschen Messungen," Phys. Z. **7**, 753–61 (1906).

28. W. Wien, *Über Elektronen*, 2nd ed.(B. G. Teubner, Leipzig, 1909), p. 32. Lorentz, who never fully accepted Einstein's relativity, was most generous in acknowledging its power and originality. But he also put his finger on a widely felt dismay with Einstein's method when he remarked: "Einstein simply postulates what we have deduced. . ." [H. A. Lorentz, *The Theory of Electrons* (1909), p. 230].

29. Phys. Z. **9**, 762 (1908). In the preface of his first edition (1911) of the first textbook on the relativity theory, *Das Relativitätsprinzip* (F. Vieweg, Braunschweig, 1911, p. i.) Max von Laue still could not point to incontrovertible experimental evidence in favor of Einsteinian relativity, but stressed the lack of persuasive falsifications, and the argument from congeniality — the two favorable criteria of a theory that Einstein also approved of. Max von Laue wrote:

    > In the five and a half years since Einstein's founding of relativity theory this theory has gathered attention in growing measure. To be sure, this attention is not always equal to adherence. Many researchers, including some with well-known names, consider the empirical grounds not sufficiently firm. Worries of this kind can of course be helped only through further experiments; in any case this book strengthens the proof that not a single empirical ground exists against the theory.
    >
    > More extensive is the number of those who cannot find the intellectual content congenial, and to whom particularly the relativity of time, with those consequences that sometimes really appear to be quite paradoxical, seem unacceptable.

30. A. Einstein, "Über das Relativitätsprinzip und die aus demselben gezogenen Folgerungen," J. d. Radioakt. v. Electr. **4**, 411–62 (dated 1907, but appeared in 1908). I discussed this response of Einstein to Kaufmann first in Holton, op. cit. in n. 26, pp. 234–6; but I include here this paragraph and the next in response to queries about Einstein's reaction, made in the question period following the public delivery of a briefer version of this paper.

31. Einstein continued to assign probabilities, e.g., in the last sentence of *The Meaning of Relativity*: "Although such an assumption is logically possible, it is less probable than the assumption that there is a finite density of matter in the universe" [Princeton University Press, Princeton, N.J., (1922, 1945), p. 108].

32. A. Einstein, *Les Prix Nobel en 1921–1922* (Stockholm, 1923), and *Nobel Lectures, 1901–1921* (Elsevier, Amsterdam, 1967), pp. 482–90.

33. The point is put in a personal way in one of Einstein's manuscripts in the Archive at the Institute for Advanced Study entitled "Fundamental Ideas and Methods of Relativity Theory, Presented in Their Development," dating from about 1919 or shortly afterward. Speaking about the fact that prior to relativity theory the theoretical interpretation of induction was quite different depending on whether the magnet or the conductor is considered in motion, he confessed that this produced in him, as perhaps in no one else, a discomfort that had to be removed: "The thought that

one is dealing here with two fundamentally different cases was for me unbearable [war mir unerträglich] . . . . The phenomenon of electromagnetic induction forced me to postulate the special relativity principle.'' In this way, ''a kind of objective reality could be granted only to the electric and magnetic field together.''

34. E.g., ''Physik and Realität,'' J. Franklin Inst. **221**, 317 (1936). Cf. also ''Über die vom Relativitätsprinzip geforderte Trägheit der Energie,'' op cit. in n. 18.

35. *Einstein, Philosopher-Scientist*, op. cit. in n. 7, p. 81.

36. ''Physik und Realität,'' op. cit. in n. 34, p. 317.

37. ''The Problem of Space, Ether, and the Field in Physics,'' in Albert Einstein, *Ideas and Opinions* (Crown, New York, 1954), p. 285.

38. In *Einstein, Philosopher-Scientist*, op. cit. in n. 7, p. 23.

39. At the Jerusalem Einstein Centennial Symposium, P. A. M. Dirac gave on 20 March 1979 a paper on ''Unification: Aims and Principles,'' in which he said:

> It seems clear that the present quantum mechanics is not in its final form. Some further changes will be needed, just about as drastic as the changes which one made in passing from Bohr's orbits to a quantum mechanics. Some day a new relativistic quantum mechanics will be discovered in which we don't have these infinities occurring at all. It might very well be that the new quantum mechanics will have determinism in the way that Einstein wanted. This determinism will be introduced only at the expense of abandoning some other preconceptions which physicists now hold, and which it is not sensible to try to get at now.
>
> So, under these conditions I think it is very likely, or at any rate quite possible, that in the long run Einstein will turn out to be correct, even though for the time being physicists have to accept the Bohr probability interpretation — especially if they have examinations in front of them.

40. ''Physik und Realität,'' op. cit. in n. 34, p. 315.

41. Einstein, letter to M. Solovine, 30 March 1952, copy in the Archive at the Institute for Advanced Study.

5. ON SOME OTHER APPROACHES TO ELECTRODYNAMICS IN 1905

Arthur I. Miller

The thrust of fundamental research in 1905 was, as it is today, toward the unification of physics within a field-theoretical framework. During the first decade of the twentieth century many physicists believed that this goal was imminent, and their researches possessed an elegance and significance that seventy-four years of relativity have tended to obscure. Albert Einstein's paper "On the Electrodynamics of Moving Bodies," the relativity paper, appeared September 26, 1905, but until 1911 many physicists referred to a "Lorentz–Einstein theory." Einstein's first paper on electrodynamics was considered to have enriched current research. This essay discusses some different approaches to those problems in electrodynamics that we now think of as posing the same puzzle — that is, relativity. The approaches to be analyzed are the theories of the electron that were proposed by Max Abraham, Paul Langevin, and H. A. Lorentz; these theories will be compared with each other and with Einstein's work. Recently discovered correspondence of Lorentz and Henri Poincaré will enable us to catch a glimpse of the drama that is science in the making.[1]

It is natural to begin with H. A. Lorentz's electromagnetic theory that he first proposed in his paper of 1892, "La théorie électromagnétique de Maxwell et son application aux corps mouvants,"[2] because his theory was central to developments in what was deemed to be basic research at the beginning of the twentieth century. Lorentz's theory was the successful result of over two decades of elaborations and purifications of James Clerk Maxwell's electromagnetic theory. One of Lorentz's fundamental premises was that the sources of the electromagnetic field were undiscovered electrons that moved about in an all-pervasive absolutely resting ether. Lorentz described the state of the ether with five fundamental equations that he considered axiomatic:

$$\nabla \times \mathbf{E} = -\frac{1}{c}\frac{\partial \mathbf{B}}{\partial t} \tag{5.1}$$

$$\nabla \times \mathbf{B} = \frac{1}{c}\frac{\partial \mathbf{E}}{\partial t} + \frac{4\pi}{c}\rho\mathbf{v} \tag{5.2}$$

$$\nabla \cdot \mathbf{E} = 4\pi\rho \tag{5.3}$$

$$\nabla \cdot \mathbf{B} = 0$$

(Equations 5.1–5.3 and $\nabla \cdot \mathbf{B} = 0$: Maxwell–Lorentz equations)

Harry Woolf (ed.), Some Strangeness in the Proportion: A Centennial Symposium to Celebrate the Achievements of Albert Einstein
ISBN 0-201-09924-1

$$\mathbf{F} = \rho\mathbf{E} + \rho\,\frac{\mathbf{v}}{c} \times \mathbf{B}, \qquad \text{Lorentz force equation} \qquad (5.5)$$

where **E** and **B** are the electromagnetic fields, ρ is the electron's charge density, and the equations are written relative to a reference system fixed in the ether that I shall call S[3]; c is the velocity of light relative to S; and **v** is the electron's velocity relative to S. Lorentz's equations are consistent with the basic requirement of a wave theory of light: namely, that relative to the ether the velocity of light is independent of the source's motion and is always c. But measuring the velocity of light in moving reference systems may give a different result. In such a case, the measured velocity of light should be direction-dependent in a manner given by Newton's addition law of velocities. However, as is so well known, despite much effort, no measurable effects of the moving earth on optical or electromagnetic phenomena had been found — that is, the velocity of light measured on the earth turned out to be c.

In his 1895 monograph entitled *Treatise on a Theory of Electrical and Optical Phenomena in Moving Bodies*,[19] Lorentz was able to explain systematically the failure of ether-drift experiments accurate to first order in the quantity v/c, where v was the velocity of neutral, nonmagnetic, nondielectric matter relative to the ether. Using Galilean spatial transformations and a mathematical "local-time coordinate" t_L, with certain "new" vectors for the electromagnetic fields, Lorentz demonstrated that the electromagnetic field equations have the same form in an inertial reference system S_r as in S (see Fig. 5.1). Thus, to order v/c, neither optical nor electromagnetic experiments could reveal the motion of S_r. Lorentz called this stunning and desirable result the "theorem of corresponding states," because if a state of optics is described in S by **E** and **B** as functions of x, y,

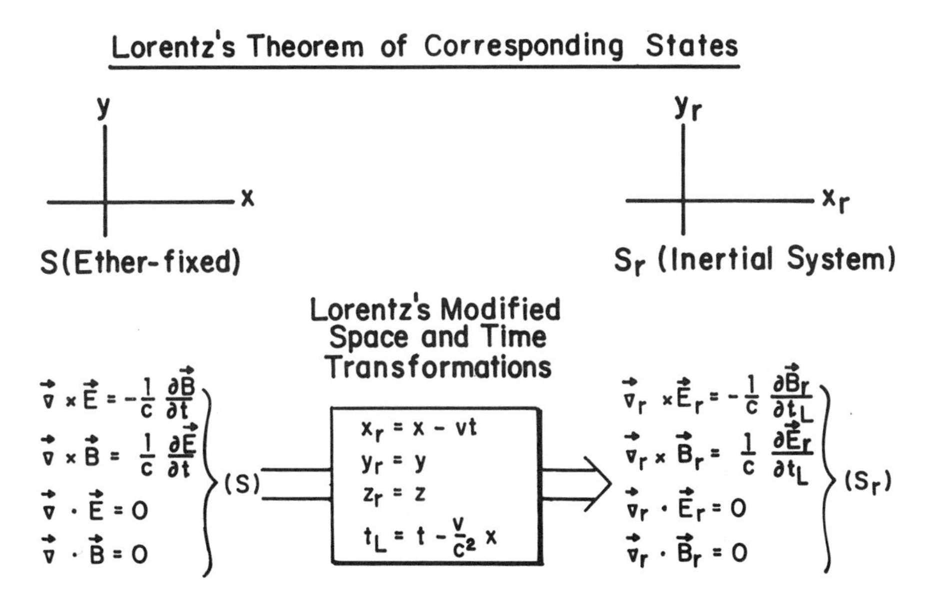

Fig. 5.1 Lorentz's modified space and time transformations contain the "local-time coordinate" t_L, and Lorentz referred to the electromagnetic field quantities $\mathbf{E}_r = \mathbf{E} + \mathbf{v}/c \times \mathbf{B}$ and $\mathbf{B}_r = \mathbf{B} - (\mathbf{v}/c) \times \mathbf{E}$ as "new" vectors.

z, and t, there is a corresponding state in S_r characterized by the "new" electromagnetic-field quantities as functions of x_r, y_r, z_r, and t_L. Needless to say, the physical time was taken to be identical with the absolute time because there was no reason to consider that time depended on the motion of a reference system.

By 1900 the electron Lorentz had hoped for had been discovered and the theory was able to explain systematically most extant empirical data. So successful had Lorentz's theory become that Henri Poincaré's assessments of it had gone from the "least defective" (in 1895) to the "most satisfactory" (1900).[4] However, Poincaré was disturbed by two blemishes on Lorentz's theory that Lorentz had been willing to overlook: 1) Lorentz's ad hoc 1892 contraction hypothesis for explaining the 1887 Michelson and Morley experiment involving an effect in second order in v/c; Poincaré scathingly criticized the contraction hypothesis; "hypotheses are what we lack least."[5] 2) But even worse, in Poincaré's opinion, Lorentz's theory violated Newton's principle of action and reaction. Lorentz had built this violation into his theory since the ether acted on bodies, but not vice versa. In his publication of 1895 Lorentz had dodged this issue by asserting that Newton's principle of action and reaction need not be universally valid. Newton's principle of action and reaction, however, was on the highest level of Poincaré's hierarchical view of a scientific theory because its generality precluded its experimental disconfirmation.[6]

The 1900 *Lorentz-Festschrift*, celebrating the twenty-fifth anniversary of Lorentz's doctorate from Leiden, included Poincaré's paper, "The Theory of Lorentz and the Principle of Reaction,"[7] in which he demonstrated the lengths to which he was prepared to go in order to save a principle. In his work, using the Maxwell–Lorentz equations, Poincaré reexpressed the Lorentz force on a charged body enclosed in a volume V, owing to charges external to V, as

$$\mathbf{F} = \nabla \cdot \overleftrightarrow{\mathbf{T}} - \frac{d}{dt}\frac{\mathbf{E} \times \mathbf{B}}{4\pi c} \tag{5.6}$$

where $\overleftrightarrow{\mathbf{T}}$ is Maxwell's stress tensor. He calculated the net force $\mathbf{F}_N$ on the charged body by integrating over the volume V, which he extended out to infinity in order to include the sources of the external electromagnetic fields. Poincaré's result is

$$\mathbf{F}_N = -\frac{d}{dt}\int \frac{\mathbf{E} \times \mathbf{B}}{4\pi c}\, dV, \tag{5.7}$$

since the contribution from the Maxwell stress tensor vanishes. Consequently, in Lorentz's theory a charged body cannot attain equilibrium. Poincaré traced this result to the absence of isolated systems in Lorentz's theory in the sense that this term was used in mechanics. Setting

$$\mathbf{F}_N = \frac{d}{dt} m_o \mathbf{v} \tag{5.8}$$

where m_o is the charged body's inertial mass, Poincaré rewrote Eq. (5.7) as

$$\frac{d}{dt}\left[m_o\mathbf{v} + \int \frac{\mathbf{E} \times \mathbf{B}}{4\pi c}\, dV\right] = \mathbf{0}. \tag{5.9}$$

The second term in Eq. (5.9) violated the principle of action and reaction as this principle was understood in Newton's mechanics. Poincaré next took the short but bold step that his philosophic view demanded: in order to rescue Newton's principle of action and reaction he compared the electromagnetic field to a "fictitious fluid" with a mass and "momentum"

$$\mathbf{G} = \frac{1}{4\pi c}\int \mathbf{E} \times \mathbf{B}\, dV. \tag{5.10}$$

Thus, Poincaré's electromagnetic momentum permitted him to simulate isolated closed systems in Lorentz's theory. In these sytems the net force $\mathbf{F}_N$ was cancelled by compensatory mechanisms in the ether arising from the electromagnetic momentum's temporal variation. But further hypotheses were necessary. Suppose, continued Poincaré, that an emitter of unidirectional radiation and an absorber were in relative inertial motion. Using Lorentz's local-time coordinate Poincaré demonstrated the insufficiency of the electromagnetic momentum for saving the principle of action and reaction separately for emitter and absorber. For this purpose he postulated an "apparent complementary force." In order to emphasize the necessity for this desperate step to save a principle, Poincaré reiterated that in conservative mechanical systems the principle of action and reaction could be considered as a consequence of the principles of energy conservation and of relative motion. According to Poincaré's view of mechanics, what he called the "principle of relative motion" could never be overthrown because it embodied the covariance of Newton's laws in inertial reference systems *and* Newton's second law;[8] furthermore, the principle of relative motion asserted the meaningfulness of only the relative motion between ponderable bodies, which is information drawn from the world of our perceptions; and so, wrote Poincaré, the "contrary hypothesis is singularly repugnant to the mind."

In a collection of Poincaré documents in Paris there is a letter of January 20, 1901 from Lorentz to Poincaré (Fig. 5.2). In this letter Lorentz expressed his admiration for Poincaré's contribution to the *Festschrift*, and then launched into an eight-page rebuttal. Lorentz's opinion of Poincaré's valiant attempt at saving the principle of action and reaction was, in his words: "But must we, in truth, worry ourselves about it?" Lorentz added in his forthright way: "I must claim to you that it is impossible for me to modify the theory in such a way that the difficulty that you cited disappears." Lorentz went on to emphasize several times that his ether acted on bodies but that there was no reaction on the ether. He explained that the "phenomena of aberration," that is, first-order effects, had "forced him" to assume a motionless ether; in fact, in Lorentz's seminal paper on electromagnetic theory,[2] second-order effects were nowhere mentioned.[10] Lorentz continued in his letter: "I deny therefore the principle of reaction in these elementary actions." In mechanics, Lorentz continued, action and reaction were instantaneous because disturbances were not mediated by an ether; however, in electromagnetic theory the reaction of an emitter of radiation was not compensated simultaneously by the action on the absorber. Poincaré had avoided this problem by attempting to satisfy the principle of reaction separately by emitter and absorber. Consistent with his desire to maintain an absolutely immobile ether, Lorentz protested Poincaré's naming the quantity in Eq. (5.10), which Lorentz compared to Poynting's vector, to be an electromagnetic momentum. To Lorentz the term momentum, of course, connoted motion. Lorentz was willing to concede only that Poincaré's electromagnetic momentum was formally "'equivalent' to a momentum." Thus, in the 1901 letter, Lorentz informed the critical Poincaré of his own sensitivity toward adding further hypotheses to an already overburdened theory, especially hypo-

Leiden, le 20 janvier 1901

Monsieur et très honoré collègue,

Permettez moi de vous remercier bien sincèrement de la part que vous avez bien voulu prendre au recueil de travaux que m'a été offert à l'occasion du 25eme anniversaire de mon doctorat. J'ai été profondiment touché de ce que tant d'illustres savants aux choisi ce jour pour me témoigner leur sympathie et l'intérêt qu'ils peuvement à mes études, malgré l'imperfection des resultats auxquels elles m'ont conduit. Cette imperfection est telle que je n'ose presque pas regarder comme aux signe d'approbation le livre qu'on m'a dédié; j'y verrai plutôt un encouragement qui m'est précieuse.

Comme votre jugement a, à mes yeux une tres grande importance, vous m'avez particulièrement obligé par le choise de votre sujet et par les paroles qui précèdent votre article. J'ai suivi vos raisonnements avec toute l'attention qu'ils demandent et je sens toute la force de vos remarques. Je dois vous avouer qu'il m'est impossible de modifier la théorie de telle façon que la difficulté que vous signalez disparaitre. Il me semble même guère probable qu'on puisse y réussir; je crois plutôt — et c'est aussi le résultat auquel tendent vos remarques — que la violation du principe de réaction est nécessaire dans toutes les théories que peuvent expliquer l'expérience de Fizeau. Mais faut-il en vérité que nous nous en inquiétions. Il y a un certain rapport entre vos considérations et une question qui a été soulevée, comme vous savez, par Helmholtz dans un de ces derniers mémoires. En effet, vos formules démontrent que l'éther contenu dans une surface fermée ne sera pas en équilibre sous l'influence des pressions de Maxwell exercées à cette surface, des que le vecteur de Poynting est une fonction du temps. De ceci, Helmholtz tire la conclusion que l'éther sera mis en mouvement dans un tel cas, et il cherche à établir les équations qui déterminent ce mouvement.

J'ai préféré une autre manière de voir. Ayant toujours en vue les phénomènes de l'aberration, j'ai admis que l'éther est absolument immobile — je veux dire que ses éléments de volume ne se déplacement pas, bien qu'ils puissent être le siège de certains mouvements internes. Or, si un corps ne se dèplace jamais, il n'y a aucune raison pour laquelle on parlerait de forces exercées sur ce corps. C'est ainsi que j'ai été amené à ne plus parler de forces qui agissent sur l'éther.

Je dis que l'éther agit sur les électrons, mais je ne dis pas qu'il éprouve de leur côté une réaction; je nie donc le principe de la réaction dans ces actions élémentaires. Dans cet ordre d'idées je ne puis pas non plus parler d'une force exercée par une partie de l'éther sur l'autre; les pressions de Maxwell n'ont plus d'existence réelle et ne sont que des fictions mathématiques qui servent à calculer d'une manière simple la force qui agit sur un corps pondérable. Evidemment, je n'ai plus à me soucier de ce que les pressions qui agisserent à la surface d'une portion le partie de l'éther ne seraient pas en équilibre.

Quant au principe de la réaction, il ne me semble pas qu'il doive être regardé comme un principe fondamental de la physique. Il est vrai que dans tous les cas où un corps acquiert une certaine quantité de mouvement *a,* notre esprit ne sera pas satisfait tout que nous ne puissons indiquer un changement simultané dans quelque autre corps, et que dans tous les phénomènes dans lesquels l'éther n'internent, ce changement consiste dans l'acquisition d'une quantité de mouvement $-a$. Mais je crois qu'on pourrait etre également satisfait si ce changement simultané ne fût pas lui même la production d'un mouvement. Vous avez déduit la belle formule

$$\Sigma M v_x + \int d\tau(\gamma g - \beta h) = \text{Const.}$$

Il me semble qu'on pourrait se borner à considerer $\int d\tau(\gamma g - \beta h)$,

$$\int d\tau(\alpha h - \gamma j), \int d\tau(\beta j - \alpha g)$$

comme des quantites dépendantes de l'état de l'éther qui sont pour ainsi dire "équivalentes" à une quantité de mouvement. Votre théoreme nous donne pour toute modification de la quantité de mouvement de la matière pondérable une modification simultanée de cette quantité équivalente; je crois qu'on pourrait bien se contenter de cela.

Je ne veux pas prétendre que cette manière de voir soit aussi simple qu'on pourrait le désirer; aussi n'aurais-je pas été conduit a cette théorie si les phénomènes de l'aberration ne m'y eussent pas force. Du reste, il va sans dire que la théorie ne doit être considerée que comme provisionare. Ce que je viens d'appeler "équivalence" pourra bien un jour nous apparaître comme une "identité"; cela pourrait arriver si nous parvenons à considérer la matière ponderable comme une modification de l'éther lui-même.

Il est presque inutile de dire qu'on pourrait aussi se tirer d'embarras en attribuant à l'éther une masse infiniment (ou très) grande. Alors les électrons pourraient réagir sur l'éther sans que ce milieu se mit en mouvement. Mais cette issue me semble assez artificielle.

Je désirerais bien vous faire encore quelques remarques au sujet de la compensation des termes en v^2, mais cette lettre devrait trop longue. J'espère donc que vous me permettrez de revenir sur cette question une autre fois. Il y a là encore bien des difficultes; vous pourriez peut être parvenir à les surmonter.

Veuillez agréer, Monsieur et tres honoré collègue, l'assurance de ma sincère considération. Votre bien devoué.

H. A. Lorentz

Fig. 5.2 Typed copy of a handwritten letter of H. A. Lorentz to H. Poincaré, 1901.

theses invented solely to save a principle whose violation permitted the theory's formulation in the first place. Lorentz concluded the 1901 letter by emphasizing that at this time, his chief concern was folding the contraction hypothesis into his electromagnetic theory.

Possibly as a result of this letter, one of Poincaré's reasons in papers of 1904 and 1908[11] for rejecting the principle of action and reaction was Lorentz's argument that the compensation between action and reaction could not be simultaneous.

To summarize: Poincaré's 1900 attempt to save the principle of action and reaction is a fine example of his application of his immense powers of mathematics, and his philosophic view, to criticizing, clarifying, and extending previously proposed physical theories. Lorentz's 1901 letter to Poincaré is typical of Lorentz's style at this point in his research on electromagnetic theory — that is, pushing ahead and leaving problems of foundations and of philosophy to others. It turned out that the road to a foundationally sound electrodynamics could be achieved only through an analysis of fundamental problems within a philosophic framework that was, to some extent, liberated from empirical data. In this way, Einstein could formulate a basis for investigating what it meant for action and reaction not to be simultaneous.[12]

We next turn to Wilhelm Wien's contribution to the *Lorentz-Festschrift* that served to change the direction of physical theory in a way that rendered superfluous Poincaré's attempts to rescue the principle of reaction, and demonstrated the conservatism of Lorentz's opinion of electromagnetic momentum. Wien was so impressed with the successes of Lorentz's electromagnetic theory and with its possibilities for unifying electromagnetism and gravitation that he suggested research toward an "electromagnetic basis for mechanics," that is, an electromagnetic world-picture.[13] In this scheme mechanics and then all of physical theory would have been deduced from Lorentz's electromagnetic theory. The oppositely directed research effort, a mechanical world-picture, had produced only increasingly complicated mechanical models in order to simulate the contiguous actions of an ether. One implication of an electromagnetic world-picture was that the electron's mass originated in its self-electromagnetic fields as a self-induction effect. Consequently, the electron's mass was predicted to be a velocity-dependent quantity.

Using high-velocity electrons from radium salts, Walter Kaufmann at Göttingen gave data for the velocity-dependence of the electron's mass.[14] Kaufmann's colleague, Max Abraham formulated the first field-theoretical description of an elementary particle.[15] Putting aside problems concerning the motion of Lorentz's ether, Abraham interpreted Poincaré's electromagnetic momentum as the electron's momentum as a result of its self-fields. Then, with a restriction on the acceleration of a rigid-sphere electron, and with new techniques of Hamilton-Lagrange mechanics applied to electrodynamics, Abraham deduced Newton's second law from Lorentz's force.[16] His principal result was that the electron's mass was a two-component quantity, which depended on whether the electron was acted on by external forces transverse or parallel to its trajectory. Abraham's transverse (m_T) and longitudinal (m_L) masses are

$$m_T = \frac{m_o^e}{2\beta^3}\left[(1 + \beta^2)\ln\left(\frac{1+\beta}{1-\beta}\right) - 2\beta\right], \tag{5.11}$$

$$m_L = \frac{m_o^e}{\beta^3}\left[\frac{2\beta}{1-\beta^2} - \ln\left(\frac{1+\beta}{1-\beta}\right)\right] \tag{5.12}$$

where $\beta = v/c$, $m_o^e = e^2/2Rc^2$ is the electron's electrostatic mass due to its self-electrostatic field, and R is the electron's radius. Kaufmann's data agreed with Abraham's equation for the transverse mass m_T. Abraham's theory agreed with every first-order experiment since it was predicated on Lorentz's electromagnetic theory, but at this point it explained no second-order data other than Kaufmann's.[17] Contrary to postrelativity myths, the first-order experiments were of primary importance to everyone.[18] The Michelson–Morley experiment was, of course, much discussed, but, like Abraham, many physicists tried assiduously to avoid the unpalatable contraction hypothesis.[19]

Therefore, by the end of 1903, it seemed that a great deal of progress had been made toward a unified description of nature that was based on the most agreeable aspects of Lorentz's electromagnetic theory. The hallmark of the new world picture, proclaimed Abraham, was "atomistic structure of electricity, but continuous spatial distribution of the ether."[20] And the electromagnetic world-picture squared with Abraham's philosophic view: since the electron's dynamics could be deduced from a single quantity — its Lagrangian, then the electromagnetic world-picture, as Abraham wrote in 1903, has an "economical meaning," a term that had been introduced by the widely read philosopher–scientist Ernst Mach.[21]

In response to Poincaré's criticisms, Kaufmann's data, the new second-order ether-drift experiments of Trouton and Noble, and those of Rayleigh and Brace, in 1904 Lorentz extended his electromagnetic theory to a theory of the electron.[22] Whereas Abraham's electron was a rigid sphere, Lorentz's was deformable and he postulated that when at rest it was a sphere, but it contracted when moving. Lorentz's electron can be likened to a balloon uniformly smeared with charge. More than ten interlocking postulates enabled Lorentz to extend the 1895 theorem of corresponding states to apply to any order of accuracy. Then, using the same approximation as Abraham, Lorentz derived transverse (m_T) and longitudinal (m_L) masses for his electron:

$$m_T = \frac{4}{3} m_o^e / \sqrt{1 - \beta^2} \tag{5.13}$$

$$m_L = \frac{4}{3} m_o^e / (1 - \beta^2)^{3/2} \tag{5.14}$$

where $\beta = v/c$, $m_o^e = e^2/2Rc^2$ is the electron's electrostatic mass due to its self-electrostatic field, and R is the electron's radius. Lorentz showed that his m_T also agreed with Kaufmann's data. This time Poincaré found no faults with the foundations of Lorentz's theory, but he quickly caught several technical errors, and his keen sense of aesthetics revealed to him certain symmetries that Lorentz's original mathematical formulation had obscured. Poincaré's completed work appeared in his classic paper of 1906, "On the Dynamics of the Electron"; a short version of this paper had appeared in June 1905.[23] The development of certain key results that survived the eventual demise of Lorentz's theory of the electron are in three letters that Poincaré wrote to Lorentz, which I discovered at the Algemeen Rijksarchief, The Hague. In order to set the stage for describing these three letters, I shall review the interpretation of Lorentz's space and time transformations of 1904, which Poincaré dubbed the "Lorentz transformations":

$$x' = klx_r \tag{5.15}$$

$$y' = ly_r \tag{5.16}$$

$$z' = lz_r \tag{5.17}$$

$$t' = \frac{l}{k} t_r - kl \frac{v}{c^2} x_r \tag{5.18}$$

$$(k = 1 / \sqrt{1 - \beta^2})$$

where the coordinates (x', y', z', t') refer to an auxiliary system Σ' in which the Maxwell–Lorentz equations have the same form as they do relative to an ether-fixed reference system; the coordinates (x_r, y_r, z_r, t_r) refer to an inertial reference system S_r; and $x_r = x - vt$, $y_r = y$, $z_r = z$, $t_r = t$, where (x, y, z, t) refer to the ether-fixed system Σ; l admits the possibility of any sort of deformation, where $l = 1$ yields Lorentz's contraction and also is the only value of l that is consistent with Lorentz's theorem of corresponding states that applies to any order in v/c. Lorentz's goal was to transform the fundamental equations of his electromagnetic theory from the inertial system S_r to the fictitious reference system Σ' in which he could perform any necessary calculations, and then transform back to S_r. For example, it was particularly useful to take Σ' to be the electron's instantaneous rest system in order to reduce problems of the electrodynamics and optics of moving bodies to problems concerning bodies at rest.

In summary (see Fig. 5.3) there were three reference systems in Lorentz's theory — S, S_r, Σ' — where Σ' was an auxiliary system possessing no physical interpretation. Consequently, there were two sorts of electrons: the "real" electron of the reference systems S and S_r, and the "imaginary" electron in Σ'. Taking Σ' to be the electron's instantaneous rest system meant that in order for Σ' to have the same properties as the ether-fixed system S, the imaginary electron had to be a sphere in Σ'. But from the spatial portion of the Lorentz transformations [Eqs. (5.15)–(5.17)], the electron in Σ' is deformed into a shape with axes (kl, l, l). Therefore in order to maintain the theorem of corresponding states, the moving electron in S_r must be deformed in a manner that compensates for the deformation in Σ' — that is, by an amount $(1/kl, 1/l, 1/l)$. Lorentz's supporting argument for this postulate of deformation included inverting the spatial part of the Lorentz transformations. In order to render the postulate of contraction further "plausible," to use Abraham's word from his subsequent criticism[24] of Lorentz's theory of the electron, Lorentz proposed further interlocking postulates; among them was a postulate that the as yet unknown intermolecular forces transformed like the electromagnetic force in order that macroscopic matter contracted when in motion. Thus, to Lorentz and Poincaré, covariance was a mathematical property that was achieved through postulated transformations, which did not possess an unambiguous physical interpretation, because the transformations contained unknown velocities relative to the ether. The reason is that in electrodynamics and optics of 1905, the ether-fixed reference system S was cavalierly taken to be the laboratory system. The tacit assumption was that the ratio of the laboratory's velocity relative to the ether with that of light was small enough to be neglected. In other words, the notion of at rest relative to the laboratory was not well defined.

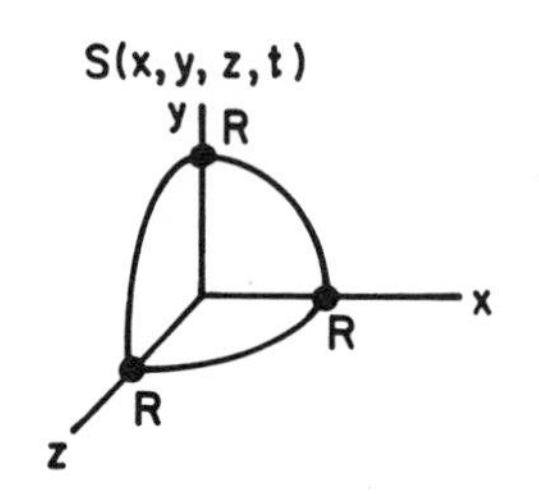

In the ether-fixed reference system S the real-sphere electron has axes: (R,R,R).

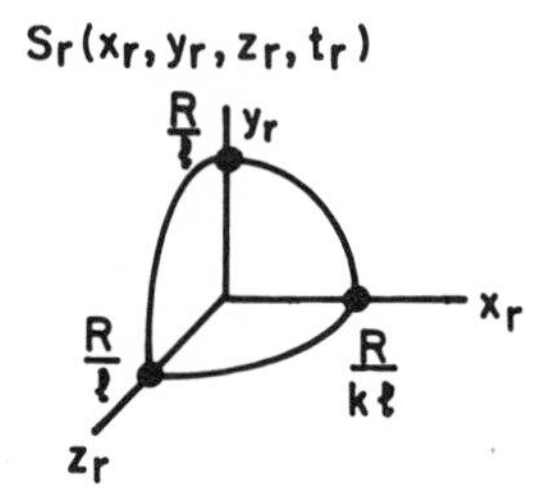

In the inertial reference system S_r the real-deformed electron has axes: $\left(\frac{R}{kl}, \frac{R}{l}, \frac{R}{l}\right)$.

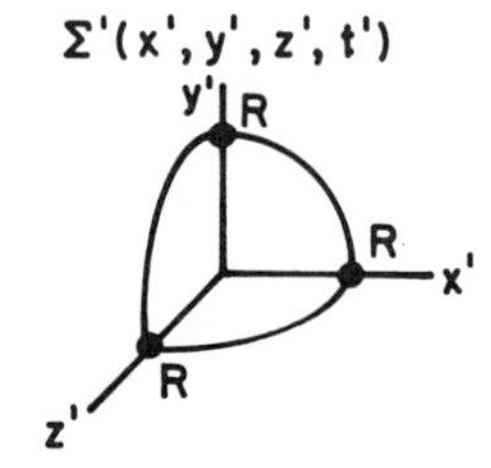

In the auxiliary reference system Σ' the imaginary-sphere electron has axes: (R, R, R).

Fig. 5.3 The three different reference sytems employed in the electromagnetic world-picture for theories of a deformable electron.

With this background, we turn to a letter (Fig. 5.4) by Poincaré written to Lorentz shortly after Poincaré had returned from the September 1904 Congress of Arts and Science at St. Louis, Missouri. Although Poincaré's letters are undated, as was customary in France, this and the next two can be set easily into the period between late 1904 and mid-1905. At St. Louis, Poincaré had discussed elatedly the main results of Lorentz's 1904 theory of the electron, and, using a term well known from fundamental studies in geometry, he had renamed Lorentz's new theorem of corresponding states the "principle of relativity," according to which "the laws of physical phenomena must be the same for a stationary observer as for an observer carried along in a uniform motion of translation; so that we have not and can not have any means of discerning whether or not we are carried along in such a motion."[25]

Poincaré began the first of this set of three letters in a familiar way, that is, by noting certain errors in Lorentz's 1904 paper. Lorentz's use of three different reference systems had led him to deduce incorrect equations for the electron's velocity, charge density, and Lorentz force as these quantities were related between S_r and Σ'. Lorentz had been aware of these errors, and he had ensured their not affecting his final results by having taken Σ' as the electron's instantaneous rest system, thereby causing offending terms to vanish. Using only S and Σ', Poincaré obtained the correct transformations for the general case where Σ' was not the electron's instantaneous rest system.

Poincaré continued by pointing out where the fundamental difficulty with Lorentz's theory lay, although he did not know yet how to resolve it: two of the quantities characterizing the moving electron (its momentum and longitudinal mass) differed when calculated from the Lagrangian and from the energy.[26] Since Lorentz had calculated the electron's mass directly from its momentum, he had missed this problem. For Poincaré, investigating the foundations of Lorentz's theory meant reformulating it within the framework of that most elegant and powerful formulation of physical theory — the mechanics of Hamilton and Lagrange. Consistent with his style of treating every problem in the most general possible manner, Poincaré's reformulation was based on a generalized version of the dynamics of the electron in which he could study every possible model of the electron, that is, all connections between k and l. It turned out as Poincaré wrote in this letter, that only the 1904 model of the electron by Poincaré's former student, Paul Langevin,[27] offered consistency regarding the electron's momentum. According to Langevin the real electron in S_r was deformed in such a way that the volumes of the real and imaginary electrons were equal; that is, $l^3k = 1$.[28]

In the conclusion of this letter, Poincaré expressed his dissatisfaction with Lorentz's method of proof for $l = 1$. Lorentz's proof[29] employed formulae for forces and accelerations valid only in the electron's instantaneous rest system, and so the proof was not general enough.

But in the next letter (Fig. 5.5), Poincaré elegantly resolved this insufficiency. By eliminating the intermediate inertial reference system, and thereby relating directly the fictitious reference system Σ' with the unobservable reference system S, Poincaré was able to write the Lorentz transformations in a form possessing a high degree of mathematical symmetry. Considering two successive Lorentz transformations along the same direction, it was easy for him to prove that the Lorentz transformations form a group, and that l must be equal to one.[30] As a bonus, Poincaré also obtained the new addition law for velocities that was independent of l.

Mon cher Collègue,

J'ai énormément regretté les circonstances qui m'ont empêché d'abord d'entendre votre conférence et ensuite de causer avec vous pendant votre séjour à Paris.

Depuis quelque temps j'ai étudié plus en détail votre mémoire électromagnetic phenomena in a system moving with any velocity smaller than that of Light, mémoire dont l'importance est extrême et dont j'avais déjà cité les principaux résultats dans ma conférence de St Louis.

Je suis d'accord avec vous sur tous les points essentiels; cependant il y a quelques divergences de détail.

Ainsi page 813, au lieu de poser:

$$\frac{1}{kl^3}\rho = \rho' \; ; \; k^2 u_x = u'_x, \; k^2 u_y = u'_y$$

il me semble qu'on doit poser:

$$\frac{1}{kl^3}\rho(1+\varepsilon v_x) = \rho' \quad \frac{1}{kl^3}\rho(v_x+\varepsilon) = \rho' u'_x$$

où $\varepsilon = -\frac{w}{c}$ ou $\varepsilon = -w$ si nous choisissons les unités de telle façon que $c = 1$.

Cette modification me semble s'imposer si l'on veut que la charge apparente de l'électron se conserve.

Les formules (10) page 813 se trouvent

alors modifiées et je trouve alors pour le dernier terme au lieu de

$$l^2 \frac{w}{c^2}(u'_y d'_y + u'_z d'_z), \; -\frac{l^2}{k}\frac{w}{c^2} u'_x d'_y, \; -\frac{l^2}{k}\frac{w}{c^2} u'_x d'_z$$

je trouve

$$l^2 \frac{w}{c^2}(u'_x d'_x + u'_y d'_y + u'_z d'_z), \; 0, \; 0$$

C'est la force de Liénard, que vous trouvez aussi mais avec des différences.

Et alors la question se pose de savoir si cette force est ou non compensée.

Ceci montre qu'entre les forces réelles X, Y, Z et les forces apparentes X', Y', Z' il y a les relations

$$X' = A(X + \varepsilon \Sigma X v_x), \quad Y' = BY, \quad Z' = BZ$$

A et B étant des coëff. et $A \varepsilon \Sigma X v_x$ représentant la force de Liénard.

Si toutes les forces sont d'origine électrique les conditions d'équilibre (ou du principe de d'Alembert modifié) donnent

$$X = Y = Z = 0$$

d'où

$$X' = Y' = Z' = 0$$

Si toutes les forces ne sont pas d'origine électrique, il y aura encore compensation pourvu qu'elles se comportent toutes comme si elles étaient d'origine électrique.

Mais il y a autre chose.

Vous supposez $l = 1$

Langevin suppose $kl^3 = 1$
J'ai essayé $kl = 1$ pour conserver l'unité
de temps, mais cela m'a conduit à des
conséquences inadmissibles.
D'un autre côté j'arrive à des contradictions
(entre les formules de l'action et de l'énergie)
avec toutes les hypothèses autres que celles
de Langevin
Le raisonnement par lequel vous
établissez que $l = 1$ ne me paraît
pas concluant, ou plutôt il ne l'est
plus et laisse l indéterminé quand
je fais le calcul en modifiant comme
je vous l'ai dit les formules de la page
813
Que pensez-vous de cela, voulez-vous que
je vous communique plus de détails ou
ceux que je vous ai donnés vous suffisent
ils
Excusez-moi en tout cas d'abuser de votre
temps.

Votre bien dévoué Collègue,

Poincaré

Fig. 5.4 Letter from H. Poincare to H. A. Lorentz, 1904.

Mon cher Collègue,

Merci de votre aimable lettre.
Depuis que je vous ai écrit mes idées se sont modifiées sur quelques points.
Je trouve comme vous $l = 1$ par une autre voie.
Soit $-\varepsilon$ la vitesse de la translation celle de la lumière étant prise pour unité.

$$k = (1 - \varepsilon^2)^{-\frac{1}{2}}$$

On a la transformation

$$x' = kl(x + \varepsilon t), \quad t' = kl(t + \varepsilon x)$$

$$y' = ly, \quad z' = lz.$$

Ces transformations forment un groupe.
Soient deux transformations composantes correspondant à

$$k, \ l, \ \varepsilon$$

et

$$k', \ l', \ \varepsilon'$$

leur résultante correspondra a

$$k'', \ l'', \ \varepsilon''$$

où:

$$k'' = (1 - \varepsilon''^2)^{-\frac{1}{2}}, \quad l'' = ll', \quad \varepsilon'' = \frac{\varepsilon + \varepsilon'}{1 + \varepsilon\varepsilon'}$$

Si nous voulons maintenant prendre

$$l = (1-\varepsilon^2)^m, \quad l' = (1-\varepsilon'^2)^m$$

nous n'aurons:

$$l'' = (1-\varepsilon''^2)^m$$

que pour $m = 0$

D'un autre côté je ne trouve d'accord entre le calcul des masses par le moyen des quantités de mouvement électromagnétique et par le moyen de la moindre action, et par le moyen de l'énergie que dans l'hypothèse de Langevin.

J'espère tirer bientôt au clair cette contradiction, je vous tiendrai au courant de mes efforts.

Votre bien dévoué Collègue,

Poincaré

Fig. 5.5 Letter from H. Poincaré to H. A. Lorentz, written between late 1904 and mid-1905.

In his next letter (Fig. 5.6), Poincaré's key result was that only Lorentz's 1904 theory of the electron offered "perfect compensation (which prevents the experimental determination of absolute motion)." Poincaré linked this result to solving the problem concerning the longitudinal mass of Lorentz's electron. Poincaré's solution was to add a term to the electron's Lagrangian that preserved the Lorentz invariance of the principle of least action. Thus, he was led to interpret the additional term as the energy due to the internal pressure that served to prevent the deformable electron from exploding in its rest system.[31] Now the theory was freed of all removable blemishes. We can well believe that Poincaré was, as he put it, in "parfait accord" with the "beaux travaux" of Lorentz. Lorentz's 1904 theory of the electron was really the culmination of over a decade of painstaking work and interaction by Lorentz and Poincaré.

Mon cher Collègue,

J'ai continué les recherches dont je vous avais parlé. Mes résultats confirment pleinement les vôtres en ce sens que la compensation parfaite (qui empêche la détermination expérimentale du mouvement absolu) ne peut se faire complètement que dans l'hypothèse $l = 1$. Seulement pour que cette hypothèse soit admissible, il faut admettre que chaque électron est soumis à des forces complémentaires dont le travail est proportionnel aux variations de son volume.

Ou si vous aimez mieux, que chaque électron se comporte comme s'il était une capacité creuse soumise à une pression interne constante (d'ailleurs

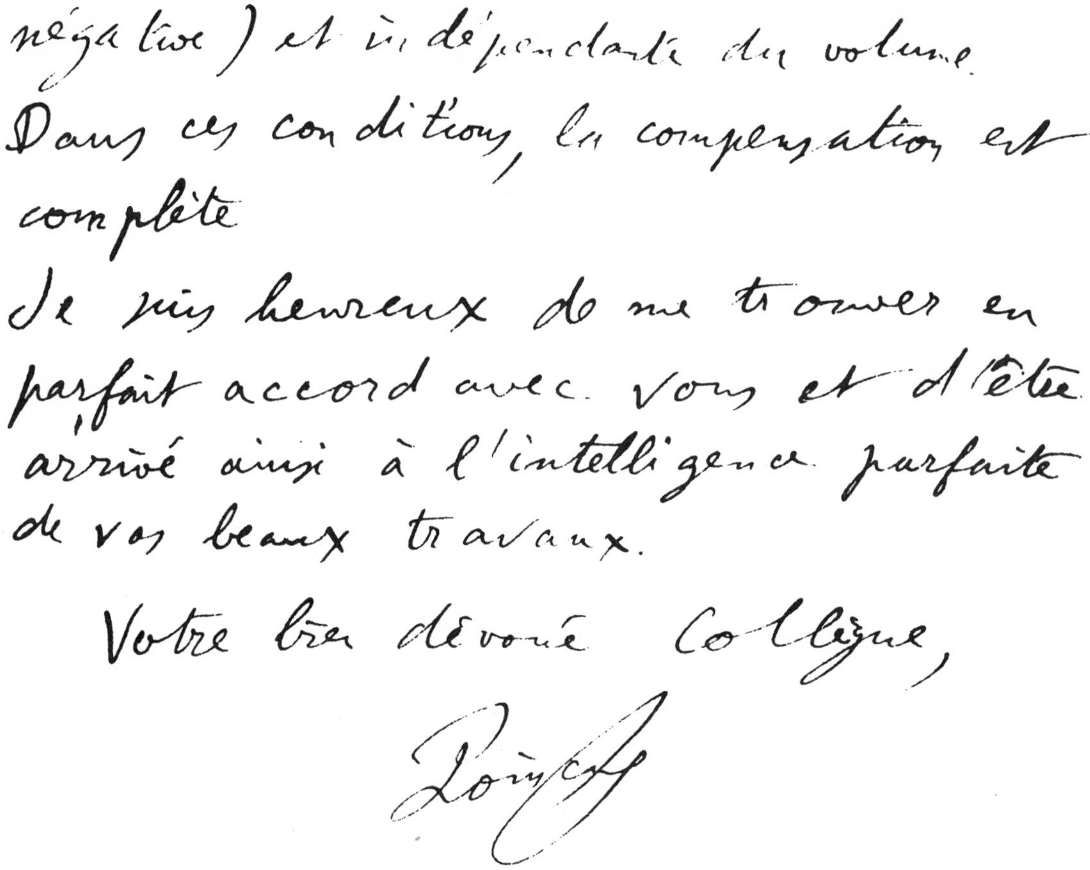

négative) et indépendante du volume.
Dans ces conditions, la compensation est
complète
Je suis heureux de me trouver en
parfait accord avec vous et d'être
arrivé ainsi à l'intelligence parfaite
de vos beaux travaux.

Votre bien dévoué Collègue,

Poincaré

Fig. 5.6 Letter from H. Poincaré to H. A. Lorentz, written between late 1904 and mid-1905.

Of the existing electron models, only Lorentz's was really adequate. It could explain optical data accurate to second order in v/c. It explained effects such as the dependence of mass on velocity, the contraction of moving objects, and the isotropy of the velocity of light in inertial reference systems, all to have been *caused* by the interaction between the electrons constituting matter and the ether. Lorentz's theory of the electron appeared to many physicists to be the most likely candidate for the cornerstone of a unified field-theoretical description of nature. As he showed in his 1905–6 publications, Poincaré had cast the theory into the Hamilton–Lagrange formalism, replete with group theory and four-dimensional spaces, and then extended it to a Lorentz-covariant theory of gravity.

From the relativity theory we have learned that it is more fruitful to consider as axiomatic that the space of inertial reference systems is isotropic for light propagation, rather than to attempt to explain this property. Then, as Einstein wrote in 1907,[19] certain postulates of Lorentz's theory become "secondary consequences," — for example, the contraction of moving bodies.[32]

Let us turn next to Kaufmann's reception of Lorentz's theory of the electron. On July 10, 1904, Kaufmann wrote to Lorentz that Lorentz's m_T agreed better with Kaufmann's most recent data than even Lorentz had imagined.[33] In order to decide the issue between the electron theories of Abraham and Lorentz, Kaufmann promised to increase his accuracy in the next series of experiments. He was as good as his word, but the results were not to please Lorentz.

In a short paper of November 30, 1905, and then in a longer one of early 1906, Kaufmann reported that his new measurements agreed best with Abraham's theory — a close second was Bucherer's — but as he put it, the "measurement results are not compatible with the Lorentz-Einstein fundamental assumption."[34] As Kaufmann wrote in 1905, "a recent publication by Mr. A. Einstein on the theory of electrodynamics . . . leads to results which are formally identical with those of Lorentz's theory" — that is, the "formal equivalence" of Einstein's 1905 relativity transformations for space and time and his prediction for the electron's transverse mass with these quantities in Lorentz's theory led to the names "Lorentz-Einstein theory" and "Lorentz-Einstein principle of relativity."[35] Just how devastated Lorentz was by Kaufmann's results is clear from his unpublished letter to Poincaré, dated March 8, 1906 (Fig. 5.7). It is a very revealing moment in the

Leiden, le 8 mars 1906.

Monsieur et très honoré collègue,

C'est déjà trop longtemps
que j'ai négligé de vous remercier de l'important mémoire
sur la dynamique de l'électron
que vous avez bien voulu m'en-
voyer. Inutile de vous dire
que je l'ai étudié avec le
plus grand intérêt et que
j'ai été très heureux de voir

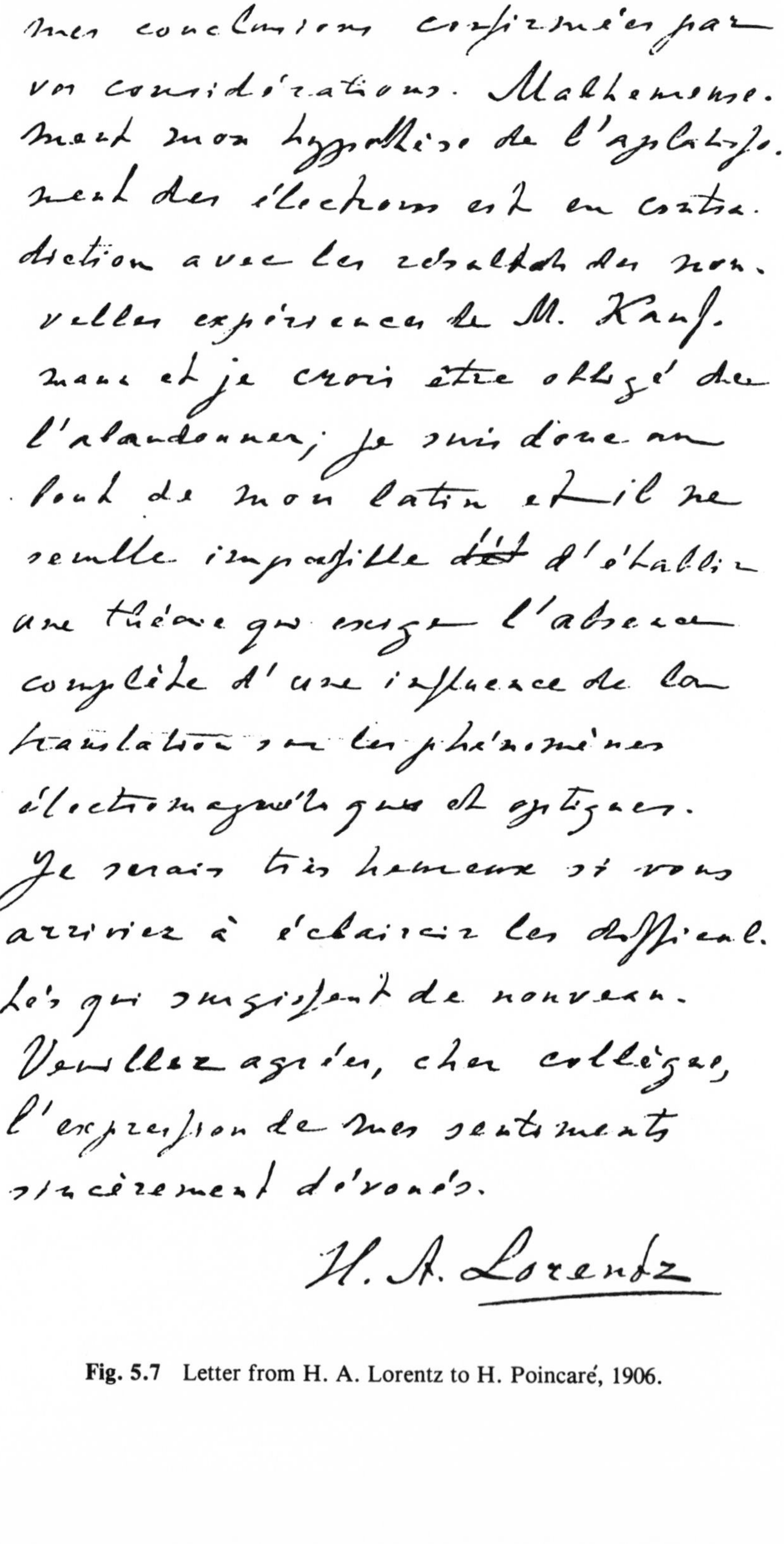

mes conclusions confirmées par
vos considérations. Malheureuse-
ment mon hypothèse de l'aplatisse-
ment des électrons est en contradiction avec les résultats des nou-
velles expériences de M. Kaufmann et je crois être obligé de
l'abandonner; je suis donc au
bout de mon latin et il me
semble impossible ~~dét~~ d'établir
une théorie qui exige l'absence
complète d'une influence de la
translation sur les phénomènes
électromagnétiques et optiques.
Je serais très heureux si vous
arriviez à éclaircir les difficul-
tés qui surgissent de nouveau.
Veuillez agréer, cher collègue,
l'expression de mes sentiments
sincèrement dévoués.

H. A. Lorentz

Fig. 5.7 Letter from H. A. Lorentz to H. Poincaré, 1906.

history of relativity theory. After congratulating Poincaré for the essay "On the Dynamics of the Electron," Lorentz wrote that, nevertheless all their work may have been for nothing:

> Unfortunately my hypothesis of the flattening of electrons is in contradiction with Kaufmann's new results, and I must abandon it. I am, therefore, at the end of my Latin. It seems to me impossible to establish a theory that demands the complete absence of an influence of translation on the phenomena of electricity and optics. I would be very happy if you would succeed in clarifying the difficulties which arise again.

What a remarkable confession, and what a clear-cut example of falsification. After all those years of work Lorentz was willing to abandon his theory because of the report of a single experiment. Lorentz repeated this message in his lectures at Columbia University in March and April 1906 that appeared in 1909 as *Theory of Electrons*: "But, so far as we can judge at present, the facts are against our hypothesis." For his part, Poincaré, too, was strongly affected by Kaufmann's new results; but his reaction was not as radical as Lorentz's. In the introductory paragraphs to the 1905 version of "On the Dynamics of the Electron," Poincaré had written that the theories of Abraham, Langevin, and Lorentz "agreed with Kaufmann's experiments." Of course he meant those based on data prior to 1905. But in the last paragraph of the introduction to the 1906 version, Poincaré wrote, concerning Lorentz's theory, that at "this moment the entire theory may well be threatened" by Kaufmann's new 1905 data.[36] Poincaré was still willing for the sake of argument to consider the principle of relativity for the moment as valid in order "to see what consequences follow from it." He called for someone to repeat Kaufmann's experiment.[37] Whereas for Lorentz, Kaufmann's results threatened a theory; for Poincaré, a philosophic view was also at stake, one that emphasized a principle of relative motion.

It turned out that Kaufmann's data were incorrect. In a letter of September 7, 1908, A. H. Bucherer in Bonn wrote Einstein in Berne of his recent experimental results concerning the mass of high-velocity electrons: "by means of careful experiments, I have elevated the validity of the principle of relativity beyond any doubt."[38] On September 23, 1908 Bucherer presented his results to the Meeting of German Scientists and Physicians at Cologne; his communication was entitled "Measurements on Becquerel-rays. The Experimental Confirmation of the Lorentz–Einstein Theory."[44] Poincaré and Lorentz welcomed Bucherer's results, and did not call for someone to repeat his experiments.[39] although, incidentally, thirty years later it turned out that Bucherer's data were inconclusive.[40] As Poincaré wrote in 1912, further weight was added to Bucherer's data because they disconfirmed Bucherer's own 1904 theory of the electron that was identical with Langevin's.[41]

The principle of relativity of Bucherer, Lorentz, and Poincaré resulted from careful study of a large number of experiments, and it was the basis of a theory in which empirical data could be explained to have been *caused* by electrons interacting with an ether. Einstein's principle of relativity excluded the ether of electromagnetic theory and did not explain anything. For example, in Einstein's relativity theory, the negative results of the ether-drift experiments were a foregone conclusion. As Lorentz wrote in his 1909 book, *Theory of Electrons*, "Einstein simply postulates what we have deduced."

In a 1907 paper[42] reviewing the status of the principle of relativity, Einstein discussed Kaufmann's new data from late 1905, those same data that, in 1906, had driven Lorentz to the "end of [his] Latin." Einstein acutely emphasized that the "systematic deviation" between the Lorentz–Einstein predictions and the data could indicate a hitherto "unnoticed source of error"; consequently, Einstein called for further experiments. Then he dismissed Kaufmann's data because in his "opinion" they supported theories that did not "embrace a greater complex of phenomena" — that is, the electron theories of Abraham and Bucherer, which could not explain optical experiments accurate to second order in v/c. Einstein's intuition served him well. Undoubtedly having the Kaufmann episode in 1907 in mind, Einstein in 1946 described one of two criteria for assessing a scientific theory thus: "The first point of view is obvious: the theory must not contradict empirical facts. However evident this demand may in the first place appear, its application turns out to be quite delicate."[43]

Abraham, Bucherer, Lorentz, and Poincaré, among others that included Max Planck[44] and Hermann Minkowski,[45] continued to cling to an ether chiefly because: 1) it was the basic ingredient to what appeared to be the realization of the long-sought-after unification of the sciences — an electromagnetic world-picture; 2) something had to mediate disturbances and explain physical effects.

The man whom Einstein held in high esteem, and to whom he referred to in correspondence of 1913 as the "fearsome Abraham"[46] refused to permit Bucherer's data to decide the issue between the competing electron theories; in other words, when his own theory was at stake, Abraham was not a falsificationist. In 1909 Abraham still pursued, as he put it, a version of the "electromagnetic mechanics" that was consistent with the rigid-sphere electron.[47]

In the part of a spirited tirade of 1914 leveled against the special relativity theory Abraham depicted well just how difficult it was to part with a philosophic position and with the great past achievements of physics — that is, the hard-won empirical data and the theories that they engendered: "the theory of relativity excited the young devoted to the study of mathematical physics; under the influence of that theory they filled the halls and corridors of the universities. On the other hand . . . the physicists of the former generation whose philosophy was formed under the influence of Mach and Kirchhoff, remained for the most part skeptical of the audacious innovators who allowed themselves to rely on a small number of experiments still debated by specialists, to overthrow the fundamental tests of every physical measurement."[48]

In conclusion, new archival materials, in conjunction with primary sources, reveal a scenario of science in the making. As was so often the case in turbulent times, the scientists' rules for assessing scientific theories were, as Poincaré wrote, "extremely subtle and delicate, and it is practically impossible to state them in precise language; they must be felt rather than formulated."[49]

Acknowledgements

This essay is based on research supported by the National Science Foundation's History and Philosophy of Science Program, by fellowships from the American Council of Learned Societies and the American Philosophical Society, and a travel grant from the Lorentz Foundation, Leiden University, The Netherlands.

The letters from Lorentz to Poincaré are in the possession of the Poincaré Estate who have granted me sole access to their materials through mid-1981. I thank the Poincaré Estate, Paris, and the Algemeen Rijksarchief, The Hague, for permission to quote from their correspondence.

It is a pleasure to acknowledge Professor Gerald Holton's comments on an early draft of this essay.

NOTES

1. For detailed analyses of the theories of the electron of Abraham, Langevin, Lorentz and Poincaré, as well as of Poincaré's philosophy of science, and for references to other secondary works see my papers listed at the end of this note and the first chapter of my book-length analysis of Einstein's 1905 relativity paper in its historic context. The book is A. I. Miller, *Albert Einstein's Special Theory of Relativity: Emergence (1905) and Early Interpretation (1905–1911)* (Addison-Wesley, Advanced Book Program, Reading, Mass., 1980). The papers are: "A Study of Henri Poincaré's 'Sur la Dynamique de l'Electron'," Arch. History Exact Sci. **10**, (3-5), 207-328 (1973); "On Lorentz's Methodology," Brit. J. Phil. Sci. **25**, 29-45 (1974); "On Einstein, Light Quanta, Radiation and Relativity in 1905," Am. J. Phys. **44**, 912-23 (1976); and "Poincaré and Einstein: A Comparative Study." Boston Studies Phil. Sci. **31** (forthcoming).

2. H. A. Lorentz, "La théorie électromagnétique de Maxwell et son application aux corps mouvants," Arch. Néerl. **25**, 363 (1892); reprinted in *Collected Papers*, Vol. 2, (Nijhoff, The Hague, 1935-39), pp. 164-343.

3. Instead of Lorentz's units I take the liberty to use the Gaussian c.g.s. system of units, whose usefulness in electromagnetic theory was first emphasized by Abraham in papers published in 1902 and 1903. Abraham also coined the term "Maxwell-Lorentz equations," which Einstein used in his 1905 relativity paper. See M. Abraham, "Dynamik des Elektrons," Nachr. Ges. Wiss. Göttingen, pp. 20–41 (1902); "Prinzipien der Dynamik des Elektrons," Phys. Z. **4**, 57-63 (1902); "Prinzipien der Dynamik des Elektrons," Ann. d. Phys. **10**, 105-79 (1903).

4. H. Poincaré, "A propos de la théorie de M. Larmor," L'Éclairage électrique **3**, 5-13, 289-95 (1895); *ibid.*, **5**, 5-14, 385-92 (1895); reprinted in *Oeuvres de Henri Poincaré*, Vol. 9 (Gauthier-Villars, Paris, 1934-1954, pp. 369-426)(this volume is referred to as *Oeuvres*), and "Sur les rapports de la Physique expérimentale et de la Physique mathématique," *Rapports presentés au Congrès international de Physique réuni à Paris en 1900*, Vol. 1 (Gauthier-Villars, Paris, 1900), pp. 1-29; translated as ch. IX and X, pp. 140–182, of H. Poincaré, *Science and Hypothesis* (Dover, New York, 1952).

5. Briefly, Lorentz invented the contraction hypothesis to explain a single experiment, and he derived its mathematical form from Newton's law for the addition of velocities, which the theorem of corresponding states was supposed to have obviated. See my 1974 paper (op. cit. in n. 1) for further discussion of the ad hocness of Lorentz's contraction hypothesis.

6. Thus Poincaré referred to Newton's principle of action and reaction as a "convention."

7. H. Poincaré, "La théorie de Lorentz et le principe de réaction" *Recueil de travaux offerts par les auteurs à H. A. Lorentz* (Nijhoff, The Hague, 1900), pp. 252–78; reprinted in *Oeuvres*, op. cit. in n. 4, pp. 464–88.

8. For brevity I use the term "covariance," which Hermann Minkowski was the first to apply effectively to the transformation properties of equations expressing physical laws. See H. Minkowski, "Die Gründgleichungen für die elektromagnetischen Vorgänge in bewegten Körpern," Nachr. Ges. Wiss. Göttingen, pp. 53-111 (1908).

9. H. Poincaré, "Sur les principes de la méchanique," *Bibliothèque du Congrès international de Philosophie tenu à Paris du 1er au 5 août 1900* (Colin, Paris, 1901), pp. 457–94; an expanded version of this paper is presented on pp. 123–39 of *Science and Hypothesis*, op. cit. in n. 4.

10. Whereas the Michelson–Morley experiment always had been of great concern to Lorentz, in 1892 it did not determine his formulation of the electromagnetic theory. This might seem surprising in the light of Lorentz's strongly empiricist message in 1886; namely, that in a question "as important" as the choice between a pure Fresnel theory, or a hybrid one containing elements of the theories of Fresnel and Stokes, one should not be guided by "considerations of the degree of probability or of simplicity of one or the other hypotheses, but to address onself to experiment" "De l'influence du mouvement de la terre sur les phénomènes lumineux," Versl. Kon Akad. Wetensch. Amsterdam **2**, 297 (1886); reprinted in *Collected Papers*, op. cit. in n. 2, Vol. 4, 153-214.

    (According to G. G. Stokes the ether at the earth's surface was dragged totally, and the dragging decreased with increasing distance from the earth's surface. In the 1886 paper Lorentz proved the inconsistency of Stokes' assuming both a velocity potential and that the relative velocity between the earth and the ether should vanish over the earth's surface. According to Augustin Fresnel a body in motion dragged the excess of ether in its interior, which he assumed to be $1 - n^{-2}$, where n was the body's refractive index.)

    To Lorentz, at that time, Michelson's interferometer experiment of 1881 was of importance since it was the only reliable means to obtain data accurate to second order in v/c. Having shown in 1886 that the Michelson interferometer experiment of 1881 was inconclusive, Lorentz urged its repetition. This was carried out in 1887, and the results apparently excluded a pure Fresnel theory. Yet, as Lorentz explained in 1892, hybrid theories, "being more complicated, are less worthy of consideration," and again in 1897 he wrote that the "Fresnel theory is without doubt simpler." Consequently, the formulation of Lorentz's electromagnetic theory ran counter to his empiricist guideline of 1886, because in the end simplicity was his guide. H. A. Lorentz, "The relative motion of the earth and the ether," Versl. Kon Akad. Wetensch. Amsterdam **1**, 74 (1892); reprinted in *Collected Papers*, op. cit in n. 2, Vol. 4, 219–23;

and "Concerning the problem of the dragging along of the ether by the earth," Versl. Kon. Akad. Wetensch. Amsterdam **6**, 266 (1897); reprinted ibid., pp. 237–44.

11. H. Poincaré, "L'état actuel et l'avenir de la Physique mathématique" delivered on 24 September 1904 at the International Congress of Arts and Science at Saint Louis, Mo. and published in Bull. Sci. Mat. **28**, 302–24; in English on pp. 91–111 of *The Value of Science*, translated by George Bruce Halsted (Dover, New York, 1958) from Poincaré's *La Valeur de la Science* (Ernest Flammarion, Paris, 1905). Poincaré, "La dynamique de l'électron," Revue générale des Sciences pures et appliquées **19**, 386–402 (1908); reprinted in *Oeuvres*, op. cit. in n. 4, pp. 551–86. Excerpts from this paper appear in Book III, pp. 199–250 of *Science and Method*, translated by Francis Maitland (Dover, New York, n.d.) from Poincaré's *Science et Méthode* (Ernest Flammarion, Paris, 1908).

12. In my book (op. cit in n. 1) analyzing Einstein's 1905 relativity paper, his thinking toward the relativity of simultaneity is analyzed in the context of the physics of 1905. This analysis is outlined in my paper "The Special Relativity Theory: Einstein's Response to the Physics of 1905," in *Jerusalem, Einstein Centennial Symposium* edited by G. Holton and Y. Elkana (Princeton University Press, Princeton, 1980). Pioneering case studies of Einstein's thinking about the special relativity theory are those of Gerald Holton: *Thematic Origins of Scientific Thought: Kepler to Einstein* (Harvard University Press, Cambridge, Mass. 1973) and *The Scientific Imagination: Case Studies* (Cambridge University Press, Cambridge, England, 1978).

13. W. Wien, "Über die Möglichkeit einer elektromagnetischen Begründung der Mechanik," *Recueil de travaux offerts par les auteurs à H. A. Lorentz* (Nijhoff, The Hague, 1900), pp. 96–107; reprinted in Ann. d. Phys. **5**, 501–513 (1901).

14. W. Kaufmann, "Die magnetische und electrische Ablenkbarkeit der Becquerelstrahlen und die scheinbare Masse der Elektronen," Nachr. Ges. Wiss. Göttingen, pp. 143–55 (1901); "Die elektromagnetische Masse des Elektrons," Phys. Z. **4**, 54–57 (1902); and "Über die 'Elektromagnetische Masse' der Elektronen," Nachr. Ges. Wiss. Göttingen, pp. 90–103 (1903).

15. In the papers cited in n. 3.

16. According to Abraham's restriction of quasi-stationary acceleration, the electron accelerates in such a way that it does not radiate. The criterion for this sort of motion is that the electron accelerates during a time that is much less than the time required for light to traverse its diameter. For details see my 1973 paper and my book cited in n. 1.

17. For example, Abraham's theory of the electron disagreed with the high-precision data of Lord Rayleigh and D. B. Brace [Lord Rayleigh, "Does Motion through the Aether cause Double Refraction?" Phil. Mag. **4**, 678–83, (1902); D. B. Brace, "On Double Refraction in Matter moving through the Aether," Phil. Mag. **7**, 317–29 (1904)]. Rayleigh and Brace sought to detect whether an isotropic substance at rest on the moving earth exhibited double refraction owing to Lorentz's postulate that bodies in motion relative to the ether became deformed — that is, whether a moving isotropic body should respond differently to light propagating through it parallel and transverse to its direction of motion. Rayleigh and Brace detected no anisotropy; Rayleigh's data were accurate to 1 part in 10^{10}, and Brace's to 1 part in 10^{13}. Abraham and Lorentz used their theories of the electron to interpret the results of Rayleigh and Brace as follows [M. Abraham, "Die Grundhypothesen der Elektronentheorie," Phys. Z. **5**, 576–79 (1904) and "Zur Theorie der Strahlung und des Strahlungsdruckes," Ann. d. Phys. **14**, 236–87 (1904); H. A. Lorentz, "Electromagnetic Phenomena in a System Moving with any Velocity Less than that of Light," Proc. Roy. Acad. Amsterdam **6**, 809 (1904), reprinted in *Collected Papers*, op. cit. in n. 2, Vol. 5, 172–97 and in part in *The Principle of Relativity: A Collection of Original Memoirs on the Special and General Theory of Relativity by H. A. Lorentz, A. Einstein, H. Minkowski and H. Weyl*, translated by W. Perrett and G. B. Jeffery (Dover, New York, n.d.), pp. 11–34; and *The Theory of Electrons* (Teubner, Leipzig, 1909); rev. ed., 1916; reprinted from the rev. ed. (Dover, New York, 1952)]. Since the index of refraction depends on the mass of the isotropic substance's constituent electrons, then the index of refraction becomes double-valued owing to the moving electrons developing transverse and longitudinal masses. Abraham's theory predicted a double refraction well within the limits of accuracy of Rayleigh and Brace, while Lorentz's 1904 theory of the electron (see below) predicted no double refraction because of hypotheses concerning how intermolecular forces transform and how moving matter contracts, among others.

18. See, for example, n. 10. Einstein mentioned explicitly only the class of first-order optical experiments ["Zur Elektrodynamik bewegter Körper," Ann. d. Phys. **17**, 891–921 (1905); a new translation is in my book, op. cit. in n. 1. The English version in *The Principle of Relativity*, op. cit. in n. 17, pp. 37–65, contains substantive mistranslations and misprints.]. R. S. Shankland reported the following interview with Einstein on 4 February 1950: "the experimental results which had influenced him [Einstein] most were the observations on stellar aberration and Fizeau's measurements on the speed of light in moving water. 'They were enough,' he said. I reminded him that Michelson and Morley had made a very accurate determination at Case in 1886 of the Fresnel dragging coefficient with greatly improved techniques and showed him their values as given in my paper. To this he nodded agreement, but when I added that it seemed to me that Fizeau's original result was only qualitative, he shook his pipe and smiled. 'Oh it was better than that!'" ["Conversations with Albert Einstein," Am. J. Phys. **31**, 47–57 (1963).]

19. Another physicist who preferred to avoid Lorentz's contraction hypothesis was Emil Cohn, whose work Einstein held in high regard [see Einstein "Über das Relativitätsprinzip und die aus demselben gezogenen Folgerungen," J. d. Radioaktivität **4**, 411-62 (1907)]. The goal of Cohn's research was not an electromagnetic world- picture, but, rather, a consistent version of the electrodynamics of moving bulk matter that was based on the best that the electromagnetic theories of Henrich Hertz and Lorentz could offer: Cohn attempted to suitably modify Hertz's electromagnetic field equations in such a way as to render them in agreement with Lorentz's 1895 theorem of corresponding states [H. A. Lorentz, *Versuch einer Theorie der elektrischen und optischen Erscheinungen in bewegten Körpern* (Brill, Leiden, 1895), reprinted in *Collected Papers*, op. cit. in n. 2, Vol. 5, 1-137; Emil Cohn, "Über die Gleichungen der Elektrodynamik für bewegte Körper," *Recueil de travaux offerts par les auteurs à H. A. Lorentz* (Nijhoff, The Hague, 1900), pp. 516-23, and "Über die Gleichungen des elektromagnetischen Feldes für bewegte Körper," Ann. d. Phys. **7**, 29-56 (1902)].

    Cohn emphasized that the term ether is a "metaphorical term [that] should not acquire an importance relative to the theory in question," and consequently in the meanwhile the ether should be taken as a "heuristic concept." The guiding theme in his research, Cohn wrote, was "scientific economy" as this term was used by Ernst Mach [Cohn, "Zur Elektrodynamik bewegter Systeme," Berl. Ber. **40**, 1294-1303 (1904); II, ibid., 1404-16 (1904)].

20. M. Abraham, "Prinzipien der Dynamik des Elektrons," op. cit. in n. 3.

21. Ibid.

22. F. T. Trouton and H. R. Noble, "The Mechanical Forces Acting on a Charged Electric Condenser moving through Space," Phil. Trans. Roy. Soc. London **A202**, 165-81 (1903); Rayleigh, op. cit. in n. 17; Brace, ibid.; Lorentz, ibid. Trouton and Noble attempted to measure the sudden torque on a parallel plate capacitor that was hung by string, owing to the capacitor's being charged or discharged; this torque is related to the earth's velocity relative to the ether. They observed no effect. The hypotheses in Lorentz's 1904 theory of particular importance for explaining this null result were the contraction hypothesis, and that all forces transform like the electromagnetic force.

23. H. Poincaré, "Sur la dynamique de l'électron," Rend. del Circ. Mat. di Palermo **21**, 129-75 (1906), reprinted in *Oeuvres*, op. cit. in n. 4, pp. 494-550; "Sur la dynamique de l'électron," Comptes rendus de l'Académie des Sciences **140**, 1504-8 (1905), reprinted in *Oeuvres*, pp. 489-93.

24. M. Abraham, "Die Grundhypothesen der Elektronentheorie," op. cit. in n. 17.

25. Poincaré, "L'état actuel et l'avenir de la Physique mathématique," op. cit. in n. 11. In geometry there is a principle of the relativity of position that asserts: 1) that the notion of distance has meaning only in terms of relations between bodies, and 2) the equivalence of all points in space. Investigations of the origins of this principle and of whether the nature of physical space could be determined empirically by distance measurements had led Poincaré to extend the principle of the relativity of position into mechanics, where he called it the "principle of relative motion" [op. cit. in n. 7]. In his widely read book of 1902, *Science and Hypothesis* (op. cit. in n. 4), Poincaré discussed this work as well as the possibility that, as he now put it, the "principle of relativity" from geometry could be extended into electromagnetic theory because thus far experiments had not revealed absolute motion. The agreement of Lorentz's theory with this principle, at least to order v/c, pleased Poincaré. Yet owing to his emphasis on empirical data, Poincaré in 1902 had not yet become a full-fledged member of the group to which he referred to as the "partisans of Lorentz." The problem was that V. Crémieu's recent experimental data contradicted Rowland's convection current, which was an essential ingredient of Lorentz's theory. By 1903 these data were found to be inaccurate [see Miller, 1973 op. cit. in n. 1].

    After Lorentz's 1904 theory of the electron, Poincaré became a partisan of Lorentz and he considered the principle of relativity to be an extension of the principle of relative motion from mechanics into the physics of the electron. Several of Poincaré's essays on the foundations of geometry are in *Science and Hypothesis*, ch. III-V. These and others of Poincaré's writings on geometry are analyzed in Miller 1973 and 1980 op. cit. in n. 1.

26. The inconsistency in Lorentz's theory concerning m_L was also brought to Lorentz's attention by Abraham in an unpublished letter of Abraham to Lorentz written from Wiesbaden on 26 January 1905 (on deposit at the Algemeen Rijksarchief, The Hague). Abraham developed the calculations in this letter in his 1905 book whose preface was also written at Wiesbaden [*Theorie der Elektrizität: Elektromagnetische Theorie der Strahlung* 1st ed. (Teubner, Leipzig, 1905); 2nd ed., (1908); 3rd ed. (1914)]. For details see Miller 1973 and 1980 op. cit. in n. 1.

27. P. Langevin, "La physique des électrons," delivered on 22 September 1904 at the International Congress of Arts and Science at Saint Louis, Missouri and published in *Revue générale des sciences pures et appliquées,* **16**, 257-76 (1905). Simultaneously and independently A. H. Bucherer in Bonn offered the same model [*Mathematische Einführung in die Elektronentheorie* (Teubner, Leipzig, 1904)].

28. See Fig. 5.3. Abraham's rigid-sphere electron does not transform according to the Lorentz transformations.

29. H. A. Lorentz, "Electromagnetic Phenomena," op. cit. in n. 17.

30. Since the ether system S never moves,, Poincaré could give only a mathematical interpretation to the Lorentz transformation's inversion symmetry: interchanging primes and changing ε to $-\varepsilon$ (where in the Lorentz transformations in Fig. 5.5, ε is the velocity of Σ' relative to S) is the mathematical operation of rotation by 180° about the common y-axes of S and Σ' ["Sur la dynamique de l'électron," 1906 op. cit. in n. 23].

31. In Abraham's opinion, nonelectromagnetic forces were necesssary to bind Lorentz's deformable electron; consequently, Lorentz's theory of the electron was not in the spirit of the electromagnetic world-picture [Abraham, "Die Grundhypothesen der Elektronentheorie," op. cit. in n. 17]. For Poincaré, on the other hand, Lorentz covariance was the guiding theme toward an electromagnetic world-picture, and the internal pressure rendered Lorentz's electron consistent with this requirement — that is, consistent with the principle of relativity. Furthermore, continued Poincaré, since we are "tempted to infer" a relation between the "causes giving rise to gravitation and those which give rise to the [internal pressure]," then Lorentz covariance permits further progress toward the unification of electromagnetism and gravitation [Poincaré, "Sur la dynamique de l'électron," 1906 op. cit. in n. 23.]

32. Historical analysis of archival materials and of primary sources permits me to conjecture that Einstein was aware of most requisite empirical data, as well as of the research toward electromagnetic and mechanical world-pictures. By 1905 his work on the fluctuations of cavity radiation and on the behavior of particles suspended in fluids had revealed to him the failure of Lorentz's electrodynamics and Newtonian mechanics in volumes the size of the electron. As Einstein recalled in his Nobel Prize acceptance address in 1923, the physicists of 1905 were "out of [their] depth." For details see my paper of 1976, op. cit., in n. 1, and especially my book, op. cit. in n. 1. A. Einstein, "Fundamental Ideas and Problems of the Theory of Relativity," delivered 11 July 1923 to the Nordic Assembly of Naturalists in acknowledgement of the Nobel Prize of 1921. Reprinted in *Nobel Lectures: 1901–1921* (Elsevier, New York, 1967), pp. 479–490.

33. Letter on deposit at the Algemeen Rijksarchief, The Hague.

34. W. Kaufmann, "Über die Konstitution des Elektrons," Berl. Ber. **45**, 949–56 (1905), and "Über die Konstitution des Elektrons," Ann. d. Phys. **19**, 487–553 (1906).

35. After pointing out a subtlety involved in defining the quantity force as "mass × acceleration," Einstein (op. cit. in n. 18) made an inappropriate choice for force that resulted in his predicting a transverse mass of $m_o/(1 - v^2/c^2)$, where for Einstein m_o was the electron's inertial mass. Kaufmann (op. cit. in n. 34) emphasized that Einstein's m_T should be the same as Lorentz's. Planck ["Das Prinzip der Relativität und die Grundgleichungen der Mechanik," Verh. D. Ges. **6**, 136–41 (1906)] showed that a consistent generalized dynamics could be obtained if force were defined as the rate of change of momentum $\mathbf{p}$, where $\mathbf{p} = m_o\mathbf{v}\sqrt{1 - v^2/c^2}$, thereby demonstrating quantitatively the mathematical equivalence of Einstein's and Lorentz's predictions for the electron's transverse mass. Kaufmann (1906 op. cit. in n. 34) complimented Einstein's approach to electrodynamics: "It is indeed remarkable," wrote Kaufmann, that from entirely different hypotheses Einstein obtained results which in their "accessible observational consequences" agreed with Lorentz's, but Einstein avoided difficulties of an "epistemological sort," that is, the inclusion of unknown velocities. Kaufmann viewed Einstein's success as based on his raising to a postulate the "principle of relative motion," which he placed at the "apex of all of physics." Then, applying this postulate to the propagation of light, Einstein deduced "a new definition of time and of the concept of 'simultaneity' for two spatially separated points, where Einstein's relationship between the time in two inertial reference systems was identical with Lorentz's local time." Kaufmann went on to emphasize that whereas Lorentz obtained his observationally accessible results through approximations (such as that of quasi-stationary motion), Einstein's kinematics of the rigid body, and his electrodynamics, were exact for inertial systems.

    Einstein (op. cit., in n. 18) had deduced the electron's m_T and m_L by relativistic transformations between the electron's instantaneous rest system and the laboratory inertial system of the electron's space and time coordinates, and of the external electromagnetic fields acting on the electron.

    In summary, Einstein was considered to have replaced Lorentz's assumptions of how electrons interact with the ether with phenomenological assumptions concerning how clocks were synchronized using light signals; that is, Einstein had improved Lorentz's theory of the electron.

36. In fact, since the first pages of the 1906 version bear the inscription, "Stampato il 14 dicembre 1905," then Poincaré must have read Kaufmann's 1905 paper (op. cit. in n. 34) and then added this final paragraph just before his paper (op. cit. in n.23) went to press in 1906.

37. Poincaré, "La dynamique de l'électron," op. cit. in n. 11.

38. Letter on deposit at the Einstein Archives, Institute for Advanced Study, Princeton, N.J.

39. To the passage in his 1908 paper (op. cit. in n. 11) where Poincaré called for others to repeat Kaufmann's 1905 measurements, he added in a footnote in his book *Science and Method* (op. cit. in n. 11), the following comment: "At the moment of going to press we learn that M. Bucherer had repeated the experiment, surrounding it with new precautions, and that, unlike Kaufmann, he has obtained results confirming Lorentz's views."

Lorentz, too, added a note to a text about to go to press. To a passage in the 1909 edition of *Theory of Electrons*, in which he discussed how well Abraham's theory agreed with the data of Kaufmann, Lorentz appended Note 87: "Recent experiments by Bucherer on the electric and magnetic deflexion of β-rays, made after a method that permits greater accuracy than could be reached by Kaufmann, have confirmed the formula [m_T], so that, in all probability, the only objection that could be raised against the hypothesis of the deformable electron and the principle of relativity has now been removed."

40. C. T. Zahn and A. H. Spees discovered that Bucherer's velocity filters were inadequate ["A Critical Analysis of the Classical Experiments on the Variation of Electron Mass," Phys. Rev. **53**, 511–21 (1938)].

41. Poincaré, however, never elevated the principle of relativity to a convention because of disagreement with the measured value of the result of his Lorentz-covariant gravitational theory for the advance of mercury's perihelion [see Poincaré, *La dynamique de l'électron* (Dumas, Paris, 1913)].

42. Op. cit. in n. 19.

43. A. Einstein, "Autobiographical Notes," in *Albert Einstein: Philosopher-Scientist*, edited by P. A. Schilpp (The Library of Living Philosophers, Evanston, Ill., 1949).

44. In the face of the spirited criticisms and hard data of Abraham and Kaufmann, Planck, at the Meeting of German Scientists and Physicians at Stuttgart, 19 September 1906, could reply only that he remained "sympathetic" toward the "Lorentz–Einstein theory" because he believed in the fecundity of its principle of relativity. ["Die Kaufmannschen Messungen der Ablenkbarkeit der β-Strahlen in ihrer Bedeutung für die Dynamik der Elektronen," Phys. Z. **7**, 753–61 (1906)]. In 1907 Planck used the principle of relativity to formulate a general dynamics that included electromagnetism and thermodynamics ["Zur Dynamik bewegter Systeme," Berl. Ber. **13**, 542–70 (1907); also published in Ann. d. Phys. **26**, 1–34 (1908)]. His goal was to continue extending the principle of relativity in these disciplines in order to determine whether it led to either theoretical inconsistencies or to empirical disagreement. Planck's empiricistic view led him to call for more experiments than Bucherer's of 1908 [A. H. Bucherer, "Messungen an Becquerelstrahlen. Die experimentelle Bestätigung der Lorentz-Einsteinschen Theorie," Phys. Z. **9**, 755–62 (1908)]. As Planck wrote in 1910: "Physical questions, however, cannot be settled from aesthetic considerations, but only by experiment, and this always involves prosaic, difficult and patient work" ["Die Stellung der neueren Physik zur mechanischen Naturanschauung," Phys. Z. **11**, 922–32, reprinted in M. Planck, *A Survey of Physical Theory*, translated by R. Jones and D. H. Williams (Dover, New York, 1960), pp. 27–44].

45. The normally taciturn Minkowski waxed enthusiastic over Bucherer's results. In the discussion session after the lecture by Bucherer (op. cit. in n. 44), Minkowski expressed his "joy" for the vindication of "Lorentz's theory" over the "rigid electron": "The rigid electron is in my view a monster in relation to Maxwell's equations, whose innermost harmony is the principle of relativity . . . the rigid electron is no work hypothesis but a hinderance to work."

46. Quoted in my 1976 paper cited in n. 1.

47. M. Abraham, "Zur elektromagnetischen Mechanik," Phys. Z. **10**, 737–41 (1909). In the preface to the 1914 edition of his *Theorie der Elektrizität*, Abraham stressed that he would not permit current data to decide the issue between his electron theory and Lorentz's. And in this unusually carefully referenced book, Bucherer's data were nowhere mentioned.

48. M. Abraham, "Die neue Mechanik," Scientia **15**, 10–29 (1914).

49. Poincaré, *Science and Method,* op. cit in n. 11.

Looking back on electrodynamics in 1905, we can conjecture that its ambiguities would have been removed without Einstein. The requisite mathematical formalism already existed. Refinements of already performed ether-drift experiments, with additional null experiments, would have given physicists such as Langevin, Lorentz, and Poincaré the license to assert that the Maxwell–Lorentz equations have exactly the same form on the moving earth as they have relative to the ether (perhaps further fine tuning of the Maxwell–Lorentz theory would also have been required). The result of these endeavors could only have been a consistent theory of electromagnetism, and not a special theory of relativity. In his obituary to Langevin, Einstein wrote: "It appears to me as a foregone conclusion that he would have developed the Special Relativity Theory, had that not been done elsewhere" ["Paul Langevin in Memoriam," in A. Einstein, *Ideas and Opinions* (Littlefield, Adams, Totowa, N.J., 1967), pp. 210–11)]. What greater honor could have been bestowed on Langevin's memory? Yet Langevin's 1911 exposition of relativity theory reveals that his view of relativity physics differed fundamentally from Einstein's [P. Langevin, "L'évolution de l'espace et du temps," Scientia **10**, 31–54 (1911)]. Langevin developed Einstein's special relativity theory with an ether whose purpose was to have been the active agent that caused physical effects such as time dilation. But general relativity could not have evolved without Einstein because its formulation required the folding together of physics and geometry. And this was a double-edged masterstroke, for Einstein broke sharply with both the physics and philosophy of his day. The year 1915 was Einstein's second *Annus Mirabilis*..

Open Discussion

Following Papers by G. HOLTON and A. I. MILLER

Chairman, C. N. YANG

Yang: I find it difficult to refrain from repeating the remark that I once read of Professor Dirac on the development of physics: he said that if he had to choose between beauty and agreement with experiment, he would choose beauty.

P. A. M. Dirac (Florida State U): You mentioned about choosing between beauty and agreement with experiment. Well that's what people ought to have done when Kaufmann obtained his results.

Holton: We know what Einstein did when he heard about Kaufmann's results — one of the foremost experimentalists of Europe disproving this unknown person's work. Einstein did not respond for nearly two years. Finally, Stark persuaded Einstein to write an article, in December 1907, in which he took this up. This is a revealing response when compared to the responses of Lorentz and Poincaré that Dr. Miller spoke about. Einstein wrote that he had not found any obvious errors in Kaufmann's article, but that the theory that was being proved by Kaufmann's data was a theory of so much smaller generality than his own, and therefore so much less probable, that he would prefer for the time being to stay with it. Actually, it took untl 1916 for a fault in Kaufmann's experimental equipment to be discovered. By that time this whole question was moot.

A. Hermann (U. of Stuttgart): It is very interesting to see what Planck thought about this contradiction between theory and experiment. Planck was a Platonist and he was always interested in finding out what he called the absolute in nature. When Planck studied Einstein's paper he was fascinated that time and space lost its absolute character, whereas the natural constants c and h were strengthened. There was additional support for the evidence that these natural constants were real absolute characters, and therefore Planck was convinced that this theory must be correct and he took a lot of time to check Kaufmann's experiments. It was Planck who pointed out in 1906 in a speech at Stuttgart that one should reject Kaufmann's experiments. Thus it was Planck who played a very important role in the acceptance of the special theory of relativity. Others who played an important role in the acceptance of the special theory were Felix Klein and Hermann Minkowski, who advanced Klein's ideas about the group character of geometry and physics.

V. Bargmann (Princeton U.): We have heard about the extremely interesting correspondence between Poincaré and Lorentz. There is one aspect of it that I always found strange. This is the fact that in the famous book by Lorentz on the theory of electrons, Poincaré is hardly mentioned. There is copious mention of Einstein's work, but hardly any of Poincaré's. Now I wonder whether Dr. Miller has any comment on this?

Miller: Lorentz mentions Poincaré in reference to the Poincaré pressure keeping the electron together and then at the end of the 1909 edition he makes mention of Einstein. His interpretation of Einstein was the same, in fact, as Max Planck's interpretation of what Einstein did: namely, to generalize the theory of the electron. One of the other points that was considered to be pleasing about Einstein's work was that he was able to derive the Lorentz transformations by assumptions that did not depend upon the ether — rather, assumptions that depended upon how clocks were synchronized. Nevertheless, people believed in an ether-based theory. There was no reason to mention Poincaré, other than the Poincaré stress and other than his making the Lorentz transformations into a more symmetrical form.

Y. Ne'eman (Tel-Aviv U.): I have a question for Professor Holton with respect to the incomprehensibility of comprehensibility. One can refine this issue at two levels. There is the comprehensibility of the universe in the sense that logic works, that it is not a mad world. And the other is that even with the logical universe, it could have been (as a physicist would describe it) so nonlinearly coupled that it would be impossible to make a scientific description. We know that science works by further gradual approximations. First we dealt with long-range forces, then we got to short-range forces, so that it was possible to tackle things one after the other. If everything would have been coupled extremely strongly, one could not have found such a description. Is it clear what Einstein meant?

Holton: No one should claim that he knows exactly what Einstein meant, but I know his letters, and in the exchange with Solovine Einstein says the fact that the order we do find has such harmony and simplicity, that is the mystery.

Ne'eman: So it's the second type?

Holton: It's the second type.

Yang: Maybe I can use the privilege of having the microphone to remark that if there were no order, there would have been no Einstein, and there would be no meeting here.

6. SPECIAL RELATIVITY THEORY IN ENGINEERING

Wolfgang K. H. Panofsky

The title of this paper may in itself be a surprise to some in that special relativity is frequently portrayed as being mysterious, abstract, and in no way related to the technological world. This is far from the truth. Special relativity plays a role in engineering applications, for it has a place in the design of certain kinds of apparatus that perform useful (or useless) functions. In fact, such devices simply would not work if special relativity were incorrect. In this sense, such devices act as a complement to experiments that are especially designed to test the validity of special relativity.

I will discuss the intellectual content of special relativity only very briefly. Special relativity resulted from a body of experimental knowledge and theoretical conjecture that accumulated over the last quarter of the previous century. Mechanics had become an exact science, and its laws exhibited the principle, known as Galilean relativity, that identical experiments that were carried out in laboratories moving uniformly with respect to one another should not exhibit different results. At the same time, Maxwell's equations had provided a full understanding of phenomena involving electricity and magnetism, including the propagation of electromagnetic radiation. Among other consequences, these equations imply that in vacuum the velocity of propagation of such radiation — that is, the velocity of light — is a constant.

A large amount of observational material relating to light emission from moving sources, light propagation in moving media, and to moving electromagnetic devices had led to a series of paradoxes that appeared to indicate a conflict between Maxwell's equations, on the one hand, and the principle of Galilean relativity, on the other. Einstein proposed that all of these paradoxes could be resolved by modifying the laws of mechanics rather than those of electromagnetism. Specifically, he postulated that the mathematical relations that described the dependence of electromagnetic phenomena on the states of motion of the observer and the physical phenomena should also apply to the laws of mechanics. Thus the velocity of light, which previously had not played a special role in mechanics, became a dominant and limiting quantity in all of the fields of physics. A more detailed mathematical formulation then flowed from two basic postulates: 1) No experiment performed within a laboratory can determine the state of absolute uniform motion of that laboratory. 2) The velocity of light in vacuum is independent of any uniform motion of the light source or observer measuring that speed. It is the analytical elaboration of these postulates that sets the framework for all the applications of special relativity.

Harry Woolf (ed.), Some Strangeness in the Proportion: A Centennial Symposium to Celebrate the Achievements of Albert Einstein
ISBN 0-201-09924-1

The application of special relativity most often cited does *not* in fact depend on the validity of special relativity: this is the exploitation of nuclear energy. There is no difference in principle between *chemical* reactions such as

$$2H_2 + 0_2 \rightarrow 2H_20 + \text{excess energy} \tag{6.1}$$

and *nuclear* reactions such as the fusion of light nuclei

$${}_1D^2 + {}_1T^3 \rightarrow {}_2He^4 + {}_0n^1 + \text{excess energy} \tag{6.2}$$

or the fission reactions of the heavy elements

$${}_0n^1 + {}_{92}U^{235} \rightarrow {}_{54}Xe^{140} + {}_{38}Sr^{94} + 2{}_0n^1 + \text{excess energy} \tag{6.3}$$

The chemical reaction (6.1) involves changes in the electron shells surrounding the nuclei, whereas the latter two reactions involve the nuclei themselves. Nuclear reactions release a millionfold more energy per atom reacting than do chemical reactions. Through his celebrated equation $E = mc^2$ Einstein demonstrated that the energy release in reactions such as these corresponds to a change in mass of the reacting constituents. But this is equally true whether we are talking about ordinary chemical energy or nuclear energy. However, because this effect is very small indeed in chemical reactions, the principle of mass-energy equivalence is much easier to verify experimentally in the case of nuclear processes. Work on nuclear energy for either peaceful or warlike purposes could have proceeded whether or not $E = mc^2$ had ever been recognized as a profound natural law. Only the existence of a sufficient energy excess and of neutron multiplication (in the case of the fission reaction) is a required condition so that a practical nuclear energy source can be built.

Einstein played a major role in ushering in the nuclear age through his famous letter persuading President Roosevelt to initiate atomic bomb research with high priority. However, since the exploitation of nuclear energy does not depend upon special relativity, I will not go into any of the enormous problems, on the one hand, and opportunities, on the other, that now face mankind in managing nuclear energy wisely.

Special relativity modifies the laws of classical mechanics if the speed of the objects involved becomes comparable to the speed of light. This happens if the total kinetic energy of such objects approaches or exceeds their so-called rest energy. What is meant by *rest energy* is the amount of energy that would be released if the objects were completely annihilated in a state of rest. As an example, if a slow negative electron and its antiparticle (called a positron), each having a rest energy of about one-half million electron volts, clash and annihilate, then the total energy of the resultant fragments is 1 million electron volts (1 MeV). One million electron volts is a unit of energy; it is the amount of energy that would be gained by a particle carrying the same electric charge as an electron when it is accelerated by an applied voltage of 1 million volts. Thus if electrons are used in engineering devices which have applied voltages approaching 0.5 million volts, then special relativity must enter into the design of such devices. Similarly, the rest energy of a proton is close to 1,000 million electron volts, or 1 GeV. Thus any device that involves protons at energies approaching or exceeding 1 GeV requires special relativity as an essential element in its design.

In modern engineering there are large families of electron devices that indeed require voltages above the hundreds of kilovolts to operate in the desired manner. Thus devices such as klystrons, high-voltage television tubes, electron microscopes, and electron accelerators for cancer therapy all require engineering analysis using special relativity in order to make them work.

A typical example is the high-powered klystron, a picture of which is shown in Fig. 6.1. The engineer designing such tubes uses the formulas of special relativity just as he does such information as the properties of materials.

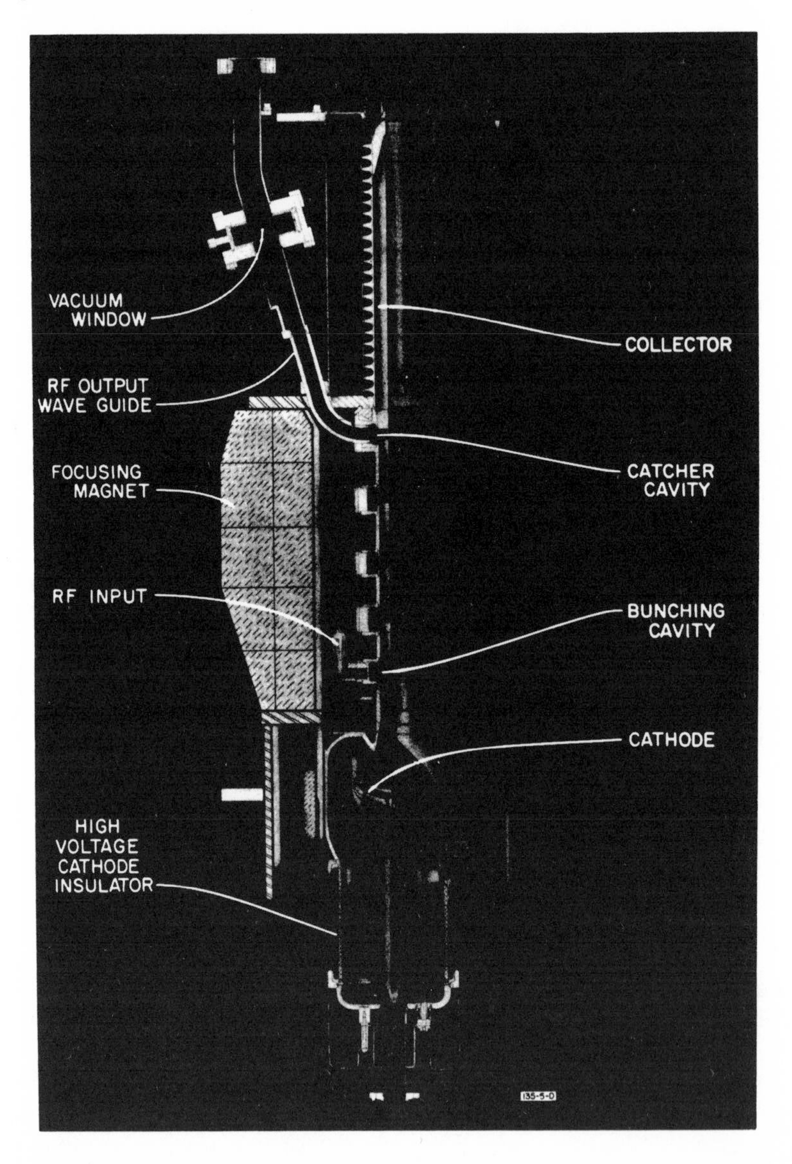

Fig. 6.1 A cutaway view of a high-power klystron. This device is about 1 meter long and operates at a frequency of about 3 GHz. It can generate a peak output power of about 30 megawatts and an average output power of about 30 kilowatts.

A klystron is basically an amplifier: it receives an electrical input signal at low voltage and low power and converts it into an output signal that is similar in shape but of much higher intensity. This is accomplished by having the weak input signal modulate the flow of a powerful beam of electrons, after which the modulated beam passes to an output structure that extracts the highly amplified information. Relativity enters into the design of this device in two fundamental ways: 1) The speed of the electron stream cannot exceed the speed of light; therefore relativistic equations are essential in calculating the timing of the information imparted to the electron beam. 2) Electrons in the beam repel one another, and for this reason there is a limit to the number of electrons that can be present in the stream without blowing it apart. Relativity ameliorates this effect in a manner that can be viewed consistently both by a real observer who analyzes the behavior of the tube in the laboratory and by a hypothetical observer who rides along with the electron beam. The observer in the laboratory would detect not only the electric *repulsion* among the electrons, which tends to explode the stream, but also a magnetic *attraction* between the parts of the beam, which behave like wires carrying parallel electric currents. These two forces approach one another in strength as the velocity of the beam electrons approaches that of light. However — and here relativity considerations are important — one can still analyze this situation by imagining that one is riding along with the beam and that the electrons thus appear to be at rest. If the equations of special relativity are correctly applied, then in the "frame" that is co-moving with the electron beam no magnetic effect is seen and the electric effect is greatly weakened. This simple conclusion is consistent with the much more complicated analysis that is necessary when the problem is treated in the "normal" reference frame, in which the device itself is at rest.

There is a fundamental reason that relativity enters more and more into engineering devices; this reason stems directly from the uncertainty principle of quantum mechanics. That principle states that the resolution in space to which an observer can observe a given object is inversely proportional to the impulse, or what is technically known as momentum transfer, which is given to the object that is to be observed. In other words, as one wishes to discern smaller and smaller objects, the "kick" that has to be given to the object to be observed must become larger and larger. Thus an optical microscope that involves light photons of energy of 1 electron volt or so is adequate to see objects in the submicron region, meaning somewhat below 0.001 millimeter. In order to see biological specimens of smaller size down to macromolecules and even individual molecules, electron microscopes having ever-increasing energies must be used. Recently electron microscopes having accelerating voltages above 1 million volts have been built and used successfully; they can discern objects to the angstrom unit size, meaning 0.0000001 millimeter (one-ten millionth of a millimeter). The ultimate microscopes are, of course, particle accelerators which I will discuss later: these look at matter down to dimensions of 10^{-15} centimeters.

Even electron microscopes developed before the war operated at voltages for which relativity is essential in order to design properly the various electromagnetic lenses that constitute the focusing elements of such instruments. It is interesting to note that the classical prewar engineering textbook on the design of electron microscopes and electron optics, by V. K. Zworykin, G. A. Morton, E. G. Ramberg, J. Hillier, and A. W. Vance gave what at that time was probably the most extensive discussion of relativistic orbit dynamics. It is quite amusing that the formulas used in that textbook were given in the typical "cookbook" fashion not uncommon in engineering texts, simply telling the designer what to do without being very explicit about the physical reasons and explanations that underlie the formulas. Accordingly, the rather extensive relativistic electron optics considerations used by electron microscope designers went unnoticed by physicists. Therefore much of what was written by Zworykin et al., and was applied by electron microscope engineers before the war, was rediscovered and rederived from fundamental principles by physicists designing particle accelerators after the war.

Special relativity is of course a totally indispensable tool for the design of high-energy particle accelerators. These come in various sizes and shapes — linear and circular — and they can and do accelerate the various charged particles that are involved in the structure of the atom: electrons, protons, and heavy ions. In particular for use in particle physics, energies are required that exceed by a very large factor the rest mass of the particle accelerated. For instance, the Stanford Linear Electron Accelerator accelerates electrons of rest energy of 0.5 million electron volts to an energy well above 20,000 million electron volts. According to the laws of special relativity these electrons are more than 40,000 times heavier when they emerge than they are when at rest. Their speed never becomes precisely equal to that of light but, again according to the relativistic rules, the final speed is within one part in 3 billion of the speed of light.

Incidentally, the fact that the speed of the accelerated electron beam closely approaches that of light but never exceeds it can be verified experimentally. An experiment has been done in which a beam of light and the electron beam traveled together along the 2-mile-long pipe, and their times of departure and arrival were measured to an accuracy of approximately 10^{-12} seconds, which is one-millionth of a millionth of a second. To within the accuracy of the observation, they arrived at the same time.

Radiation of Electromagnetic Energy from Fast Charged Particles

In general, when charges are accelerated, that is, if they do not move with uniform speed in a fixed direction, then they radiate electromagnetic energy. A radio antenna is essentially a carrier of such charges that go back and forth, thereby producing radio waves. One would assume that when particles are speeded up in an accelerator the losses that would occur due to just such radiation would have to be reckoned with. This is true, although the severity of the problem depends drastically on what particle is being accelerated and how the acceleration is done. The heavier the particle the less is the radiated power. For this reason, the matter of radiated energy does not as yet have to be taken into account for machines accelerating protons to energies thus far attainable or in view. On the other hand, the matter is indeed serious for the much lighter electrons. However we have an interesting phenomenon: If you impart energy to electrons by accelerating them in a straight line, then once the speed of light is approached most of the energy becomes an increase in mass and the velocity changes only very little. Therefore the acceleration force effectively pushes against what appears to be a rapidly increasing inertia, and there is little acceleration. Radiation from electrons accelerated in a linear accelerator is therefore negligible. This is not the case when electrons are confined by magnets to circular orbits and swing around as their energies increase. In that case, the acceleration is the ordinary centripetal acceleration that applies to any objects moving in a circle, be they people on a merry-go-round, microbes in a centrifuge, or electrons in a circular orbit. Hence the radiation from electrons in circular machines is a serious problem indeed, and as the energy increases more and more radio frequency power is required to compensate for this radiation loss. I will discuss this problem in somewhat greater detail later. This is another example of a relativity consideration playing a controlling role in the design of equipment; it controls the design of accelerators and even the basic choice as to which accelerator serves the need in question.

The linear accelerator at Stanford is shown in Fig. 6.2. The particles are injected at low energy and are then carried along in synchronism with a traveling electric field, gaining energy as they go. The inner diameter of the accelerating pipe is about 2 centimeters, whereas the length, as already mentioned, is about 2 miles, or 3 kilometers.

A question that comes to mind is: How it is possible to keep the electron beam within this narrow pipe for such a long distance? The answer can again be provided in either of two ways, using relativistic arguments. The first is as follows: As the particle accelerates, its longitudinal momentum is being increased continually. You may recall that momentum is the product of mass times velocity. In this case the velocity is almost constant and nearly equal to the speed of light, whereas the mass continues to increase in accordance with the energy-mass equivalence principle of Einstein. It turns out that, on the average, there are no radial forces acting on the electrons in the pipe;

Fig. 6.2 Aerial photograph of the Stanford Linear Accelerator Center (SLAC). The 2-mile-long accelerator extends from background to foreground. After acceleration, the electron beam can be directed to any of several different experimental areas located in the larger buildings in the foreground.

therefore the transverse momentum does not increase but stays constant throughout the acceleration process, as shown in the next diagram (Fig. 6.3). The angle made by the beam with respect to the axis continually decreases. Thus the beam becomes increasingly collimated or confined, and for this reason it does not spread out by a substantial amount during acceleration.

There is another, and more striking, way to visualize this situation by using the arguments of special relativity. Suppose we could observe this process while traveling along with the accelerating electrons. From this vantage point we would observe that the length of each of the individual segments of the accelerator became progressively shorter as our speed relative to the pipe approaches the speed of light. In fact, the 2-mile-long accelerator would appear to have a length of only about 30 inches as we rode along with the electrons! This means that the task of properly injecting electrons into the 2-mile-long Stanford machine is simply to aim them accurately enough so that they will hit a target 2 centimeters in diameter at a distance of only 30 inches, and any fool can do that!

Whether you are looking at this problem from the point of view of an observer at rest relative to the accelerator structure or from the point of view of an observer traveling along with the electrons, the ease of confining such an electron beam can be understood in a consistent manner. Thus to qualify as a designer for an electron linear accelerator you have to understand something about special relativity, and this is true also for designers of circular accelerators, which are used, among other things, for accelerating protons to the highest energies yet achieved by artificial means.

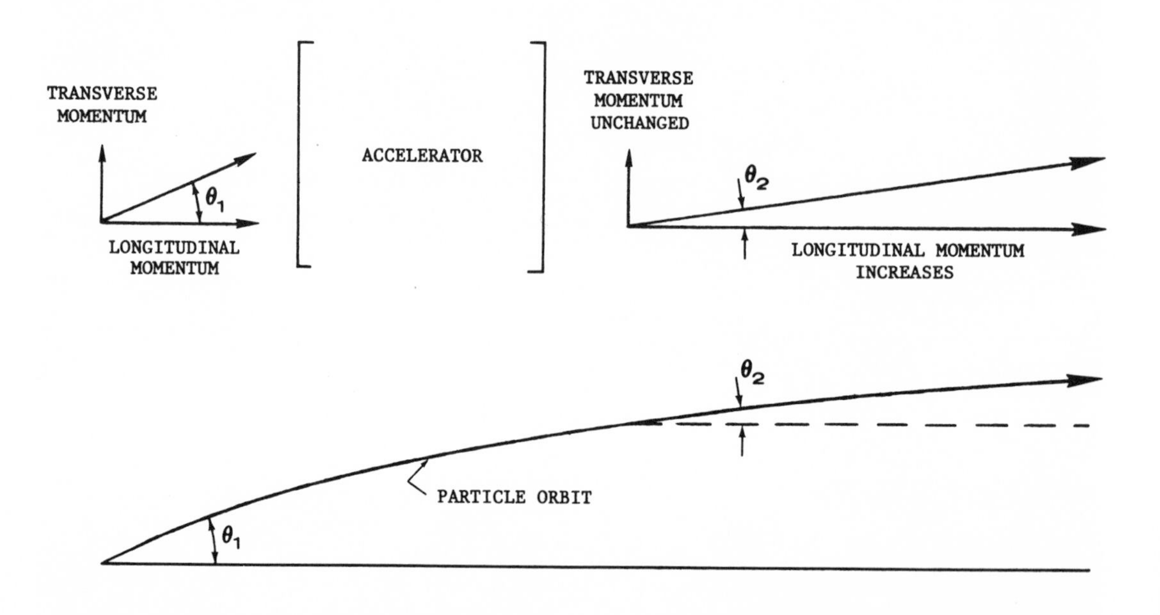

Fig. 6.3 During the acceleration process, the longitudinal momentum of the beam electrons increases continuously, while the transverse momentum remains the same. The net effect is to decrease the angle made by the beam particles with respect to the accelerator axis from an initial value of θ_1 to a final, much smaller value of θ_2.

A recent development in the accelerator field has been the confinement of high energy electrons in circular storage rings after they have previously been accelerated by an accelerator. Figure 6.4 shows one such installation. Such storage rings have become powerful tools for high-energy elementary particle physics, and they have proved to be very useful in other applications. In particular, the circulating beam of electrons has the characteristic that it emits high-energy X rays of unprecedented intensity. Although this process is a serious limitation faced by the designer of circular machines for accelerating or storing electrons, one man's pain can be another's pleasure: this radiation can be used for whatever useful purposes people may have for using X rays. This type of radiation goes under the name *synchrotron radiation.* It might be interesting to consider — again using the arguments of relativity — why such high-energy X rays are emitted.

Fig. 6.4 The SPEAR storage ring at the Stanford Linear Accelerator Center is shown at the right in this aerial photograph. In this device, counterrotating beams of electrons and positrons of energies up to 4 GeV per beam collide at two different locations around the ring circumference.

This radiation process is shown schematically in Fig. 6.5. You will remember that traveling around in a circle is equivalent to being accelerated continuously toward the center of the circle. In other words, if you are not accelerated at all you travel in a straight line, but if you are continuously accelerated from a straight line toward a fixed point, then the result is that you travel in a circle. It is also characteristic of the laws of electricity that if a charge such as an electron is accelerated, it radiates electromagnetic radiation. For instance, a radio antenna is simply a piece of wire in which electrons are alternately accelerated and decelerated, and this motion of electric charge produces radio waves that can be received at long distances. Thus an electron traveling in a circle should behave like a little antenna that emits radio waves in a pattern that is exactly the same as would be produced by a short-wire antenna pointed toward the center of the circle. But does it?

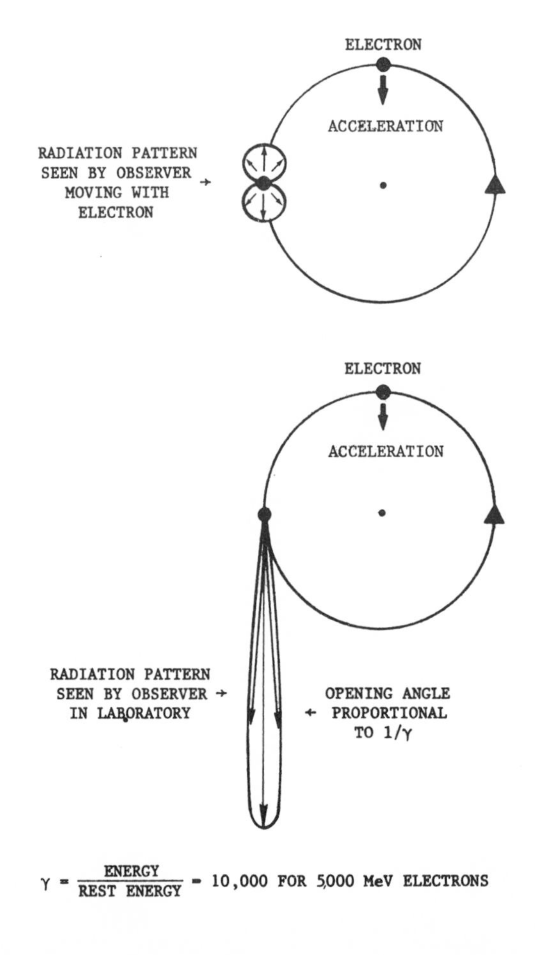

Fig. 6.5 The radiation pattern emitted by an electron is a storage ring as seen by an observer moving with the electron (*above*) and by an observer in the laboratory (*below*). In the latter case, the opening angle of the emitted radiation is proportional to $1/\gamma$, where γ is the ratio of the electron's total energy to its rest energy. For 5,000 MeV electrons, $\gamma \simeq 10{,}000$.

In practical electron storage rings, the frequency with which the electron goes around is roughly 1 million times per second, and from this one might surmise that the frequency of the radiation produced would also be 1 million cycles per second, or 1 MHz, which is in the middle of the AM radio broadcast band of the electromagnetic spectrum. However, you may also remember that radio waves have a very much longer wavelength and lower frequency than X rays. Why, then, does an electron traveling around in a circle at nearly the speed of light emit X rays rather than radio waves?

The reason can be understood as follows: First, it is indeed true that *if* you were traveling along with the electrons you would see a radiation pattern that corresponds to that of an ordinary wire antenna. This is shown by the upper sketch in Fig. 6.5. However, from the point of view of an observer who is not traveling with the electron but is at rest relative to the storage ring, the radiation pattern is thrown forward into a very narrow cone. The angle of this cone is just the inverse of the ratio of the total energy of the electron to its rest energy. This ratio is given by the symbol gamma in relativity. Thus the wave striking an observer is not continuous but is a short pulse emitted for only that fraction of the period of revolution of the electrons that corresponds to the ratio of these two energies, that is, the fraction 1/gamma.

But this is not all. We also have to consider the fact that during the time when the observer receives this short pulse of radiation the electron itself has moved. As is shown in the next diagram (Fig. 6.6), the electron has moved through this small angle 1/gamma as the radiation cone sweeps

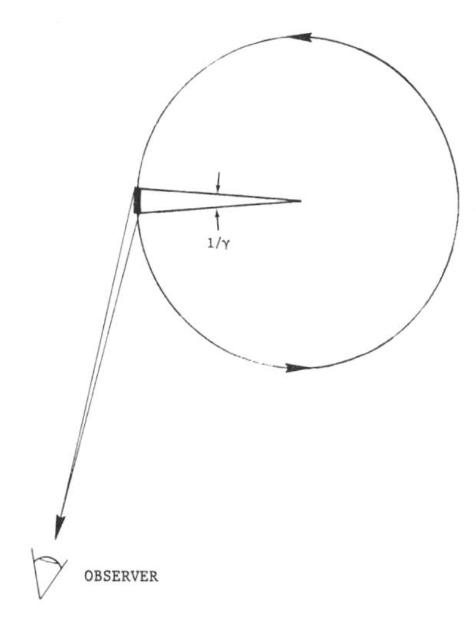

Fig. 6.6 As the radiation cone sweeps past the observer, the emitting electron moves through the small angle $1/\gamma$. If the electron were traveling at exactly the speed of light, then the radiation emitted at the beginning and at the end of the short time interval would arrive at the observer at exactly the same time. However, since the electron does not travel at quite the speed of light, there is a small difference, and special relativity can be used to show that the time of arrival of the two radiation pulses at the observer is shortened relative to the time interval of emission by the ratio $1/\gamma^2$. This means that the time interval during which the observer receives radiation is shorter by a factor of $1/\gamma^3$ than the time it takes for the electron to go once around its circular orbit. Consequently, the wavelength of the radiation is reduced by the same factor, and its frequency is correspondingly increased by the factor of γ^3.

past the observer. If we consider light pulses that are emitted at the beginning and at the end of that time interval, we find that these two light pulses will arrive at the observer at almost the same time. In fact, if the electrons traveled exactly at the speed of light the pulses would arrive exactly at the same time because of the extra distance that the first light pulse had to travel to reach the observer. However, since the electron does not quite travel at the speed of light, there is a small difference, and special relativity can be used to show that the time of arrival of the two light pulses at the observer is shortened relative to the time interval of emission by the ratio 1/gamma^2. This means that the time during which the observer receives light is shorter by a factor of 1/gamma^3 than the time it takes for the electron to go once around its circular orbit. In consequence, the wavelength of the radiation is reduced by the same factor, and its frequency is correspondingly increased by the factor gamma^3.

For practical electron storage rings the value of gamma can be very high. As an example, if we have electrons of energy 5,000 MeV going around in a circle, then the mass or energy of such particles is 10,000 times the rest energy: gamma = 10,000, and gamma^3 = 10^{12} or 1 million million. This huge factor shifts the emitted frequency from the radio broadcast band into the X-ray region. Electron storage rings are thus very practical devices for producing extremely intense, high-energy X rays, which are being used in a large variety of applications: research in biology and medicine, production of integrated circuits and other semiconductor devices, determination of the structures of complex atoms and molecules, and so on. So the design of X-ray sources based on synchrotron radiation again clearly requires the engineer to be fluent in the principles of special relativity.

The Twin Paradox

A great deal of folklore has developed around what is commonly called "the twin paradox" of special relativity. If two clocks keep identical time when they are at rest relative to one another, then the predicted result of relativity is that if one of them is set in motion and then returns it will show less elapsed time than the one that remained at rest. In human terms, if two identical twins were truly identical and if one of them went on a trip and came back, then he would be younger than his stay-at-home brother. This prediction raises two questions: First, is it true? And second, is it really a paradox, and if so, what is the paradox? We can answer these questions unequivocally: Yes, it is true; And no, it is not a paradox because we understand it. Indeed, the twin taking the two-way trip would age less rapidly as measured on the biological time scale of his brother, and the amount of this difference depends on the ratio of his travel speed to the speed of light. The effect is very small indeed when one considers ordinary speeds. For example, if the twin's trip proceeds at 600 miles per hour, then the difference in aging rate between the twins will differ by about 1 part in 1 million million or 0.002 second in a human lifetime, hardly an observable amount. However, what is most puzzling is the principle of the thing: If the twin comes back and if the effect *were* observable, then the twins would no longer be identical, because one of them has aged less than the other. Therefore the final outcome is in principle different if A has gone on a trip and B has stayed at home, or if B has gone on a trip and A has stayed at home. This seems to violate what one naively believes to be the essence of special relativity, namely that you cannot determine absolute motion, since there is no fixed reference frame that defines a state of rest. This conclusion, however, does not apply to this particular instance. Special relativity indicates only that reference frames moving at a *constant* relative velocity cannot be distinguished. It does *not* say that you cannot distinguish between accelerated and nonaccelerated states of motion. For one twin to take a trip and to return means that he has to be accelerated during part of his travel because during his turn-around his speed obviously has to be reversed. Therefore the one who has aged less is the one who has been in accelerated motion. Thus the twin paradox is not actually a paradox but constitutes only one of the many experiments that can tell the difference between what professionally are called inertial and noninertial frames (an inertial frame is one that is in uniform motion relative to that "inertial" frame defined by the preponderance of the masses in the universe).

In order to describe what is going on here from the point of view of either one twin or the other, we have to go beyond the boundary of special relativity and turn to general relativity, which offers a consistent method for dealing with accelerated states of motion also. However, special relativity does give the "right answer" if a phenomenon is *described* in the inertial frame, and in such a frame the stay-at-home twin has aged more than his itinerant brother.

This may all be interesting enough, but why do I mention it in a talk on relativity and engineering? The answer is that although the twin paradox is not observable with actual human twins, there is a very strong and demonstrable effect when one talks about phenomena on an atomic scale. We do indeed have clocks that can be put into rapid motion and sent around circles and come back home. Such clocks are radioactive atoms or particles that can decay radioactively into lighter fragments. Such unstable particles have built into them a basic clock that determines the probability that the particle will disintegrate. For example, the mu-meson, one of the particles in modern physics that occurs copiously in secondary cosmic rays reaching the earth, has a "mean life" of 2×10^{-6} seconds. That is, on the average it will decay after 1/500,000 of a second. This lifetime has been measured very precisely. If one takes a beam of such particles and captures it in a ring-shaped electromagnet, then the particles will go around and around until they decay, and we can thus keep track of their numbers in time.

Such mu-meson beams can be produced with a velocity approaching the speed of light. For instance, if one stores 3 GeV (3 billion electron volts) mu-mesons in a ring magnet as has been done during the past few years in experiments at CERN in Geneva, then the so-called time dilation factor that is observed is about 30. Thus a mu-meson at rest will have an average lifetime 30 times shorter than those traversing circular orbits at 3 GeV. This effect agrees very precisely with the prediction of special relativity theory. The twin paradox is thus both real and precisely measureable in practical situations.

These "practical situations" have thus far been confined to radioactive clocks. There has been much speculation in science fiction writing about space travel in which passengers go out into space and then return, having outlived their contemporaries. Can this be made to work? In practical terms the answer is no. Although space vehicles are useful platforms from which to mount highly sensitive experiments to examine various features of the general theory of relativity, space travel by human beings is one field for which the engineer does not have to worry about special relativity.

7. COMMENTS ON "SPECIAL RELATIVITY THEORY IN ENGINEERING"

Edward M. Purcell

Professor Panofsky's masterful review of what may become known as *high-gamma* engineering needs no amplification or embellishment by me. I do want to state for the record what many here know to be true: that the Stanford Linear Accelerator is an example of superb engineering from which many other lessons can be learned, including lessons in management.

There is another direction in which one can look for "practical" consequences of special relativity. Navigation is a practical problem nearly as old as civilization. It was just over 200 years ago that John Harrison's Chronometer No. 4 was certified by the Board of Longitude for the famous £20,000 prize. With an accuracy approaching 1 second per day, it met the requirement that position in longitude should be determinable to within 30 miles after a long voyage. Since then chronometric accuracy has improved by nearly nine powers of 10, from 1 second per day to less than (before long, much less than) 1 microsecond per year. Moreover — and justifying a comparison with Harrison's superlative device — these modern clocks, too, can travel. As you are aware, these developments have opened up astonishing possibilities for fundamental experiments. One is tempted to redefine a *Gedankenexperiment* as one that is conceivable but not yet funded. The implications of such far-reaching technical advances for experiments on gravitation are to be found in Professor Shapiro's paper (Chap. 8).

Returning to navigation, one might wonder whether an improvement by 10^9 over Harrison's chronometer is not, as a practical matter, superfluous by several powers of 10. For if clock accuracy remained the only limit on determination of longitude by that procedure, it would imply a positional accuracy in longitude of, *quite literally,* a hair's breadth. Clearly such an extrapolation is ridiculously unrealistic. Inaccuracy in celestial observation, not to mention irregularity in the rotation of the earth, would swamp any clock error long before such a limit was approached. A modern navigation system, however, can use traveling clocks in a different way. The Global Positioning System (GPS/NAVSTAR), now in an early testing stage, will employ twenty-four satellites in 12-hour orbits, each carrying an atomic clock and broadcasting time signals continuously on two wavelengths. Receiving signals from any four satellites, an observer with relatively modest equipment will be able to determine his own position on the globe, and his altitude as well, with an

Harry Woolf (ed.), Some Strangeness in the Proportion: A Centennial Symposium to Celebrate the Achievements of Albert Einstein ISBN 0-201-09924-1

uncertainty approaching 10 meters. Absolute timing accuracy measured in nanoseconds makes that feasible. (See Table 7.1, which appeared in a recent article on the NAVSTAR system.)

Table 7.1
Relativistic clock frequency shifts

- Receiving station clock frequency f_r
- Transmitting satellite clock frequency f_t

$$\frac{f_r}{f_t} = 1 + \frac{1}{c^2}(\phi_t - \phi_r) + \frac{1}{2}\left(\frac{v_r^2}{c^2} - \frac{v_t^2}{c^2}\right) + \frac{\bar{k}}{c}\cdot(\bar{v}_t - \bar{v}_r) + \ldots = 1 + \delta$$

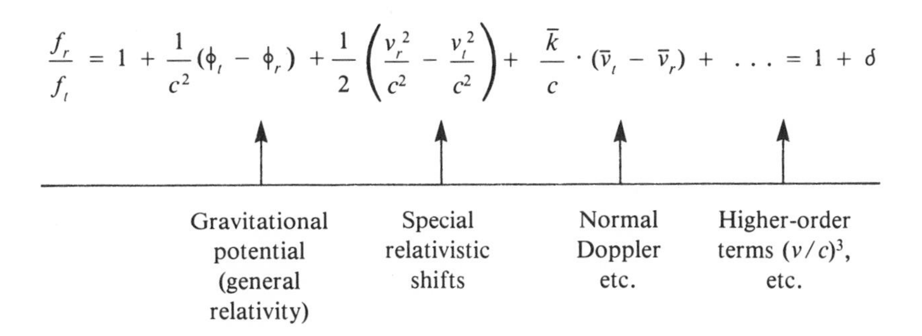

- Observed satellite increase in frequency (observed at receiver) $\delta \simeq 447.9 \times 10^{-12}$ or 448 μsec/sec of time offset. (38.7 μsec/day)
- Earth oblateness and solar perturbation: varies from 445.8 to 450.2×10^{-12}
- Satellite clock is purposely set low to 10.22999999545 MHz or 4.45×10^{-10} low in frequency relative to a nominal 10.23 MHz.

Source: J. J. Spilker, Jr., "GPS Signal Structure and Performance Characteristics," Navigation **35** (2) (1978).

It is essential to take into account the difference in clock rate according to special relativity, the difference being, even for the moderate speeds of aircraft, greater than the expected drift in clock rate. The table reminds us that at this level too, differences implied by the equivalence principle (here labeled as *general relativity*) also have to be incorporated. Indeed, as you may know, a remarkably accurate verification of the gravitational red shift by means of a rocket-borne clock has been achieved by R. F. C. Vessot (to whom, incidentally, I am indebted for suggesting the example of the Global Positioning System). Already in the routine maintenance of our standard of time, atomic clocks are transported by air from Washington to Boulder, Colorado, for comparison; the accuracy is such that it is necessary to allow for time dilation en route, as well as for the extra altitude of Boulder. Evidently the modern navigation systems engineer has to understand special relativity — and even, unlike the high-gamma electronic engineer, a little general relativity.

There is another effect, strictly relativistic, second order in v/c, and important in engineering — namely, *magnetism*. Two copper wires 1 centimeter apart, carrying a few hundred amperes of electric current, interact with a palpable force, a few ounces per foot of wire. That force can be traced back to the mean drift velocities of the electrons in the wire, which in this example would hardly exceed 1 centimeter per second. If the same electrons were made to stream through the wires with speeds approaching c, the force would be increased by something like 10^{21} to quadrillions of tons per foot of wire. Equally gigantic would be the ordinary electrostatic repulsion of the electrons in the wires, were it not for the presence of almost precisely compensating positive charges. This is just to emphasize that the magnetic forces of our everyday experience are indeed as feeble as we should expect relativistic effects to be that are the manifestation of such small velocities. And that is true even though such forces, ingeniously multiplied in an ancient device called an electric motor, can hoist an elevator or propel a locomotive. At least that is so in the universe we know about, which appears to contain only electric charges, and no magnetic charges whatever.

I bring up this very elementary physics to introduce a reference to a paper now largely forgotten, one of the early American responses to Einstein's 1905 article. The paper was by Leigh Page, then a young instructor at Yale; it appeared in the *American Journal of Science* in 1912. As far as I know it was the first paper to show explicitly and simply how one can derive from Coulomb's Law of electrostatic attraction and the kinematics of special relativity, and nothing more, both Ampere's law for the magnetic interaction of currents and Faraday's law of induction. Of course Page did not claim to have gone beyond Einstein. Early in the paper, which is entitled "A Derivation of the Fundamental Relations of Electrodynamics from those of Electrostatics," Page wrote:

> The object of this paper is to show, that if the principle of relativity had been enunciated before the date of Oersted's discovery, the fundamental relations of electrodynamics could have been predicted on theoretical grounds as a direct consequence of the fundamental laws of electrostatics, extended so as to apply to charges relatively in motion as well as to charges relatively at rest. Of course, only that part of the theory derived from the principle of relativity that is independent of any a priori knowledge of the electrodynamic equations, will be made use of. That is to say, we will use only the kinematics of relativity: to use the dynamics of relativity, which is derived from the electrodynamic equations, would be to reason in a circle.

Leigh Page, incidentally, served as one of the two judges in the *Scientific American*'s famous Einstein Prize essay contest of 1921. The other judge was Professor E. P. Adams of Princeton, translator in that same year of Einstein's four Princeton Lectures (not to be confused with N. I. Adams, later coauthor with Page of a widely used text in electromagnetism).

Recognizing the essential relation of electromagnetism to special relativity puts our assigned topic, "Special Relativity in Engineering," in a rather odd light — a bit like "Euclidean Geometry in Engineering." This the way the world *is*. And it does not really take gigavolts or nanoseconds to demonstrate it; stepping on the starter will do it!

Finally, we should remind ourselves that the influence on modern technology of Einstein's work is by no means limited to manifestations of special relativity, even if we subsume thereunder all of classical electromagnetism. Take the revolution in modern optics, beginning with everything called laser or maser. Its roots are in Einstein's great paper in 1917 on the quantum theory of radiation, where first appeared those now familiar coefficients A_n^m and B_n^m. Van der Waerden says of this paper in his *Sources of Quantum Mechanics*: "All subsequent research on absorption, emission and dispersion of radiation was based on it." Based on it even more directly is every working laser. Professor Panofsky remarked that $E = mc^2$ does not have to be invoked to explain nuclear energy. But one cannot begin to explain the machinery now being constructed for laser fusion, a colossal engineering enterprise, without talking about B_n^m.

Open Discussion

Following Papers by W. K. H. PANOFSKY and E. M. PURCELL

Chairman, C. N. YANG

R. Sexl (U. of Vienna): Now one can base an introduction to relativity on engineering aspects. For instance, the old problem of synchronizing clocks is still very often based on trains moving through thunderstorms, whereas at the moment, it would be much better to base it on the actual clock network on earth. There are certain timekeeping stations that are synchronized the very way that Einstein said they should be, not with light signals really, but with radio signals, so that you can actually, just from the working of the earth's clock network, read off the fact that the speed of light is constant in all directions. Even the present navigation system for ships, the Lawrence sea network, operates on the same principle: a captain receives radio signals traveling with the speed of light. If these signals were moving according to classical ether theory, than all the ships would sink, for they would be off their courses by a few kilometers. Maybe that some of the ships do sink is a disproof of relativity, but I am not quite sure!

L. van Hove (CERN, Geneva): In addition to the points mentioned by Professor Panofsky and going a little beyond special relativity, adding to it field theory and quantum theory, also subjects very central in Einstein's thinking, we come to another aspect, which is the existence of antiparticles, of antimatter, and the existence of annihilation between a particle and an antiparticle into radiation. There seems to be a trend to using positrons, which are the antiparticles of electrons, for an increasing amount of what might be called applied work — namely, the study of electron distributions in material, through the two-photon annihilation.

W. Rindler (U. of Texas, Dallas): Professor Panofsky remarked about the remarkable coherence of electron beams and gave a very nice explanation that they act somewhat like parallel currents that attract each other. There is another rather beautiful explanation of the same fact, which is also very beautifully relativistic: You can think of a small cloud of electrons at rest in a laboratory as a clock that would run rather fast so the electrons would repel one another; they would expand rather fast. As the electrons move, the clock is slowed down, and the electron beam is held together, which is a nice manifestation of the relativistic time dilation effect.

I wonder if I might make a rather frivolous comment also on engineering, since this is an engineering session, and report to you the invention by the Polish relativist Bielnik of a relativistic refrigerator. You put a chicken in this refrigerator and you take it out a year later, and it's quite fresh because it was kept fresh by fast motion. It has aged only about a few seconds while you aged a year!

M. Goldhaber (Brookhaven): Perhaps Professor Panofsky is not old enough to remember that $E = mc^2$ did play an important role in the development of nuclear physics, especially about 1930 after Bohr had the idea that energy might not be conserved in β-decay, with the number of β particles just petering out with increasing energy. From an important collection of data, C. D. Ellis and N. F. Mott showed, however, that the upper limit of the β spectrum agreed with the mass differences of parent and daughter atoms if you used the equation $E = mc^2$. Thanks to this equation a linear table of masses soon replaced a square table of reaction energies.

Yang: I would like to raise a question for discussion, especially perhaps with the physicists and historians in the audience. It is commonly said that of the two revolutions in physics for which Einstein was responsible, special relativity and general relativity, general relativity theory might not have been discovered for a long time without Einstein, but special relativity certainly would have been discovered early. Listening to what was said this morning by Professor Miller and realizing the brilliance of Poincaré and the steadiness and great solidity of the knowledge of Lorentz, it is not at all clear to me that indeed it would have been that simple a matter to shake off the stationary ether concept if Einstein had not come along. I am not a historian of science; I do not know whether this is a thesis that is sustainable, but if there is anybody who would like to remark on it, I think it would be a very interesting topic to discuss.

A. I. Miller (Harvard U.): Certainly the work of Poincaré and Lorentz would have been fixed up eventually to the extent that what they were waiting for was further ether-drift experiments. These certainly would have failed had they been done, and therefore, one could more than cavalierly take the ether-fixed system to be the one on the earth. And one would have then a very exact principle of relativity that is based on an ether. Whether one could have obtained the mass-energy equivalence is doubtful, however, but if Einstein had not come along, one would have had a viable electrodynamical theory. General relativity theory is something else. There, one had to make the step to combine physics and geometry, something that Poincaré was very much against, as were most other philosopher–physicists of that time. But just to reiterate, the Lorentz–Poincaré theory of the electron is mathematically identical to special relativity theory, but it is not observationally identical.

G. Holton (Harvard U.): Einstein himself said that not Poincaré or Lorentz but Langevin might have developed the special theory of relativity.

A. Hermann (U. of Stuttgart): Max Born told us several times that he was educated at Göttingen and took part in a course with Hermann Minkowski and that he had the feeling when the Einstein paper came out that Minkowski was very, very near to a special theory of relativity and this, of course, is connected with Felix Klein's *Erlanger* program. So Max Born felt that the special theory of relativity would have come without Einstein, perhaps one or two years later. However, it is quite different with quantum theory and of course with the general theory of relativity.

Yang: To that, I would remind you that Whitehead had said, in his analysis of the origin of scientific thought, that to come very close to a thing and to understand it are two quite distinct things.

P. A. M. Dirac (Florida State U.): In one respect Einstein went far beyond Lorentz and Poincaré and the others, and that was in asserting that the Lorentz transformation would apply to the whole of physics and not merely to phenomena based on electrodynamics. Any other physical forces that may be introduced in the future will have to conform to Lorentz transformations, which is going far beyond what the people who were working with electrodynamics were thinking about. And, of course, in a way Einstein was wrong, because the Lorentz transformation does not apply to everything. There is the microwave radiation, which does provide an absolute velocity. It provides an ether, but the real importance of Einstein's work was to show how Lorentz transformations dominate physics, and that is especially important in atomic physics; it led to De Broglie's connection between waves and particles, and it led to the relativistic theory of the electron and antimatter. These are all consequences of the dominant position that the Lorentz transformations hold in physics. I am inclined to think that this is really a more important discovery than general relativity.

B. Hoffman (Queens College, City U. of New York): I am wondering whether people would have discovered the special theory of relativity without Einstein. It is true that Poincaré had all the mathematics and somewhat more than Einstein had in his 1905 paper, but in Poincaré's work there was always the implication that there was a rest system — something at rest in the ether — and so you get the impression that Poincaré and any followers would have said, yes, if something is moving relative to the ether, it is contracted. But, of course, people who believe this would think that our stationary rods were expanded, instead of contracted, and Poincaré would have had one clock going slower, but the other going faster. This reciprocity was a very subtle point, and it is quite likely that people might never have realized that it was a reciprocal relationship.

IV. Developments in Relativity (continued)

8. EXPERIMENTAL CHALLENGES POSED BY THE GENERAL THEORY OF RELATIVITY

Irwin I. Shapiro

Throughout the first half century of its existence, the general theory of relativity remained largely in the domain of theoretical physics. Meaningful tests of the theory were, for the most part, beyond the reach of experimental techniques. In many other branches of physics, measurements were disclosing inadequacies in theory; in gravitation virtually no experimental progress was being made. This imbalance is now beginning to be redressed, but just barely. Although there has been an enormous increase in experimental activity during the last decade, progress has not been rapid; the obstacles are too great.

In this paper we review the progress made in experimental tests of the general theory of relativity and discuss the prospects for further improvement and for some new tests. We consider, in turn, the following aspects of gravitation:

1. Principle of equivalence for massive bodies. (How the mighty fall)
2. Clock comparisons. (The ticks that count)
3. Light deflection. (The photons that barely streak by the sun)
4. Signal retardation. (The echoes that arrive home late)
5. Perihelion precession. (The orbits with that little bit extra)
6. Time variation of the gravitational "constant". (Is gravity losing its touch?)
7. Lense–Thirring effect. (The drag that refreshes)
8. Gravitational radiation. (The waves that pass gently in the night)

Tests of these aspects constitute in part the experimental challenges for the next decade.

1. Principle of Equivalence for Massive Bodies

Laboratory experiments on small bodies to test the (weak) principle of equivalence were the concern even of nonphysicists at least back as far as the fifth century.[1] At that time, Ioannes Grammaticus recorded the fact that gravitational and inertial mass were equivalent although, to be sure, he did not use those names. Recording of more formal tests began with the work of Galileo

Harry Woolf (ed.), Some Strangeness in the Proportion: A Centennial Symposium to Celebrate the Achievements of Albert Einstein ISBN 0-201-09924-1

and continued with Newton, Bessel, Eötvös, Dicke, and, most recently, Braginsky. This latest result[2] indicates that any dependence on composition of the ratio of inertial to (passive) gravitational mass for laboratory-sized bodies is no more than about 2 parts in 10^{12}:

$$\frac{m_i}{m_g} = 1 \pm 2 \times 10^{-12} . \tag{8.1}$$

This result implies that, to high accuracy, the nuclear, the electromagnetic, and even the weak interactions contribute equally to gravitational and inertial mass.[3] For the gravitational interaction, the picture is quite different. The ratio of the gravitational binding energy to the total energy varies with the square of the characteristic length of the body. For a laboratory-sized body of radius 1 meter, only about 1 part in 10^{23} is contributed by the gravitational binding energy — some eleven orders of magnitude smaller than even the impressive Braginsky–Panov experiment could detect. In particular, for a spherically homogeneous body, we have

$$\frac{\text{Gravitational binding energy}}{\text{Total energy}} \equiv \Delta = \frac{4\pi}{5}\frac{G\rho R^2}{c^2} \sim 10^{-23};\ R \sim 1\ m, \tag{8.2}$$

where G is the gravitational constant, ρ ($\sim 5\ g\ cm^{-3}$) the density, and c the speed of light.

For massive bodies, we can write any possible deviations of (m_i/m_g) from unity as $\eta\Delta$, where Δ is given in Eq. (8.2) and η, as first calculated by Nordtvedt and Will in the parametrized post-Newtonian (PPN) formalism,[4] is given for "fully conservative" theories by

$$\eta = 4\beta - \gamma - 3 . \tag{8.3}$$

Here β and γ are two PPN parameters. Possible preferred frame and location effects have been ignored; a more general expression is given by Will.[5] General relativity, of course, requires

$$\eta = 0 . \tag{8.4}$$

How can η be determined experimentally? Obviously the sizes of the test bodies must be increased to a planetary scale to be useful. But three or more are needed. Were only two available, a violation of the principle of equivalence could not be distinguished from a rescaling of the mass of one of the bodies with respect to that of the other since no independent measurements of the masses are feasible. The first experiment tried with such massive bodies utilized the moon-earth-sun system, first suggested for this purpose by Nordtvedt.[6] The optical retroreflectors left on the moon by the Apollo astronauts made this experiment possible. If η were positive, one would see the moon's geocentric orbit shifted in the direction of the sun. Measurements of the earth-moon distance would then disclose an additional monthly variation, with an amplitude of about 8η meters. Although 8 meters ($\eta = 1$) may seem fairly large compared to the (current) uncertainty of about 10 centimeters in the laser measurements of the distance between a telescope on earth and a retroreflector on the moon, one must remember the competing classical effect. This effect has the same period but an amplitude of 110 kilometers. Fortunately, the *uncertainty* in our knowledge of the classical effect is only about 1 centimeter. Thus an accurate test is possible. Two groups of scientists analyzed these laser measurements; the results showed η to differ from zero by an amount insignificant compared with the estimated uncertainties. These are given as 0.015 by Shapiro et al. and as 0.03 by Williams et al.[7] How could different uncertainties be obtained from analysis of the same data? The answer is simple: The uncertainty is dominated by systematic errors and the allowance for such errors in the overall uncertainty is partly a matter of judgment. Shapiro et al. assigned an error four times the statistical standard deviation.

Future improvements in this limit on any deviation of η from the null value predicted by general relativity are certainly possible. The accuracy of measurement of distances to the lunar retroreflectors can be expected to improve, but only rather slowly with time. Greater improvement in the near future in the estimation of η can be obtained through more accurate auxiliary measurements of the variations in the rotation of the earth and through refinement of the theory presently used to describe the motion of and about the center of mass of the moon.[8]

One other system that deserves mention is the Mars-sun-Jupiter system, where the effect of any violation of the principle of equivalence would be nearly one thousandfold larger than for the moon-earth-sun system. However, the accuracy of the relevant measurements[9] is nearly one hundredfold lower and the periodicity of the effect about twentyfold longer, being basically the ratio of Mars' orbital period to the moon's. Any non-null effects on the two systems should therefore be comparable in their detectability.

We expect the uncertainty in the estimate of η to be reducible to about 0.5 percent or less in the next decade from the analysis of the measurements on both systems.

One can conceive of an enormous improvement in this test were pulsars to be discovered in a tightly bound ternary system, since, for a neutron star, Δ is of the order of 0.1 to 0.2, some eight orders of magnitude larger than for the earth. Unfortunately, only one pulsar out of more than 300 known is in a binary system, with the companion not yet detected. The probability of detecting three pulsars in a single system must, therefore, be considered very low indeed.

2. Clock comparisons

Some of the more famous of the predictions of general relativity concern the effects of changes in gravitational potential on the rate of a clock and on the frequency of an electromagnetic signal. For example, such a signal is predicted to be red shifted if the gravitational potential is of larger magnitude at the point of transmission than at the point of detection. The fractional change, $\Delta f/f$, in frequency is predicted to be

$$\frac{\Delta f}{f} \simeq \frac{1}{c^2}[\phi_1 - \phi_2] \; ; \; \phi_i << c^2 \; (i = 1,2) \; , \tag{8.5}$$

where ϕ_1 and ϕ_2 are, respectively, the (Newtonian) gravitational potentials at the points of transmission and detection. The most precise laboratory verification of Eq. (8.5) was obtained in an elegant series of experiments by Pound, Rebka, and Snider.[10] Laboratory experiments are sensitive only to the difference in altitude of the transmitter and detector; hence $\phi_1 - \phi_2$ in Eq. (8.5) can be replaced by gh where g is the acceleration of gravity on the earth and h is the altitude difference between the positions of transmitter and detector. The experiments of Pound et al. utilized the Mössbauer effect in which the recoil momentum upon emission or absorption of 14.4 kev gamma rays from the Fe^{57} nucleus is shared by the whole crystal, thus allowing the formation of extremely narrow spectral lines. In particular, at the top of a 25-meter-high tower, they placed a crystal emitter that could be moved sinusoidally, and at the bottom they placed a fixed crystal absorber. The absorption was measured as a function of the vertical velocity of the emitter, which could thereby compensate for a gravitational effect on the frequency of the rays emitted at the top. Of course, emitter and absorber were also reversed to reduce systematic errors. By careful attention to detail, they reduced their uncertainty to a level corresponding to about 5 parts in 10^5 of the natural width of the resonance line in their experiment. The measured effect, divided by the predicted effect, was unity to within their estimated uncertainty of 1 percent.

Experiments have also been carried out to test directly the effect on atomic clocks of differences in gravitational potential. The first such experiment was conducted by Hafele and Keating,[11] who flew four cesium-beam clocks around the world in opposite directions and compared their readings before and after the flights with clocks that remained at home. The results confirmed the Einsteinian predictions but, as expected, did not still the doubters of "the twin paradox." In particular, Hafele and Keating found the fractional differences between the measurements and the predictions for the combined gravitational and kinematic effects to be about -0.01 ± 0.08 for the westward and -0.4 ± 0.5 for the eastward flight. The fractional error was higher for the latter flight primarily because, for it, the gravitational and kinematic contributions were of opposite sign, thus reducing the magnitude of the differences in the readings between the clocks that flew and those that did not.

Alley and his collaborators also used a plane to fly sets of atomic clocks for comparison with similar clocks on the ground (C. O. Alley, private communication, 1977). In these experiments, laser signals were transmitted from the ground and reflected by corner cubes on the plane so as to realize in practice the Einstein *Gedanken* experiment of synchronization of separated clocks by exchange of light signals. The path of the plane, which circled at an altitude of about 10 kilometers, also had to be monitored. The three cesium clocks on board had instabilities at the level of about 2 parts in 10^{14} for averaging times of 30 hours. The cumulative time differences between the clocks on the plane and those on the ground for a typical flight were about 50 nanoseconds. From three clocks and five flights of 15 hours each, Alley and his coworkers deduced a ratio of measured to predicted values of 0.987 ± 0.016, about one-third of this uncertainty being an allowance for possible systematic errors. These experiments were thus not quite so accurate as that of Pound and Snider, but the two groups did perform conceptually different experiments.

By far the most impressive clock experiment was performed by Vessot and Levine[12] who in 1976 placed a hydrogen maser frequency standard on a spacecraft that traveled on an orbital arc with a 10,000 kilometer maximum altitude. The height of the rocket that launched this hydrogen maser equaled the height of Pound's tower. The communication link, used for comparison of the hydrogen maser standard in orbit with the corresponding masers on the ground, had to involve radio signals; a clever three-frequency scheme was employed to virtually eliminate the otherwise deleterious effects of the ionosphere. A two-way, two-frequency link, controlled from the ground, and a one-way, one-frequency link, controlled from orbit, accomplished this goal: the two frequencies of the former were chosen to surround the single frequency of the latter so as to provide nearly complete cancellation of the effects of the ionosphere. Final results have not yet been obtained from this experiment because of difficulties with the determination of the trajectory which was apparently affected in an important manner by nongravitational accelerations. Preliminary results show that the ratio of the measured red shift to the predicted value differs from unity by 50 parts in 10^6 with an uncertainty of about 100 parts in 10^6. The uncertainty in the final result is expected to be somewhat lower (Vessot, private communication, 1978).

None of these clock experiments is nearly sensitive enough to yield useful information on either the γ or the β parameter of the PPN formalism.

What of the future? The most dramatic advance that could soon be made involves the placement of a hydrogen maser frequency standard on a spacecraft that passes near the sun. An efficient technique is the use of a gravity assist from Jupiter to transfer the spacecraft into an extremely eccentric orbit that almost grazes the sun. The gravitational red shift for a clock passing that close to the sun is of the order of 10^{-6} to 10^{-7}, depending on the exact periapse of the orbit of the spacecraft. Clock stabilities of the order of one part in 10^{15}, for averaging times of the order of 100 seconds or more, have already been achieved in the laboratory, and could yield an uncertainty in the ratio of measured to predicted red shift values of between 10^{-8} and 10^{-9} in such a solar-probe experiment. This experiment would be sensitive to second-order (β) effects in the Schwarzschild metric. However, the requirements for the experiment are by no means trivial. A "drag-free" satellite is needed to overcome the rather severe nongravitational accelerations due primarily to radiation pressure. The development of a drag-free system with the appropriate characteristics for operation in the hostile environment near the sun is a difficult task.[13] It is also necessary to develop a hydrogen-maser standard with the required lifetime.

A solar probe is currently under study by the U.S. National Aeronautics and Space Administration. When such a probe might be launched is unclear; even less clear is whether it will have an appropriate drag-free system; and least clear is whether a hydrogen maser or equivalent frequency standard will be included.

Further laboratory experiments, some suggested by Nordtvedt and Will, may be carried out to compare different kinds of clocks, for example clocks based on a hydrogen maser and those based on a cavity resonance, which have different fundamental length scales. Such comparisons may provide useful bounds on, or measurements of, possible nonmetric effects.[14]

3. Light Deflection

Measurements of the deflection of starlight by the gravitational field of the sun have been undertaken with varying degrees of vigor since 1919. According to the PPN formalism, the deflection for a small impact parameter can be expressed as

$$1''.75 \left[\frac{1+\gamma}{2}\right] \frac{R_\odot}{d}, \tag{8.6}$$

where $R_\odot$ is the radius of the sun, d the impact parameter of the light rays, and γ a PPN parameter that is identically unity in the general theory of relativity.

Eclipse expeditions from 1919 to 1952 gave a variety of results; data from the same eclipses in the hands of different analysts also occasionally gave inconsistent results. Overall, these results yield a value for γ consistent with unity to within an uncertainty of order 50 percent. The most recent attempt at eclipse observations was made in 1974 by a team from the University of Texas; their result, too, was consistent with unity to within the estimated one-standard-deviation uncertainty of about 10 percent.[15]

The real advance in this test has come about through the use of radio interferometry, first suggested over a decade ago.[16] With the technique of very-long-baseline radio interferometry (VLBI), the angular accuracies achievable can be much higher than can now be achieved on the surface of the earth with optical techniques. The essence of this radio technique is the absence of a direct electrical connection between the different elements of the interferometer array. At each site, the radio signals are mixed with local-oscillator signals, usually derived from a hydrogen maser frequency standard, and filtered. The resultant video signals, suitably digitized, are recorded on magnetic tapes. The tapes from all elements of the interferometer are brought together and cross-correlated pairwise.

The first accurate deflection experiment utilizing VLBI was carried out in 1972 by Counselman et al.[17] They used four antennas, two each at the National Radio Astronomy Observatory in Green Bank, West Virginia, and at the Haystack Observatory in Massachusetts, 845 kilometers distant. Two compact extragalactic radio sources, designated 3C 273 and 3C 279, were observed, each by an antenna at each observatory. The latter source is occulted by the sun on October 8. The "fringe spacing" for this interferometer, which operated at a radio frequency of 8 GHz, was about $0\overset{''}{.}01$. Two sources were observed simultaneously so that use of difference techniques could reduce the effects of both the atmosphere and the relative drift of the clocks at the two sites. Through the use of two antennas at each site, it was possible to monitor continuously the fringe phase for each source. The fringe phase varies so rapidly, and so erratically, for signals that pass close to the sun, that any "break" in the monitoring of the phase could be serious. In general, the fringe phase can not be "connected" across breaks without the introduction of 2π ambiguities. Thus, for each break, a new integration constant, in effect, has to be estimated in the analysis, often causing a substantial loss of accuracy in the estimate of γ.

The amount of magnetic tape recorded in this experiment was rather extensive; if placed end to end, the tapes would stretch from Princeton nearly to Ulm. The number of bits recorded was almost 10^{12}. Luckily, they were almost all noise and, with about five months of compression, yielded a few bits of signal, namely the result

$$\gamma = 0.98 \pm 0.06 \,. \tag{8.7}$$

The uncertainty in this experiment was not dominated by signal-to-noise problems, nor by the solar corona, but rather by a lack of proper calibration; suitable equipment was not then available.

A better experiment was done subsequently, with a shorter baseline and a direct electrical connection between the antennas; this technique is dubbed connected-element radio interferometry. This experiment was carried out in 1974, and repeated in 1975, by Fomalont and Sramek.[18] They utilized a nearly linear array of four antennas at the National Radio Astronomy Observatory; three spanned a distance of only 2.7 kilometers, but the fourth yielded a baseline of 35 kilometers. (Use of the three closely spaced antennas instead of only one provided a certain amount of redundancy and introduced little extra complication since the four-element array was fully instrumented for other astronomical purposes.) To reduce the effects of the earth's atmosphere, they observed three sources that also formed a nearly linear array. These sources were almost equally spaced on the sky and spanned an arc of about 10°, with the central source being occulted by the sun on April 11. All observations were carried out simultaneously at 2.7 and 8.1 GHz. Thus, the dispersive property of a weak plasma ($n^2 - 1 \propto f^{-2}$, where n denotes the index of refraction and f the radio frequency) could be used with these dual-band observations to reduce the effects of the corona. The three sources could not be observed simultaneously but, rather, were observed sequentially. The breaks in the measurements of the fringe phases for each source were here not so serious as in the VLBI experiment, mainly because the much shorter baselines involved yield correspondingly reduced fringe rates which allow phase connection to be accomplished unambiguously.

The main uncertainty in the results from these connected-element interferometer experiments was attributable to the atmosphere. Use of a three-source array and linear interpolation was only partially successful in the reduction of the atmospheric effects. Combining the data from the two years, Fomalont and Sramek obtained

$$\gamma = 1.01 \pm 0.02\,, \tag{8.8}$$

a result about three times more accurate than was obtained in the earlier, VLBI, experiment.

What future improvements are feasible in the use of interferometry to measure the gravitational deflection of radio waves? Connected-element interferometry experiments will probably be performed with the very large array (VLA), now being constructed in Socorro, New Mexico. This array will contain twenty-seven telescopes, each 25 meters in diameter, and will span a figure *Y* with maximum dimension about 25 kilometers. The accuracy achievable will probably be limited by the neutral atmosphere to a standard error, $\sigma(\gamma)$, of about 0.006. The improvement would be brought about by the presumably more stable atmospheric conditions in New Mexico and the use of more antennas, which should tend to average out the errors introduced by the atmosphere.

The first VLBI experiments were limited by lack of proper calibration equipment and of the means for dual-band observations. These limitations are, of course, not fundamental and can be overcome. Further, the effects of the atmosphere are largely uncorrelated over distances greater than several tens of kilometers and will be proportionately less important for longer baselines. An analysis carried out by M. Ratner (private communication, 1976) shows also that the effects on accuracy of turbulence in the corona will decrease with increasing baseline length, although more slowly than in an inverse proportion. For baselines on the earth longer than about 5,000 kilometers, however, the lack of mutual visibility of the sources prevents any additional improvement in achievable accuracy.

Several types of improved VLBI experiments seem feasible. All would use fully calibrated equipment. In one approach, observations of each source would be made simultaneously in two frequency bands, one each centered at about 2 and 8 GHz. With available wideband receivers, group-delay measurements could be made at both bands with sufficient accuracy to eliminate the 2π ambiguity in the fringe phase, or phase delay, measurements.[19] The use of this technique would obviate the need to follow fringe phase continuously and would enable the experiment to be carried out with the use of only a single antenna at each site; the observations of two or more sources could be carried out cyclically. A second approach would be to use a single, but much higher, radio frequency, about 25 GHz, for which the coronal effects would be less than 0.1 percent of the predicted gravitational deflection, except for signals that pass within about two solar radii from the sun's surface. With this approach only one radio frequency need be employed, but present equipment limitations would require the fringe phase to be monitored continuously and hence would require the use of two antennas at each site.

A third, and most unusual, approach could be tried in the fall of 1979 when two spacecraft at greatly different distances from the sun will be nearly aligned as they pass through superior conjunction. Both spacecraft have dual-band (2 and 8 GHz) transmitters; one, Voyager 2, will be in the vicinity of Jupiter; the other, Pioneer Venus, will be in orbit about the planet Venus. Comparison of the measurements of the signals from the two spacecraft would allow significant reduction in the errors that would otherwise be introduced by the relative (erratic) drifts in the frequency standards at the different observing sites and by the differences in the propagation medium encountered by the signals received at these sites. The smaller the angular separation between two sources, in general the greater will be the cancellation of the propagation medium effects. Of course, the smaller the angular separation, the smaller will be the sought-for difference in gravitational deflection. However, for sources *not* at distances from the sun large compared with the observer's distance from the sun, this lack of angular separation is not always a very serious impediment. The angle α of apparent displacement of the position of the source as seen from earth, due to the gravitational deflection, is predicted to be[20]

$$\alpha \simeq \frac{4r_0}{d}\left(\frac{1+\gamma}{2}\right)\left(\frac{r_s}{r_s+r_e}\right); \; d \ll r_e, r_s \,, \tag{8.9}$$

where r_s and r_e are, respectively, the distances of the source and the earth from the sun. Thus, even for precise alignment of the two spacecraft, the predicted differential deflection would drop only to 40 percent of the full deflection for a source "at infinity." The centennial of Einstein's birth offers the first opportunity to measure the deflection in this novel manner.

Overall, one can reasonably expect the uncertainty in the estimate of γ to be reduced to about 0.001 from VLBI experiments.

Activity in the optical part of the spectrum has also far from ceased. To perform the deflection experiment, Henry Hill and his coworkers have been developing a very sophisticated optical astrometric telescope that uses laser interferometric techniques, among others. This group intends to use the sun's diameter as the unit of length and to use planets, rather than stars, as targets. One reason for this choice is that the difference between the celestial, or ecliptic, latitude of a given star and that of the sun remains virtually constant in the vicinity of the close passage. Thus, the estimate of this difference is highly correlated with the estimate of the deflection. With the target a planet, the closest angular separation from the sun varies from conjunction to conjunction and, with data from several conjunctions, one can solve simultaneously for both the relevant orbital elements and the parameter γ that describes the deflection in the PPN formalism. Hill anticipates that the uncertainty in the estimate of γ will be of order 0.002. However, the predicted relativistic deflection for light reflected from planets is severalfold lower than for light from distant stars, as shown by Eq. (8.9). In addition, planets are not point targets; in measuring the deflection the topography on the planet as well as the albedo variations near the limb of the planet become possibly important sources of systematic error. Hill hopes to overcome such difficulties with the use of a finite Fourier transform technique to define the position of the limb.[21] The time scale required to obtain a useful result is of course unclear, as in all of these experiments.

The future looks brightest when one considers optical interferometry from earth orbit. Here the potential exists, although the funds may not, to obtain values for $\sigma(\gamma)$ as low as 10^{-6} or perhaps lower. A preliminary examination of the characteristics of such an instrument is in progress,[22] and calculations of the predicted second-order relativistic deflection effects have been carried out.[23] Although there are many small effects, previously ignored, that must be considered, there appears to be no fundamental problem that would prevent the detection of second-order relativistic effects.

4. Signal Retardation

General relativity predicts that the transit times of light signals traveling between two "fixed" points would be increased if a massive body were placed near the path of these signals.[24] How could this prediction be tested? A natural method is to utilize two planets as the "fixed" points and the sun as the massive body. Light signals can be transmitted from the earth toward a target planet and the echoes detected and their transit time measured. According to the PPN formalism, the "extra" time delay, $\Delta\tau$, of the echo, attributable to the direct effect of the gravitational potential of the sun, is given by

$$\Delta\tau \simeq \frac{4r_o}{c}\left[\frac{1+\gamma}{2}\right]\ln\left[\frac{r_e + r_p + R}{r_e + r_p - R}\right]. \tag{8.10}$$

Here, in addition to quantities already defined, we have $r_o \equiv GM_\odot/c^2 \approx 1.5$ km; $M_\odot$ the mass of the sun; r_e and r_p the heliocentric coordinates of the earth and target planet, respectively; and R the distance between the planets. There is a logarithmic singularity in $\Delta\tau$ such that the maximum delay is of the order of 250 microseconds for rays that just graze the limb of the sun.

Experiments to measure $\Delta\tau$ were carried out first with radar observations of the planets[25] and more recently with the radio tracking of the Mariner 6 and 7 spacecraft[26] in heliocentric orbit and of the Mariner 9 spacecraft in Mars orbit.

Radio tracking of spacecraft has the potential for much higher accuracy than do passive radar measurements which, in addition to lower signal-to-noise ratios, are hampered by the wondrously complicated topography of planetary surfaces, from scales of millimeters and below to kilometers and above. In 1976, the Viking spacecraft provided the opportunity to make a substantial improvement in the accuracy of this experiment. A number of factors contributed. For example, the round-trip delays of radio signals were measured with uncertainties down to the 10-nanosecond level, severalfold lower than was obtainable with apparatus available previously. Even that limit was not set by the attainable signal-to-noise ratio, but by the difficulty in calibration of the time delays through the spacecraft and through the antenna-transmitter-receiver combination on the earth. The signal-to-noise limit for most of the measurements was only a few nanoseconds. Thus, under some conditions, it was possible to measure the distance from a point on the earth to a point on Mars with an uncertainty of under 2 meters (perhaps a feat worthy of entry in the Guinness book of world records).

It is not enough to make accurate measurements. One must be able to *interpret* them properly, a nontrivial task. Recall that the round-trip time of radio signals sent to Mars when it is near superior conjunction is about 2,500 seconds and that one wishes to account for all contributions of a few nanoseconds or larger. The Viking spacecraft afforded the possibility for a vastly improved accuracy of interpretation, and that is where the real advantage lay. How did this improvement come about? First, one must know the orbits of the bodies involved. The Viking landers allowed us to deal with the earth and Mars as the relevant orbiting bodies; each has an area-to-mass ratio some millionfold smaller than those of spacecraft, and hence these planets are correspondingly immune to nongravitational accelerations, which are difficult to model. Furthermore, the orbits of the earth and Mars are affected in only a minor way by the gravitational accelerations of the smaller asteroids and the comets that are ignored in the model. Hence these orbits can be determined accurately from the Viking data. Second, one must know the nonrelativistic effects on the signal propagation. The principal such effect is caused by the plasma in the solar corona. If the echo time-delay measurements can be made simultaneously at two widely separated frequency bands, then the effect of the plasma on the signal propagation can be deduced, as mentioned before. Unfortunately the two Viking landers were equipped with only a single-frequency "transponder" that operated only at a radio frequency of 2.2 GHz. At this frequency, the coronal contribution to the delay reaches at least 100 μsec for ray paths that pass within a few solar radii of the center of the sun. Luckily, the two Viking orbiters were equipped with a transponder that operated at two radio frequencies, 2.2 and 8.4 GHz. But this dual-band capability was available only on the downlink, spacecraft-to-earth path and not on the uplink, earth-to-spacecraft path. Thus, the space-time path over which a reliable calibration of the plasma effects could be obtained was different from the space-time path over which the calibration was needed. Nevertheless this partial ability to calibrate still represented a vast improvement over the conditions for prior spacecraft experiments, where only the single 2.2 GHz frequency was available for both links.

To determine the plasma effects as reliably as possible one had to observe a lander and an orbiter simultaneously. Two different ground installations were required because each could track signals from only one spacecraft at a time. A large number of these simultaneous observations were made surrounding the superior conjunction of Mars that occurred on November 25, 1976. For most of the days of observation, the data, after calibration for the plasma effects, were consistent with each other to within 200 nanoseconds. However, for about 25 percent of the days on which data were obtained, the delay measurements, although internally consistent on each day, differed from those on other days in an inexplicable, and not systematic, manner by up to 15 microseconds. These deviant data have never been explained.

If one treats the deviant data as would a good experimentalist, by carefully sweeping them under the celestial carpet, and analyzes the remainder, the results are in remarkable accord with the predictions of general relativity. Because some useful data were obtained for ray paths that passed within four solar radii of the center of the sun, the characteristic logarithmic signature of $\Delta\tau$ was distinct enough for the estimate of γ to be only weakly correlated with the estimates of the orbital parameters of the earth and Mars.

A preliminary analysis was carried out in which the delay measurements obtained near superior conjunction were compared with the predictions of general relativity, the latter based on orbits, and lander and tracking-station locations, determined from data obtained far from superior conjunction. This analysis[27] yielded results in excellent agreement with general relativity, any deviation from $\gamma \equiv 1$ being unlikely to exceed the estimated uncertainty $\sigma(\gamma) = 0.01$. A more recent analysis[28] yielded

$$\gamma = 1.000 \pm 0.002 , \tag{8.11}$$

with the quoted uncertainty, about twice the formal standard error, reflecting a modest factor-of-2 allowance for the possible contribution of unknown systematic errors.

The Viking experiment has clearly yielded the most accurate verification yet obtained for the predictions of general relativity on the influence of massive bodies on the propagation of light signals.

What are the prospects for further improvements in the measurement of the gravitational retardation of light signals? It is certainly within the grasp of present technology to place landers on the surface of an inner planet. Viking has already proved this assertion for Mars. Landers equipped with transponders operable simultaneously at four widely separated frequency bands, say from about 8 to 35 GHz, would allow improvement of several orders of magnitude to be made in the estimate of γ. Not only would the effects of the mean corona be virtually eliminated, via a suitable two-frequency calibration technique, but, with a third frequency, the scattering contribution, which is proportional to f^{-4}, could also be reduced. (The angular deviations vary with f^{-2}, and, for small angles, the corresponding path-length changes will vary as stated.) A fourth band is suggested for the confidence rendered by redundancy. Transmitters have been developed that are adequate to meet the requirements of this experiment at all the proposed frequencies. Such an experiment might yield an accuracy nearly sufficient to detect second-order (post-post Newtonian) effects. My calculations and those of R. Epstein showed that this second-order contribution to the delay is only some tens of picoseconds. Thus improvement in the accuracy of measurement of echo delays would also have to be attained. Reaching that level of accuracy when the signals pass close to the sun may well be possible; however, obtaining high enough accuracy in the measurements and in the interpretation to distinguish reliably the second-order contribution seems beyond the present state of the art of such radio experiments. In any event such an interplanetary experiment would be enormously expensive. Recall that Viking was a billion dollar project. Only in conjunction with a project driven by many other goals could such a relativity experiment conceivably be carried out.

An alternative technical approach would be to use laser tracking, instead of radio tracking, to virtually eliminate coronal problems. It would, of course, be necessary to place a laser receiver, transmitter, and accurate clock on the planet's surface. From both a technical and an economic viewpoint, the laser method is in a quite distinct class from the radio-tracking method. But there is no fundamental impediment; it is purely a matter of time and money.

5. Precession of Perihelia

Astronomers have attempted to follow the precession, or advance, of the perihelia of planetary orbits for several hundred years. Until the mid-nineteenth century, the results were always in agreement with Newtonian theory to within the uncertainties of the observations and the theoretical calculations. The latter were limited primarily by imprecise knowledge of the masses of the perturbing bodies. Leverrier was the first to note a serious discrepancy between observations and calculations relating to the changes in the elements of the orbit of Mercury.[29] He could not assign this discrepancy uniquely to one element but, on the basis of auxiliary arguments, concluded that the conflict was due to an advance of the perihelion position about 36 seconds of arc per century greater than expected. This discrepancy was definitively assigned to the perihelion advance and its value refined to about 43 seconds of arc per century by Newcomb at the end of the nineteenth century. It was, of course, this excess that was dramatically and naturally explained by Einstein's general theory of relativity. In the PPN formalism, the relativistic contribution $\dot{\omega}_R$ to the secular advance is given by

$$\dot{\omega}_R = \frac{6\pi r_o}{a(1 - e^2)} \left[\frac{2 + 2\gamma - \beta}{3} \right] \frac{\text{rad}}{\text{rev}}, \tag{8.12}$$

where, in addition to the previously introduced quantities, we define a as the semimajor axis and e as the eccentricity of the orbit.

At present, optical observations spread over more than 200 years yield an uncertainty in the determination of the rate of precession of approximately 0.4 seconds of arc per century. Radar observations, on the other hand, which span only 10 years yield an uncertainty half that of the optical — about 0.2 seconds of arc per century. The real problem is the interpretation of these measurements. The solar system is a dirty laboratory, although, admittedly, rather clean by usual astrophysical standards. If we assume that the sun is rotating uniformly, then we can conclude from the radar observations that the combination of parameters $(2 + 2\gamma - \beta)/3$ deviates from unity by no more than 0.005.[30] The question of the rotation of the sun is still somewhat controversial. The experiment of Dicke and Goldenberg[31] purportedly detected an unexpectedly large visual oblateness of the sun, which required a rapidly rotating inner part to explain it, as previously predicted by Dicke and Peebles.[32] But the interpretation of the "measured" oblateness seemed, in fact, to be flawed because the distinction between a true visual oblateness of the sun and a variation of brightness very near the limb could not be made reliably. Hill's later observations of the sun's visual oblateness utilized a different optical technique[33] and failed to confirm the results of Dicke and Goldenberg. However, these observations are very difficult to make at the required level of accuracy and probably should be repeated before definitive conclusions are drawn.

Are there any other suitable objects for which the relativistic contribution to the advance can be determined accurately without any appreciable uncertainty about the contribution of the sun's gravitational oblateness? Results have, in fact, been obtained for several other objects, for example Mars and the asteroid Icarus.[34] But the uncertainties are of the order of 10 to 20 percent and far too large to provide useful results. Satellites of the outer planets provide even poorer prospects. The orbit of Mercury remains the best natural testing ground of the predictions of general relativity regarding perihelion precession. Table 8.1 contains a list of the nonrelativistic contributions to the precession, the magnitudes and uncertainties in our knowledge of these contributions, and possible means for reduction of the uncertainties. Even if we accept uniform solar rotation, as was done in preparation of the table, we find that the largest uncertainty is still from the solar oblateness. Based on a standard model for the sun, the contribution of the oblateness to the precession would be about 0.015 seconds of arc per century. Our estimate of the uncertainty, however, matches the size of the effect. A direct method for reduction of this uncertainty, but an expensive one, involves the placement of a "drag-free" spacecraft in an orbit that passes within several radii of the surface of the sun, as was discussed in Sec. 2. By this means, the uncertainty in our knowledge of the solar oblateness could probably be reduced by an order of magnitude.[35] Surprisingly to many, the next limitation reached is due to the coupling between the spin and orbital motions of Mercury. The spin of Mercury is coupled to its orbital motion in a 3:2 resonance. The lack in our knowledge of the fractional difference between the equatorial principal moments of inertia of Mercury introduces a corresponding uncertainty in the contribution of the spin-orbit coupling to the perihelion advance of Mercury's orbit. The estimated magnitude of the effect is of the order of 0.003 seconds of arc per century (Shapiro, unpublished, 1969), and the uncertainty is at least twice that figure, but could be reduced by radio tracking of a Mercury orbiter. Next on the list is the uncertainty in the mass of Jupiter. The accuracy of measurement of the round-trip propagation times of the radio signals sent from the earth to the Viking landers is, however, sufficient to allow an improved value of the mass of Jupiter to be obtained soon. The last entry on the table, the uncertainty in our knowledge of the mass of Venus, is already so small that for many years it will not pose any hindrance to the interpretation of the observations of Mercury.

There is also another, relativistic contribution to the perihelion advance of Mercury's orbit, probably comparable in magnitude to the nonrelativistic contribution from the spin-orbit coupling, but of opposite sign. This relativistic effect is directly proportional to the angular momentum of the sun[36] and the uncertainty in its magnitude stems mostly from the possibility that the sun's interior may be rotating more rapidly than its surface. Only if 1) the uncertainty in the determination of the total advance were reduced below 0.001 arcseconds per century, 2) the classical contributions to the advance were determined with comparable or lower uncertainty, and 3) the general theory of relativity were valid to this same order, could the angular momentum of the sun be determined from the perihelion advance of Mercury's orbit.

Aside from the question of the sun's angular momentum, have not measurements of the signals from the binary pulsar[37] made obsolete any further concern with the advance of the perihelion of Mercury's orbit? The periastron advance of the orbit of the binary pulsar, after all, is about $4\overset{\circ}{.}2$ per year, about 35,000 times larger than the advance for the orbit of Mercury. Unfortunately measurement of the periastron advance per se serves primarily to determine a single function of the individual masses of the binary system. At present these quantities cannot be determined independently with any useful accuracy. Thus, although this binary system is not now useful as a laboratory to test the predictions of general relativity for the periastron advance (see,

Table 8.1

Major uncertainties in the nonrelativistic contributions to the perihelion advance of Mercury's orbit

Source	Magnitude (arc sec/century)	Uncertainty (arc sec/century)	Cure
Solar oblateness	~0.015	+0.015 −0.008	Solar Probe
Spin-orbit coupling	~0.003	+0.006 −0.003	Mercury Orbiter
Mass of Jupiter	~132	0.002	Mars Landers
Mass of Venus	~240	0.0005	Not needed

however, Sec. 8), it has for the first time allowed the determination of a parameter of an astrophysical system from a general relativistic effect.

6. Time Variation of the Gravitational "Constant"

Dirac[38] formulated a "large numbers hypothesis," whose roots can be found in Milne's and Eddington's works. This hypothesis is based on the fact that the ratio of the electrostatic to the gravitational force between an electron and a proton, about 10^{39}, is approximately equal to the age of the universe expressed in atomic units. Dirac assumed that this (near) equality was of fundamental importance and time-independent. Since the age of the universe is clearly time-dependent, so must be some other quantity. Dirac proposed therefore that the gravitational constant would vary with (atomic) time. For discussion, consider the ad hoc relation:

$$G(t) = G_o + \dot{G}_o (t - t_o) , \tag{8.13}$$

where G_o is the gravitational "constant" at some recent epoch t_o, and $\dot{G}_o$ is the corresponding time rate of change. To determine whether or not $\dot{G}_o$ vanishes, we can measure the time kept by gravitational clocks on an atomic time scale. Since the fractional rate of change of G is certainly small compared to orbital angular velocities of solar-system objects, one can easily show that

$$2 \frac{\dot{G}_o}{G_o} \simeq -2 \frac{\dot{a}}{a} \simeq - \frac{\dot{P}}{P} \simeq \frac{\dot{n}}{n}, \tag{8.14}$$

where a, P, and n are, respectively, the semimajor axis, the orbital period, and the mean motion of an object orbiting under the control of the gravitational interaction.

Any change in the orbital mean motion of a body is manifested most noticeably in its longitude $L(t)$. Any linear change in n with (atomic) time would cause a quadratic change ΔL in the deviation of L from its nominal ($\dot{G}_o = 0$) value: $\Delta L = (2\pi/P)(\dot{G}_o/G_o)(t - t_o)^2$. To determine $\dot{G}_o/G_o$ from its contribution to ΔL, two approaches have been used: one is based on radar observations of the inner planets and the other on optical observations of the moon. Both of these techniques are somewhat intricate[39] and both share a serious difficulty, the masking of the quadratic dependence of ΔL on t by other terms. Consider, for example, the unavoidable linear term $a + b(t - t_o)$, whose coefficients a and b depend on the initial conditions that describe the orbit of the body. Because we lack sufficiently accurate independent knowledge of these conditions, we must estimate them simultaneously with $\dot{G}_o/G_o$. Thus, the presence of a linear term alone reduces the sensitivity of the data to the coefficient of any quadratic term by a factor of about 6, a measure of the accuracy with which a straight line can match a parabola. A further difficulty that affects importantly only the lunar technique is due to dissipative tides, which also produce a quadratic change in longitude. This effect is, however, not necessarily fatal in practice, because estimates of the tidal contribution to ΔL can be obtained by other means, for example from monitoring of the orbits of suitable artificial earth satellites. The results obtained so far from the radar observations of the inner planets are dominated by systematic errors, primarily associated with the topography of the planetary surfaces. In fact, although the formal (statistical) standard error, $\sigma(\dot{G}_o/G_o)$, in the estimate of $\dot{G}_o/G_o$ is of the order of a few parts in 10^{12} per year, a reasonable estimate of the contribution of systematic errors allows us to conclude only that:[40]

$$-0.5 \times 10^{-10}\ \text{year}^{-1} < \dot{G}_o/G_o < 1.5 \times 10^{-10}\ \text{year}^{-1} . \tag{8.15}$$

Van Flandern analyzed lunar observations and concluded that there has been a significant decrease with time in the gravitational interaction.[41] His value was $\dot{G}_o/G_o = -(3.6 \pm 1.8) \times 10^{-11}\ \text{year}^{-1}$, although during the past few years this number, and its uncertainty, have fluctuated by amounts somewhat larger than their values because of changes in his analysis. In view of the intrinsic weakness of the statistical evidence for a change in G as well as the seriousness of systematic errors affecting the analysis,[42] one cannot conclude that any variation of gravitational time with respect to atomic time has been uncovered.

Continued observations and analysis using these techniques alone are bound to lead to an improved sensitivity for the detection of any change in the gravitational constant. For radar observations of the planets or for radio-tracking of planetary orbiters or landers, this sensitivity will increase as $t^{5/2}$, where t is the interval over which the observations extend. Over the next few years the most accurate bound, perhaps smaller than 1 part in 10^{11} per year, may result from analysis of the echo-delay measurements obtained with the Mariner 9 and Viking spacecraft.

A limit on any bound that can be placed on $\dot{G}_o/G_o$ from planetary observations is posed by the uncertainty in our knowledge of the mass loss of the sun, currently about 1 part in 10^{14} per year; this limit will clearly not become a serious concern for quite some time. For lunar observations, one is presently limited by lack of accurate knowledge of the tidal contributions, and the time scale for improvement is not at all clear. Indeed, the dissipation accompanying tides may even be a function of time as well as of the frequency of the driving forces.

7. The Lense–Thirring Effect

In 1918 Lense and Thirring pointed out that the plane in which a small body orbits a large rotating mass will be "dragged" in the direction of the rotation.[43] In particular, the rate of change $\dot{\Omega}_{LT}$ of the longitude of the ascending node of the orbit is predicted from the PPN formalism to be

$$\dot{\Omega}_{LT} = \frac{2GI\omega}{c^2a^3(1 - e^2)^{3/2}} \left(\frac{1 + \gamma}{2}\right), \tag{8.16}$$

where, in addition to the quantities previously defined, I and ω denote, respectively, the moment of inertia and angular velocity of the central body. No experimental support for this prediction exists. However, techniques are available that could be used to test it.

In 1970, Miller and Shapiro studied several possibilities for the detection of the Lense–Thirring effect, including the radio tracking from the earth of two drag-free spacecraft placed in counterorbiting polar orbits about the sun.[44] These possibilities are, unfortunately, quite impractical to pursue for economic and other reasons. A much more practical scheme, independently conceived, was proposed by Van Patten and Everitt, who suggested the placement of such spacecraft in orbit about the earth.[45] They considered, further, a radio tracking link established between the two spacecraft so as to achieve high accuracy during the close passages, which would be arranged to occur over the poles.

For a satellite orbiting 800 kilometers above the earth's surface, $\dot{\Omega}_{LT}$ is predicted by the general theory of relativity to be about 0.2 seconds of arc per year. To discern $\dot{\Omega}_{LT}$ requires extreme care, especially in view of the far larger contribution to the nodal precession of the second zonal harmonic of the earth's gravitational field. However, by keeping the orbital inclinations very nearly polar, and determining their values accurately, one can estimate this contribution with sufficient accuracy. (Recall that the effect of the second zonal harmonic on nodal precession vanishes for a strictly polar orbit, and by keeping the effect small one lessens the sensitivity of the result to the uncertainty in our knowledge of the magnitude of this harmonic.) Radio tracking via earth-satellite links allows the total nodal precession rate to be determined; the difference then allows the value of $\dot{\Omega}_{LT}$ to be deduced. Of course, there are many other factors to consider, including other harmonics of the gravity field, tidal accelerations, geodetic precession, tracking-system limitations and so on. Van Patten and Everitt have considered all of these factors and have concluded that this approach could be used to detect the dragging of inertial frames predicted by the general theory of relativity.[46]

Another, related experiment was proposed earlier by Schiff and, independently, by Pugh.[47] Each suggested the placement of a gyroscope in earth orbit to measure the effect of the earth's rotation on the direction of spin of the gyroscope. Schiff's calculations show that for a gyroscope in an elliptical orbit about a massive spherically symmetric central body, the average rate of change $\dot{\mathbf{n}}_s$ of the gyroscope's spin axis will be

$$\dot{\mathbf{n}}_s = \frac{GI\omega}{2c^2a^3(1 - e^2)^{3/2}}[(\mathbf{n}_c \times \mathbf{n}_s) + 3(\mathbf{n}_0 \cdot \mathbf{n}_c)(\mathbf{n}_o \times \mathbf{n}_s)] , \tag{8.17}$$

where the new quantities are $\mathbf{n}_s$, $\mathbf{n}_c$, and $\mathbf{n}_o$, the unit vectors in the directions of the spin vector of the gyroscope, the spin vector of the central body, and the normal to the orbital plane of the gyroscope, respectively. There is also an average contribution, due to the geodetic precession, given by

$$\dot{\mathbf{n}}'_s = \frac{3(GM)^{3/2}}{2c^2a^{5/2}(1 - e^2)^{3/2}}\mathbf{n}_o \times \mathbf{n}_s , \tag{8.18}$$

where M is the mass of the central body.

For an artificial satellite in a polar orbit about the earth at an altitude of 800 kilometers, Eqs. (8.17) and (8.18) predict that the magnitudes of $\dot{\mathbf{n}}_s$ and $\dot{\mathbf{n}}'_s$ will be 0.05 and nearly 7 arcseconds per year, respectively, with $\mathbf{n}_s$ precessing about the earth's spin vector and $\mathbf{n}'_s$ about the normal to the orbital plane. In the PPN formalism for "fully conservative" theories without preferred-frame effects, these predictions are modified by the factors $(1 + \gamma)/2$ and $(1 + 2\gamma)/3$, respectively.[48]

The experimental detection of $\dot{\mathbf{n}}_s$, and even of $\dot{\mathbf{n}}'_s$, presents formidable technical problems. Required are gyroscopes with unaccountable drift rates well below 0.05 seconds of arc per year, and a means to track a suitable reference object with a comparably low uncertainty. The Stanford researchers, led by Fairbank and Everitt, have been actively developing the necessary apparatus for about fifteen years. They have made extremely clever use of cyrogenic physics in this development and have devised sophisticated solutions for many of the technical problems. Their design of the gyroscope consists of a very precisely spherical, homogeneous fused quartz ball, 4 centimeters in diameter, coated with a thin film of niobium, a superconductor. The ball is to be placed in a spherical quartz container and suspended electrically by suitable voltages applied across three mutually perpendicular axes. The ball will be initially spun up by gas jets, kept in a region of extremely low magnetic fields, under about 2×10^{-7} gauss, and have its spin axis direction monitored through its London moment by a three-axis magnetic readout system, based on the use of superconducting quantum interference devices (SQUIDs). Such a method is required because of the highly spherical and homogeneous properties of the ball, which, in turn, are needed to reduce the level of unwanted gravitational torques to benign levels.

Awesome progress has been made in the development of complete laboratory models of both the gyroscope and the novel telescope system, the latter to track reference stars. Nonetheless, it is difficult to predict reliably when this experiment might be performed in space.

8. Gravitational Radiation

General relativity predicts the existence of gravitational radiation. Its direct detection is another matter. There are two main problems: 1) the difficulty in developing instrumentation of sufficient sensitivity to detect quadrupole radiation, the lowest-order possible in general relativity, and 2) the lack of secure knowledge of the characteristics of gravitational radiation from astrophysical sources.

Basically two types of gravitational-radiation detectors are under development: acoustically coupled and electromagnetically coupled antennas. An example of the former is the resonant bar, pioneered by Weber; examples of the latter are optical interferometers and the radio-tracking of spacecraft.[49]

The amplitudes of gravitational waves are conventionally described in terms of h, the dimensionless perturbation of the background metric, usually twice the fractional strain induced in the detector. This quantity h is not to be confused with Planck's constant; it is hoped in fact that the former is numerically many orders of magnitude larger when the latter is expressed in cgs units.

Sources of gravitational radiation are conventionally divided into three categories: burst, periodic, and stochastic. Bursts of gravitational radiation may be produced in discrete, violent, and short-lived astrophysical events such as in the direct collapse of very massive objects that may form black holes. Periodic gravitational radiation may be produced by such phenomena as oscillating stars and binary star systems. Stochastic gravitational radiation is, by definition, composed of the aggregate of all such radiation from distant sources in the universe, which produce an "incoherent" background.

The uncertainties accompanying the theoretical predictions for burst sources are vast. Not only are the rates of occurrence and the maximum amplitudes of the gravitational radiation from such events uncertain by orders of magnitude,[50] but the shape of any pulse is also mostly a matter of speculation. Unless the most optimistic predictions of magnitudes and rates of occurrence are correct, the probability of the reliable detection of such radiation will likely remain very small for a decade or more. The characteristics of the radiation from periodic sources, on the other hand, can be predicted rather more reliably from the known properties, for example, of nearby binary star systems. But, due to the perversity of nature, the gravitational radiation from these sources is expected to be quite weak: $h \sim 10^{-20}$ or less. The sensitivity of detectors would have to improve over present versions by six or more orders of magnitude in order for them to be able to detect such radiation. The predictions for the stochastic background, like those for the burst sources, are fraught with uncertainty. Models "likely" to be valid[51] yield values for power spectral densities well below a reasonable extrapolation of the capabilities of current detection equipment. It appears that the capability may exist within the next decade to detect this stochastic background only if it contains sufficient energy to close the universe, and if most of this energy is contained in the low end of the power spectrum, below, say, about 1 Hz. (These somewhat vague statements about gravitational radiation detection are necessitated by the myriad factors, both known and unknown, on which more precise statements would depend.)

Although the prospects for the *direct* detection of gravitational radiation in the near future may be somewhat bleak, the prospects for convincing *indirect* detection are quite bright, due to the pulsar discovered in a "tight" binary system.[52] Gravitational radiation from this system should cause a detectable decrease in its ~8-hour orbital period. A sensitive measure of such a decrease is provided by the orbital position of the pulsar. The pulsar should "gain" in its orbit quadratically with time compared to predictions based on the assumption of a constant orbital period.

Taylor, Fowler, and McCulloch[53] have estimated the orbital properties of the pulsar with high accuracy. Their basic measurements are the times of arrival t_a of the pulses from the pulsar. The interpulse period corresponds to the remarkably constant value, ~60 milliseconds, of the spin period of the pulsar; its estimated spin-down time is about 200 million years. By use of template-matching techniques, Taylor and his colleagues have been able to estimate t_a's with uncertainties as low as 50 microseconds, despite the full width of the main "spike" in the shape of the pulse being about 2 milliseconds at half power. These measurements of t_a were used, after correction for various effects, to estimate, by weighted least squares, twenty parameters of a theoretical model, consistent with general relativity, that describes the physical properties of the pulsar system, including the rate of change, $\dot{P}$, of its orbital period, and certain characteristics of the pulses.

The formal solution obtained by Taylor et al. for $\dot{P}$ was $(-3.2 \pm 0.6) \times 10^{-12}$, whereas the predicted value, based on contributions from gravitational radiation alone, was $(-2.40 \pm 0.05) \times 10^{-12}$. This theoretical value was based on the usual quadrupole formula[54] and incorporates the necessary information on the orbit and on the masses of the two bodies, as derived from the same analysis of the times of arrival of the pulses.

A more pictorial representation of the significance of the estimate of $\dot{P}$ was obtained by Taylor and his colleagues as follows. They separated all of their useful measurements of t_a into seven sets; each set contained observations obtained near a different epoch. The epochs were spread between late 1974 and late 1978 with a gap of about two years between the first and second epochs. They analyzed each of the seven sets separately and, for each, estimated the time, T_p, of passage of the pulsar through periastron for the orbit encompassing the chosen epoch. In Fig. 8.1, we show the differences between these estimated values of T_p and the predicted values, the latter based on an assumed constant orbital period, P_0, obtained from the first set of data. The curvature determined by the change with time of these seven differences is a direct measure of $\dot{P}$. The solid curve in the figure represents the best-fit parabola, but with the curvature fixed by setting $\dot{P} = -2.4 \times 10^{-12}$, the value predicted by general relativity. The agreement between observation and theory is manifestly excellent. However, for the last six points, which are confined to a two-year interval, the distinction between a "best-fit" parabola and a "best-fit" straight line is of marginal significance. The accumulated effect of curvature over two years, without consideration of masking (see Sec. 6), amounts to no more than about 200 microseconds, only severalfold larger than the smallest uncertainties in the individual estimates of t_a. The determination of significant curvature thus rests very heavily on the first point.

To interpret properly any result for $\dot{P}$, one must also investigate the "cleanliness" of this binary system. The measurements themselves and a variety of astrophysical arguments indicate that the companion to the pulsar is very likely to be a neutron star. If so, tidal and other Newtonian contributions to $\dot{P}$ should be insignificant. Furthermore, the plasma density along the line of sight within the system has been shown by simultaneous multifrequency measurements of t_a's to be lower than 10^6 centimeter^{-3}, which indicates that drag and mass loss contribute negligibly to $\dot{P}$. All in all, it appears that the entire value of $\dot{P}$ may well be attributable to gravitational radiation.

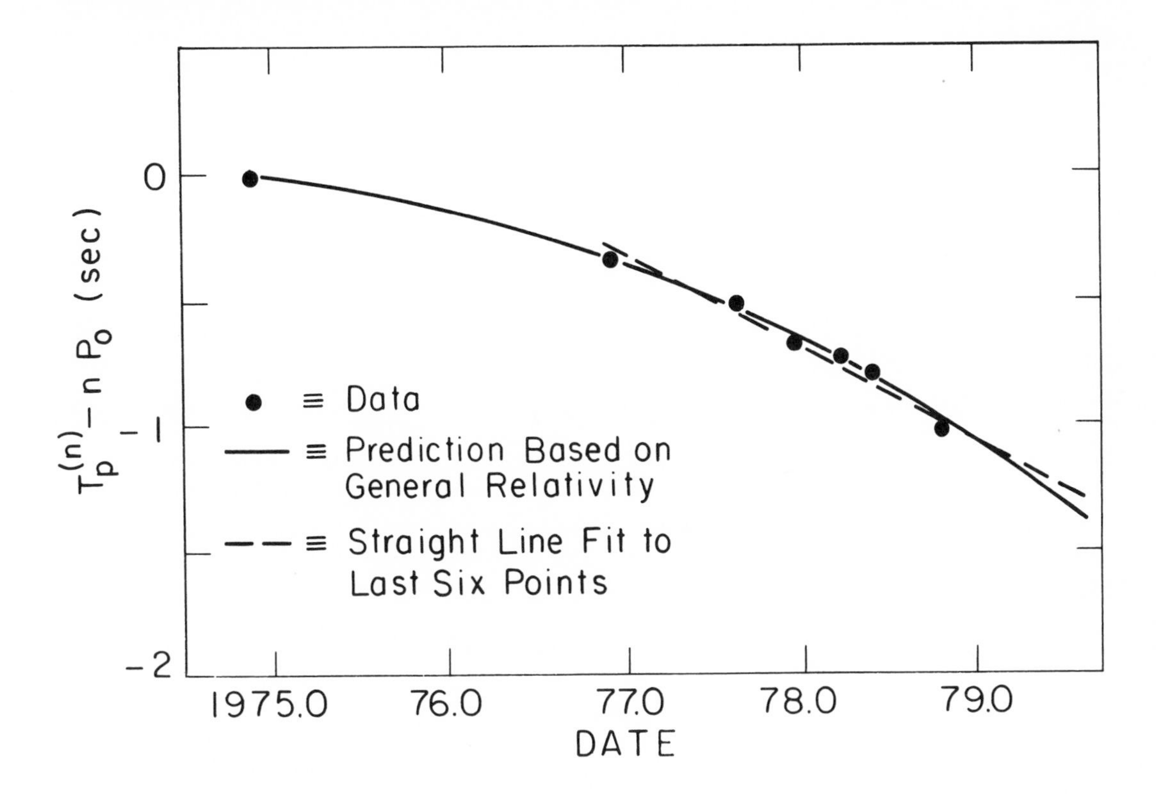

Fig. 8.1 Illustration of the observed decrease in orbital period of the pulsar in a binary system (PSR 1913 + 16). The ordinate represents the calculated difference between the time of periastron passage, $T_p^{(n)}$, and the time predicted, based on a constant orbital period, P_0. The symbol n denotes the number of orbits traversed by the pulsar since the late 1974 epoch to which P_0 corresponds. Each datum shown represents the analysis of a large set of pulse-arrival-time measurements accumulated in the immediate vicinity of the epoch of the datum. The curvature of the solid line represents the predicted effect of gravitational radiation on the orbital period and, hence, on the time of periastron passage; the intercept and mean slope of the solid line are based on the data and have no theoretical significance. The dashed straight line, fit to the last six points only, illustrates that the significance of the curvature defined by the data depends heavily on the earliest datum. (After Taylor *et al.*[55])

Should we now conclude that the existence of gravitational radiation has been established? Probably not. But the significance of the curvature of $T_p^{(n)} - nP_0$ versus n (see Fig. 8.1) will increase approximately as $t^{5/2}$, so that after a few more years of observation, Taylor et al. will be able to obtain a much more accurate result for $\dot{P}$. It seems unlikely that geodetic precession,[56] or any change in the pulse mechanism, will soon stop the beaming of steady pulses towards the earth and thus prevent the accumulation of the needed data. In a few more years the relevant astrophysical questions may also be settled more definitively. As a result, the present very tantalizing evidence for the existence of gravitational radiation may grow in a few years to a rather reliable quantitative demonstration.

9. Conclusions

What conclusions can be drawn from the ensemble of solar-system experiments? Combining the various estimates of the parameters in the PPN formalism yields the following values for γ and β for fully conservative theories in the absence of preferred frame and location effects:

$$\begin{aligned} \gamma &\cong 1.000 \pm 0.002 \\ \beta &\cong 1.000 \pm 0.004 \end{aligned} \quad , \tag{8.19}$$

with the correlation in these estimates being under 0.5. These results provide a remarkable verification of the predictions of the general theory of relativity. One might well wonder which alternative theories of gravitation have thereby been proven invalid. Unfortunately, none. All other widely discussed and analyzed theories of gravitation, such as the theory of Brans and Dicke and that of Rosen, have a special property: their predictions either agree or can be made to agree with those of general relativity in the post-Newtonian regime.[57] In many of these theories the agreement is secured by the selection of an appropriate value of an adjustable parameter, such as γ (no such parameters exist in general relativity). For example, the scalar-tensor theory of Brans and Dicke contains only one additional parameter, a dimensionless constant $\omega(>0)$ related to γ by

$$\gamma = (\omega + 1) / (\omega + 2) . \tag{8.20}$$

A limit of 0.2 percent on the deviation of γ from unity restricts ω to being greater than about 500, since $\omega \geqslant (1 - \gamma)^{-1}$ when $0 < 1 - \gamma \ll 1$. But no matter how accurate the experiment, it could not prove such a theory invalid, provided that general relativity is a correct description of gravitation to that level. Only if general relativity is proven wrong, or if experiments can be made sensitive to "post-post-Newtonian" effects, will there be a substantial winnowing of competing theories.

What are the prospects in the next decade for further progress? It is conceivable that post-post-Newtonian effects might be detected, but such experiments, as well as lesser ones, would be very expensive. Perhaps the main experimental advance of the next decade will be provided by a "clean" laboratory discovered far from the solar system. Detection and study of the companion to the binary pulsar could conceivably establish this system as the sought-for laboratory for testing the general theory of relativity. Even without such detection, accurate measurement of $\dot{P}$ may allow the elimination of theories of gravitation that predict violations of the principal of equivalence for massive bodies. These theories predict dipole gravitational radiation whose effects on $\dot{P}$, if present, should be distinguishable[58] provided the mass and structure of the pulsar are not virtually identical with those of its companion. Although Taylor et al. found such a near equality in mass, analysis of the data in a manner consistent with a theory that violates the principle of equivalence would likely lead to a substantial disparity in the estimates of the two masses and, hence, to a prediction of discernible dipole gravitational radiation. Of course, the *direct* detection of gravitational waves not only would provide another tool for testing theories of gravitation, mainly through studies of the polarization properties of the waves, but would also open a new window for further exploration of the universe.

For the long term, we can confidently expect the appearance of experimentally verified deviations from the predictions of the general theory of relativity. The ultimate synthesis of gravitational and quantum phenomena will surely point to experimental consequences not envisioned or predicted by Einstein's theory. Einstein himself would doubtless have welcomed such progress.

NOTES

1. M. R. Cohen and I. E. Drabkin, *A Source Book in Greek Science* (McGraw-Hill, New York, 1948), p. 217.
2. V. B. Braginsky and V. I Panov, Zh. Eksp. Teor. Fiz. **61**, 873 (1971).
3. L. I. Schiff, Proc. Nat. Acad. Sci. **45**, 69 (1959); M. Haugan and C. Will, Phys. Rev. Lett. **37**, 1 (1976).
4. C. M. Will, in *Experimental Gravitation*, edited by B. Bertotti (Academic Press, New York, 1973), p. 1.
5. Ibid.
6. K. Nordtvedt, Phys. Rev. **169**, 1017 (1968).
7. I. I. Shapiro et al., Phys. Rev. Lett. **36**, 555 (1976); J. G. Williams et al., Phys. Rev. Lett. **36**, 551 (1976).
8. R. W. King et al., J. Geophys. Res. **83**, 3377 (1978).
9. Shapiro et al, op. cit. in n. 7.
10. R. V. Pound and J. L. Snider, Phys. Rev. B **140**, 788 (1965).
11. J. C. Hafele and R. E. Keating, Science **177**, 166, 168 (1972).
12. R. F. C. Vessot and M. W. Levine, in *Gravitazione Sperimentale* (Accademia Nazionale dei Lincei, Rome, 1977), p. 371.
13. C. W. F. Everitt and D. DeBra, in *A Close-up of the Sun*, edited by R. W. Davies and M. Neugebauer, Jet Propulsion Laboratory Publication 78-70, 1978, p. 60.
14. C. M. Will, in *General Relativity: An Einstein Centenary Survey*, edited by S. W. Hawking and W. Israel (Cambridge University Press, Cambridge, 1979), p. 24.
15. Texas Mauritanian Eclipse Team, Astron. J. **81**, 452 (1976).
16. I. I. Shapiro, Science **157**, 806 (1967).
17. C. C. Counselman et al., Phys. Rev. Lett. **33**, 1621 (1974).
18. E. B. Fomalont and R. A. Sramek, Phys. Rev. Lett. **36**, 1475 (1976).
19. A. E. E. Rogers et al., J. Geophys. Res. **83**, 1 (1978).
20. Shapiro, op. cit. in n. 16.
21. H. A. Hill and R. T. Stebbins, Astrophys. J. **200**, 471 (1975).
22. R. D. Reasenberg and I. I. Shapiro, in *Proceedings of the 5th International Symposium on Space Relativity*, Acta Astronautica, in press.
23. R. Epstein and I. I. Shapiro, submitted to Phys. Rev. Lett.
24. I. I. Shapiro, Phys. Rev. Lett. **13**, 789 (1964).
25. I. I. Shapiro et al., Phys. Rev. Lett. **20**, 1265 (1968).
26. J. D. Anderson et al., Astrophys. J. **200**, 221 (1975).
27. I. I. Shapiro et al., J. Geophys. Res. **82**, 4329 (1977).
28. R. D. Reasenberg et al., Astrophys. J. (Lett.) **234**, L219 (1979).
29. U. J. J. Leverrier, Ann. Obs. Paris **5**, 1 (1859).
30. Shapiro et al., op. cit. in n. 7.
31. R. H. Dicke and H. M. Goldenberg, Astrophys. J. Suppl. **27**, 131 (1974).
32. R. H. Dicke and P. J. E. Peebles, Nature **202**, 432 (1964).
33. Hill and Stebbins, op. cit. in n. 21.
34. I. I. Shapiro et al., Astron. J. **76**, 588 (1971).

35. J. D. Anderson et al., in *Gravitazione Sperimentale* (Accademia Nazionale dei Lincei, Rome, 1977), p. 393; R. D. Reasenberg and I. I. Shapiro, in *A Close-up of the Sun*, edited by R. W. Davies and M. Neugebauer, Jet Propulsion Laboratory Publication 78-70, 1978, p. 19.

36. S. Weinberg, *Gravitation and Cosmology: Principles and Applications of the General Theory of Relativity* (Wiley, New York, 1972).

37. R. A. Hulse and J. H. Taylor, Astrophys. J. (Lett.) **191**, L59 (1974).

38. P. A. M. Dirac, Proc. Roy. Soc. Lond. **A165**, 199 (1938).

39. R. D. Reasenberg and I. I. Shapiro, in *Gravitazione Sperimentale* (Accademia Nazionale dei Lincei, Rome, 1977), p. 143.

40. R. D. Reasenberg and I. I. Shapiro, in *Atomic Masses and Fundamental Constants* 5, edited by J. H. Saunders and A. H. Wapstra (Plenum Press, New York, 1976), p. 643.

41. T. C. Van Flandern, Scientific American **234**, 44 (1976).

42. See Reasenberg and Shapiro, op. cit. in n. 39, for discussion.

43. J. Lense and H. Thirring, Phys. Z. **19**, 156 (1918).

44. L. Miller, S.B. thesis, Massachusetts Institute of Technology (1971).

45. R. A. Van Patten and C. W. F. Everitt, Celestial Mechanics **13**, 429 (1976).

46. Ibid.

47. G. E. Pugh, Research Memorandum No. 11, Weapon System Evaluation Group, Department of Defense (1959); L. I. Schiff, Phys. Rev. Lett. **4**, 215 (1960).

48. Will, op. cit. in n. 4.

49. See, for example, R. Weiss, in *Sources of Gravitational Radiation*, edited by L. Smarr (Cambridge University Press, Cambridge, England, 1979).

50. K. S. Thorne and V. B. Braginsky, Astrophys. J. (Lett.) **204**, L1 (1976).

51. L. A. Rosi and R. L. Zimmerman, Astrophys. Space Sci. **45**, 447 (1976).

52. Hulse and Taylor, op. cit. in n. 37.

53. J. H. Taylor, L. A. Fowler, and P. M. McCulloch, Nature **277**, 437 (1979).

54. Ibid.

55. Ibid.

56. Ibid.

57. C. Brans and R. H. Dicke, Phys. Rev. **124**, 925 (1961); N. Rosen, Ann. Phys. **84**, 455 (1974).

58. C. M. Will and D. M. Eardley, Astrophys. J. (Lett). **212**, L91 (1977).

9. COMMENTS ON "EXPERIMENTAL RELATIVITY"

David T. Wilkinson

Irwin Shapiro's excellent paper is so comprehensive and authoritative that it would be presumptuous of me to try to comment on it directly. Instead I have chosen to make a few remarks about the history, character, and shortcomings of experimental relativity. In his letter of invitation Harry Woolf suggested that "the symposium be an occasion where scientists can reflect on what science is, and how at its best it is carried out." My comments respond more to this charge than to Irwin's paper.

1. Slow Progress

What a marked contrast one finds between the developments of quantum physics and relativity physics during the first three-quarters of this century. As opposed to the rapid growth of quantum mechanics through the exemplary give-and-take interaction of theory and experiment, one sees in relativity Einstein's glittering general theory and two early experimental successes, followed by decades of painfully slow experimental progress. It has taken sixty years finally to achieve empirical tests of general relativity at the 1 percent level. Why should this be? Such an elegant theory, so rich in predictions, might be the answer to an experimentalist's dreams; we are told exactly what to look for. The answer, of course, lies in the extreme weakness of the gravitational force compared to the electrical force, making the predicted effects always small and buried in noise and competitive systematic effects.

Progress beyond the early experimental tests, using Mercury's perihelion precession and light deflected by the sun, required development of technology and experimental techniques well beyond those available in the early 1920s. The magnitude of these developments is best illustrated by a partial list of the main tools of today's experimental relativist.

Microwave technology
Large radio dishes
4 to 5-meter-class optical telescopes
High-speed computers
Cryogenic technology

Harry Woolf (ed.), Some Strangeness in the Proportion: A Centennial Symposium to Celebrate the Achievements of Albert Einstein
ISBN 0-201-09924-1

Ultralow-noise electronics
Rockets, satellites, and planetary probes

Clearly, these are not developments made in university laboratories by graduate students. Millions of dollars and expert engineering were required, and one notices that the motivations for most of these developments were economic or political, rather than scientific. So, often the experimental relativist must be an opportunist, keeping a watchful eye on technology and, these days especially, the space program. Of course, not all relativity experiments involve huge, expensive facilities. The Eötvös-class null experiments, the gravitational red shift tests and the search for gravity waves are examples of important laboratory experiments. But all relativity experiments seem to have one common feature: they are hard.

2. Character of Relativity Experiments

To better appreciate the relativity experimentalist's predicament, one needs to look more closely at his main problem: weak signals buried in noise. As my experimenting colleagues know all too well, weak signals inevitably force the experimentalist to confront two often separable problems: statistical errors and systematic errors.

Statistical Errors

Noise superimposed on the signal one seeks is an unavoidable consequence of the quantum mechanical world we live in. The random, unpredictable nature of noise ultimately limits the accuracy achievable in an experiment. Modern signal-processing techniques often allow the experimenter to improve the signal-to-noise ratio (S/N) of his measurement by averaging for longer and longer periods of time, but the returns are slow; S/N improves only as the square root of the averaging time. However, there are limits, other than the experimenter's patience, to how long one can time-average. Some of the phenomena used in testing relativity (eclipses, occultations, etc.) are transient in nature, but usually even more restrictive are the limiting systematic errors.

Systematic Errors

These are not really errors, but false signals generated by the apparatus or by competing phenomena. They hide in the noise, look like the signal (unless discriminated), and keep experimentalists awake at night. The good experimenter anticipates systematic effects and minimizes them in designing the apparatus or isolates them in the measuring procedure. But the checks for systematic errors take time, and one usually must compromise between averaging time (to reduce statistical errors) and measuring or searching for small false signals (systematic errors).

Because systematic errors hide in the noise, one often cannot evaluate them or discover them (if they were not anticipated) until the measurement is over and the data are being reduced. Only then comes the painful realization that the checks for systematic errors were inadequate, or that something unexpected may have happened that cannot be separated from the relativity effect. This is not a serious problem if the apparatus or observing procedure can simply be modified and the measurements taken again — common practice in most physics laboratories, especially with "table-top" experiments. Unfortunately, since relativity tests often deal with transient phenomena or require scheduled instruments or weeks of integration time, reobservation is sometimes very difficult. Then too, some relativity experiments are just so hard that no one has had the mettle to try to improve and repeat them. Consequently, many experiments that have resulted in interesting tests of relativity have a shortcoming, an unchecked systematic effect or some worrisome unexplained glitch that could not be reproduced and understood. At best such effects are fully explained and qualify or limit the accuracy of the result; at worst they are underemphasized and not included in the error estimates.

A new, and potentially serious, problem has emerged from the relativist's natural interest in space. As we rely more and more on instruments in space, limited accessibility of the apparatus becomes a major problem in planning precision experiments. All systematic errors must be anticipated and appropriate checks designed well in advance of any data taking. Many experienced practitioners have serious misgivings about how far one can go with precision measurements without being able to "kick the apparatus." Unfortunately, some of the most interesting relativity tests may have to be done this way.

Data Analysis

Another characteristic, not unique to relativity experiments, but common to most, is the need to process large quantities of data. For example, one may be required to integrate the raw data over long times to reduce noise effects or to compute and remove large known effects (such as nonrelativistic orbit terms). Often it is necessary to choose and fit a model to the data to extract the small relativistic parameters from other effects present in the signal. The analysis techniques are usually too complex to describe in detail in the literature, and there is no common agreement about the best way to treat every situation. Consequently, the experimenter has considerable latitude, and the reader is often left wondering about how sensitive the result is to the choice of a reduction procedure.

Let us look at a few ways that choices in data processing can affect the final result of an experiment. The most obvious one is bad luck. When noisy data are averaged, statistical errors can be large; necessary choices like how to bin the data, or what procedure to use in filtering the data, can change the final result. Fortunately, there are standard ways of dealing with these problems and other scientists can estimate the possible effects on the results, if enough information is given in the paper. However, this is often not the case; even the basic statistics of the noise component of the data are sometimes not reported in experimental papers.

It is sometimes necessary to deal with data that are less than ideal, especially when it is difficult or impossible to repeat the measurements. There may be known causes of intermittent noise or glitches in the data, so editing is required. Often the causes are unknown (for example, see Irwin Shapiro's discussion of a large excess delay observed in 25 percent of the Viking data). Clearly, how one decides which data to analyze can have a significant effect on the result. Unfortunately, there are no standards for how to deal with this problem. Ideally, one reports the results of all editing procedures used, but experimenters should at least describe the procedure adopted and indicate how much of the data (including "preliminary data") were not used in the final analysis. Even this is not currently the practice in journal literature.

Here I should mention the too apparent tendency for experimentalists to apply poor or incorrect statistical procedures to their data. Straightforward procedures such as least-squares model fitting, statistical hypothesis tests, and the computation of error correlations should be part of every experimentalist's basic training. The literature indicates that they are not. When reading the experimental literature of a century ago one is frequently awed by the sophistication of the equipment and the ingenuity of the old-timer's techniques, but appalled by their treatment of data and errors. I am afraid that the same remark will be appropriate a hundred years from now.

Ultimately, the experimentalist, plagued by noise systematic errors and too many choices in data analysis, must still come up with a result. (After all that's what the theorists really care about, not all of these details, caveats, and excuses.) How can the aforementioned effects propagate to the results? What mistakes must the experimentalist avoid, above all?

3. **Pitfalls of Experimental Relativity**

Undiscovered Systematic Errors

Well-designed experiments minimize systematic errors by eliminating or measuring competing effects. But such measures are never perfect, and uncertainty in the residual systematic effects often limits experimental accuracy. This is, however, not the main problem because here the experimenter has recognized the systematic effects and included their uncertainty in the reported error limits. Much more serious are any underestimated or undiscovered systematic effects. Underestimation happens when false signals have been attributed to the relativistic effect and the final error limits do not include this bias; thus the result is wrong and misleading. Theorists will waste a lot of time worrying about it (or not enough time if the result happens to agree with theory), and it may take years to find and correct the experimental problem. From the perspective of progress in the field, this is a very costly error, but, unfortunately, it is also one of the easiest to make. Good experimentalists learn early to put most of their thought and effort into problems connected with systematic effects. Unfortunately, there are no standard ways of approaching systematic errors, as there are for statistical errors. Each experiment has its own set of systematic effects and it is the experimentalist's job to think of and evaluate them all. If there is an art in experimental physics, this is it.

Objectivity in Data Analysis

Many of the experiments discussed by Irwin Shapiro would not have been possible without the help of modern, high-speed computers. There is too much data to reduce and analyze in any other way. But the ability to manipulate large volumes of data often presents the experimenter with some freedom in choosing data analysis procedures, and these choices usually affect the final results. It is becoming more difficult for the experimenter to remain completely objective in processing data. For example, one particularly seductive procedure is to analyze the data several ways and *then* build a case to justify and report only one of the results. This problem is particularly acute in relativity, where the experimenter knows the "right" answer and everyone knows that he knows.

Don't Stop When You Get the "Right" Answer

Another problem with knowing the "right" answer is that one relaxes, or stops, when he gets it. With systematic effects to think about, and choices and checks to make in data analysis, the decision about when to stop and publish is sometimes more subjective than objective. Ideally, the current value of the result (compared to expectations) does not enter into this decision, but we all know that it does. It is well known that experimenters work harder on experiments that give unexpected results; for one thing, they draw more fire. One would like to think that the experiments that agree with theory have received as much scrutiny as those that do not, but they have not.

4. Conclusions

The demand for experimental results in relativity still greatly exceeds the supply and, as I have tried to indicate, there are more reasons to suspect the quality of the product than in many other areas of physics. It is hard to see this situation improving much for the next decade. Expected trends in scientific manpower and budgets do not extrapolate to the levels required to make major changes in the productivity of experimental relativity. So, if one assumes that experiments continue at about the current level, what might be done to minimize the harmful effects of incorrect results? There seem to be two obvious recommendations, not easily implemented: 1) Experimentalists should support each new result with a long paper or report dealing with experimental details. Not only does doing so permit experts to evaluate the experiment, but it stimulates others to think about possible systematic effects. 2) Theorists should develop a better understanding, as well as a more healthy skepticism, of their experimental colleagues' work. Perhaps the overly pessimistic tone of these Comments will be excused if it helps theorists to better appreciate the problems of experimentalists.

Open Discussion

Following Papers by I. I. SHAPIRO and D. T. WILKINSON

Chairman, R. DICKE

R. Dicke (Princeton U.): An important thing that none of the speakers so far has pointed out is that all the other tests of general relativity have involved the gravitational field as a quasi-static field, and it is only in the gravitational waves that you see gravity for the first time as a dynamical field.

J. H. Taylor, Jr. (U. of Massachusetts): That was a superb review by Irwin Shapiro of the experimental state of affairs in relativity, and I think there is no necessity for me to say anything further about the experimental work that my group and I have been doing on the binary pulsar. For those of you who might be interested in the nature of the first point on the graph that Irwin showed (since a considerable fraction of the credibility of the curvature in the line drawn through the points must be assigned to the first point) I will make the following two brief remarks. First, the number of independent measurements that went into that first point is some 300. One could further subdivide the data, making the first point into two or three points instead. There is no discrepancy between the individual values measured that way. So, unless we have been trapped by one of the kinds of systematic errors that Wilkinson mentioned, I have good confidence in the measurements as presented. Second, if one does a formal solution for the curvature of that same line, based on all the data except the earliest measurements, one gets the same result to within statistical uncertainty. The result is decreased from a very believable 6 σ kind of result to a rather more cautiously stated 2.5 σ result.

Dicke: Let us turn very briefly to some discussion of these two papers.

D. Sciama (U. of Texas and Oxford): Maybe in the next few years there is the possibility of a strong field test of general relativity. If Cygnus X-1 does contain a black hole and if the X rays from Cygnus X-1 are linearly polarized, which is quite possible, then there will be general relativity effects on the plane of polarization that will rotate the plane as one goes up in energy from 2 kilovolt X rays, say, to 50 kilovolts and the calculations show that the effect of the rotation is very large — not seconds of arc as we have been hearing so far, but on a particular model, a rotation of 100°. If the black hole is not rotating, the effect would be due entirely to the bending of the X rays in the gravitational field that drags the polarization direction around. But if it is a fast-rotating black hole, there would also be a very large dragging of inertial frames effect that would contribute to the net rotation of the plane of polarization. I am hoping that in the next few years X ray polarimeters will be flown that will be able to make this test.

Anonymous: I would like to take up a remark that was made in the Comments by Dr. Wilkinson that seems to me can be reversed with certain right. I think that experimentalists should not take too seriously what sometimes is reported to be a proof of a certain theorem within a theory. It has been stated that in the beautiful measurements of Dr. Taylor and his associates one tests the prediction of general relativity. This is not the time to take up controversial questions of a detailed nature, but I think it would be better to say for the moment that what is tested there is whether Einstein's quadrupole formula, together with certain additional assumptions, is used in order to compare the theory with the measurements. I think it is an open question at the moment whether that particular formula has actually correctly been deduced from general relativity theory; that is a separate issue. Along similar lines, another point: Usually when one compares measurements of the beta-gamma parameters and even more detailed things with the theory, one bases these comparisons on parametrized post-Newtonian formalism or some such similar formalism. These are weak field approximation methods and one applies them without much criticism immediately to pulsars. Whether or not, if you look at the weak field approximations, you see that first of all, even if they are asymptotic expansions, they converge very slowly, it is more a theoretical simplicity assumption than a theoretically justified result to say that the compactness of bodies does not affect the overall motion. Perhaps one also should have some reservations as to how precisely certain things really have been deduced within the theory.

Dicke: Thank you. I appreciate your remarks. I guess there's a certain degree of healthy skepticism on both sides.

C. W. Misner (U. of Maryland): Perhaps the foregoing discussion has overlooked a major part of the observational support for Einstein's theory — a part that I think explains the theory's widespread acceptance in the years before the space program and other technological progress brought increasing support from new experiments. The older body of experimental evidence for Einstein's theory of gravity is the vast number of quite commonplace experiments in which both gravitation and light are employed simultaneously, and where the usual computations employ Newtonian gravitation theory and the electrodynamics of special relativity.

Physicists do not believe that physics has been perfected already, and therefore some schizophrenia is a normal way of life. We have heard here about Einstein's balancing the demands of particle and wave properties of light when these had not yet been resolved reasonably. But the schizophrenic state is not the aim of physics. We should be resolute in attacking it, as Einstein was. We should rejoice and accept as confirmation of the theory any piece of experimental data that newly becomes explicable in a nonschizophrenic way. Therefore I want to do an experiment that is evidence for general relativity. [Holds up a notepad, then drops it on the floor.] The experiment to be explained here is not merely the Newtonian free fall of a mass, but the fact that it was also *seen* to fall. Gravitation and electromagnetism act in a single experiment. For another example consider a cyclotron. Its design requires special relativity to calculate the motions of the particles that are being accelerated. But gravitation is not ignored in the design either, as without gravity the cyclotron magnets would float away and go out of alignment. General relativity is prepared to comfortably describe both the particle trajectories and the gross mechanical stability of the magnets. It does not require that after calculating the charged particle orbits, we go into the next office, open a new book, and change our worldview before engineering the magnet alignment system. What are the alternatives to this schizophrenia? The only scientific view I know that allows us sanely to do both a Newtonian calculation (magnet support design) and an electrodynamics calculation (particle orbits) is to require that these theories be seen as limiting cases by correspondence principles either from general relativity or from one of the many competing theories that use the same Einsteinian framework of ideas. Modern experiments are important for distinguishing the variants among relativistic theories of gravity, but when we count evidence for the common central ideas of these theories, the major lode of observational data already existed in Einstein's time and newly became intelligible physics with the advent of general relativity. It consists of the everyday experiments in which both light and gravity occur in the same laboratory and have to be thought of by the same physicist.

10. THEORETICAL ADVANCES IN GENERAL RELATIVITY

Stephen W. Hawking

It is impossible to describe in a paper of this length all the advances that have been made in our understanding of the general theory of relativity, or to mention all the names of those who have made significant contributions. I shall merely pick out a few of what I consider to be the highlights. I apologize in advance for what will be a personal and rather biased selection.

From the start general relativity was recognized to be a major intellectual revolution that transformed our ideas of space and time from a fixed background to a dynamical entity influenced by the events taking place in it. The theory, however, was thought to be almost impossibly complicated mathematically; in the 1920s there were said to be only a dozen people in the world who understood it. It was also regarded as having little practical importance because it was believed that the gravitational field would never become sufficiently strong that there would be much difference between the predictions of general relativity and those of the much simpler Newtonian theory of gravity. These views have changed only in the last twenty years or so. We now recognize that very strong gravitational fields must occur in astrophysical objects such as pulsars and compact X-ray sources and that all the evidence points to the conclusion that the universe has evolved from an earlier state in which conditions were extreme. We have also found that general relativity is not such a difficult theory as was first supposed. We now have a fairly complete qualitative understanding (some aspects of which I shall describe), and the advent of computers has enabled us, at least in principle, to obtain quantative predictions to any order of accuracy that our research budget will support. General relativity is now taught to most theoretical physics graduates and the theory has become a service subject for astrophysics, like electrodynamics.

The general theory of relativity makes two logically independent assumptions. The first, and really the most revolutionary, is that the effects of gravitation on matter fields can be described by replacing the flat Minkowski metric of special relativity by a curved metric g_{ab}. This is the mathematical expression of what Einstein called the "principle of equivalence." In modern language we would say that general relativity is a metric theory. This assumption has been tested to very high accuracy by the Eötvös experiment, and its refinements by Dicke and Braginskii and their collaborators. The second assumption is that the metric g_{ab} is related to the distribution of matter and energy in space-time by the famous Einstein equations:

$$R_{ab} - \tfrac{1}{2}g_{ab}R + \Lambda g_{ab} = 8\pi G T_{ab}$$

Harry Woolf (ed.), Some Strangeness in the Proportion: A Centennial Symposium to Celebrate the Achievements of Albert Einstein ISBN 0-201-09924-1

Here R_{ab} is the Ricci tensor of the metric g_{ab}, R is the Ricci scalar, T_{ab} is the energy momentum tensor, Λ is the cosmological constant (probably zero), and G is Newton's constant of gravitation.

It is not immediately obvious that these very geometrical field equations have the properties that would be required to describe gravity, and indeed it took Einstein several years and a number of false attempts to find them. In his original 1915 paper he showed that they would reproduce the results of Newtonian gravitational theory in the limit of low masses, large separations, and low velocities. He also showed that in the linear approximation, data on an initial surface would determine the gravitational field up to the freedom to make coordinate transformations. Subsequent work by a number of authors has extended this result to the full nonlinear case: we now can prove that the initial data determines a unique evolution of the gravitational field at least for some finite interval of time. One cannot prove that it does so for all time because there are examples in which singularities occur and provide an endpoint for the evolution and for time itself. I shall discuss singularities further.

The fact that the Einstein equations form a hyperbolic set implies that changes in the gravitational field produced by changes in the distribution of matter will propagate at a finite velocity, which, in fact, is the same as the velocity of light. Einstein himself has shown in 1916 that such "gravitational waves" occurred in the linear approximation. However, there was for a long time disagreement about whether these waves had any physical significance and, in particular, whether they would carry energy away from a system that emitted them. The problem was made harder by the fact that the mass of an isolated system such as a double star is normally defined by the limiting values near infinity of certain metric components on a spacelike hypersurface. Later spacelike surfaces will intersect any gravitational radiation emitted by the system, however, and thus the mass measured on them will be the same as the original mass. This difficulty was finally overcome in 1962 by Bondi, Metzner, and Van den Burg who used a new definition of mass related to the asymptotic behavior of the metric on null hypersurfaces of constant retarded time rather than on spacelike hypersurfaces of constant time. They showed that the mass associated with successive null hypersurfaces decreased by an amount that could be interpreted as the energy of the gravitational radiation that escaped to infinity between the surfaces of constant retarded time. It remains an unresolved problem, though, to relate this mass to that defined on spacelike hypersurfaces. One would expect that the spacelike mass would be the limit of the null mass at infinite negative retarded time. Another problem that has not been completely cleared up is to relate the decrease of mass to the motion of the bodies producing the gravitational radiation. One would expect some sort of radiation reaction force as in electrodynamics, but attempts to derive this from various approximation schemes have produced conflicting results.

It has long been an outstanding problem to show that the mass of a bounded system is necessarily positive. This is because in general relativity there is no local measure of gravitational energy. Bondi's result showed that gravitational waves carry positive energy. On the other hand, gravitational potential energy is negative because gravity is attractive. One might therefore imagine that a body such as a star could have such a high negative gravitational binding energy that its total energy or mass was negative. Recently, however, Schoen and Yau have shown that the mass measured on a spacelike hypersurface is always positive. It seems that what happens is that when a body becomes so tightly bound gravitationally that it is in danger of having a negative mass, an event horizon forms and the body collapses to make a black hole that has positive mass. I shall return to black holes later on in this paper.

Up to about 1960, most work in general relativity was performed in terms of metric components in a single coordinate system. This was alright for studying local questions, but it led to confusion and misunderstanding about the properties in the large. A particular example of this was the Schwarzschild solution discovered only one year after the formulation of the theory. This is the unique spherically symmetric vacuum solution and represents the gravitational field of an isolated body of mass M in asymptotically flat space. The metric components, in the normally used coordinate system, become singular at the "Schwarzschild radius" $r = 2M$. Although Eddington and Oppenheimer found coordinate systems in which this singularity did not occur, its fictitious character was not generally recognized until 1959 when Fronsdal, Finkelstein, and Kruskal obtained the maximal analytic extension, which consisted of two asymptotically flat regions connected by a "throat" at $r = 2M$.

One reason that the apparent singularity at $r = 2M$ was not investigated more thoroughly was that it was generally thought to be unphysical: no "real" body would ever become so compressed that it would be inside its Schwarzschild radius and so would produce a qualitative change in the structure of space-time. However, it was recognized from the beginning that gravitational forces could produce global effects on the scale of the whole universe. Because gravity is observed always to be attractive, the very weak gravitational interactions between individual particles all add up in a large conglomeration of matter and can be dominant over all other forces. This caused a severe problem in the Newtonian theory of gravity because it implied that the bodies in the universe should fall together, contrary to the accepted belief that the universe was essentially unchanging in time. Einstein still held this static view when he proposed his first cosmological model in 1917. In it he overcame the problem of gravitational attraction by introducing a repulsive (in both senses of the word) cosmological term into the field equations. This had the effect of changing the three-dimensional sections of constant time from ordinary three-dimensional Euclidean space into 3-spheres, thus introducing the revolutionary concept that space could be finite but unbounded.

Apart from the rather ad hoc introduction of the cosmological term, Einstein's static model suffered from the fact that it was unstable to small perturbations that could make it either collapse or expand indefinitely. It was abandoned when observations of distant galaxies by Slipher and Hubble in the early 1920s showed that they were moving away from us with velocities proportional to their separation from us. This was interpreted as indicating that the universe was expanding. The first cosmological model to describe such an expansion was the solution discovered by Friedmann in 1922. It contained pressure-free matter and was spatially homogeneous and isotropic. Friedmann's models were generalized by Robertson and Walker in the 1930s, but the assumption of spatial homogeneity and isotropy was retained. At the time this was simply because it was thought to be too difficult to consider models that were anisotropic or inhomogeneous (anisotropic models were first examined by Taub in 1951 and by Heckmann and Schucking in 1958). The very high degree of isotropy of the microwave background radiation discovered by Penzias and Wilson in 1965 indicates that the large-scale structure of the universe is very close to that of the Friedmann model, at least back to the time at which the radiation was last scattered. For earlier times we have indirect evidence from the fact that the observed cosmological abundances of helium and deuterium agree very well with nucleosynthesis calculations in a Freidmann model first performed by Gamow and his collaborators and later refined by Hyashi, Hoyle, Taylor, Peebles, Fowler, Wagoner, and others. Indeed it is very difficult to account for so much helium in any other way. By contrast, calculations performed in anisotropic or inhomogeneous models produce very different abundances of these elements.

The agreement between these calculations and observations indicates that a Friedmann model was a good approximation to the universe, at least back to the time of nucleosynthesis, which begins in these models at about 100 seconds after the initial singularity. But what of the singularity itself? All the Friedmann models (except those with a large repulsive cosmological term, and they are in conflict with observations) have an initial singularity, the Big Bang. Although most of the early world pictures had involved a Moment of Creation, the progress of scientific thought in the nineteenth and early twentieth centuries had made the idea that time had a beginning rather unfashionable. There were therefore a number of attempts to overcome the problem posed by the singularities of the Friedmann model.

One idea, proposed by Bondi, Gold, and Hoyle in 1948, was that the universe was in a "steady state" and presented the same general appearance at all times as well as at all points in space. As the galaxies receded from each other, new matter was continually created in between to maintain the average density of the universe at a constant value. This theory was aesthetically very attractive and had the great merit of making definite testable predictions. Unfortunately, these predictions were inconsistent with observations of radio sources and of the microwave background radiation, so the theory had to be abandoned.

Another attempt to avoid singularities was the suggestion that maybe they were simply a consequence of the high degree of symmetry of the Friedmann models. The assumption of isotropy implied that the relative motion of any pair of particles had to be purely radial, while the assumption of spatial inhomogeneity ruled out the possibility of pressure gradients. It was therefore not surprising that the world lines of all particles should intersect each other at some instant of time. In a realistic solution one might expect transverse velocities and irregularities. As one evolved the solution back in time, these would grow large and might prevent the occurrence of a singularity of infinite density and produce a "bounce" that connected the present expanding phase of the universe to a previous contracting one. In 1963 Lifshitz and Khalatnikov claimed that a fully general solution of the Einstein equations would not contain a singularity. They based this on an analysis of a solution with a singularity that they expanded in a power series about the singular point. They found it contained one fewer arbitrary function than a fully general solution and therefore would, presumably, be of measure zero in the space of all solutions. They later realized that there was a more general class of solutions containing singularities that had the full number of arbitrary functions and so concluded that singularities *could* occur in general solutions.

Their techniques did not enable them to show that singularities necessarily *would* occur in cosmological models. This was done by a completely different approach initiated by Penrose in 1965. He used global geometrical arguments to show that singularities were inevitable in the collapse of a star if certain global conditions were satisfied. Penrose's methods were extended and applied to cosmology by Hawking and Geroch. This work culminated in a general all-purpose singularity theorem of Hawking and Penrose in 1970. The theorem showed that a singularity was inevitable under the following conditions:

1. General relativity is correct.
2. The energy momentum tensor of matter satisfies the inequality

$$\left(T_{ab} - \tfrac{1}{2} g_{ab} T\right) V^a V^b \geqslant 0$$

for any timelike vector V^a.
3. There are no closed timelike curves.

4. Any timelike or null geodesic contains some point at which

$$V_{[a} R_{b]cd[e} V_{f]} V^c V^d \neq 0.$$

5. There is some point p such that all the past directed (or future directed) null geodesics from p start converging again.

Condition 2 is a very reasonable inequality, which is satisfied by any matter with positive energy density and pressure. Condition 3 is the requirement of causality, that one should not be able to travel into one's own past; it seems very reasonable physically, but there are other singularity theorems that do not need this condition. Condition 4 is a generality requirement that states roughly that every timelike or null geodesic encounters some matter or randomly oriented curvature; it could be weakened considerably. Condition 5 demands that there be sufficient concentration of matter or energy in the universe to focus every past directed light ray from some point p. Observations of the microwave background radiation reveal that it has a thermal spectrum indicating that it must have been scattered many times. One can show that this implies that there must be sufficient matter in the universe to cause condition 5 to be satisfied.

The theorem shows that time will have a beginning for some timelike or null geodesic. It does not prove that this will be so for every geodesic, and indeed there are some solutions in which this is not the case. However, it seems clear that these are special cases and are unstable; in the general case there will be a curvature singularity that will intersect every world line. Thus general relativity predicts a beginning of time.

The theorems also imply that singularities will occur if a star shrinks to less than its Schwarzschild radius. At first people did not believe that this could ever happen, but around 1930 Chandrasekhar and Landau showed that in Newtonian theory a cold body could not be supported by degeneracy pressure if it had a mass greater than one and one-half times that of the sun. In general relativity the mass limit would be even lower because of the nonlinearity of the interaction. This raised the question of what would happen to a star of more than the limiting mass after it had exhausted its nuclear fuel and had cooled. In some cases the star might explode or throw off enough matter to reduce its mass below the limit, but it was difficult to believe that this happened in every case, and, even if it did, one could always consider a *Gedanken* experiment in which one collapsed two solar masses of iron. The answer was provided by Oppenheimer and his collaborators Volkoff and Synder in 1939: the star would collapse until it shrank to a singularity of zero volume and infinite density.

Many people, including Eddington and Einstein himself, did not believe Oppenheimer's conclusion and his work was generally ignored until interest in strong gravitational fields was stimulated by the discovery first of quasars in 1963 and then of pulsars in 1967. Oppenheimer's work was reappraised in the light of the better understanding that had been obtained of the global structure of the Schwarzschild solution and of a new solution, discovered by Kerr in 1963, that represented the gravitational field of a stationary rotating object. Oppenheimer's conclusion that a singularity would occur in the collapse of a star depended on the star being exactly spherically symmetric. However, the theorems mentioned previously showed that singularities would still occur even if the collapse was not exactly spherical. These singularities were even more disturbing than the cosmological ones; it was bad enough for time to have a beginning, but now it seemed that time would have an end as well, at least for an observer foolish or unfortunate enough to follow the collapse of the star.

At the singularity the space-time continuum or manifold just comes to an end. One loses the ability to predict the future because we have no laws to govern what comes out of a singularity. This is, of course, true of the cosmological singularity in our past. We can, however, regard what came out of it as the initial data for the universe. But we might be worried that new unpredictable information might enter the universe every time a star collapsed. Fortunately, it seems that this does not happen, at least at the classical level, because it appears that the singularities formed by gravitational collapse always occur in regions of space-time, called black holes, in which there is such a strong gravitational field that no light or information can escape to an external observer. This is called the "Cosmic Censorship Hypothesis" and forms the basis for all theoretical work on black holes. It remains the major unproved conjecture in classical general relativity, but it is supported by perturbation and computer calculations and by the failure of a number of attempts to establish inconsistencies among the results that can be derived from it.

One of these results is that the surface area of the event horizon, the boundary of the black hole, can never decrease with time (at least when quantum effects can be ignored) and increases whenever matter or energy falls into the black hole. This suggests that when a black hole is formed by gravitational collapse, a certain amount of energy will be radiated to infinity in the form of gravitational waves and the rest will cross the event horizon into the black hole, which will then settle down to a stationary state. The combined work of Israel, Carter, Hawking, and Robinson showed that these black hole stationary states all belong to the Kerr–Newman family of solutions which depend on only three parameters — the mass, the angular momentum, and electric charge — but which are otherwise independent of the details of the body that collapsed. This result, called the "no-hair theorem" by Wheeler, is the basis for nearly all theoretical work on black holes. It enables one to construct detailed models of astrophysical objects thought to contain black holes and to compare their predictions with observations. This has been done with considerable success in a number of cases, notably that of Cygnus X-I, so that we are now reasonably confident that we have detected black holes.

The increase of the area of the event horizon of a black hole is very reminiscent of the behavior of entropy according to the second law of thermodynamics. This led Beckenstein in 1972 to suggest that one could regard some multiple of the area of the event horizon as the entropy of the black hole. The entropy would measure the very large amount of information about the collapsing body that was lost when the black hole settled down to a stationary state characterized by only three parameters. In a purely classical theory this amount of information would be infinite because a black hole of given mass could have been formed out of the collapse of a body containing an indefinitely large number of particles of indefinitely small mass. However the uncertainty principle of quantum mechanics would imply that these particles would have a wavelength inversely proportional to their mass. One would not expect that the body would collapse if this wavelength were longer than the radius of the black hole that it would have formed. It would therefore seem reasonable that the relation between the area of the event horizon and the entropy would be essentially quantum mechanical and would involve Planck constant h. From h, c the velocity of light, and G Newton's constant of gravity, one can form a combination, the Planck length, which is about 10^{-33} centimeters. One would therefore expect that the entropy would be some numerical factor times the area in units of Planck length squared.

There was, however, an apparently insuperable obstacle to this identification of the area of the event horizon with entropy: If black holes had a finite entropy, they should also have a finite nonzero temperature and should be able to remain in equilibrium with thermal radiation at the same temperature. On the other hand, this would be impossible if, as was thought at the time, black holes could absorb radiation but not emit anything in return. The difficulty was overcome in 1974 when investigations of the behavior of quantized matter fields in a black hole background geometry revealed that the gravitational field near the black hole would cause creation and emission of particles just as if the black hole were a hot body with a temperature inversely proportional to the mass (in the nonrotating case).

One can visualize this process heuristically as follows: The quantum mechanical uncertainty principle implies that the whole of space is filled with pairs of "virtual" particles and antiparticles that are constantly materializing in pairs, separating, and then coming together again and annihilating each other. These particles are called virtual because, unlike "real" particles, they cannot be observed directly with a particle detector. Their indirect effects can nonetheless be measured, and their existence has been confirmed by a small shift (the Lamb shift) they produce in the spectrum of light from excited hydrogen atoms. In the presence of a black hole one member of a pair of virtual particles may fall into the hole, leaving the other member without a partner with which to annihilate itself. The forsaken particle or antiparticle may fall into the black hole after its partner, but it may also escape to infinity, where it appears to be radiation emitted by the black hole.

Another way at looking at the process is to regard the member of the pair of particles that falls into the black hole — the antiparticle, say — as being really a particle that is traveling backward in time. Thus the antiparticle falling into the black hole can be regarded as a particle coming out of the black hole but traveling backward in time. When the particle reaches the point at which the particle-antiparticle pair originally materialized, it is scattered by the gravitational field so that it travels forward in time.

The particle or antiparticle that escapes to infinity carries away positive energy. By conservation of energy this implies that the other member of the pair carries negative energy into the hole. (Classically, a particle outside the event horizon must have positive energy; however, inside the hole there are orbits that have negative energy as seen from infinity.) The influx of negative energy causes the mass of the hole to go down and the area of the event horizon to shrink. This increases the temperature of the hole and the rate of emission. Eventually it seems that the black hole will disappear completely in a tremendous explosion. Stellar mass black holes would have a lifetime much longer than the present age of the universe, but there might be a population of much smaller primordial black holes that might have been formed by the collapse of inhomogeneities in the very early dense stages of the universe. Observational searches for an isotropic background of gamma rays emitted by such primordial black holes or for gamma ray or radio burst produced by their final explosion have placed the very low upper limit of about 200 per cubic light years on the present density of primordial black holes in the mass range around 10^{15} grams. This indicates that the early universe must have been fairly smooth and nonturbulent, with a reasonably stiff equation of state.

The quantum mechanical emission from black holes violates the classical result that the area of the event horizon can never decrease. However, it can be replaced by a generalized second law of thermodynamics, according to which the sum of a certain multiple of the area of black hole event horizons plus the entropy of matter and radiation outside black holes never decreases. This is because the emission from black holes is truly thermal; that is, the particles have random phases and all configurations of the emitted particles with the same energy have equal probabilities. One can regard this as a reflection of the fact that the particles come from inside the black hole, a region of which an external observer has no knowledge apart from its total energy (and angular momentum and charge). The randomicity or intrinsic entropy that quantum gravity seems to introduce is a completely new feature, which is not shown by other quantum field theories such as quantum electrodynamics or Yang–Mills theory. It seems to be connected with the fact that general relativity allows space-time to have nontrivial topologies.

One consequence of quantum gravitational randomicity is that even supposedly stable particles like baryons should decay. One can see this in the following way: Suppose that a star composed of baryons collapses to form a black hole. For most of its lifetime the temperature of the black hole will be very low, and, so, it will emit mainly zero-rest mass particles such as neutrinos, protons, and gravitons. In the final stages the temperature will become sufficiently high to emit baryon–antibaryon pairs. However, even if there is some asymmetry that favors the emission of baryons over antibaryons, the black hole will by then have insufficient rest-mass energy to emit as many baryons as there were in the original star. Thus when the black hole disappears completely (assuming that it does), baryon number conservation will have been violated. If this can happen in the presence of large black holes, one would expect that it would also occur spontaneously because of "virtual" black holes produced by quantum fluctuations of the vacuum.

In order to treat these fluctuations one needs a proper quantum theory of gravity. This is also required to discuss the final stages of the evaporation of black holes and the behavior of space-time near the singularities, especially the original big bang singularity, that are predicted by the classical theory. In my opinion this is the most outstanding problem in theoretical physics at the present time. A considerable amount of effort is being expended on attempting to combine general relativity and quantum mechanics and on the closely related problem of unifying gravity with the other interactions in physics. Although there have been some partial successes, I think it would be fair to say that we are still some way from the hoped for synthesis. The problem is that the perturbation techniques that have been so successful with quantum electrodynamics and Yang–Mills theory do not seem to work in gravity. The divergences that arise in closed loop Feynman diagrams are not of the same form as that of the original action and so cannot, as in electrodynamics or Yang–Mills theory, be absorbed into a redefinition of the coupling constant and the masses. This seems to be true both for "pure" gravity and for the various supergravity theories in which the gravitational field is combined with various other bosonic and fermionic fields, which are related to each other by fermionic supersymmetry transformations. My personal view is that the nonrenormalizability of quantum theories of gravity indicates, not a failure of the theories themselves, but a breakdown of the standard perturbation techniques. This should come as no surprise to those who have worked in classical general relativity and who know that perturbation expansions have only a limited range of validity; one cannot treat a black hole as a perturbation of flat space, yet summing Feynmann graphs around flat space amounts to just that. One might expect that on the scale of the Planck length there would be fluctuations of the metric and of the topology endowing space-time with what Wheeler has called a "foamlike" structure. How to handle this is not yet clear, but the most promising approach seems to be to use path integrals evaluated in the Euclidean regime — that is, over positive definite or positive semidefinite metrics. Such path integrals tend to involve infinities. Most of these, however, cancel out in supersymmetric theories, leading to the hope that one could give a well-defined mathematical meaning to the path integral. The ultimate aim is a complete and consistent quantum theory that would incorporate gravity and all other interactions of physics. This would realize the cherished dream of Einstein's later years, although in a form that he might not recognize.

11. COMMENT ON S. HAWKING'S "THEORETICAL ADVANCES IN GENERAL RELATIVITY"

William G. Unruh

Stephen Hawking has given an admirable, albeit of necessity brief, survey of the history of the theoretical development of general relativity. To those of us educated in the field in the last twenty years or so, the fumblings and misunderstandings of the years from 1915 to 1950 seem remote and strange. The emphasis on asking questions about relations between physical observables, rather than coordinate relations, that we have received in our training makes the long disputes over questions like the reality of the gravitational wave, or the physical reality of the $r = 2M$ Schwarzschild singularity hard to understand.

I would like to leave the history of general relativity to those far more able to deal with it, and to examine briefly what I feel the present theoretical significance of general relativity is. As Hawking has mentioned, we now have, I believe, a reasonably deep understanding of the classical theory, and have developed a number of mathematical tools to deal with various problems therein. Although I am sure that the theory still holds some physical surprises, and that our understanding is in places more superficial than we imagine, the theoretical activity in this field is at a low ebb. The main difficulty is that we need some real physical situations to which we can apply our insights and techniques. Fortunately the astronomers seem to be working hard at trying to find such systems for us.

The main theoretical interest has therefore shifted toward the problem that Hawking has just called the most important facing theoretical physics today — namely, the elucidation of the relation between gravity and quantum mechanics. I have purposely stated the problem in this manner as it is not yet clear whether the solution to this problem will be the quantization of gravity (or rather, the development of a theory of gravity that can be quantized) or some sort of transformation or relativization of quantum mechanics.

It is ironic that just at the time when the experimental tests are becoming possible, are being performed, and are verifying the predictions of the theory, the most determined assault on structure of the theory itself is taking place. In the attempt to make general relativity compatible with quantum field theory, many feel that the theory must be altered in some fashion. The most notable example of this is, of course, supergravity, with its addition of the gravitino. Although many of these extensions to general relativity have great mathematical beauty, my views tend to echo those

Harry Woolf (ed.), Some Strangeness in the Proportion: A Centennial Symposium to Celebrate the Achievements of Albert Einstein
ISBN 0-201-09924-1

of Eddington in his 1925 book on relativity: "The theory is intensely formal as indeed all such action-theories must be, and I cannot avoid the suspicion that the mathematical elegance is obtained by a short cut which does not lead along the direct route of real physical progress. From a recent conversation with Einstein I learn that he is of much the same opinion. Nevertheless when the path of progress is uncertain it would be unwise to ignore advance along any open route."

The alternative view, which has probably been expressed most consistently by J. Wheeler, is that we must first find the essential physical fact and philosophical outlook that relate quantum mechanics and gravity, and that the proper mathematical formulation will then suggest itself. Unfortunately, as the proper viewpoint has yet to be found, this approach has not even left us any beautiful mathematics to play with.

In a sense the two approaches represent the very different histories of quantum mechanics and general relativity. In the history of quantum theory, our understanding of the physical world has time and again been led by the mathematics. From Schroedinger's equation, to the Dirac equation, to SU(3), to the Yang–Mill's unified theories, people have developed theories for their mathematical appeal, and then found to their amazement (or at least mine) that they actually applied to the real world. On the other hand, in general relativity, progress was time and again held up because people asked mathematical, rather than physical, questions of the theory.

To most people, at least sometimes, the investigation of the role of quantum mechanics in gravity seems premature. Order-of-magnitude estimates suggest that quantum effects should become important only over distances of order 10^{-33} centimeters, energies of order 10^{22} Mev, and so on. As no experiments exist, or have any possibility of existing, at this level, this theoretical endeavor seems doomed to failure for lack of any guidance in our speculations by the real world. The answer to this objection is twofold.

The first point is that general relativity provides a radically new arena in which our traditional concepts in quantum theory can be examined and tested. Our physical interpretation of, especially, quantum field theory are heavily based on the background existence of a flat space-time metric and the Poincaré group. In a curved space-time even the elementary concepts of a vacuum state and of a particle interpretation become extremely difficult. Furthermore, the existence of things like variable topologies, or of the possibility of universes with closed timelike curves, can force us to radically reexamine some of the basic tenets of our interpretation of quantum mechanics. In this sense, general relativity can be seen as continuing the role of gadfly to quantum mechanics that Einstein himself played.

A recent example of this, closely connected with the black hole evaporation process mentioned by Hawking, is the behavior of accelerated particle detectors. Since in general relativity one often has to worry about the behavior of accelerated observers, the behavior of such detectors even in flat, Minkowski space-time is of interest. By building a simple model of a particle detector, and accelerating it in the vacuum state in flat space-time, it is possible to show that such a detector behaves exactly as if it were immersed in a *thermal* sea of particles of temperature

$$T = \frac{8\pi\hbar a}{kc} .$$

To give an idea of the magnitude of this effect, note that 1 °K requires an acceleration of 10^{23} cm/sec^2.

Furthermore, what such an accelerated observer regards as the detection of particles, a noninertial observer regards as the emission of such particles. This surprising result appears to unify kinematics, thermodynamics, and quantum mechanics.

The second reply to such criticism is that we have already a powerful "experimental" fact that needs to be explained, and that is the logical inconsistency of the two theories. The basic equations of general relativity, $G_{\mu\nu} = T_{\mu\nu}$, contain incompatible quantities on the two sides of the equals sign. Furthermore, this incompatibility has consequences even at the present experimental scale.

Because an unstable process such as the collapse of a star is quite possibly driven by fundamentally quantum events, such macroscopic processes will have a quantum nature associated with them. The gravitational field must therefore also share this quantum nature even on macroscopic levels. An example is a steel ball suspended in a closed box by an electromagnet with the state of the electromagnet controlled by a low-level radioactive source. (The similarity with the Schroedinger cat-in-the-box is obvious.) Because the position of the iron ball depends on the fundamentally quantum radioactive source, its position will have a large quantum uncertainty. Will one be able to determine the position of the ball in a quantum sense by looking only at its gravitational field?

If we look at the history of general relativity, a very similar situation existed at about 1905: special relativity and Newtonian gravity were also incompatible. Einstein's resolution of this incompatibility came not by trying to find experimental evidence of relativistic effects in gravity, effects that even today we have great difficulty in measuring. Rather, he seized on the apparently irrelevant equivalence of inertial and gravitational mass. Our hope is that we today can find the relation between the two great accomplishments of twentieth-century physics, general relativity and quantum mechanics.

Open Discussion

Following Papers by S. HAWKING and W. G. UNRUH

Chairman, R. DICKE

P. Bergmann (Syracuse U.): I very largely share Stephen Hawking's expectations as to the future of relativity and quantum theory, but I would like to comment on one particular aspect. It seems Einstein always was of the opinion that singularities in a classical field theory are intolerable. They are intolerable from the point of view of classical field theory because a singular region represents a breakdown of the postulated laws of nature. I think one can turn this argument around and say that a theory that involves singularities and involves them unavoidably, moreover, carries within itself the seeds of its own destruction — or putting it perhaps more optimistically, the seeds of its further evolution in a radical direction leading away from the present theory. I would like to comment on the violation of the conservation laws that Steve Hawking already mentioned that is unavoidable if a black hole first imbibes baryons and then disgorges a different number of baryons as a result of what most of us call the Hawking effect, that is, a quantum radiation. A different model that illustrates the same point is the following: Supposing that you submerge a black hole in an environment with which it is in thermodynamic equilibrium, but where the environment carries a chemical potential, so that, for instance, the density of neutrinos and antineutrinos is not equal in the environment. We can then maintain a stationary, though not a stable interaction between environment and black hole, such that the black hole absorbs a permanently different composition of neutrino-antineutrino mixture as it spews out, because no chemical potential can be associated with the black hole. This is one of the implications of the no-hair condition. There is, I think, one possible way out of the occurrence of singularities that deserves some further investigation, and that is the following: If, near the black hole, conditions are very exotic, it is conceivable that the energy condition, which is one of the five conditions enumerated by Hawking, could be violated. It seems, in some respects, that if you are going overboard and admit serious exceptions to what we consider the conventional behavior of nature, a violation of the energy condition might be swallowed just as much as the occurrence of a singularity. I am not terribly optimistic, but I think this possibility should be excluded. I would like to end by saying that I agree with Stephen Hawking that the whole situation looks like one in which a completely new idea is required, one that will enable us to integrate the present fragmentary theories into a new organic whole.

Y. Ne'eman (Tel-Aviv U.): I would like to discuss that issue of the nonconservation of baryon number that was so well explained in Stephen Hawking's paper, but, as you may know, I spent most of my life as a particle physicist and I have only a kind of visiting relationship with the gravitational region. Among my colleagues in particle physics, there is a tendency now to go to theories that are extremely fundamental and are called grand unification theories, in which the weak, the strong, the electromagnetic interactions all get unified at some limit, which happens to be a limit where the intermediate bosons are of Planck mass level. These physicists do not take any note of the existence of gravitation, and my criticism to them is that the conclusions that they draw with respect to events in the Planck mass region should consider the fact that we are not alone:

there is also gravitation, and therefore when you get to Planck mass level, you should consider gravitation as an important force that must also be taken into consideration. I would like, hesitantly, to make the same kind of remark here: We owe a lot to baryon number. We owe our existence to the conservation of baryon number; otherwise we would be floating in the universe as $E = mc^2$. So the topic deserves at least some consideration on our part. For instance, if whoever made the laws decided that baryon number should indeed be conserved, there would be nothing easier for him than to say that gravitation stops at the point where the baryon number is threatened. That means that the black hole would not dissolve at the moment when it might do away with baryon number conservation. There might be some special state; but maybe it will turn into a particle with baryon number 10-to-the-I-don't-know-what, which exists somewhere in the spectrum and which has a certain mass. You can stay at that level and you will not find empty space at that position because gravitation is not alone, and you can have an additional law that will stop it from doing away with the black hole at that level. My position at least is that of an agnostic with respect to this issue, and I would not do away with this important law just on the kind of evidence that gravitation gives us. I do admit, certainly, that the arguments derived from gravitation by itself are very compelling, but it is not by itself.

S. Hawking (through Alan S. Lapedes): Stephen's comment was that he finds it interesting that people have such an emotional attachment to baryon conservation. This may be because most people do not believe in eternal life. They would like to hope that the particles which make up their bodies would live forever.

J. Stachel (Boston U.): Just a follow-up on Peter Bergmann's comment — a historical note that might prove at some future date to be of more than historical interest. At least once Einstein faced a singularity question in his own published writings — not in the black hole context, of course, but in the context of the cosmological singularity. In the appendix to the "Meaning of Relativity" on cosmology, he made a comment on this subject that is of some interest to report. It perhaps has not received the significance it should, at least in the English-speaking world, because the paragraph is marred by two mistranslations in the English version. But essentially what he said is that one should not take the singularity too seriously. One should not worry about it too much because it is quite likely that, under the conditions of ultra-high matter and field density that would prevail near the "singularity," — "das Raumerfuellende," that which fills space — will not be correctly described by the concepts we use in ordinary general relativity. I interpret that to mean that, in other words, the very concept of a $T^{\mu\nu}$ or a T^{ik} will break down and that from the point of view of somebody who is searching for a unified field theory of quite a different kind, where there would be no $T^{\mu\nu}$, the problem would pose itself in quite a different way. I guess nobody has a very good idea how to do such a thing, but at least it is wise to bear in mind that it might be a possible way forward at some future date.

I. Shapiro: What happens to the charge in the evaporation of a black hole with nonconservation of baryons? How does the charge get dissipated?

Hawking (through Lapedes): Charge is absolutely conserved. This is because it is coupled to a long-range field, the electromagnetic field. And similarly, energy and angular momentum are coupled to a long-range field, the gravitational field. Therefore, they are conserved. However, baryon number is apparently not coupled to any long-range field, or at least if it were, it would violate the Dicke–Braginsky equivalence experiment, unless the coupling would be very much weaker than gravity.

Anonymous: I would like to know how to treat the so-called naked singularity — that is, the singularity outside the black hole.

Hawking (through Lapedes): Professor Hawking does not believe there are any apart from the big bang itself. This is the cosmic censorship hypothesis in which he believes. This remains the unsolved problem of classical general relativity, to prove the cosmic censorship conjecture.

Dicke: Any questions for Dr. Unruh?

D. Sciama (Oxford U.): It is a comment rather than a question, but perhaps Dr. Unruh won't agree with it. In his beautiful paper he was too modest to tell us that he was the one who discovered that the accelerating observer sees a heat bath. It is a further property of that heat bath I want to mention because it relates so closely to the discussions we had yesterday about Einstein's work on fluctuations in a heat bath. Not only is the mean spectrum that the accelerating detector sees a Planck one, but it also sees all the energy fluctuations that Einstein discovered in the 1900s. Moreover, in the Hawking radiation process, the same thing is true. The mean flux from a black hole is thermal, but the energy fluctuations in that radiation field are again precisely the same as Einstein fluctuations in a black body radiation bath. Therefore, in this set of remarkable works of just the last three or four years, for the first time there have been brought together the two major themes of Einstein's working life — fluctuation theory and special and general relativity.

W. G. Unruh: I think I agree.

V. Quantum Theory

12. NO FIRM FOUNDATION: EINSTEIN AND THE EARLY QUANTUM THEORY

Martin J. Klein

Einstein knew what was important on an occasion like this. "If we would honor his memory fittingly," he wrote on the three-hundredth anniversary of Johannes Kepler's death, "we must get as clear a picture as we can of his problem and the stages of its solution."[1] That was what Einstein wanted most to know about his great predecessors — what those physicists of the past "were aiming at, how they thought and wrestled with their problems."[2]

There is no doubt about what Einstein's problem was: to construct a new theory that could serve as the foundation for all of physics. His "Autobiographical Notes"[3] make it clear that Einstein saw his life's work as an unremitting struggle to establish such a new, unified foundation, and that his "fifty years of conscious brooding"[4] on the nature of quanta formed an integral part, in fact a large part, of that struggle. Problems less weighty than that of the foundations of physics might occupy him temporarily, but Einstein would not, could not, escape the demands of his "passion for comprehension"[5]; it was the true grand passion of his life.

Einstein's education in physics was largely self-education. When he was a student in Zürich he "gradually learned to live in peace with a somewhat guilty conscience" and to arrange his studies in a way that corresponded to his own interests and his own rate of intellectual digestion. This meant giving up any hopes of being a "good student"; it meant working in the laboratory but cutting most of the lectures so as to spend the time in his own room "studying the masters of theoretical physics with sacred passion."[6] What did he learn from those books by Kirchhoff, Helmholtz, Boltzmann, and Hertz, among others?[7] How did those long hours of reading and thinking lead him to the problem that dominated his scientific work?

The treatises Einstein studied set forth the mechanical world view as it was conceived and employed in the late nineteenth century.[8] This was, in effect, a highly developed form of the program Newton had formulated for natural philosophy: "from the phenomena of motions to investigate the forces of nature, and then from these forces to demonstrate the other phenomena."[9] Newton had laid down the "mathematical principles of natural philosophy," the "rational mechanics" that made such investigations possible and had given one superb example of their use, his "explication of the System of the World." He saw this derivation of the law of universal gravitation

Harry Woolf (ed.), Some Strangeness in the Proportion: A Centennial Symposium to Celebrate the Achievements of Albert Einstein
ISBN 0-201-09924-1

from the phenomena of the heavens and its application to deducing "the motions of the planets, the comets, the moon, and the sea" as only a beginning. It was his hope that "we could derive the rest of the phenomena of Nature by the same kind of reasoning from mechanical principles," since he had "many reasons to suspect that they may all depend upon certain forces by which the particles of bodies . . . are either mutually impelled towards one another and cohere in regular figures, or are repelled and recede from one another."

What Einstein found in his books was a partial realization of Newton's hopes, and he was struck by the power of this mechanical physics. "What made the greatest impression upon the student," he wrote in his "Autobiographical Notes," "was less the technical construction of mechanics and the solution of complicated problems than the achievements of mechanics in areas that apparently had nothing to do with mechanics: the mechanical theory of light, which conceived of light as the wave motion of a quasi-rigid elastic aether, but above all the kinetic theory of gases." He listed some of its successes and then added: "These results supported mechanics as the foundation of physics and at the same time of the atomic hypothesis, already firmly established in chemistry. . . . Apart from this it was also of profound interest that the statistical theory of classical mechanics was capable of deducing the laws of thermodynamics, something that was in essence already accomplished by Boltzmann."[10]

The deep impression created by this mechanical world picture never left Einstein. It inspired his first paper, which was a study of intermolecular forces and capillary phenomena,[11] written in the fall of 1900, shortly after his graduation from the Polytechnic in Zürich. Einstein had taken up one of the classical problems of mechanical physics — posed by Newton, studied in detail by Laplace and by Gauss, and taken up again by J. D. van der Waals at the beginning of his career.[12] Although Einstein never got very far in his attempt to relate the surface energy of a liquid to its chemical structure, and had no success in trying to relate intermolecular forces to the force of gravitation, he did get a taste of "the magnificent feeling" that comes from recognizing "the unity of a complex of phenomena which appear to be completely separate things according to our direct sense perceptions."[13]

When Einstein put this work aside he turned to another problem well within the mechanical tradition.[14] After a thorough study of Boltzmann's *Lectures on the Theory of Gases*, Einstein decided to write an article that would "fill the gap" that he thought Maxwell and Boltzmann had left in the mechanical foundation of thermodynamics. He would derive the second law "using only the equations of mechanics and the theory of probability."[15] This article, submitted to the *Annalen der Physik* at the end of June 1902, just as its author started work at the Swiss patent office in Bern, became the first of a series of three.[16] Although Einstein began with Boltzmann's problem of the statistical mechanical basis for thermodynamics, by the time he completed the last of the series early in 1904 he had gone well beyond this problem and the mechanical tradition to which it belonged. He was now working on his own fresh problems; he had found his characteristic style of doing physics, and was ready to try to deal with some of the deep difficulties he saw arising from the divided foundations of physics.

Even while he was still studying the great expositors of a physics based on mechanics, Einstein was aware that the Newtonian vision had been found unattainable and that the mechanical world picture was no longer held by many of his contemporaries. But he had had a glimpse of what a unified physics based on one sound theoretical foundation might be like, and that image never left him. If mechanics could not provide the basis, a replacement would have to be found. To many, such a replacement seemed to be already at hand in the new electrodynamics.

"The most fascinating subject at the time I was a student," Einstein wrote, "was Maxwell's theory. What made it seem revolutionary was the transition from forces acting at a distance to fields as fundamental quantities. The incorporation of optics into electromagnetic theory, with its connection between the speed of light and the absolute electrostatic and electromagnetic systems of units, as well as the relation of the index of refraction to the dielectric constant, and the qualitative connection between the reflection coefficient and the metallic conductivity of a body — it was like a revelation."[17] Maxwell himself described his work as a dynamical theory of the electromagnetic field: he called it an electromagnetic field "because it has to do with the space in the neighborhood of the electric or magnetic bodies," and dynamical "because it assumes that in that space there is matter in motion, by which the observed electromagnetic phenomena are produced." This matter was of a special sort, the "aethereal substance." The field had to be thought of as "a complicated mechanism capable of a vast variety of motion," yet "a mechanism . . . subject to the general laws of Dynamics."[18]

For Maxwell and for his immediate successors establishing the nature of that "complicated mechanism" was one of the crucial problems.[19] Without that one would still not have a real explanation of electromagnetism in mechanical terms. Many models or mechanical analogies were proposed for this purpose, but none of them carried any real conviction. The question of the mechanical representation of the field remained unresolved. In the 1890s H. A. Lorentz deepened the level of understanding of electromagnetism by carefully separating the roles of matter and field in the theory.[20] Although he began in 1892 with a mechanical basis for the field equations, it played no part in his work, and he quietly dropped it from all his subsequent discussions. The aether, which carried the fields, became in effect an immaterial medium. It interacted with matter only by the forces it exerted on the electric charges carried by material particles, and the charged particles, in turn, acted as the sources of the field.

By the beginning of this century the impressive successes of Lorentz's electron theory (as it came to be called) suggested to a number of physicists that electrodynamics might be the fundamental theory on which all physics could be based. Electrodynamics had resisted all attempts to reduce it to mechanics; perhaps it was time to reverse the process. Electrodynamics might itself be the deeper theory, and mass, that most characteristic concept of mechanics, might always be of electromagnetic origin. A new electromagnetic world view to replace the mechanical one seemed a very real possibility to Wilhelm Wien, Max Abraham, and a number of other physicists.[21]

We know that Einstein studied the Maxwell theory before 1900 during his years at Zürich, probably through the expositions of Hertz, Boltzmann, and August Föppl.[22] It seems likely that he did not study Lorentz's clarification of the theory until several years later, for he wrote to Michele Besso in January 1903 that he planned to make "comprehensive studies in electron theory" as soon as he had finished the work he had on hand.[23] Einstein later described Lorentz's work as an "act of intellectual liberation," and to the end of his life went on admiring its "rare clarity, logical consistency, and beauty."[24] His study of Lorentz would certainly have indicated to Einstein that "mechanics was being abandoned as the basis of physics, almost without this being noticed, because adapting it to the facts finally showed itself to be a hopeless task."[25] But it was also clear that this withering away of mechanics had not yet been completed. Lorentz's theory was marked by a dualism in its foundations since it used both the electromagnetic field, governed by Maxwell's equations, and the material particle, governed by Newton's laws of motion. The idea of explaining everything on the basis of electromagnetism was still only a hope. It could not be accomplished until the existence of charged particles could be deduced from a field theory, and the linear equations of Maxwell's theory did not include such a possibility.

Einstein had been aware that there were serious doubts about mechanics as the foundation for all of physics long before he studied the electron theory. In 1897, during his first year at the Polytechnic, Einstein had met Michele Besso, an Italian-Swiss engineer six years older than he was. The two hit it off at once in a friendship that would last for over a half century. Soon after their first meeting Besso recommended Ernst Mach's books on mechanics and the theory of heat to the "young, ardent devotee of science."[26] Both books made a great impression on Einstein, and he later credited Mach's *Mechanics* in particular with having "shaken this dogmatic faith" in the mechanical world view.[27]

Much has already been written about the ways in which Mach's work influenced Einstein, and I do not intend to go over the arguments about the effect of his positivism on Einstein's thinking.[28] However important this may or may not have been, one should not overlook the fact that Mach's *Mechanics* does conclude with an explicit denial of "the view that makes mechanics the basis of the remaining branches of physics and explains all physical phenomena in mechanical terms." This view, writes Mach, is "a prejudice." He recognizes that, looked at historically, the mechanical view of the world is "both intelligible and pardonable," and "may also, for a time, have been of much value," but he finds it an "artificial conception." "We have no means of knowing, as yet, which of the physical phenomena go deepest, whether the mechanical phenomena are perhaps not the most superficial of all, or whether all do not go equally deep."[29] One major theme of Mach's book is that the historical priority of mechanics is no ground for giving it logical priority. He makes this point again equally vigorously in his *Wärmelehre*. Mach's arguments for this position must have accounted for at least part of the "profound influence" his books had on the young Einstein, along with his famous analysis and explicit criticism of the basic concepts of Newton's *Principia*, to which Einstein referred explicitly.[30]

I think there is another feature of Mach's work that may have impressed him. Mach points out clearly that the structure of a developed science like mechanics owes something to historical accident. Mechanics, for example, might look quite different if it had been developed along lines suggested by Huygens, which would have been a logical possibility, rather than along the Newtonian lines actually followed.[31] Statements that now have the status of basic laws would then be derived theorems, and vice versa. Mach goes even further when he emphasizes that one should not think of a principle like that of the impossibility of perpetual motion as merely a theorem of mechanics. "Since its correctness was perceived long before the completion of mechanics, and since it even contributed in an essential way to the establishment of mechanics, it is surely probable that it does not really depend on mechanical knowledge, but rather that its roots are to be sought in more general and deeper convictions."[32] This result of Mach's "historical-critical exposition" suggests a flexible approach to physics in which certain general principles are seen as more significant, more reliable, than the logical structures on which they seem to be based. Whether or not Einstein learned to think this way from Mach cannot be said, but learn it he did; nothing is more characteristic of his early work than his ability to select such principles.

The discussion up to this point has only suggested some of the background and the interests that Einstein brought to the problems that fascinated him in those early years at Bern. To see how he put this background to use we must turn to his work, and particularly to the papers in which he began to struggle with the question — the "incredibly important and difficult" question, as he would soon come to call it[33] — of quanta. And we must begin, not with the masterwork of 1905 in which light quanta were introduced, but rather with the paper I referred to earlier as the one in which Einstein first broke new ground and begin to write physics in his own characteristic style.[34] This is his paper of 1904, usually grouped with its two predecessors as a trilogy on the statistical foundations of thermodynamics. It is, however, also the first work in which Einstein refers to Max Planck's theory of blackbody radiation. I now think it was written as a consequence of his reading of Planck and that it represents the first results of his thinking about that strange theory.[35]

Describing the paper in a brief introductory section Einstein wrote: "First an expression for the entropy of a system is derived which is completely analogous to the one which Boltzmann found for ideal gases and which is assumed by Planck in his theory of radiation. Then a simple derivation of the second law is given. Following this we investigate the significance of a universal constant which plays an important role in the general molecular theory of heat. Finally, there follows an application of the theory to blackbody radiation which results in a most interesting relation between the universal constant mentioned (which is determined by the sizes of the elementary units of matter and electricity), and the order of magnitude of the wavelengths of radiation, a result obtained without having recourse to special hypotheses."

In this passage Einstein refers to the starting point of Planck's derivation of the distribution law for blackbody radiation, the relationship between entropy and "probability." He also alludes to Planck's most striking result — his evaluation of the constant he called k, and with it the mass of an atom and the natural unit of electric charge, from the experimentally determined constants in the radiation law.[36] These were evidently the features of Planck's work that impressed Einstein first, the features he began to think about before he tried to deal with Planck's law or its derivation. His previous work on statistical mechanics had prepared him for handling these questions his own way. He certainly knew what entropy meant in a statistical theory, and if there were a link between radiation and the basic constant of statistical mechanics he ought to be able to find that, too. While his results were not the same as Planck's they certainly suggested that it would be worthwhile thinking more about the problem of radiation.

Einstein had already shown on very general grounds that if a system is in equilibrium with a heat bath at temperature T, then the probability dW that the energy of the system has a value in the interval E to $E + dE$ is given by the equation,

$$dW = C \exp(-E/kT)\, \omega(E)\, dE. \tag{12.1}$$

In this equation C is fixed by the condition that the probability distribution is normalized, and $\omega(E)$ is the function, now known as the structure function, which characterizes the system in question. It is defined by the condition that $\omega(E)dE$ is the volume of that part of the system's phase space between the energy surfaces corresponding to energies E and $E + dE$, in other words the volume of the energy shell. He had already shown that for all systems $\omega(E)$ is related to the temperature through the equation,

$$\frac{d\ln\omega(E)}{dE} = \frac{1}{kT}, \tag{12.2}$$

from which it follows directly that the entropy S of the system must be related to $\omega(E)$ by the equation,

$$S = k\ln[\omega(E)]. \tag{12.3}$$

This is the expression for the entropy to which Einstein referred in his introductory remarks, but how it was "fully analogous" to Planck's starting point he did not say. He would return to that in due time.

As for the constant k, it was well known from the kinetic theory of gases that one could relate it to the constant R in the ideal gas equation of state by the equation

$$k = R/N_0 \tag{12.4}$$

where N_0 is Avogadro's number, the number of molecules in a gram molecular weight. Einstein went on to show that it had a more general significance. From the distribution law, Eq. (12.1), it is easy to see that the energy fluctuations satisfy the equation

$$\overline{E^2} - (\bar{E})^2 = kT^2 \frac{d\bar{E}}{dT}. \tag{12.5}$$

The left-hand side can also be expressed as $\overline{(\Delta E)^2}$, where ΔE is defined as $(E - \bar{E})$, and so the equation can be rewritten in the form,

$$\overline{(\Delta E)^2} = kT^2 \frac{d\bar{E}}{dT}. \tag{12.6}$$

Since the energy fluctuations are a measure of the degree to which the system departs from its average or thermodynamic behavior, Einstein described the universal constant k as determining the "thermal stability" of systems. He considered this to be its real significance. Einstein found this last equation particularly interesting because "it no longer contains any quantity that suggests the assumptions underlying the theory" from which it was derived. Such results always alerted him to the possibility that they were more general than the theory that produced them, and this case was no exception.

If only it were possible to measure the energy fluctuations for some system, Eq. (12.6) would then allow a direct determination of k. Einstein had to admit that no such possibility had yet been found, but he could point to one system where experimental information could be linked to the fluctuations. That system was thermal radiation in an empty enclosure, blackbody radiation. The fluctuations would be due to interference of the waves making up the radiation field. Although one would expect the relative fluctuations in energy to be negligibly small when the linear dimension of the enclosure is large compared to the wavelength at which the spectral distribution has its peak, the relative fluctuations should be of the order of unity when that dimension is approximately equal to the wavelength at the maximum of the distribution.

Einstein then boldly applied Eq. (12.6) to calculate the energy fluctuations of thermal radiation, a system that certainly did not satisfy the conditions of its derivation. Using the Stefan–Boltzmann law for the average energy, he readily estimated the linear dimension of a volume in which the relative fluctuations would be of the order of one. The condition he found required this dimension to be inversely proportional to the temperature with a coefficient that could be evaluated from k and the Stefan–Boltzmann constant. Both the form of this result and the order of magnitude of the coefficient agreed with the way the wavelength of the maximum in the distribution law was known to depend on temperature, from experiment and from Wien's displacement law.

It was this remarkable connection between k and the wavelength that Einstein referred to at the start of his paper and again in a letter to Conrad Habicht: "I have now found the relationship between the size of the elementary units of matter and the wavelengths of radiation in an extremely simple way."[37] And as he said at the end of the paper: "I believe that because of the great generality of our assumptions this agreement ought not be ascribed to chance." Einstein was now convinced that studying the fluctuations in those properties of a system whose average values are given by thermodynamics provided a new and valuable way of getting at some important but subtle physical relationships. He was the first to take fluctuations — which are, after all, the peculiarly statistical part of a statistical theory — completely seriously, and he would continue to rely on the remarkable insights they could offer in the course of his next twenty years of exploring the problems of radiation. [38]

For a year after he submitted this paper to the *Annalen* Einstein wrote nothing, but on March 17, 1905, he sent another paper to the same journal which showed how very far his thinking had gone during the interval.[39] Even he referred to it as "very revolutionary," a word he would never apply to his work on relativity.[40] It offered a new "heuristic viewpoint" for considering the emission of light and some of its interactions with matter. Einstein showed that a number of phenomena of this sort including fluorescence, photoionization, and the photoelectric effect, previously quite obscure, could be more readily understood if one gave up the idea that radiant energy is continuously distributed in space. He proposed instead that one think of the energy of light as consisting of a finite number of localized energy quanta, units of energy that can be produced or absorbed only as a whole. This proposal of light quanta was utterly heretical, and Einstein was well aware of it. He knew the mass of evidence for the wave theory of light that had been built up through the nineteenth century, evidence that culminated in Hertz's experimental confirmation of Maxwell's theory of light as an electromagnetic wave. What could have prompted Einstein to make his extraordinary suggestion of light quanta? Most of his paper was devoted to answering precisely this question.

Einstein began by pointing to the dichotomy in the foundations of physics, his deepest concern. "There is a profound formal difference," he wrote, "between the theoretical ideas which physicists have formed concerning gases and other ponderable bodies and the Maxwell theory of electromagnetic processes in so-called empty space." The discrete mechanics of material bodies composed of atoms and the continuous field theory of electromagnetism employed very different sets of variables and very different mathematical formalisms. This was the dualism between particle and field, between mechanics and electromagnetism, in which neither set of concepts had been reducible to the other. It was the dualism brought out most sharply by Lorentz's electron theory, and Einstein had discovered that it had some disturbing aspects. When these two disparate fundamental theories had to be brought to bear together in the same situation there could be serious problems. The first section of Einstein's paper was concerned with one of these, one so serious that Paul Ehrenfest would later call it "the ultraviolet catastrophe."[41]

Einstein had been reflecting some more on what Planck had done. Planck's derivation of the radiation spectrum purported to be based on electrodynamics, the electron theory, and Boltzmann's statistical mechanics, but Einstein (and some of Planck's other readers) had trouble disentangling the various strands in his argument. Einstein took up the problem in his own way and showed that the combination of classical theories on which Planck had drawn would lead, not to Planck's results, but to a disaster.

He considered a volume, enclosed by reflecting walls, that contained electromagnetic radiation, a gas, and also a number of charged particles (electrons) bound harmonically to various points within the enclosure. These oscillating electrons would exchange energy with the freely moving gas molecules by collisions, and they would also emit and absorb electromagnetic radiation. If there were oscillators of essentially all relevant frequencies present, then at equilibrium the distribution law should be that of blackbody radiation. The oscillating electrons (Planck's resonators) served as links between the material system — the gas, described by mechanics — and the electromagnetic radiation, described by Maxwell's equations. Both theories could be used to find the average energy u_ν of an oscillator of frequency ν when the system is in equilibrium at temperature T. For equilibrium with the gas the oscillator must have an average energy given by the equipartition theorem,

$$u_\nu = kT. \tag{12.7}$$

For equilibrium with the radiation the rates of absorption and emission must be equal, and Planck had shown in 1899 that this leads to the condition,

$$u_\nu = (c^3/8\pi\nu^2)\, \rho(\nu,T), \tag{12.8}$$

where c is the velocity of light and $\rho(\nu,T)d\nu$ is the energy per unit volume of blackbody radiation at temperature T having frequencies in the interval ν to $\nu + d\nu$.[42] Since both equations must hold at equilibrium, the radiation spectrum seems to be determined, with

$$\rho(\nu,T) = (8\pi\nu^2/c^3)kT. \tag{12.9}$$

But, as Einstein immediately pointed out, this result is not only in conflict with experiment, it is intrinsically unacceptable. It does not really give a definite distribution of energy between field and matter since the integral of $\rho(\nu,T)$ over all frequencies is infinite.

Einstein was the first to point out this unfortunate consequence of classical physics and to recognize its implications. Planck had never mentioned the equipartition theorem, nor had he ever written the equivalent of Eq. (12.9) in any of his papers on the radiation problem. Lord Rayleigh did arrive at the essential features of this equation in June 1900 using a very different argument, but he had immediately supplied an exponential convergence factor and never discussed the basic difficulty that Einstein emphasized.[43]

The distribution law that accounted for all experimental results was the one Planck had arrived at in October 1900 and then derived two months later.[44] Einstein was still not ready to analyze that derivation, but he did comment on one important aspect of Planck's work. As I mentioned earlier the most impressive consequence of that work — the one that may well have convinced Planck himself to take it seriously — was the new determination of the basic constants. Einstein showed that this result did not really depend on Planck's special assumptions, or even on his complete distribution law, and that Eq. (12.9) did have some connection with that law.

Planck's distribution had the form,

$$\rho(\nu,T) = \frac{\alpha\nu^3}{\exp(\beta\nu/T) - 1}, \tag{12.10}$$

where α and β are constants. In the region of long wavelengths and large radiation densities (high temperatures) this goes over into an equation of the same form as Eq. (12.9),

$$\rho(\nu,T) = (\alpha/\beta)\nu^2 T. \tag{12.11}$$

By identifying the coefficients of the two equations one could find k, and the basic constants that follow from it, from Planck's radiation constants α and β. The results were, of course, the same as Planck's. "The greater the energy density and the wavelength," Einstein concluded, "the more serviceable the theoretical foundations we used prove to be; but for short wavelengths and low radiation densities they fail completely."

That failure of the foundations was the great problem, but there was no way Einstein could find to attack it directly. He had apparently not yet seen what enabled Planck to reach the new distribution law for radiation; he simply did not mention the introduction of "energy elements" $h\nu$, which Planck had originally called "the most essential point of the whole calculation."[45] Einstein turned instead to a consideration of "the connection between blackbody radiation and experiment without assuming any picture of the way radiation is produced and propagates." In other words he would not try to argue *to* the distribution law, since there seemed to be no sound basis on which to derive it, but would try instead to argue *from* the known distribution to any of its consequences capable of being explored by experiment. What other properties must radiation have if its spectral distribution at thermal equilibrium has the known form?

Einstein took as the basis of his discussion the older and simpler distribution law, first given by Wien, according to which

$$\rho(\nu,T) = \alpha\nu^3 \exp(-\beta\nu/T). \tag{12.12}$$

This is the limiting form of Planck's distribution, Eq. (12.10), in the region of high frequencies and low temperatures, just the region in which the classical foundations had failed. It was also the region in which Wien's law had shown itself completely adequate for describing experimental results. Einstein emphasized repeatedly in his paper that the conclusions he drew about the nature of radiation could only be expected to hold in the region where Wien's distribution law was valid.

Using the relation among temperature, entropy, and energy given by the second law, Einstein derived from Wien's $\rho(\nu,T)$ an expression for the entropy S of that part of the blackbody radiation having frequencies in the interval ν to $\nu + d\nu$. If we write the energy E of that part of the radiation in the form

$$E = V\rho(\nu,T)d\nu, \tag{12.13}$$

where V is the volume of the enclosure, then S is given by the equation

$$S = (E/\beta\nu)\{1 + \ln (V\alpha\nu^3 d\nu/E)\}. \tag{12.14}$$

Although this expression is not particularly transparent, one can now readily calculate the change in the entropy of this radiation when the volume changes from V_1 to V_2 at fixed energy E:

$$S_2 - S_1 = (E/\beta\nu) \ln (V_2/V_1). \tag{12.15}$$

The entropy of this radiation shows the same volume dependence as "the entropy of an ideal gas or a dilute solution," where one has the equation,

$$(S_2 - S_1)_{\text{gas}} = Nk \ln (V_2/V_1), \tag{12.16}$$

if the gas is composed of N molecules.

Was this striking similarity a mere coincidence, or did it suggest something essential about the nature of radiation? To answer this question Einstein turned to Boltzmann's relation between entropy and probability, a relation that in his opinion neither Planck nor Boltzmann had given its full physical significance nor used to its full potentialities. Promising a detailed discussion in a later paper, Einstein restricted himself to using the principle in the form

$$S_2 - S_1 = k \ln W, \tag{12.17}$$

where the left-hand side is the entropy difference between two states of a thermodynamic system and W is the relative probability of the occurrence of these two states. If one considers a gas contained in a volume V_1 and asks for the probability W_g that the N molecules of the gas are all to be found in the subvolume V_2 at some moment, the answer will evidently be that

$$W_g = (V_2/V_1)^N. \tag{12.18}$$

The point to be stressed is that "no assumption needs to be made about the law of motion of the molecules in order to deduce this equation"; it requires only that the particles move independently of one another and show no preference for one part of the volume compared to another.

Einstein then turned the argument around and applied it to the radiation. Equations (12.15) and (12.17) show that the probability of finding all the blackbody radiation (at frequency ν) in the subvolume V_2 of volume V_1 must be given by the equation

$$W = (V_2/V_1)^{N'}, \tag{12.19}$$

where the exponent N' is given by

$$N' = (E/\beta k\nu). \tag{12.20}$$

Einstein drew the conclusion that he found inescapable: "Monochromatic radiation of low density (within the region of validity of the Wien radiation formula) behaves with respect to thermal phenomena as if it consisted of independent energy quanta of magnitude $\beta k\nu$." This was the basis for his suggesting that revolutionary "heuristic viewpoint" on the nature of radiation — the hypothesis of light quanta — which, he hoped, might "prove useful to some investigators in their researches."[46]

We know how seriously Einstein took this proposal, how effectively he used it in predicting the dependence of the energy of photoelectrons on the frequency of the light that freed them from a metal surface and in explaining the lack of dependence of that energy on the intensity of the incident light. But he was virtually alone in giving his confidence to light quanta. His contemporaries were not about to consider a corpuscular model of light proposed by an unknown young man on the strength of an argument based on the theory of fluctuations, a theory that not even the founders of statistical mechanics had ever previously put to any use. It would be a long time before the light quantum won much support from anyone other than its proposer.

Einstein wrote nothing more on quanta in 1905. He was concentrating on several other problems, as we know. But in March 1906, exactly a year later, he completed another study, "On the Theory of the Production and Absorption of Light."[47] He had at last penetrated the obscurities of Planck's theory of the spectrum of blackbody radiation, and was evidently surprised at what he found. Anyone who still thinks that Einstein's use of light quanta in 1905 was a generalization or extension of Planck's theory need only read this paper of 1906 to be disabused of that idea. For Einstein began by remarking that when he wrote his paper of the previous year it had seemed to him that Planck's work was a "contrast" to his own. Planck had indicated that his theory was built on Maxwell's theory, the electron theory, and statistical mechanics, and yet it had led to the distribution in Eq. (12.10). Einstein was convinced, however, that if one started with precisely those basic theories the only possible outcome was the unacceptable distribution given by Eq. (12.9). How could Planck have arrived at a different answer if he had argued from the same premises? That was what puzzled Einstein, even as it puzzled Lord Rayleigh, who asked the same question in *Nature* in the spring of 1905, and Paul Ehrenfest, who asked it again a year later.[48]

The answer to this puzzle consisted of two parts, as Einstein proceeded to show. The first was related to that "expression for the entropy of a system . . . assumed by Planck" which Einstein had already noticed in 1904. As he had shown then, and now rederived, the entropy of a system having energy E could always be written in the form $S = k \ln [\omega(E)]$, as given in Eq. (12.3), where $\omega(E)dE$ is the phase space volume of the energy shell. If the system consists of a collection of N harmonic oscillators obeying ordinary mechanics, then $\omega(E)$ is proportional to E^{N-1}, and there is no escape from equipartition and the disastrous distribution law that goes as $\nu^2 T$. To arrive at Planck's distribution, however, one must forbid the oscillators to have arbitrary values of their energies, and require instead that these energies assume only integral multiples of ε, where ε has the value,

$$\varepsilon = k\beta\nu. \tag{12.21}$$

With this assumption, calculating the volume of an energy shell of thickness ε for the system of N oscillators of frequency ν is equivalent to counting the number of ways W of distributing the given number of units ε over these oscillators. This is just how Planck determined the quantity W that he called the number of complexions and related to the entropy S by an equation resembling Boltzmann's,

$$S = k \ln W. \tag{12.22}$$

Since Planck had in effect made this restrictive assumption on the energies of his oscillators he must also have made another assumption, if only implicitly, and this was the second part of Einstein's answer to the puzzle. To go from the average energy u_ν of an oscillator to the spectral distribution function $\rho(\nu, T)$, Planck had used Eq. (12.8), which he had derived from electrodynamics. That derivation was no longer of any use, however, since it assumed a continuously varying oscillator energy. This meant that Planck had implicitly assumed the continuing validity of the classical relation between oscillator energy and spectral density even for his oscillators with a discrete set of energy values. This was far from an obvious assumption because in the region of validity of Wien's radiation formula the average energy u_ν of the oscillator is small compared to its energy quantum $k\beta\nu$.

Einstein's conclusion from all this was not that Planck's theory should be rejected, but rather that one had to recognize that Planck "had introduced a new hypothetical element into physics — the light quantum hypothesis."

When Einstein reviewed Planck's *Lectures on the Theory of Thermal Radiation* a few months later, he was even more impressed with this work.[49] Planck's book brought out the structure of his theory more clearly than his many papers had, and Einstein found it "a wonderfully clear and unified whole," "admirably suited for making the reader completely familiar with the material." Despite Planck's "use of a hypothesis based only on analogy," Einstein thought that "every unprejudiced reader" would find the result "a very probable one."

The more Einstein reflected on the implications of Planck's work, the more far-reaching he found them to be. By November 1906 he had seen an entirely new dimension of the problem, and he reported on it in his paper, "Planck's Theory of Radiation and the Theory of Specific Heat."[50] He now had an even simplier way of showing just how Planck had modified the kinetic-molecular theory of heat. This time Einstein worked directly with the structure function $\omega(E)$ of a single harmonic oscillator of frequency ν. If the oscillator is described by the standard equations of motion, then $\omega(E)$ is a constant; the average energy of the oscillator is then necessarily kT. To obtain Planck's distribution, the constant structure function had to be replaced by an $\omega(E)$ that differed from zero only when E had the values, $0, \varepsilon, 2\varepsilon, \ldots$, and that gave equal weight to all these integral multiples of the energy unit ε. With this new structure function the average energy is easily shown to be given by the equation,

$$\bar{E} = \frac{\varepsilon}{\exp(\varepsilon/kT) - 1}, \tag{12.23}$$

using the distribution function of Eq. (12.1). This expression for $\bar{E}$ replaces kT, but approaches it when the temperature is sufficiently large compared to (ε/k). The energy unit ε must be taken to be

$h\nu$, where h is Planck's notation for the constant Einstein was still denoting by $k\beta$. The distribution law then follows at once with the help of Eq. (12.8), which must be assumed to remain valid despite the limitation imposed on the oscillator's energy.

Einstein emphasized the seriousness of the step Planck had taken. "While up to now molecular motions have been supposed to be subject to the same laws that hold for the motions of the bodies we perceive directly . . . we must now assume that, for ions which can vibrate at a definite frequency and which make possible the exchange of energy between radiation and matter, the manifold of possible states must be narrower than it is for the bodies in our direct experience." The laws of motion obeyed by Planck's oscillators could not be the familiar laws of mechanics. But if that were so, one could not stop there. "I now believe that we should not be satisfied with this result. For the following question forces itself upon us: if the elementary oscillators used in the theory of the energy exchange between radiation and matter cannot be interpreted in the sense of the present kinetic-molecular theory, must we not also modify the theory for the other oscillators that are used in the molecular theory of heat?" Einstein had no doubt about the answer: "If Planck's theory strikes to the heart of that matter, then we must expect to find contradictions between the present kinetic-molecular theory and experiment in other areas of the theory of heat," contradictions that could also be resolved by restricting the possible values of the oscillators' energy. The rest of Einstein's paper described the contradictions between theory and experiment concerning the specific heats of solids and showed how the new way of treating oscillators removed these contradictions. Einstein had seen that there would have to be a quantum theory of matter as well as radiation and had taken the first step toward creating one.

This paper showed the power of such a theory as Einstein established unexpected connections between the optical properties of solids (their infrared absorption frequencies) and their thermal behavior (specific heats). It also contained a new law for the specific heats of solids to replace the Dulong–Petit rule: at temperatures high compared to the characteristic oscillation frequency ν of the atoms in the solid — or rather compared to $(h\nu/k)$ — the specific heat will approach the classical, or Dulong–Petit, value; but at low temperatures compared to $(h\nu/k)$ the specific heat of any solid will be small, approaching zero as the temperature approaches absolute zero. This general theorem was confirmed by the measurements made by Walther Nernst and his coworkers a few years later in the course of experiments initially undertaken for quite another reason. Nernst's quick recognition of "the mighty logical power of the quantum theory" helped bring that theory into a prominent position that it might not easily have attained otherwise. Einstein's quantum theory of matter was accepted much more readily than were his ideas on the quantum structure of radiation.

By 1909 Einstein was ready to present some of the results of his continual reflection on the nature of radiation. He did so in a pair of papers devoted to that "extraordinarily important problem," both published in the *Physikalische Zeitschrift* that year.[51] The second of these was the text of the lecture he gave in September to the annual meeting of German scientists, held that year at Salzburg. No less severe a critic than Wolfgang Pauli has called this paper "one of the landmarks in the development of theoretical physics."[52] Even among Einstein's papers it stands out as one of the masterworks.

"It is undeniable," Einstein told his colleagues in Salzburg, "that there is an extensive group of facts concerning radiation which show that light has certain fundamental properties that can be understood much more readily from the standpoint of the Newtonian emission theory than from the standpoint of the wave theory. It is my opinion, therefore, that the next phase of the development of theoretical physics will bring us a theory of light that can be interpreted as a kind of fusion of the wave and emission theories. The purpose of the following arguments is to give a foundation for this opinion, and to show that a profound change in our views of the nature and constitution of light is absolutely necessary.''

Few, if any, of the physicists in Einstein's audience were prepared to follow him when he called for that ''profound change'' in thinking about the nature of radiation.[53] He was by this time no longer the unknown patent examiner who had proposed a possible granular structure to radiation four years earlier. Younger physicists like Max von Laue, Jakob Laub, and Rudolf Ladenburg had already traveled to Berne to meet, talk, and even to work with Einstein.[54] He was in correspondence with Planck and Lorentz, exchanging views on the radiation problem in particular.[55] The ''secular cloister'' of the Swiss patent office had been left behind forever at just about the time of the Salzburg meeting as Einstein took up his first academic position at the University of Zürich. Nevertheless his ideas on radiation were still far too extreme to win much acceptance. They would require giving up Maxwell's equations and admitting that light waves themselves have a corpuscular structure, as Planck pointed out in the discussion period. ''That seems to me,'' he added, ''to be a step that, in my opinion, is not yet called for.''[56]

In putting forward his new ideas, Einstein, who was addressing a major scientific gathering for the first time, expressed himself with his usual modesty, recognizing the degree to which he had been isolated from his contemporaries. What he was reporting,he said, was in large part ''the result of reflections which had not yet been sufficiently checked by others.'' He was talking about them despite that, not through ''excessive confidence'' in his own insight but in the hope of inducing one or another of those in the audience to concern himself with these questions. (He had expressed the same thought in a letter to Laub, writing: ''This quantum question is so incredibly important and difficult that everyone should busy himself with it.[57]) Despite his qualifying remarks Einstein did have confidence in the structure of argument he had created to support his assertion that the theory of radiation was only at the beginning of ''a development whose outcome could not yet be foreseen, but that would without any doubt be of the greatest significance.''

Einstein began his lecture by reviewing the successes and the difficulties of the electromagnetic theory of light, with particular emphasis on the clarification brought to this subject by Lorentz's work. He characterized his own ''so-called relativity theory'' as a reconciliation of the relativity principle, which had to be assumed valid for all physical phenomena on the basis of experiments like Michelson's, with the essential feature of Lorentz's theory, the principle that the speed of light is independent of the speed of its source. Einstein called attention to two features of the theory of relativity relevant to his subject, ''our views on the nature of light,'' two features that relativity shared with an emission or corpuscular theory rather than with the wave theory. Since the theory of relativity abolished the aether, the electromagnetic fields that constitute light were no longer to be thought of as states of a hypothetical medium, but rather as independently existing entities sent out from their source in much the same way as in the Newtonian emission theory. Furthermore, because of the equivalence of mass and energy, when light is emitted from a body A and then absorbed by another body B, there is a transfer of mass from A to B. But these were the only ways in which the theory of relativity modified the mode in which one had to think about the nature of radiation. The principal support for Einstein's opinion came from other quarters, and he proceeded to marshal that support.

He reminded his audience that there were many well-known properties of light that could not be accounted for by the wave theory. Why, for example, did the occurrence or nonoccurrence of a particular photochemical reaction depend on the color of the incident light, and not on its intensity, when the intensity ought to determine the available energy? Why should light of short wavelengths be more effective in producing chemical reactions than light of longer wavelengths? Why are higher temperatures, and therefore higher molecular energies, needed if there is to be a larger component of short wavelength radiation in the radiation emitted by a body? How does a single photoelectron acquire so much energy from a light source whose energy is presumably distributed at a very low density? Einstein suggested that all these difficulties might have a common origin in one essential feature of the wave theory of light: the basic emission process does not have a simple inverse. If the emission of light consists of the production of an expanding spherical wave by an oscillating charge, then its inverse, the absorption of a contracting spherical wave, would require "an enormous number" of emitters for its approximate realization; it would surely not be an elementary process.

A corpuscular theory of light would evidently be better in this respect, since the ejection of a particle of light and its absorption would possess the kind of symmetry Einstein saw as desirable. As usual with his points there was a simple physical argument involved. The example Einstein cited was the production of secondary cathode rays by X rays. Since the velocities of these secondaries are of the same order of magnitude as the velocities of the primary electrons that produce the X rays, it is hard to see how the X rays can be expanding spherical waves. A particle theory, on the other hand, would make all the radiant energy available at absorption. "The elementary process of emission of radiation seems to be direct," Einstein concluded.

He found the strongest support for his opinion — that radiation is not adequately characterized by the wave theory — in the implications of Planck's theory of blackbody radiation, a theory he described briefly because he "probably should not assume it to be generally known." Einstein emphasized that, despite any appearances to the contrary, Planck's use of energy elements ε, where ε is equal to $h\nu$, meant a real departure from previous theory. The discreteness of these energy elements would be innocuous only if the average energy u_ν of one of Planck's oscillators was always large compared to $h\nu$. This condition, however, was flagrantly violated in commonly studied circumstances, so that Planck's count of complexions gave a value for the entropy essentially different from the result of applying Boltzmann's methods. "Accepting Planck's theory means, in my opinion, plainly discarding the foundations of our theory of radiation," Einstein concluded. But that was not a reason for rejecting Planck's theory, which had just had another striking success as recent experiments confirmed Planck's value for the unit of electric charge.

Einstein then raised the crucial questions, the questions that determined the difference between his own views and those of even his boldest colleagues: "Would it not be conceivable that Planck's radiation formula were indeed correct, but that it could be derived by some method that did not depend on such an apparently monstrous assumption as Planck had used? Would it not be possible to replace the hypothesis of light quanta by some other hypothesis by means of which one could do equal justice to the known phenomena? If it is necessary to modify the principles of the theory could one not at least retain the equations for the propagation of radiation and interpret only the elementary events of emission and absorption in a way different from that used up to now?"

His answer to all these questions was negative. The hypothesis of light quanta was not merely a possible basis for arriving at Planck's radiation law, not merely a sufficient condition. By assuming the correctness of that law and arguing from it, exploring its implications, Einstein had now shown that the quantum structure of light was a necessary consequence of Planck's law. This was, after all, the direction he had taken in his 1905 paper, but then he had limited himself to the high-frequency region by starting with the limiting form of Planck's law, the Wien distribution. His new arguments led to something even more interesting — a direct indication of that "fusion of the wave and emission theories" that the future would have to bring if, as he had written earlier, "Planck's theory strikes to the heart of the matter."

There were two independent arguments that Einstein referred to in his lecture, leading to the same conclusion by distinctly different methods, both of them uniquely his own. He did not give the details in his lecture since both methods had been published in the paper he had written at the beginning of the year. [58] Einstein's first argument involved a calculation of the energy fluctuations in blackbody radiation, something he had been concerned with since 1904. He had now refined his analysis of Boltzmann's principle and could rederive the basic fluctuation equation, Eq. (12.6), without needing to refer to any underlying statistical *mechanics*. (This was part of his creation of a statistical thermodynamics, which I cannot go into here.[59]) Einstein applied the fluctuation equation to that part of the blackbody radiation, in a partial volume V at temperature T, having frequencies in the interval ν to $\nu + d\nu$. This system has energy E (I omit the average sign for convenience) given by the equation

$$E = V\rho(\nu,T)\,d\nu\ , \tag{12.24}$$

where $\rho(\nu,T)$ is the Planck distribution,

$$\rho(\nu,T) = \left(\frac{8\pi\nu^2}{c^3}\right)\frac{h\nu}{\exp(h\nu/kT) - 1}\ . \tag{121.25}$$

(This is Eq. (12.10) with the constants now written in Planck's notation.) A direct application of Eq. (12.6) leads to the result

$$\overline{(\Delta E)^2} = \{h\nu E + \frac{c^3}{8\pi\nu^2 V d\nu}E^2\}, \tag{12.26}$$

which can also be written in the form

$$\overline{(\Delta E)^2} = \{h\nu\rho + \frac{c^3}{8\pi\nu^2}\rho^2\}\ V d\nu. \tag{12.27}$$

If the fluctuations in energy were due only to interfering waves, as electromagnetic theory would imply, only the second term in either of these equations could be present. Einstein argued this on dimensional grounds and his assertion was confirmed several years later when Lorentz carried out the rather lengthy calculations in detail.[60] This second term would alone be present if the Planck law were replaced by its low-frequency ("Rayleigh–Jeans") limit.

Suppose, on the other hand, that the radiation were composed of classical independent particles, each of energy $h\nu$, so that there would be $E/h\nu$ such particles in V, on the average. Then a familiar statistical argument would lead to just the first term in Eq. (12.26) for the fluctuations. (The mean square fluctuation in the number of particles would be the average number $E/h\nu$ itself.) This would also be the result if the high-frequency or Wien limit of Planck's law had been used in Eq. (12.24).

Einstein remarked that it was as though there were two independent causes producing the fluctuations. The same kind of fusion of wavelike and particlelike fluctuations resulted from his second argument, an analysis based on the method he had developed for his theory of Brownian motion. He considered a flat plate, a mirror, suspended in an enclosure containing an ideal gas as well as thermal radiation. The plate is free to move in the direction perpendicular to its plane. The collisions of the gas molecules with the plate will set it into an irregular motion, a particular case of Brownian motion. When the plate moves, the forces on its front and back due to radiation pressure become unequal, and it will suffer a "radiation friction," approximately proportional to its velocity. If this were the only effect of the radiation, all the energy of the gas would eventually be transformed into energy of the radiation. Since there is an equilibrium between gas and radiation, however, the frictional loss of energy must be made up by the fluctuations of the radiation pressure. The kinetic energy of the moving plate must, on the average, be equal to $kT/2$.

Let v be the velocity of the plate at time t, and let P be the resistive force per unit velocity due to the radiation pressure. Suppose that during the time interval τ the plate acquires momentum Δ due to fluctuations in radiation pressure, and loses momentum $Pv\tau$ due to "radiation friction." The condition of stability is that

$$\overline{(mv + \Delta - Pv\tau)^2} = \overline{(2mv)^2} . \tag{12.28}$$

Since the fluctuations are irregular, $\overline{v\Delta}$ will vanish. The term $\overline{(Pv\tau)^2}$ can be neglected as small, so that we obtain

$$\overline{\Delta^2} = 2P\tau m\overline{v^2}. \tag{12.29}$$

We now introduce the condition that the plate be in equilibrium with the gas,

$$\frac{m\overline{v^2}}{2} = \frac{kT}{2} . \tag{12.30}$$

It is still necessary to evaluate the coefficient of "radiation friction," P, in terms of the spectral distribution function for the radiation $\rho(\nu,T)$. Einstein considered the special case in which the plate is a mirror only for radiation in the frequency interval ν to $\nu + d\nu$, and is transparent otherwise. A relatively lengthy calculation that Einstein published in 1910,[61] led to the result

$$P = \frac{3}{2c}\left[\rho - \frac{\nu}{3}\frac{\partial\rho}{\partial\nu}\right]Ad\nu , \tag{12.31}$$

where A is the area of the plate. If one now evaluates P for the case that ρ is the Planck distribution, and works out the mean square fluctuation $\overline{\Delta^2}$ of the plate's momentum, the result has the form,

$$\overline{\Delta^2} = \{h\nu\rho + \frac{c^3}{8\pi\nu^2}\rho^2\} \frac{A\tau}{c} d\nu . \tag{12.32}$$

The structure of this equation exactly matches that of Eq. (12.27) for the energy fluctuations, and the result can be interpreted in a similar fashion.

Convincing as these results were to Einstein, they left his colleagues unpersuaded. While he saw the analysis of fluctuations as a powerful instrument for exploring the structure of radiation, most other physicists were barely convinced of the existence of such fluctuations.[62] Einstein himself was so certain that he had found the right direction for making progress on the radiation problem that he was "ceaselessly occupied" with it at this time, trying to develop that "fusion of the wave and emission theories."

There are only a few remarks in his 1909 papers and some evidence from his letters about the nature of his efforts.[63] He apparently tried to find a suitable modification of the wave equation, and considered both linear and nonlinear possibilities of various orders. Early in 1909 he was still hopeful enough to remark: "The variety of possibilities does not seem to be so great that one has to be scared off by the problem."[64] In September he was almost apologetic about being unable to offer a solution to the problem. "The principal difficulty," he explained, "lies in the fact that the fluctuation properties of radiation, as expressed in the above equations, present small foothold for setting up a theory. Just suppose that the phenomena of diffraction and interference were still unknown, but that one knew that the average value of the irregular fluctuations of the radiation pressure were given by the second term of the above equation, where ν is a parameter of unknown significance that determines the color. Who would have enough imagination to construct the wave theory of light on this foundation?"[65]

Einstein's intense efforts to construct a new fundamental theory that would unify particle and field, a modified electron theory, seem to have gone on through 1910, judging by remarks in his letters. By May 1911, however, he seems to have abandoned that search, as he explained in a letter to Michele Besso: "I no longer ask whether these quanta really exist. Nor am I trying any longer to construct them, because I now know that my brain is incapable of accomplishing such a thing. But I am searching through the consequences [of the quantum hypothesis] as carefully as possible in order to learn the range of applicability of this concept."[66] The spring of 1911 was also the time when Einstein began to devote himself to the problem of gravitation, the problem that would grip him more and more tightly during the next five years.

He did not put the quantum theory completely aside, as his letter to Besso indicates. In Prague, in Zürich, and in Berlin after his move there in 1914, Einstein went on exploring the connections between the concept of quanta and the rest of physics. His considerations went from the specific heats of solids to the law of photochemical equivalence. He used his favorite methods, those of statistical thermodynamics, to study the question of zero-point energy, and to conclude from another investigation that: "A quantum type of change in the physical state of a molecule seems to be no different in principle from a chemical change."[67] But interesting and important as this work may have been, Einstein's real efforts were being directed elsewhere — to the construction of his general theory of relativity. This was the period he would later describe as "the years of anxious searching in the dark, with their intense longing, their alternations of confidence and exhaustion, and the final emergence into the light."[68]

Einstein resumed his reflections on the problem of radiation in 1916. During that summer he wrote to Besso that he had a "splendid idea" about the emission and absorption of light. "An astonishingly simple derivation of the Planck formula, I might even say *the* derivation. Everything completely quantal."[69] Einstein sent a short paper on this subject to the German Physical Society in July[70] and then wrote a longer one, containing even more striking results, some months later.[71] The first paper is almost never discussed, since its basic argument and results are repeated in Einstein's later more complete and more accessible paper, but it is worth looking at more closely for the insight it gives into Einstein's approach to the problem.

He began with some remarks about Planck's original derivation which "had called the quantum theory into being 16 years ago." Despite the "unprecedented boldness" of its derivation, everything about Planck's radiation law had been "brilliantly confirmed" in a variety of ways. What was still unsatisfactory, however, was the incompatibility between the *classical* analysis of absorption and emission, that had allowed Planck to relate the spectral distribution to the average energy of an oscillator, and the *quantized* energy of that oscillator itself. As Einstein pointed out, any residual doubts about the idea of quantization must surely have been dispelled by the great success of Bohr's theory of spectra. A unified discussion of Planck's radiation law would therefore require the replacement of the classical parts of his derivation by quantum theoretical considerations.

Einstein described the classical theory of an oscillator in the radiation field in a way "first employed in the theory of Brownian motion." If E represents the energy of the oscillator at some moment, one then asks how E will be changed after a time τ, where τ is long compared to the period of the oscillator but short enough so that the fractional change in energy during τ is very small. There are two ways in which E changes: the oscillator will lose energy $\Delta_1 E$ by radiating, where

$$\Delta_1 E = -AE\tau . \tag{12.33}$$

The oscillator's energy will also change by an amount $\Delta_2 E$ due to the work done on it by the electric field of the surrounding radiation, where, on the average,

$$\Delta_2 E = B\rho\tau . \tag{12.34}$$

"This second change increases with increasing density of radiation and has a magnitude that depends on 'chance,' and also a sign that depends on 'chance.'" (The sign depends on the phase relationship between oscillator and radiation.) The constants A and B can be calculated from electromagnetic theory. Since the average value of E taken over many oscillators must be stationary, it follows that

$$\overline{E + \Delta_1 E + \Delta_2 E} = \overline{E} , \tag{12.35}$$

or,

$$\overline{E} = (B/A)\rho . \tag{12.36}$$

When the values of A and B are supplied from the electromagnetic calculations, Eq. (12.36) reduces to Eq. (12.8), the result that Planck had derived in 1899. Einstein's new description of Planck's work served as his guide to modifying it in a way appropriate to the quantum theory.

Instead of restricting his attention to harmonic oscillators, Einstein considered a gas of molecules in equilibrium with thermal radiation. Each molecule could exist only in certain states, Z_1, $Z_2, \ldots$, where it would have the corresponding energy values $\varepsilon_1, \varepsilon_2, \ldots$. In thermal equilibrium at temperature T, the number of molecules in state Z_n must be proportional to the quantity, $p_n \exp(-\varepsilon_n / kT)$, where p_n is the statistical weight of state Z_n, a number characteristic of the state and independent of temperature. A molecule in state Z_n can make a transition to a state Z_m of higher energy ε_m by absorbing radiation of frequency $\nu = \nu_{nm}$. In this process energy $(\varepsilon_m - \varepsilon_n)$ is absorbed. Transitions in the opposite direction are also possible by emission of radiation of the same frequency ν. Einstein required that these two processes exactly balance at equilibrium, so that only a single pair of states need be considered in finding the condition for equilibrium.

He took the transitions to be of two types, choosing his hypotheses on the analogy of the classical discussion. If a molecule is in the upper state it can radiate and go to the lower state whether or not external radiation is present. This process, Einstein commented, is just like a sort of radioactive decay. The number of such transitions per unit time is set equal to $A_m{}^n N_m$ where N_m is the number of molecules in state Z_m and $A_m{}^n$ is a constant characteristic of the pair of states involved. The second type of transition is effected by the radiation field in which the molecule is placed. Just as a classical oscillator can be forced to absorb or to emit energy by radiation of its own frequency depending on the relative phase, so there can be transitions induced between states Z_m and Z_n in both directions. The numbers of such transitions per unit time are, respectively, $B_m{}^n N_m \rho$ and $B_n{}^m N_n \rho$, for induced transitions from Z_m to Z_n and from Z_n to Z_m.

The equilibrium condition is the equality of the total transition rates in the two directions,

$$A_m{}^n N_m + B_m{}^n N_m \rho = B_n{}^m N_n \rho \, . \tag{12.37}$$

The exponential law for the equilibrium population of the states requires that

$$\frac{N_n}{N_m} = \frac{p_n}{p_m} \exp\{(\varepsilon_m - \varepsilon_n) / kT\} \, . \tag{12.38}$$

If these two equations are combined, one obtains

$$A_m{}^n p_m = \rho [B_n{}^m p_n \exp\{(\varepsilon_m - \varepsilon_n) / kT\} - B_m{}^n p_m] \, . \tag{12.39}$$

The requirement that ρ must go to infinity as T does leads to the relationship

$$B_n{}^m p_n = B_m{}^n p_m \, . \tag{12.40}$$

Using this in Eq. (12.39) and writing α_{mn} for the ratio of $A_m{}^n$ to $B_m{}^n$ we finally obtain the result,

$$\rho = \frac{\alpha_{mn}}{\exp\{(\varepsilon_m - \varepsilon_n) / kT\} - 1} \, . \tag{12.41}$$

Despite appearances the quantities α_{mn} and $(\varepsilon_m - \varepsilon_n)$ cannot depend on the characteristics of the molecules considered because ρ must be a universal function of ν and T according to Kirchhoff's law. Furthermore, the frequency dependences of α_{mn} and of $(\varepsilon_m - \varepsilon_n)$ are fixed by Wien's displacement law[72]: α_{mn} must be proportional to ν^3 and $(\varepsilon_m - \varepsilon_n)$ to ν,

$$\varepsilon_m - \varepsilon_n = h\nu \,, \tag{12.42}$$

where h is a universal constant. This result, as Einstein remarked, is identical to one of the postulates of Bohr's theory of spectra. The quantities $A_m{}^n$ and $B_m{}^n$ would be directly calculable "if we possessed an electrodynamics and a mechanics modified according to the quantum hypothesis."

Even without supplying those new theories Einstein was able to take another decisive step. "It can be conclusively shown," he wrote to Besso on August 24, 1916, "that the elementary processes of emission and absorption are directed events. One need only investigate the (Brownian) motion of a molecule . . . in the radiation field."[73] And on September 6 he emphasized to Besso what was really involved: "The essential thing is that the *statistical* considerations that lead to Planck's formula have now become *uniform* and as general as possible, with nothing assumed any longer about the mediating molecules except the most general idea of quantization. It then follows . . . that in every elementary transfer of energy between radiation and matter there is a transfer of momentum $h\nu/c$. It follows from this that every such elementary process is a *completely directed* event. And with that light quanta are as good as certain."[74]

In the longer version of his paper Einstein emphasized the need to consider the elementary processes of absorption and emission as completely directed, calling this the "principal result" of his analysis. His calculation of the Brownian motion of a molecule in the radiation field followed the lines he had laid down in 1909. The mean square momentum $\overline{\Delta^2}$ a molecule acquires from the fluctuating radiation in time τ again satisfies the equation

$$\overline{\Delta^2} = 2P\tau kT \,. \tag{12.43}$$

This time, however, Einstein calculated both $\overline{\Delta^2}$ and P (the "friction coefficient") for the assumed mode of interaction between field and matter. The results were consistent — that is, Eq. (12.43) was satisfied — only for directed processes. Even though the theory left "the time and direction of the elementary processes to 'chance'," Einstein expressed his "full confidence in the reliability of the course taken."

But Einstein's colleagues were no more to be persuaded in 1917 by the fluctuation arguments he found so convincing, than they had been in 1905 or 1909. Einstein tried hard to find a crucial experiment that would distinguish between the wave and quantum theories. When he proposed one in 1921,[75] and it seemed for a short while as though the wave theory were ruled out by the observations of Hans Geiger and Walther Bothe, the result was startling. "If your light experiment really turns out anticlassically," Paul Ehrenfest wrote to Einstein, "I mean after both theoretical and experimental criticism, then, you know, you will have become really *uncanny* to me." That experimental proof did not, in fact, hold up under Ehrenfest's sharp critical analysis of its premises.[76] But by 1923 the Compton effect was well established and that discovery, as Arnold Sommerfeld put it, "sounded the death knell of the wave theory of radiation."[77]

For Einstein, however, there would be no simple answer to the problem of radiation. The duality he had found in 1909 was there, and it was not to be explained away. "We now have two theories of light," he wrote in 1924, "both indispensable, but, it must be admitted, without any logical connection between them, despite twenty years of colossal effort by theoretical physicists."[78]

Only a few months earlier Einstein had recorded a statement about his work for the collection of recordings in the Preussischen Staatsbibliothek in Berlin.[79] He had begun with a general characterization of his interests: "From the earliest years all my scientific efforts have been directed to deepening the foundations of physics. The viewpoints and requirements of philosophy in the narrower sense had only a secondary effect on me." After describing his work on the special and general theories of relativity Einstein concluded with the following paragraph:

> The other great problem that I have been concerned with since about 1900 is that of radiation and the quantum theory. Stimulated by the work of Wien and Planck I recognized that mechanics and electrodynamics were in irresolvable contradiction with the experimental facts, and so I helped in creating that complex of ideas known by the name of quantum theory, which has been developed so fruitfully particularly by Bohr. I shall probably devote the rest of my life to the fundamental clarification of this problem, however slight the prospects for attaining this goal may be.

NOTES

1. A. Einstein, "Johannes Kepler," reprinted in his *Ideas and Opinions*, edited by C. Seelig (Dell, New York, 1973), p. 256.
2. R. S. Shankland, "Conversations with Albert Einstein," Am. J. Phys. **31**, 50 (1963).
3. A. Einstein, "Autobiographical Notes," in *Albert Einstein: Philosopher-Scientist*, edited by P. A. Schlipp (Library of Living Philosophers, Evanston, Ill., 1949) pp. 2–95.
4. A. Einstein, letter to M. Besso, 12 December 1951, in *Albert Einstein, Michele Besso, Correspondance 1903–1955*, edited by Pierre Speziali (Hermann, Paris, 1972), p. 453.
5. Einstein, *Ideas and Opinions*, op. cit. in n. 1, p. 333.
6. A. Einstein, "Autobiographische Skizze," in *Helle Zeit-Dunkle Zeit*, edited by C. Seelig (Europa Verlag, Zürich, 1956), p. 10.
7. In addition to references *Einstein, Philosopher-Scientist*, op. cit. in n. 3, and *Helle Zeit-Dunkle Zeit*, op. cit. in n. 6, see also P. Frank, *Einstein. His Life and Times*, translated by G. Rosen (Knopf, New York, 1947), pp. 18–21, and G. Holton, "Influences on Einstein's Early Work," in his *Thematic Origins of Scientific Thought. Kepler to Einstein* (Harvard University Press, Cambridge, Mass., 1973), pp. 197–217.
8. For a general discussion see M. J. Klein, "Mechanical Explanation at the End of the Nineteenth Century," Centaurus **17**, 58–82 (1972).
9. I. Newton, *Mathematical Principles of Natural Philosophy*, translated by A. Motte, revised by F. Cajori (University of California Press, Berkeley, 1966), pp. xvii–xviii.
10. Einstein, *Einstein, Philosopher-Scientist*, op. cit. in n. 3, pp. 18–21.
11. A. Einstein, "Folgerungen aus den Capillaritätserscheinungen," Ann. d. Phys. **4**, 513–23 (1901).
12. See M. J. Klein, "The Historical Origins of the Van der Waals Equation," Physica **73**, 28–47 (1974).
13. A. Einstein, letter to M. Grossmann, 14 April 1901, in C. Seelig, *Albert Einstein und die Schweiz* (Zürich, 1952), p. 52. Einstein's letter is marked only April 14. Seelig dates it in 1902, but the contents of the letter seem to show clearly that it was written in 1901.
14. He described his current work in an undated letter to Marcel Grossmann written during the summer of 1901. This letter is in the Einstein Archive at the Institute for Advanced Study at Princeton.
15. A. Einstein, "Kinetische Theorie des Wärmegleichgewichtes und des zweiten Hauptsatzes der Thermodynamik," Ann. d. Phys. **9**, 417–33 (1902).

16. The others were A. Einstein, "Eine Theorie der Grundlagen der Thermodynamik," Ann. d. Phys. **11**, 170–87 (1903), and "Zur allgemeinen molekularen Theorie der Wärme," Ann. d. Phys. **14**, 354–62 (1904).
17. *Einstein, Philosopher-Scientist*, op. cit. in n. 3., pp. 32–33.
18. J. C. Maxwell, "A Dynamical Theory of the Electromagnetic Field," Trans. Roy. Soc. **155**, 459 (1865); reprinted in *The Scientific Papers of James Clerk Maxwell*, edited by W. D. Niven (Cambridge University Press, Cambridge, England, 1890; also Dover, New York, 1965), p. 526, see especially pp. 527, 533.
19. See Klein, op. cit. in n. 8 and also the works cited by R. McCormmach, "H. A. Lorentz and the Electromagnetic View of Nature," Isis **61**, 459–97 (1970).
20. T. Hirosige, "Origins of Lorentz's Theory of Electrons and the Concept of the Electromagnetic Field," Hist. Studies Phys. Sci. **1**, 151–209 (1969).
21. See McCormmach, op. cit. in n. 19.
22. See Frank, op. cit. in n. 7 and Holton, ibid.
23. Einstein, letter to M. Besso, undated, January 1903, *Einstein, Besso Correspondance*, op. cit. in n. 4, p. 4.
24. A. Einstein, "H. A. Lorentz, His Creative Genius and His Personality," in *H. A. Lorentz. Impressions of His Life and Work,* edited by G. L. de Haas-Lorentz (North Holland, Amsterdam, 1957), pp. 5–9.
25. *Einstein, Philosopher-Scientist* op cit. in n. 3, pp. 26–27. This was an observation he had already made almost forty years earlier: A. Einstein "Über die Entwicklung unserer Anschauungen über das Wesen und die Konstitution der Strahlung," Phys. Z. **10**, 817 (1909).
26. M. Besso, letters to A. Einstein, 12 October to 8 December 1947 and A. Einstein, letter to M. Besso, 6 January 1948, *Einstein, Besso Correspondance*, op. cit. in n. 4, pp. 386, 391.
27. *Einstein, Philosopher-Scientist*, op. cit. in n. 3, pp. 20–21.
28. See G. Holton, "Mach, Einstein, and the Search for Reality," pp. 219–59 in his book, *Thematic Origins*, op. cit. in n. 7. Also see J. T. Blackmore, *Ernst Mach. His Work and Influence* (University of California Press, Berkeley, 1972), pp. 247–85.
29. E. Mach, *Die Mechanik in ihrer Entwicklung. Historish-kritisch dargestallt* (Leipzig, 1883); translated as *The Science of Mechanics,* by T. J. McCormack (Open Court, La Salle, Ill., 1960) pp. 596–97.
30. E. Mach, Die Principien der Wärmelehre. Historisch-kritisch dargestellt (Leipzig, 1896). Einstein refers explicitly to the *Wärmelehre* in his letter to Besso, 6 January 1948, *Einstein, Besso Correspondance*, op. cit. in n. 4, p. 39.
31. Mach, op. cit. in n. 29, pp. 305–16.
32. Mach, op. cit. in n. 30, p. 316.
33. A. Einstein, letter to J. J. Laub, undated, probably 1909, quoted in C. Seelig, *Albert Einstein. A Documentary Biography,* translated by M. Savill (Staples Press, London, 1956), p. 87.
34. Einstein, 1904, op. cit. in n. 16.
35. For an alternative discussion of this and other papers by Einstein on the early quantum theory see T. S. Kuhn, *Black-Body Theory and the Quantum Discontinuity*, 1894–1912 (Oxford University Press, New York, 1978), pp. 170–87, Einstein's early papers on statistical mechanics are also discussed by J. Mehra, "Einstein and the Foundation of Statistical Mechanics," Physica **79A**, 447–77 (1975). See also my paper, "Thermodynamics in Einstein's Thought," Science **157**, 509–16 (1967).
36. M. Planck, "Über das Gesetz der Energieverteilung im Normalspektrum," Ann. d. Phys. **4**, 553–63 (1901); "Über die Elementarquanta der Materie und der Elektricität," Ann. d. Phys. **4**, 564–66 (1901). Einstein was almost surely unacquainted with Planck's first publication of his results in December 1900. For discussion of the role of the natural constants in Planck's thinking, see M. J. Klein, "Thermodynamics and Quanta in Planck's Work," Physics Today **19** (11), 23–32 (1966), and Kuhn, op. cit. in n. 35, pp. 90–91, 110–3.
37. A. Einstein, letter to C. Habicht, 14 April 1904, quoted in Seelig, *Einstein und die Schweiz*, op. cit. in n. 13, p. 64.
38. Einstein's search for an observable fluctuation phenomenon led to his "invention" of the Brownian motion the following year; see A. Einstein, "Über die von der molekularkinetischen Theorie der Wärme geforderte Bewegung von in ruhenden Flüssigkeiten suspendierten Teilchen," Ann. d. Phys. **17**, 549–60 (1905). See also M. J. Klein, "Einstein and the Wave-Particle Duality," Natural Philosopher **3**, 1–49 (1964).

39. A. Einstein, "Über einen die Erzeugung und Verwandlung des Lichtes betreffenden heuristischen Gesichtspunkt," Ann. d. Phys. **17**, 132–48 (1905). See M. J. Klein, "Einstein's First Paper on Quanta," Natural Philosopher **2**, 57–86 (1963).

40. M. J. Klein, "Einstein on Scientific Revolutions," Vistas Astron. **17**, 113–20 (1975).

41. P. Ehrenfest, "Welche Züge der Lichtquantenhypothese spielen in der Theorie der Wärmestrahlung eine wesentliche Rolle?," Ann. d. Phys. **36**, 91–118 (1911).

42. M. Planck, "Über irreversible Strahlungsvorgänge," Berl. Ber., 440–80 (1899). Einstein refers to Planck's summary publication of this work under the same title, Ann. d. Phys. **1**, 69–122 (1900). See Kuhn, op. cit. in n. 35, pp. 72–84.

43. Lord Rayleigh, "Remarks upon the Law of Complete Radiation," Phil. Mag. **49**, 539–40 (1900). See also M. J. Klein, "Max Planck and the Beginnings of the Quantum Theory," Arch. Hist. Exact Sci. **1**, 459–79 (1962), and Kuhn, op. cit. in n. 35, pp. 144–47.

44. M. Planck, "Über eine Verbesserung der Wien'schen Spectralgleichung," Verh. D. Phys. Ges. **2**, 202–04 (1900); "Zur Theorie des Gesetzes der Energieverteilung im Normalspectrum," ibid. **2**, 237–45 (1900).

45. Planck, "Zur Theorie des Gesetzes der Energieverteilung im Normalspectrum," p. 239.

46. For another analysis of Einstein's argument see J. Dorling, "Einstein's Introduction of Photons: Argument by Analogy or Deduction from the Phenomena?" Brit. J. Phil. Sci. **22**, 1–8 (1971).

47. A. Einstein, "Zur Theorie der Lichterzeugung und Lichtabsorption," Ann. d. Phys. **20**, 199–206 (1906).

48. Lord Rayleigh, "The Dynamical Theory of Gases and Radiation," Nature **72**, 54–55 (1905), and "The Constant of Radiation as Calculated from Molecular Data," ibid. **72**, 243–44 (1905), P. Ehrenfest, "Zur Planckschen Strahlungstheorie," Phys. Z. **7**, 528–32 (1906).

49. A. Einstein, "Review of M. Planck, *Vorlesungen über die Theorie der Wärmestrahlung*," in *Beiblätter zu den Annalen der Physik* **30**, 764–66 (1906). This is discussed in M. J. Klein and A. Needell, "Some Unnoticed Publications by Einstein," Isis **68**, 601–604 (1977).

50. A. Einstein, "Die Plancksche Theorie der Strahlung und die Theorie der spezifischen Wärme," Ann. d. Phys. **22**, 189–90, 800 (1907). For discussion of this work and its consequences see M. J. Klein, "Einstein, Specific Heats, and the Early Quantum Theory," Science **148**, 173–80 (1965).

51. A. Einstein, "Zum gegenwärtigen Stand des Strahlungsproblems," Phys. Z. **10**, 185–93, 323–24 (1909), and "Über die Entwicklung unserer Anschauungen über das Wesen and die Konstitution der Strahlung," Phys. Z. **10**, 817–26 (1909). For discussion of these see M. J. Klein, op. cit. in n. 38.

52. W. Pauli, "Einstein's Contributions to Quantum Theory," in *Einstein: Philosopher-Scientist*, op. cit. in n. 3, pp. 149–60.

53. See the discussion of this and a number of related points in the chapter by A. Pais in this volume.

54. L. Pyenson, "Einstein's Early Scientific Collaboration," Hist. Studies Phys. Sci. **7**, 83–123 (1976).

55. R. McCormmach, "Einstein, Lorentz, and the Electron Theory," Hist. Studies Phys. Sci. **2**, 41–87 (1970).

56. M. Planck, Phys. Z. **10**, 825 (1909).

57. Einstein, letter to J. J. Laub, op. cit. in n. 33.

58. Einstein, "Zum gegenwärtigen Stand des Strahlungsproblems," op. cit. in n. 51.

59. See, in particular, A. Einstein, "Uber die Gültigkeitsgrenzen des Satzes vom thermodynamischen Gleichgewicht," Ann. d. Phys. **22**, 569–72 (1907), and "Theorie der Opaleszenz von homogenen Flüssigkeiten und Flüssigkeitsgemischen in der Nähe des kritischen Zustandes," Ann. d. Phys. **33**, 1275–98 (1910).

60. H. A. Lorentz, *Les théories statistiques en thermodynamique* (Teubner, Leipzig, 1916), pp. 114–120. This is a convenient place to remark on the later history of the Einstein fluctuation formula, Eq. (12.26). In 1919 Ornstein and Zernike argued that Einstein's result was invalid for the energy fluctuations in a subvolume containing only radiation (free of matter), and that a correct calculation would result in only the second or wave term in Eq. (12.26). Their essential point was that the entropy of radiation is not an additive function of the volume due to the correlations between different partial volumes. Only when matter is present to ensure thermal equilibrium and destory phase relations is the entropy of the whole enclosure equal to the sum of the entropies of its parts, and only then does Einstein's formula hold. See L. S. Ornstein and F. Zernike, "Energiewisselingen der zwarte straling en licht-atomen," Versl. Akad. Amsterdam **28**, 281–92 (1919). This point was discussed further by P. Ehrenfest, "Energieschwankungen im Strahlungsfeld," Z. Phys. **34**, 362–73 (1925); **35**, 316 (1925). When the quantum mechanics was developed, one of its

major early achievements seemed to be the confirmation of the correctness of Einstein's result, Eq. (12.26), by a direct calculation analogous to Lorentz's classical calculation. See M. Born, W. Heisenberg, and P. Jordan, "Zur Quantenmechanik II," Z. Phys. **35**, 557-615 (1926). This result was challenged by Heinsenberg himself, however, in his article, "Über Energieschwankungen in einem Strahlungsfeld," Ber. Sächs. Akad., Math. Phys. Kl. **83**, 3-9 (1931). Heisenberg's calculation was then criticized by M. Born and K. Fuchs, "On Fluctuations in Electromagnetic Radiation," Proc. Roy. Soc. **A 170**, 252-65 (1939); **A 172**, 465-66 (1939). The whole problem has recently been reviewed and clarified by J. J. Gonzalez and H. Wergeland, "Einstein-Lorentz Formula for the Fluctuations of Electromagnetic Energy," Arkiv. for Det. Fysiske Seminar i Trondheim No. 6 (1972).

61. A. Einstein and L. Hopf, "Statistische Untersuchung der Bewegung eines Resonators in einem Strahlungsfeld," Ann. d. Phys. **33**, 1105-15 (1910). In this paper the authors also show that one cannot blame the unsatisfactory classical or Rayleigh-Jeans distribution law for radiation on the application of the equipartition theorem to the radiation itself, or even to the motion of oscillators of arbitrarily high frequencies. Their method is to calculate $\overline{\Delta^2}$ and P independently from electromagnetic theory, thereby obtaining a differential equation for $\rho(\nu,T)$ whose unique solution is the Rayleigh-Jeans law. The equipartition theorem is applied only to the translational motion of molecules, an incontestably valid use (perhaps the only one) of that long-disputed theorem.

62. See, for example, Planck's views on the need to exclude fluctuations from physics as given in his lecture at Leyden in 1908. M. Planck, "Die Einheit des physikalischen Weltbildes," Phys. Z. **10**, 62-75 (1909).

63. See McCormmach, op. cit. in n. 55, and M. J. Klein, op. cit. in n. 35.

64. Einstein, "Zum gegenwärtigen Stand des Strahlungsproblems," op. cit. in n. 51.

65. Einstein, "Über die Entwicklung unserer Anschauungen über das Wesen and die Konstitution der Strahlung," op. cit. in n. 51.

66. Einstein, letter to M. Besso, 13 May 1911, *Einstein, Besso Correspondence*, op. cit. in n. 4, p. 19.

67. A. Einstein, "Beiträge zur Quantentheorie," Verh. D. Phys. Ges. **16**, 820-28 (1914).

68. Einstein, *Ideas and Opinions*, op. cit., p. 283.

69. Einstein, letter to M. Besso, 11 August 1916, *Einstein, Besso Correspondance*, op. cit., pp. 78-79.

70. A. Einstein, "Strahlungs-Emission und -Absorption nach der Quantentheorie," Verh. D. Phys. Ges. **18**, 318-23 (1916). I am grateful to Kathryn Olesko for pointing out to me the importance of this earlier version of Einstein's argument.

71. A. Einstein, "Zur Quantentheorie der Strahlung," Physik. Ges., Zürich, Meitteilungen **16**, 47-62 (1916). Also published in Phys. Z. **18**, 121-28 (1917). The latter reference is usually quoted in the literature.

72. This law requires that the spectral distribution for blackbody radiation must have the form

$$\rho(\nu,T) = \nu^3 f(\nu/T),$$

where $f(\nu/T)$ is a function of the single variable (ν/T).

73. A. Einstein, letter to M. Besso, 24 August 1916, *Einstein, Besso Correspondance*, op. cit. in n. 4, p. 80.

74. A. Einstein, letter to M. Besso, 6 September 1916, *Einstein, Besso Correspondance*, op. cit. in n. 4, pp. 81-82.

75. A. Einstein, "Über ein den Elementarprozess der Lichtemission betreffendes Experiment," Berl. Ber., 882-83 (1921).

76. P. Ehrenfest, letter to A. Einstein, 11 January 1922. See my article, "The First Phase of the Bohr-Einstein Dialogue," Hist. Studies Phys. Sci. **2**, 1-39 (1970) for discussion.

77. A. Sommerfeld, letter to A. H. Compton, quoted by Compton in J. Franklin Inst. **198**, 70 (1924). See R. H. Stuewer, *The Compton Effect: Turning Point in Physics* (Science History Publications, New York, 1975) for a full discussion of this work and its impact.

78. A. Einstein, "Das Comptonsche Experiment," Berliner Tageblatt, 20 April 1924, l. Beiblatt.

79. A. Einstein, recording made on 6 February 1924; text given in F. Herneck, "Zwei Tondokumente Einsteins zur Relativitätstheorie," Forschungen und Fortschritte **40** (5), 133-135 (1966).

13. EINSTEIN'S CRITIQUE OF PLANCK

Thomas S. Kuhn

No paper so rich and authoritative as Professor Klein's provides an opening for useful commentary, and I shall attempt none. Instead, in further celebration of Albert Einstein, I shall provide one more example of his extraordinary critical sense, his ability to discover in existing theory both difficulties and consequent rewards unseen by most or all of his contemporaries. In the story I shall tell, it is Einstein rather than Planck who first quantized the Planck oscillator, and I must therefore begin by emphasizing that this new version of the origin of the quantum discontinuity only changes the nature but in no way devalues the contribution due to Planck. On the contrary, Planck's derivation of his famous blackbody distribution law becomes better physics, less sleepwalking, than it has been taken to be in the past. Furthermore, Planck himself appears far less the conservative, attempting to deny his own offspring. In particular, his so-called second theory becomes the first theory from his pen to call upon discontinuity at all. In the development of his own thought it is a step toward, not a retreat from, the new physics of quanta.

Since my present concern is with Einstein rather than Planck, I shall here have to premise most of that vaguely heretical position. But, as needed background, let me first make clear of what the position consists. As is well known, Max Planck first presented his famous distribution law for blackbody radiation in October 1900. At that time the law lacked a derivation, but Planck supplied two closely related ones in papers published during December 1900 and January 1901. Those derivations have for years been read as introducing the concept of energy quantization. To carry through his derivations, it is said, Planck was forced to restrict to integral multiples of the quantum $h\nu$ the energy of the resonators he had introduced to equilibrate the radiation field. I have elsewhere argued, however, that Planck had no conception of a discrete energy spectrum when he presented the first derivations of his blackbody distribution law. Nor had he yet developed one when he published, six and one-half years later, the first edition of his *Lectures on the Theory of Thermal Radiation*. Rather, he thought his derivations classical, compatible with both Maxwell's equations and Newton's laws.[1]

In those lectures, Planck emphasized that, with one notable exception, his treatment of radiation precisely paralleled Boltzmann's treatment of the ideal gas. Both men determined entropy as the logarithm of an appropriate combinatorial probability, and both then derived distribution laws from knowledge of an equilibrium entropy function. In each case, furthermore, computing

Harry Woolf (ed.), Some Strangeness in the Proportion: A Centennial Symposium to Celebrate the Achievements of Albert Einstein ISBN 0-201-09924-1

the relevant probability required first dividing the continuum of total available energy into small units of size ε and then counting the number of ways in which a given distribution of that total energy could be achieved, the distribution being over molecules for Boltzmann, over resonators for Planck.[2] Among all possible distributions, the one that could be achieved in the largest number of ways was the most probable, and the entropy of that distribution was proportional to the logarithm of the number of ways it could be achieved.

In all these respects Boltzmann's and Planck's derivations of distribution laws were the same, but in another they differed. Boltzmann's result, it turned out, was independent of the size of the energy element ε.[3] To satisfy the laws of thermodynamics, Planck's ε had, on the other hand, to be equal to $h\nu$, with h a universal constant and ν the resonator frequency. Planck was forced, that is, to restrict the size of the cell used in probability computations. Within and between those cells, however, resonators emitted and absorbed continuously, governed in full by Maxwell's equations. A physically structured phase space rather than discontinuous energy levels was, Planck thought, the novelty his derivation required.[4] Though obscurely expressed, that is the viewpoint that underlies Planck's derivation papers of 1900 and 1901. It remained Planck's viewpoint, furthermore, but no longer obscure, in the published *Lectures* of 1906. And it is precisely this viewpoint that Einstein attributes to Planck in his generally laudatory review of the book. Sketching the key elements of Planck's theory, Einstein says nothing of energy quanta or of a restriction on resonator energy. Instead he notes that, for radiation "in contrast to the previously standard gas-theoretical hypothesis of infinitely small elementary regions — the elementary region is chosen of finite magnitude ($= h\nu$)."[5]

When he wrote that description of Planck's viewpoint, Einstein was already aware that Planck's distribution law could not be derived in so classical a way. Apparently he had not yet identified precisely where the derivation went astray, but he would shortly do so, and anticipating his conclusion will clarify what is presently at issue. When one subdivides a continuum for purposes of computing probabilities, one substitutes an approximate for an exact count of the number of ways of achieving a given distribution. The approximation is adequate only if ε is small compared with the average energy of a resonator or, to put the same condition differently, only if $h\nu$ is everywhere small compared with kT (T being the absolute temperature and k the so-called Boltzmann constant). That condition is not fulfilled everywhere in the range of experimental blackbody measurements, and, where it is not, Planck's derivation does not apply. To preserve it, the energy of a resonator must be restricted to cells far smaller than Planck's ε. It must, that is, be very close or equal to some integral multiple of $h\nu$. The resonator must be quantized.

The dependence of his argument on an approximation that required (but could not have received) physical justification is the point Planck missed. Today, no competent physicist would do so. But in 1900, though statistical techniques were quite standard in physics, the combinatorial or probabilistic approach was scarcely known and intuition was uninhibited by subsequent sophistications. A paper written by Boltzmann in 1877 supplied Planck's single significant precedent, and he followed that paper closely. Its argument, like Planck's later one, required that sums over the subdivided energy continuum be replaced at key points by integrals over the continuum itself. That transition, again like Planck's, can be justified only under specified physical conditions, but Boltzmann explicitly regarded it as "self explanatory," a mathematical manipulation required by any mathematical gas theory.[6] Both in late 1900 and in 1906, Planck followed that lead, overlooking the need for physical justification of an apparently mathematical step and therefore also failing to recognize that such justification could be provided only in the case of gases, not of radiation. Among his readers Einstein alone was acute enough to identify this subtle shortcoming of Planck's argument, and his understanding was reached only by stages. It is to those stages that I now turn, sketching a path the points on which have already been touched by Professor Klein at one or another part of his paper. My excuse for the duplication must be that the scenery is different when the points are placed on a single path, especially if that path starts from a revised view of Planck's own work.

As Professor Klein has already pointed out, Einstein's first allusion to Planck's radiation theory occurs in the third and last of the series of papers on statistical mechanics that he published between 1902 and 1904. There is no reason to suppose that Einstein had studied Planck's work with much care when that paper was written,[7] but the paper itself provides clear reason for his beginning to do so. Immediately after introducing his new conception of energy fluctuation and an equation governing it, Einstein wrote: "The equation just found would permit an exact determination of the universal constant *k* if it were possible to determine the energy fluctuation of a system. In the present state of our knowledge, however, that is not the case. Indeed, for only one sort of physical system can we presume from experience that an energy fluctuation occurs. That system is empty space filled with thermal radiation."[8] Reformulating the foundations of statistical theory had led Einstein to fluctuations, and fluctuations had by 1904 led him to the blackbody problem on which Planck was a leading authority. If Einstein had not already examined Planck's papers closely, he would now have begun to do so.

In any case, Einstein knew Planck's theory well by the time he wrote his famous light-quantum paper in the following year. Evidence of the care and depth of Einstein's study is provided by an extended footnote subtly explicating Planck's obscure concept of natural radiation.[9] Evidence for some part of his attitude towards that theory is provided by his sketch of a derivation for the alternative distribution law ultimately to be known by the names of Rayleigh and Jeans. To model the radiation problem, Einstein employed resonators in a reflecting enclosure. Although he did not say so, that model was Planck's, but Einstein's result was not. Even if one rejected — as most physicists then would have[10] — Einstein's claim that his result was the only one compatible with classical theory, a clear puzzle remained, one also emphasized by Professor Klein. How could Planck and Einstein, apparently using the same model and the same theories, have derived different distribution laws? That puzzle had a special urgency because Planck's law worked; it satisfied experiment.

Not surprisingly, the man who first posed the puzzle was also the first to solve it. In a paper on "The Theory of the Emission and Absorption of Light," submitted for publication in March 1906, Einstein used his earlier formulation of statistical mechanics to develop the following formula for W, the probability corresponding to the most likely distribution of energy over a collection of n resonators:

$$W = \int_{E}^{E+\Delta E} \cdots \int dE_1\, dE_2 \ldots dE_n$$

Here E is the total energy of the collection, the resonators in which have individual energies E_i, and ΔE is a small increment sufficient to include most fluctuations. If the energy of individual resonators varies continuously, Einstein stated, then straightforward integration leads to the form for W corresponding to the Rayleigh–Jeans law. If, on the other hand, resonator energy is restricted to integral multiples of $h\nu$, then Einstein's multiple integral becomes a multiple sum that, on inspection, yields Planck's combinatorial form for W and hence his distribution law. Starting from a single statistical-mechanical formula, Einstein had shown that the Wien law follows for continuous resonator spectra, the Planck law for discrete spectra with energy levels $nh\nu$.

It is immediately after introducing this result that Einstein wrote the passage already discussed by Professor Klein. We must recognize, he said, two propositions as basic to the Planck theory: first, the energy of elementary resonators must be restricted to integral multiples of $h\nu$; second, Maxwell's equations, though not applicable to the detailed behavior of individual resonators, must nevertheless yield the correct value for average resonator energy. Those two propositions entered physics together, and Planck had not imagined that his theory might imply either one.[11] That is what Einstein had in mind when, at the beginning of his paper, he said that, although he had previously thought his theory in conflict with Planck's, he now realized that "the Planck theory implicitly makes use of the recently mentioned light-quantum hypothesis."[12] Einstein's text is the first to display the essential role of quantization in blackbody theory. Only with the introduction of a discrete energy spectrum could Planck's argument evade the Rayleigh–Jeans law.

Nevertheless, although Einstein early in 1906 saw how Planck's argument must be revised to give the Planck distribution law, he apparently did not yet see where Planck's original version of the derivation had gone astray. If Planck had reached a distribution law different from that of Rayleigh and Jeans, then there must be a mistake in his argument, one that Einstein had not yet found. Perhaps that is why, when Einstein reviewed Planck's *Lectures,* published later in the same year, he said nothing about any difficulty with the derivation, about the Rayleigh–Jeans law, or even about quantization. Instead, he sketched Planck's continuum version of the blackbody derivation, singling out finite cell size as the significant novelty. About that derivation he said: "Though it has not yet been possible to reach its goal by recourse only to theoretical tools firmly supported by experiment — the author instead resorting to an hypothesis based solely on analogy — still every unprejudiced reader will recognize that the result arrived at is extremely probable."[13]

When Einstein spoke of "an hypothesis supported merely by analogy," he probably had in mind the analogy, emphasized by Planck, between Planck's way of deriving the entropy of resonators and Boltzmann's way of deriving the entropy of the molecules of a gas. That, in any case, is the place where Einstein, at the beginning of 1909, first located the problem with Planck's derivation. "Neither Herr Boltzmann nor Herr Planck," he wrote, "have given a definition of [the quantity] W. In a purely formal manner they [instead] set W equal to the number of ways of achieving the distribution in question [*Anzahl der Komplexionenen des Betrachteten Zustandes*]."[14] That procedure is justified, he continued, only if each way of achieving the distribution can be shown to be equally probable, a problem that Boltzmann at least recognized, but to which Planck pays no attention at all. If he had, Einstein concluded, "he would have reached the formula defended by Jeans. As overjoyed as every physicist must be that Planck has overlooked this requirement in so fortunate a way, it would be correspondingly unsuitable to forget that the Planck radiation formula is incompatible with the theoretical foundations from which Herr Planck began."[15]

About the presence of incompatibility between Planck's premises and conclusions, Einstein is, of course, correct, but his diagnosis of the incompatibility's nature is not. Planck had not justified his choice of equiprobable configurations, but neither did Einstein, nor has any physicist since. In quantum theory, unlike classical theory, the equiprobability of specified configurations is a postulate, not subject to derivation from nonprobabilistic principles. More to the point, the particular choice of equiprobable elements made by Einstein is the closest available parallel to Planck's. What differentiates the two treatments is not the statistical weight attributed to particular configurations but, rather, the way in which a configuration is itself defined: for Planck each resonator is placed in a finite region, for Einstein at a point. That is the difference that Einstein finally emphasized in a second, very different diagnosis of Planck's original continuum derivation, pointing out that in this form the "derivation could make the Planck radiation formula appear to be a consequence of contemporary electromagnetic theory."[16] To show that it was not, Einstein developed the argument anticipated at the start of this brief essay. Planck's derivation would go through if only the element ε were, at all frequencies, small compared with the mean energy of a resonator. Failing that, the derivation could be preserved only by quantization. "In my opinion," Einstein concluded, "accepting the Planck theory means flatly rejecting the foundations of our radiation theory."[17]

By "the Planck theory," Einstein of course meant his own revised version. Only that version called for a break with classical theory, a break that Einstein had believed to be unavoidable since at least 1905. By 1909, when he wrote the words just quoted, a few others had reached the same conclusion: Laue and Ehrenfest in 1906; Lorentz, followed by Wien and Planck, in 1908. Except with respect to light particles, Einstein was not long alone. But, at each step in the development of a nonclassical blackbody theory, Einstein had been ahead: in 1905 he was the first to see that Planck's premises ought to have led him to the Rayleigh–Jeans law; in 1906 he was the first to recognize the new premises that Planck's argument required; and in 1909 he was the first to isolate the oversight that had made possible Planck's original derivation. The version of radiation theory that tradition has long attributed to Planck was introduced by Einstein. Once again the profundity of his concern with the foundations of his field had served both him and physics well.

NOTES

1. Thomas S. Kuhn, *Black-Body Theory and the Quantum Discontinuity, 1894–1912* (Clarendon Press, Oxford, 1978, and Oxford University Press, New York, 1978), provides an extended account of this position. The first five chapters describe the development of Planck's theory through 1906; Ch. 7 deals with Einstein's reformulation of that theory; and Ch. 8 and 10 recount Planck's conversion to discontinuity and his early responses to it.
2. Strictly speaking, the parallelism or analogy sketched here is between Boltzmann's derivation of a distribution law for gas molecules and Lorentz's familiar reformulation of Planck's derivation for radiation. Planck himself considered the different ways of distributing energy, not over a collection of resonators at a single frequency, but over collections of resonators at different frequencies. Failure to distinguish between these two sorts of arguments has led to much misunderstanding of Planck's early papers. See Kuhn, op. cit. in n. 1, pp. 102–10, for an extended discussion.
3. Boltzmann's ε must be large enough to include many molecules but small enough so that the distribution function changes very little between neighboring cells. Within these limits the value of ε affects only an additive constant in the entropy and may therefore be disregarded.
4. The phrase "physical structure of phase space" was first used by Planck only in 1916 ["Die physikalische struktur des Phasenraumes," Ann. d. Phys. **50,** 385–418 (1916)]. But the conception can be traced at least from the *Lectures* of 1906, where Planck randomly distributes resonators over a phase plane divided into elliptical annuli of area h. Note that in the *Lectures* Planck explicitly allows the resonators to lie within the annuli, not just at their elliptical boundaries (Kuhn, op. cit. in n. 1, p. 129), so that the presence of the phase-plane boundaries has a physical effect on resonators even though the boundaries do nothing to influence resonator position. That way of conceiving phase space is basic also to all versions of Planck's second theory. My claim is that something very like it is implicit in his first derivation papers of 1900 and 1901.
5. A. Einstein, "[Review of] *M. Planck. Vorlesungen über die Theorie Wärmestrahlung,*" *Beiblätter zu den Ann. d. Physik* **30,** 764–6, (1906), quotation from p. 766.
6. L. Boltzmann, "Über die Beziehung zwischen den zweiten Hauptsatze der mechanischen Wärmetheorie und der Wahrscheinlichkeitsrechnung respektive den Satzen über die Wärmegleichgewicht," Wiener Ber. II, **76,** 373 (1877), where the relevant discussion is on p. 48. See also Kuhn, op. cit. in n. 1, pp. 59f.

7. Professor Klein may well be right — there is little relevant evidence — that this paper was written as a consequence of Einstein's reading Planck and thinking about his strange theory. But it seems to me more likely that Einstein, like Gibbs, was led to fluctuation phenomena by the logic of his theory and that the possibility of applying his results to blackbody measurement was an unexpected dividend. Einstein's 1904 paper seems an entirely natural extension of the work he had published in 1903, probably prompted in part by an awkwardness in the way he had defined the function $\omega(E)$ in the earlier paper. Compare Kuhn, op. cit. in n. 1, pp. 177f.

8. A. Einstein, "Zur allgemeinen molekularen Theorie der Wärme," Ann. d. Phys. **14,** 354–62 (1904), quotation on p. 360f. Instead of the now standard constant k, Einstein refers to his own constant χ, which is half as large ($k = 2\chi$).

9. A. Einstein, "Über einen die Erzeugung und Verwandlung des Lichtes betreffendren heuristischen Gesichtspunkt," Ann. d. Phys. **17,** 132–48 (1905), footnote on p. 135. Though Einstein does not say so, his definition of disorder for radiation is very different from Planck's. Furthermore, its introduction at this place is gratuitous, since Einstein's argument makes no use of it. I think it likely that he had spent much time trying to understand Planck's obscure formulation and that he supplied the result for the benefit of readers who he believed would share his puzzlement.

10. On the status of the equipartition theorem see S. G. Brush, "Foundations of Statistical Mechanics, 1845-1915," Arch. Hist. Exact Sci. **4,** 145–83 (1967), especially pp. 162–68.

11. I am uncertain to what extent Professor Klein agrees with this position, but I believe he will be read as rejecting it. In his paper, immediately after the introduction of Eq. (12.22), he writes: "Since Planck had in effect made this restrictive asumption [quantization] on the energies of his oscillators he must also have made another assumption, if only implicitly. . . .Planck had implicitly assumed the continuing validity of the classical relation between oscillator energy and spectral density even for his oscillators with a discrete set of energy values." Those sentences would ordinarily be read as saying something about Planck's state of mind in the period before Einstein described the two assumptions as basic to the Planck theory. On that reading, I believe the sentences are false. Planck made neither assumption, not even "implicitly" or "in effect." But Professor Klein may mean only that the logic of Planck's position demands these assumptions, that they are somehow implied by it. In that case the sentences are not false, but it is unclear what they can possibly mean. What can "Planck's postion" be if it is not a position actually held by Planck?

12. A. Einstein, "Zur Theorie der Lichterzeugung und Lichtabsorption," Ann. d. Phys. **20,** 199–206 (1906), quotation on p. 199.

13. Einstein, op. cit. in n. 5, p. 765.

14. A. Einstein, "Zur gegenwartigen Stand des Strahlungproblems," Phys. Z.**10,** 185–93 (1909), quotation on p. 187.

15. Ibid., p. 187f.

16. A. Einstein, "Über die Entwicklung unserer Anschauungen über das Wesen und die Konstitution der Strahlung," Phys. Z. **10,** 817–25 (1909), quotation on p. 822.

17. Ibid.

Open Discussion

Following Papers by J. KLEIN and T. S. KUHN

Chairman, JULIAN SCHWINGER

Y. Ne'eman (Tel-Aviv U.): To Professor Klein. The names of Rutherford and Bohr were mentioned by you, and when one looks at their years and asks what were the important developments in physics at that time, a principal answer is to be found in the deciphering of the atom. Was it in the nature of Einstein's curiosity, which was somehow blind or not interested in something that was closer to phenomenology and only interested in turning existing theory into more profound theory and more unifying and fundamental that diminished his interest in this area, or was there something else that you detect in following his work during these years? I mean, what is the amount of his interest in what is happening with the structure of the atom? He was so versatile; you have mentioned interests from capillarity to cosmology. Did he interact in any way with atomic structure, and is this the beginning of his lack of connection to this part of physics, which was, perhaps, crucial in the failure of the unified field theory twenty or thirty years later?

Klein: I think that is a difficult question to answer. It seems to me that in the early years, in the years before he moved to Berlin, his contact with the whole complex of things that goes with atomic physics was relatively slight. I think that at least part of the reason for that, again going back still further to his years at the patent office, was his limited time and exposure to the literature. He did after all spend forty-eight hours a week in the patent office, and he did have available to him for most of that time only the *Annalen* for reading. In the years in Berlin I think his contact with what was going on was strong. The Berlin Collquium was a great center. Several of his papers during the First World War are comments on the current state of what we now call the old quantum theory. (There are two very important papers along that line.) He is reported to have reacted favorably, understandingly, and glowingly to Bohr's work in 1913 and similarly to Sommerfeld's extension of Bohr's ideas in 1916. Those are the years when he was primarily occupied in developing the general theory, which precluded any serious activity of any other kind. If you ask why having developed a new theory of radiation in 1916–17 he did not throw himself into questions of atomic physics, I am not sure a complete answer can be found. Again, I would think it had a lot to do with deep parental concern, appropriate of course, for his other ideas, but beyond that I am not really sure. When he responded to several of the developments in the early 1920s it was again sympathetically and understandingly to de Broglie, to Schroedinger. He was apparently in close touch, probably through Born, with what was going on at Göttingen. The record of his comments and reactions at the Solvay conferences in the twenties again suggests an awareness of what was happening and a close interest. But why he did not prosecute the ideas himself, I do not know.

Schwinger: Was it not Einstein who at the time of the Sommerfeld quantization conditions gave it the form that did not depend upon the separation of coordinates, as everybody else's had done, which was a deep bit of insight?

Klein: Yes, that was one of the papers I alluded to a few minutes ago. It was in 1917, exactly. And that is one that Schroedinger said had influenced him greatly, almost ten years later.

R. Jost (E. T. H., Zurich): In this paper, if I am not mistaken, Bohr is not mentioned.

Klein: You may well be right. But, again it is a comment on the current state of the literature, where the dominant names were Epstein and Schwarzschild.

J. L. Ehlers (Max Planck Inst., Munich): Professor Klein, you have emphasized the importance which the formula for the energy fluctuation played for these developments. I think the first derivation of that fluctuation formula from quantum field theory was in the last part of the famous three-man paper Born–Heisenberg–Jordan published in 1926. Since Einstein had great interest in these fluctuations and also was in close contact at the time with Born, I wonder whether any reaction of Einstein's is known to this derivation within quantum field theory, a theory that did not correspond to his view of physics.

Klein: All I can say is, I do not remember that it is commented on explicitly in the Born–Einstein correspondence. They discuss many things and there are a number of letters from those years; but I do not remember any mention of that point in the three-man paper. But my memory may be wrong.

H. D. Smyth (Princeton U.): The point I wish to put before Professor Klein is based on my understanding of his comment that the community of physics did not really accept the idea of the particle nature of light until the Compton-effect experiment. I was taught atomic physics here at Princeton in 1918 and 1919, and there was not any question about this. We knew the Millikan experiment; we knew the various other experiments that had tried to get the maximum energy of photoelectrons; we knew of the experiments of Franck and Hertz. The whole experimental program under Karl Compton really was based on the idea of the particle nature of light at least as a working hypothesis.

Klein: Well, I cannot quarrel with your memory. But, I can only comment that the literature of the period you are talking about, the literature of the late teens and the early twenties, does not show any support for the notion of free light quanta, as far as I am aware of it.

Smyth: I do not think my memory is defective, but I will remind you there was a war that did have an effect on the publication program. If anyone else learned physics at other progressive institutions in that period, I would be glad to have them support my point of view.

Klein: Apropos of that remark, I might point out that in 1921 Einstein was sufficiently concerned with the lack of direct evidence for free light quanta to propose an experiment that he thought would distinguish between the effects of light as consisting of quanta, or as at least showing direct quantum properties and the classical theory. The experiment turned out ultimately to be based on a misconception, but there was enough interest in it at the time that Bothe and Geiger set out to test his idea (this is not the famous Bothe–Geiger experiment of a few years later) and again, the result was inconclusive. All I can say is that both the periodical literature and the texts of the time that I have read simply do not bear it out. If that was so at Princeton, then Princeton was an enclave of light quanta in that period!

E. Wigner (Princeton U.): In spite of all my admiration of Professor Kuhn, I would like to contradict him a little bit. I think Planck did wonderful work, and even the discovery of his equation is wonderful. I do not believe he believed in the details of any derivation of it, and this was natural since the physics of that time was full of contradictions. In Bohr's atom the electrons wandered around on their orbits following the laws of classical mechanics, but then they decided to jump around. The motion on the orbit was governed by the classical equations, and it was very unclear how they realized that they had to jump away from the equations. That Planck believed in the quantum jump is evident: he postulated the emission and the absorption as definite processes and postulated for both of them the energy $h\nu$. I believe it is clear that he believed that the intensity of the radiation of frequency ν is always a multiple of $h\nu$. And this, I think, was in contradiction with many, many basic concepts of classical theory — and we felt that very strongly in the days before quantum mechanics — a little bit even after — that physics as we knew it then was not self-consistent. And, therefore, the fact that there was a deviation from the fluctuation theory was true, but we felt that we had to swallow that fact. And perhaps I should also remind you how little Planck's ideas were accepted in the beginning; at the Solvay Congress the authorities said: "Oh, the present radiation intensity is just a temporary thing. It is because the blackbody radiation is not yet in equilibrium with matter. Eventually the Rayleigh–Jeans law will be put into effect." So that I feel that the quantum was discovered, at least from what I read, by Planck and even though we admire Einstein, this is not for what we admire him most.

Kuhn: Thank you, Dr. Wigner. I think it would be a mistake for me to try here and now to give the full argument that leads me to this conclusion. It is developed at very great length in the first half of a book that appeared only in the middle of November 1978 so that very few people have had a chance to see it yet. There is just one thing that I would like to say now. After 1906, and more particularly after 1909, the situation I have been describing changes quite drastically. Particularly after 1909, when people talk about the Planck theory, they do mean the version that involves quantization. On the other hand, the people who are deeply involved with the blackbody problem are not, by the time of the first Solvay congress in 1911, still saying that quantization will go away or that we will go back to the Rayleigh–Jeans law. I had rather expected the sort of thing you describe, but I find, going over the conference and the discussion, that although people are still deeply puzzled as to what form the quantum theory will take — whether it will be like Planck's second theory, whether it will be a more nearly continuum theory — that a large proportion of the participants are deeply convinced, by the end of 1911, that there is going to be a fundamental break, something that they had really not seen two years before, and that Jeans' theory is not going to work.

N. Razinsky (Fusion Energy Foundation): I would like to follow up on the point of Einstein's participation in the question of the structure of the atom and also follow up on Dr. Wigner's comment on the electron jumping. The question is for Dr. Klein. It is my understanding that during the Solvay conferences in the late 1920s — early 1930s — there was quite a battle going on between Schroedinger and Niels Bohr on the question of atomic structure.

Klein: I didn't hear of any battle that went on between Bohr and Schroedinger at the Solvay conferences.

Razinsky: Did not. Well, Einstein was . . .

Klein: In Copenhagen, but not at the Solvay conferences.

Razinsky: Then maybe it was in Copenhagen. I think Einstein at that point was very involved in the question of whether the Bohr model was correct or whether Schroedinger's approach was a better idea, and brought up the point many times during these conferences or during sessions at Copenhagen, so he did have an interest in it. I think that what happened eventually was that Schroedinger was beaten down in the controversy and Einstein may have gone down with him. I wonder if that is your impression or if you have any comment on that? Let me add one more thing. I think that the point that Einstein was trying to make was that the ad hoc asumptions that Bohr had made could possibly be replaced by a more causal theory, which was what Schroedinger was getting at and this was the point of contention. I want to know if that is your perception of the matter.

Klein: The first answer that I would make to that is that your questions really deal with a later period, and I think Professor Pais's paper is likely to answer them. The second is that we have present some people who were around at some of these epic debates, and they are certainly in a better position to comment than I am.

Jost: I am very glad that we finally decided that Planck's law was by Planck. In addition, I fully agree that in a way, if you read Planck's papers, you get the impression that he never knew what he did. [Klein: Who does?] If he discovers something, yes. If he doesn't discover something, he usually knows what he did. I would, however, like to put in a good word in favor of Planck. On the W — on his W. I think Planck says in his paper of 1901 that of course the correctness of the W which he writes down can only be proved by the experimental check. This shows at least, that at this point, he knows that he did not know what he did.

A. Hermann (U. of Stuttgart): I would like to make a remark on the connection between Einstein's first ideas on quantum theory and the atom. As you of course know, Dr. Klein, Einstein tried to make a connection between Planck's constant h and the elementary constant of electricity. In 1909, he found that the dimension of e^2 divided by c has the same dimension as h and he so thought that there must be a connection, and then in the same year, Wilhelm Wien said that he could not believe that there is a connection between Planck's constant and the elementary constant of electricity, but a connection between h and the atom, and that you must derive the h from the atom. Then Sommerfeld turned it the other way and said you must derive the nature of the atom from the h, and this was the idea that Bohr picked up in 1913. So there was some connection, some concern of Einstein with links between the two questions.

Wigner: I would like to point out one more fact even though I know that I am a little bit out of order. And this is that it was clear in those days, and in fact it was clear even very much later to everybody, that the physics that we then had was not consistent. It was not based on some simple, straightforward principle, the kind of beautiful, simple, and self-consistent principle Einstein said we had to postulate. Hence the fact that the fluctuations did not come out right, and many other things, does not invalidate the fact that a step toward the discovery of new laws of nature was made just the same. I remember so well that when I attended even very much later in the 1920s the colloquia in Berlin, it appeared to me that many people felt that perhaps man is not bright enough to understand microscopic physics fully. And to my surprise, very much later, I heard the same view expressed by the person whom I call my famous brother-in-law (Dr. P. A. M. Dirac). He felt that way even though he never was present at any of the colloquia I referred to. You see, the consistency did not exist at that time, and this had a very great influence on all thinking.

J. Stachel (Boston U.): I would like to comment on three topics that have come up. First, on the Bohr atom: Not only was Einstein enthusiastic when he first heard about Bohr's work, but he remarked that he had had similar ideas a while before. This seems to indicate that he was already interested before 1913 in the problem of atomic structure and was amazed that Bohr had been able to get someplace with this difficult problem. Second, on the photon concept, and its acceptance by the physics community. If you read the recently published volumes of Bohr's works, which include his published and unpublished writings from the late teens and early twenties of this century, you see that he sometimes refers to the photon concept of Einstein almost condescendingly. Even after the Compton effect was discovered, he persisted in his attempts to do without it. Even more interesting, perhaps, if you read Rosenfeld's comments and the Bohr–Rosenfeld paper, you see, I think, that Bohr never accepted the idea that the particulate structure of the radiation field had a reality comparable to the reality of the electron. In that sense, one could say that Bohr never really fully accepted the photon concept. Third, on the question of Einstein's reaction to the three-man paper of Born, Heisenberg, and Jordan: Unfortunately, the correspondence between Einstein and Jordan and Heisenberg exists only in one-sided form. With the exception of one postcard to Jordan, the Einstein letters to both appear to have been destroyed during the war — I am not sure exactly how. But a number of letters from Heisenberg and Jordan to Einstein have been preserved in the Einstein Archive. From these letters it is clear that they were in active correspondence with Einstein all during this period, and one can try to reconstruct the discussion from the remaining letters. (In particular, I have been in correspondence with Jordan about his letters.) For example, Jordan was in correspondence with Einstein before the Dreimännerarbeit appeared, and he outlined the calculations of the derivation of the fluctuation phenomena in the last section of that paper well before it appeared. Einstein was apparently not to happy with the treatment of blackbody fluctuations, and wrote to Heisenberg about this question, among a number of others. Heisenberg gave Jordan (who wrote the section on statistics of wave fields) the job of replying to Einstein. Jordan could not really satisfy Einstein's objection, and his comments spurred Jordan to continue to work on the problem, until he came to a satisfactory resolution of Einstein's problem within the quantum-mechanical framework a couple of years later. During this period he continued to correspond with Einstein, and the paper in which he published his results explicitly mentions that some of the things in it were done in response to comments by Einstein. On this question, as well as several others, Einstein was certainly in dialogue with Jordan and Heisenberg and in this sense was also closely involved in the early stages of development of matrix mechanics. Unfortunately, we do not have the texts of the Einstein letters to see exactly what he said.

14. EINSTEIN ON PARTICLES, FIELDS, AND THE QUANTUM THEORY

Abraham Pais

1. Einstein, the Quantum and Apartness

Apart. . . . 4. Away from others in action or function; separately, independently, individually . . .

Oxford English Dictionary

1.1 *Introduction*

In 1948 I undertook to put together the Festschrift in honor of Einstein's seventieth birthday.[1] In a letter to prospective contributors I wrote: "It is planned that the first article of the volume shall be of a more personal nature and written by a representative colleague, shall pay homage to Einstein on behalf of all contributors."[2] I then asked Robert Andrews Millikan (1868–1953) to do the honors as the senior contributor.[3] He accepted and his article is written in his customary forthright manner. On that occasion he expressed himself as follows on the equation $E = h\nu - P$ for the photoelectric effect: "I spent ten years of my life testing that 1905 equation of Einstein's and, contrary to all my expectations, I was compelled in 1915 to assert its unambiguous verification in spite of its unreasonableness since it seemed to violate everything we knew about the interference of light."[4]

Physics had progressed, and Millikan had mellowed since the days of his 1915 paper on the photo effect, as is evidenced by what he had written at that earlier time: "Einstein's photoelectric equation . . . appears in every case to predict exactly the observed results . . . Yet the semicorpuscular theory by which Einstein arrived at his equation seems at present wholly untenable"[5]; and in his next paper he mentioned "the bold, not to say the reckless, hypothesis of an electromagnetic light corpuscle."[6] Nor was Millikan at that time the only first-rate physicist to hold such views, as will presently be recalled. Rather, the physics community at large had received the light quantum hypothesis with disbelief and with skepticism bordering on derision. As one of the architects of the pre-1925 quantum theory, the old quantum theory, Einstein had quickly found both enthusiastic and powerful support for one of his two major contributions to this field: the quantum theory of specific heat. (There is no reason to believe that such support satisfied any particular need in him.) By sharp contrast, from 1905 to 1923 he was a man apart in being the only one, or almost the only one, to take the light quantum seriously.

Harry Woolf (ed.), Some Strangeness in the Proportion: A Centennial Symposium to Celebrate the Achievements of Albert Einstein
ISBN 0-201-09924-1

If I had to characterize Einstein by one single word I would choose "apartness." This was forever one of his deepest emotional needs. It was to serve him in his singleminded and single-handed pursuits, most notably on his road to triumph from the special to the general theory of relativity. It was also to become a practical necessity for him, in order to protect his cherished privacy from a world hungry for legend and charisma. In all of Einstein's scientific career, this apartness was never more pronounced than in regard to the quantum theory. This covers two disparate periods, the first one of which (1905–23) I have just mentioned. During the second period, from 1926 until the end of his life, he was the only one, or again nearly the only one, to maintain a profoundly skeptical attitude to quantum mechanics. I shall discuss Einstein's position on quantum mechanics in Secs. 6 and 8 but cannot refrain from stating at once that Einstein's skepticism should not be equated with a purely negative attitude. It is true that he was forever critical of quantum mechanics. But at the same time he had his own alternative program for a synthetic theory in which particles, fields, and quantum phenomena all would find their place. Einstein pursued this program from about 1920 (before the discovery of quantum mechanics!) until the end of his life. Numerous discussions with him in his later years have helped me gain a better understanding of his views. Some personal reminiscences of my encounters with Einstein are found in Sec. 8.1.

But let me first return to the days of the old quantum theory. Einstein's contributions to it can be grouped under the following headings:

1. *The light quantum.* In 1900 Planck had discovered the blackbody radiation law without using light quanta. In 1905 Einstein discovered light quanta without using Planck's law.
2. *Specific heats.* In 1907 Einstein published the first paper on quantum effects in the solid state, which showed the way out of the paradoxes arising from the application of the classical equipartition theorem beyond its domain of validity.

These two topics have been discussed elsewhere in this volume. The present paper begins with Einstein's next contribution.

3. *The photon.* The light quantum as originally defined was a parcel of energy. The concept of the photon as a particle with definite energy and *momentum* emerged only gradually. Einstein himself did not discuss photon momentum until 1917. Relativistic energy momentum conservation relations involving photons were not written down till 1923. Einstein's role in these developments is discussed in Sec. 2 of this paper.

The reader may wonder why the man who wrote down the relation $E = h\nu$ for light in 1905 and who propounded the special theory of relativity in that same year would not have stated sooner the relation $p = h\nu/c$. I shall comment on this question in Sec. 8.3.

4. *Quantum statistics.* Einstein's work is treated in Sec. 4.
5. *Wave mechanics.* Einstein's role as a key transitional figure in its discovery will be discussed in Sec. 5.

Before continuing the outline of this paper (in part 4 of this section), I should like to take leave of our main character for a brief while in order to give an overview of the singular role of the photon in the history of the physics of particles and fields. In so doing I shall interrupt the historical sequence of events in order to make some comments from today's vantage point. This will naturally lead to an understanding of Einstein's own singular position in the days of the old quantum theory (see the third part of this section).

1.2 *Particle Physics: The First Fifty Years*

Let us leave aside the photon for a while and ask how physicists reacted to the experimental discovery or the theoretical prediction (whichever came first) of other new particles.[7]

The discovery in 1897 of the first particle, the electron, was an unexpected experimental development that brought to an end the ongoing debate: Are cathode rays molecular torrents or etherial disturbances? The answer came as a complete surprise: They are neither, but rather, a new form of matter. There were some initial reactions of disbelief. Joseph John Thomson (1856–1940) once recalled the comment of a colleague who was present at the first lecture Thomson gave on the new discovery: "I [J.J.T.] was told long afterwards by a distinguished physicist who had been present at my lecture that he thought I had been 'pulling their legs.'"[8] Nevertheless the existence of the electron was widely accepted within the span of very few years. By 1900 it had become clear that beta rays are electrons as well. The discoveries of the free electron and of the Zeeman effect (1896) combined made it evident that a universal atomic constituent had been discovered and that the excitations of electrons in atoms were somehow the sources of atomic spectra.

The discovery of the electron was a discovery at the *outer* experimental frontier. In the first instance this finding led to the abandonment of some earlier qualitative concepts (of the indivisibility of the atom), but it did not require, or at least not at once, a modification of the established corpus of theoretical physics.

During the next fifty years three other particles entered the scene in ways not so dissimilar from the case of the electron, namely via unexpected discoveries of an experimental nature at the outer frontier. They are the proton, (or, rather, the nucleus), the neutron[9] and — just a half century after the electron — the muon, the first of the electron's heavier brothers. As to the acceptance of these particles, it took little time to realize that their coming was in each instance liberating. Within two years after Rutherford's nuclear model, Niels Bohr (1885–1963) was able to make the first real theoretical predictions in atomic physics. Almost at once after the discovery of the neutron, the first viable models of the nucleus were proposed, and nuclear physics could start in earnest. The muon is still one of the strangest animals in the particle zoo, yet its discovery was liberating too since it made possible an understanding of certain anomalies in the creation and absorption of cosmic rays. (Prior to the discovery of the muon, theorists had speculated about the need for an extra particle to explain these anomalies.)

To complete the particle list of the first half century there are four more particles[10] that have entered physics, but in a different way: initially they were theoretical proposals.

The first neutrino was proposed in order to save the law of energy conservation in beta radioactivity. The first meson (now called the pion) was proposed as the conveyer of nuclear forces. Both suggestions were ingenious, daring, innovative, and successful — but they did not demand a radical change of theory. Within months after the public unveiling of the neutrino hypothesis the first theory of the weak interaction was proposed, which is still immensely useful. The meson hypothesis immediately led to considerable theoretical activity as well.

The neutrino hypothesis was generally assimilated long before this particle was actually observed. The interval between the proposal and the first observation of the neutrino is even longer than the corresponding interval for the photon. The meson postulate found rapid experimental support from cosmic ray data — or so it seemed. More than a decade passed before it became clear that the bulk of these observations actually involved muons instead of pions.

Then there was the positron, "a new kind of particle, unknown to experimental physics, having the same mass and opposite charge to an electron."[11] This particle was proposed in 1931, after a period of about three years of considerable controversy over the meaning of the negative energy solutions of the Dirac equation. During that period one participant expressed fear for "a new crisis in quantum physics."[12] The crisis was short-lived however. The experimental discovery of the positron in 1932 was a triumph for theoretical physics. The positron theory belongs to the most important advances of the 1930s.

And then there was the photon, the first particle to be predicted theoretically.

Never, either in the first half century or in the years thereafter has the idea of a new particle met for so long with such almost total resistance as the photon. The light quantum hypothesis was considered somewhat of an aberration even by leading physicists who otherwise held Einstein in the highest esteem. Its assimilation came afer a struggle more intense and prolonged than for any other particle ever postulated, because never, to this day, has the proposal of any particle but the photon led to the creation of a new *inner* frontier. The hypothesis seemed paradoxical: Light was known to consist of waves, hence it could not consist of particles. Yet this paradox alone does not fully account for the resistance to Einstein's hypothesis. Let us look at the situation more closely.

1.3 *On the Assimilation of the Light Quantum Hypothesis*

1.3.1 *Einstein's Caution.* Einstein and Michele Angelo Besso (1873–1955) had met in Zürich in about 1897 and continued their friendship by correspondence from 1903 until the year of their death. Einstein's letters provide a rich source of his insights into physics and people. His struggle with the quantum theory in general and with the light quantum hypothesis in particular is a recurring theme. In 1951 he wrote to Besso: "Die ganzen 50 Jahre bewusster Grübelei haben mich der Antwort der Frage 'Was sind Lichtquanten' nicht näher gebracht."[13]

Throughout his scientific career quantum physics remained a crisis phenomenon to Einstein. His views on the nature of the crisis would change, but the crisis would not go away. In this paper I intend to follow the evolution of Einstein's position on the quantum theory. The period up to 1917 has been discussed by M. Klein in the preceding paper. I shall therefore be quite brief about those early days.

Einstein's sense of the crisis led him to approach quantum problems with great caution in his writings — a caution evident from the very first line ever written on light quanta, the title of his 1905 paper,[14] in which he referred to his hypothesis that light is emitted and absorbed in quanta as a heuristic viewpoint. In the earliest years following this proposal Einstein had good reasons to regard the light quantum hypothesis as provisional. He could formulate it clearly only in the domain $h\nu/kT >> 1$, where Wien's blackbody radiation law holds.[15] Also, he had used this law as an experimental fact without explaining it. Above all, it was obvious to him from the start that grave tensions existed between his principle and the wave picture of electromagnetic radiation — tensions which, in his own mind, were resolved neither then nor later. A man as perfectly honest as Einstein had no choice but to emphasize the provisional nature of his hypothesis. He did this very clearly in 1911, at the first Solvay congress, where he said: "I insist on the provisional character of this concept [light quanta] which does not seem reconcilable with the experimentally verified consequences of the wave theory."[16]

This statement seems to have created the belief in several quarters that Einstein was ready to retract. In 1912 Arnold Sommerfeld wrote: "Einstein drew the most far-reaching consequences from Planck's discovery [of the quantum of action] and transferred the quantum properties of emission and absorption phenomena to the structure of light energy in space without, as I believe, maintaining today his original point of view [of 1905] in all its audacity."[27] Referring to the light quanta, Millikan stated in 1913 that Einstein "gave . . . up, I believe, some two years ago"[18]; and in 1915 he wrote, "Despite . . . the apparently complete success of the Einstein equation [for the photo effect] the physical theory of which it was designed to be the symbolic expression is found so untenable that Einstein himself, I believe, no longer holds to it."[19]

It is my impression that the resistance to the light quantum idea was so strong that one almost hopefully mistook Einstein's caution for vacillation. However, judging from his papers and letters there is no evidence that at any time he retracted any of his statements made in 1905.

1.3.2 *Electromagnetism: Free Fields and Interactions.* Einstein's 1905 paper on light quanta is the second of the revolutionary papers on the old quantum theory. The first one of course was the paper of December 1900 by Max Planck (1858–1947) in which the quantum was first introduced.[20] Both papers contained proposals that flouted classical concepts. Yet the resistance to Planck's ideas, while certainly not absent, was much less pronounced and vehement than in the case of Einstein. Why?

First a general remark on the old quantum theory. Its main discoveries concerned quantum rules for stationary states of matter and of pure radiation. By and large no comparable breakthroughs occurred in regard to the most difficult of all questions concerning electromagnetic phenomena: the interaction between matter and radiation. There, advances became possible only after the advent of quantum field theory when the concepts of particle creation and annihilation were formulated.

Now when Planck introduced the quantum in order to describe a property (the black body spectral function) of essentially a pure radiation system he did so by a procedure of quantization applied to matter (the so-called Planck oscillators). He was unaware of the fact that his proposal implied the need for a revision of the classical description of the radiation field itself. I shall not discuss the manifest lack of consistency of his procedures.[21] Allegedly, Planck's reasoning involved a proposed modification in the interaction between matter and radiation — not a modification of the radiation field itself. This did not seem too outlandish, since the interaction problem was full of obscurities in any event. By contrast, when Einstein proposed the light quantum he had dared to tamper with the Maxwell equations for free fields, which were believed (with good reason) to be much better understood. Therefore it seemed less repugnant to accept Planck's extravaganzas than Einstein's.

This difference in assessment of the two theoretical issues, one raised by Planck, one by Einstein, is quite evident in the writings of the leading theorists of the day. Planck himself had grave reservations about light quanta. In 1907 he wrote to Einstein: "I am not seeking for the meaning of the quantum of action [light quanta] in the vacuum but rather in places where absorption and emission occur and [I] assume that what happens in the vacuum is rigorously described by Maxwell's equations."[22] A remark by Planck at a physics meeting in 1909 vividly illustrates his and others' predilections for "leaving alone" the radiation field and for seeking the resolution of the quantum paradoxes in the interactions: "I believe one should first try to move the whole difficulty of the quantum theory to the domain of the interaction between matter and radiation."[23] In that same year Hendrik Antoon Lorentz (1865–1928) expressed not only his belief in "Planck's hypothesis of the energy elements" but also his strong reservations regarding "light quanta which retain their individuality in propagation."[24]

Thus by the end of the first decade of the twentieth century many leading theorists were prepared to accept the fact that the quantum theory was here to stay. However, the Maxwell theory of the free radiation field, pure and simple, provided neither room for modification (it seemed) nor a place to hide one's ignorance, in contrast with the less transparent situation concerning the interaction between matter and radiation. This position did not change much until the 1920s, and it remained one of the deepest roots of resistance to Einstein's ideas.

1.3.3. *The impact of experiment.* The first three revolutionary papers on the old quantum theory were those by Planck (1900), Einstein (1905), and Bohr (1913). All three contained proposals that flouted classical concepts. Yet the resistance to the ideas of Planck and Bohr, while certainly not absent, was much less pronounced and vehement than in the case of Einstein. Why? The answer: because of the impact of experiment.

Physicists, good physicists, enjoy scientific speculation in private, but tend to frown upon it when done in public. They are conservative revolutionaries, resisting innovation as long as possible at all intellectual cost, but embracing it when the evidence is incontrovertible — if they do not, physics tends to pass them by.

I often argued with Einstein about reliance on experimental evidence for confirmation of fundamental new ideas. In Sec. 8 I shall have more to say about Einstein's position on this issue. Meanwhile, I shall discuss the influence of experimental developments on the acceptance of the ideas of the three men just mentioned.

Planck's proximity to the first-rate experiments on blackbody radiation being performed at the Physikalisch Technische Reichsanstalt in Berlin was beyond doubt a crucial factor in his discovery of 1900 (though it would be very wrong to say that this was the only decisive factor). In the first instance, experiment also set the pace for the acceptance of the Planck formula. One could (and did and should) doubt his derivation as, among others, Einstein did in 1905. But at that same time neither Einstein nor any one else denied the fact that Planck's highly nontrivial universal curve admirably fitted the data. Somehow he had to be doing something right.

Bohr's paper[25] of April 1913 about the hydrogen atom was revolutionary and certainly not at once generally accepted. But there was no denying that his expression $2\pi^2e^4m/h^3c$ for the Rydberg constant of hydrogen was remarkably accurate (to within 6 percent, in 1913). When, in October 1913, Bohr was able to give an elementary derivation[26] of the ratio of the Rydberg constants for singly ionized helium and hydrogen, in agreement with experiment to five significant figures, it became even more clear that Bohr's ideas had a great deal to do with the real world. When told of the helium/hydrogen ratio, Einstein is reported to have said of Bohr's work, "Then it is one of the greatest discoveries."[27]

Einstein himself had little to show by comparison. To be sure, he had mentioned a number of experimental consequences of his hypothesis in his 1905 paper. But he had no curves to fit, no precise numbers to show. He had noted that in the photoelectric effect the electron energy E is constant for fixed light frequency ν. This explained Lenard's results. But Lenard's measurements were not so precise as to prevent men like J. J. Thomson and Sommerfeld from giving alternative theories of the photo effect, of a kind in which Lenard's law does not rigorously apply.[28] Einstein's photoelectric equation $E = h\nu - P$ predicts a linear relation between E and ν. At the time that Einstein proposed his heuristic principle no one knew how E depended on ν beyond the fact that one increases with the other.[29] Unlike Bohr and Planck, Einstein had to wait a decade before he saw one of his predictions vindicated, the linear E-ν relation, by Millikan in Chicago for the photo effect[30] and by William Duane (1882–1935) and his assistant Franklin Hunt, at Harvard, for the inverse photo effect[31] (Volta effect). One immediate and salutary effect of these experimental discoveries was that alternative theories of the photo effect vanished from the scene.

Yet Einstein's apartness did not end even then.

I have already mentioned that Millikan relished his result on the photo effect but declared that, even so, the light quantum theory "seems untenable." In 1918, Ernest Rutherford (1871–1937) commented as follows on the Duane–Hunt results: "There is at present no physical explanation possible of this remarkable connection between energy and frequency".[33] One can go on. The fact of the matter is that, even after Einstein's photoelectric law was accepted, almost no one but Einstein himself would have anything to do with light quanta. This went on until the early 1920s, as is best illustrated by quoting the citation for Einstein's Nobel prize in 1921: "To Albert Einstein for his services to theoretical physics and especially for his discovery of the photoelectric effect."[34] This is not only a historic understatement but also an accurate reflection on the consensus in the physics community.

To summarize, the enormous resistance to light quanta found its roots in the particle-wave paradoxes (see M. Klein in this volume). The resistance was enhanced because the light quantum idea seemed to overthrow that part of electromagnetic theory that was believed to be best understood: the theory of the free field. Moreover, experimental support was long in coming and, even after the photoelectric effect predictions were verified, light quanta were still largely considered unacceptable. Einstein's own emphasis on the provisional nature of the light quantum hypothesis tended to strengthen the reservations held by other physicists.

1.4 *The Quantum Theory: Lines of Influence*

The skeleton diagram given in Fig. 14.1 is an attempt to reduce the history of the quantum theory to its barest outlines. At the same time this figure will serve as a guide to the rest of this paper. $X \rightarrow Y$ means: the work of X was instrumental to an advance by Y. Arrows marked M and R indicate that the influence went via the theory of matter and of radiation, respectively.

If Planck, Einstein, and Bohr are the fathers of the quantum theory, then Gustav Robert Kirchhoff (1824–87) is its grandfather. Insofar as he is the founder of optical spectra analysis [in 1860, together with Robert Bunsen (1822–99)],[35] an arrow leads to Johann Jakob Balmer (1825–98), the inventor of the Balmer formula (in 1885).[36] From Balmer we move to Bohr, the founder of atomic quantum dynamics. Returning to Kirchhoff, as the discoverer of the universal character of blackbody radiation[37] his influence goes via Wien to Planck.

The arrow from Bose to Einstein refers to Bose's work on electromagnetic radiation and its impact on Einstein's contributions to the quantum statistics of a material gas. See Sec. 4, where also Einstein's influence on Dirac is briefly mentioned.

The arrow from Wien to Planck refers to the latter's formulation of his blackbody radiation law, while the triangle Wien–Planck–Einstein represents the mutual influences that led to the light quantum hypothesis.

The triangle Einstein–de Broglie–Schroedinger refers to the role of Einstein as the transitional figure in the birth of wave mechanics (Sec. 5). The h marking the arrow from Planck to Bohr serves as a reminder that not so much the details of Planck's work on radiation as the very intro-

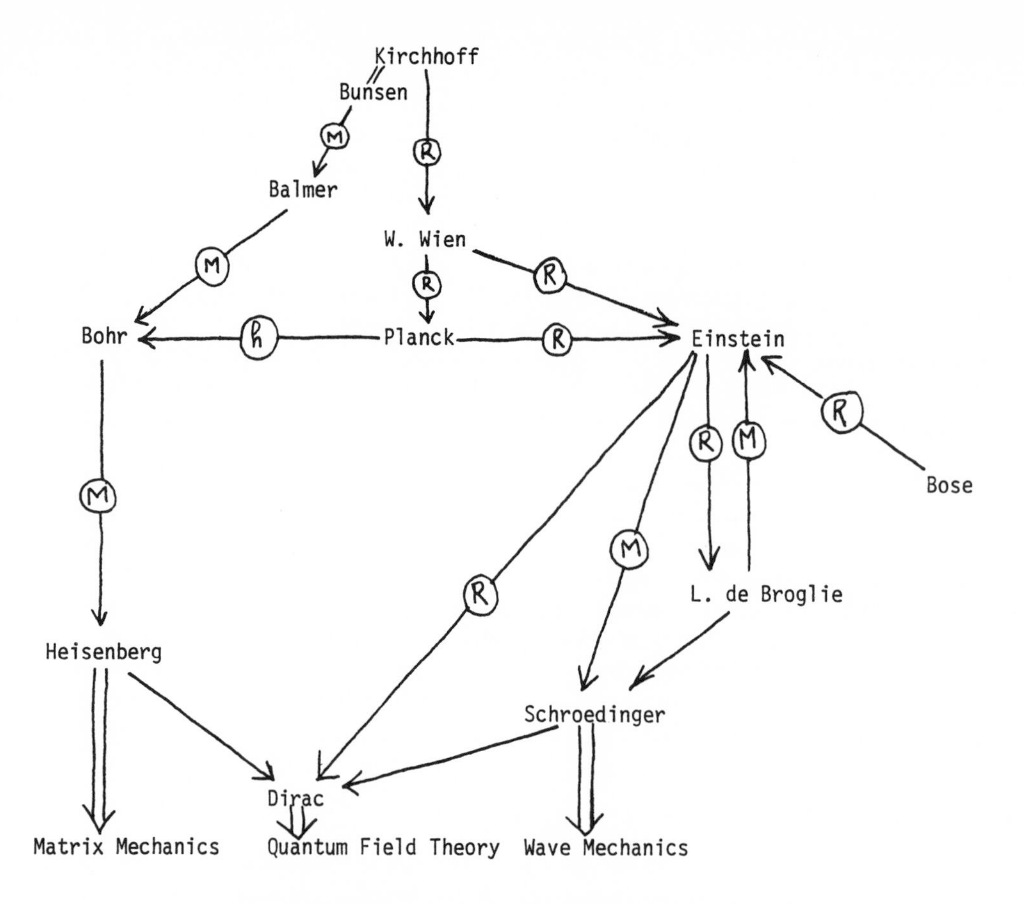

Fig. 14.1 The quantum theory: lines of influence.

duction by Planck of his new universal constant *h* was decisive for Bohr's ideas about atomic stability. An account of Bohr's influence on Heisenberg and of Heisenberg's and Schroedinger's impact on Dirac is beyond the scope of the present paper.

In the case of Einstein and Bohr, it cannot be said that the work of one induced major advances in the work of the other. Therefore the simplified diagram does not and should not contain links between them. Nevertheless, for forty years there were influences at work between Einstein and Bohr, and these were in fact intense, but they were on a different plane. In a spirit of friendly and heroic antagonism these two men argued about questions of principle. Sec. 3 deals with Bohr's resistance to Einstein's idea of the photon. This was but a brief interlude. It ended with the detailed experimental vindication of the photon concept to which Bohr fully subscribed from then on. Their far more important debate on the foundations of quantum mechanics began in 1927. On these issues the intellectual resistance and opposition of one against the most basic views held by the other continued unabated until the end of Einstein's life. At issue were the criteria by which one should judge the completeness of the description of the physical world. Their discussions have not affected the evolution of physical theory; yet theirs will be remembered as one of the great debates on scientific principle between two dominant figures of their day. The dialogue between Bohr and Einstein had one positive outcome: it forced Bohr to express the tenets of complementarity in increasingly precise language. (See Sec. 6.)

After 1930 Einstein considered quantum mechanics to be consistent and successful but incomplete. At the same time he had his own aspirations for a future theory of particles and fields. I shall try to make clear in Sec. 8 what these were. I do not believe that Einstein presented valid arguments for the incompleteness of the quantum theory. But neither do I think that the times are ripe to answer the question whether the quantum mechanical description is indeed complete, since the physics of particles and fields is a subject beset with many unresolved fundamental problems. Among these there is one that was most dear to Einstein and with which he (and all of us, to date) struggled in vain: the synthesis of quantum physics with general relativity. Since we still have far to go, any assessment of Einstein's views must necessarily be tentative. In order to stress this I have prefaced Sec. 8, on Einstein's vision, with a very brief overview (Sec. 7) of the current status of particle physics.

2. From the Light Quantum to the Photon

2.1 *Momentum Fluctuations*

A photon is a state of the electromagnetic field with the following properties.

1. It has a definite frequency ν and a definite wave vector $\mathbf{k}$.
2. Its energy E:

$$E = h\nu \tag{14.1}$$

and its momentum $\mathbf{p}$:

$$\mathbf{p} = h\mathbf{k} \tag{14.2}$$

satisfy the dispersion law

$$E = c\ |\mathbf{p}| \tag{14.3}$$

characteristic for a particle of zero rest mass.[38]
3. It has spin one and (like all massless particles with nonzero spin) two states of polarization. The single particle states are uniquely specified[39] by the properties (1)–(3).

The number of photons is in general not conserved in particle reactions and decays. I shall return to the nonconservation of photon number in the section on quantum statistics (Sec. 4) but would like to note here an ironical twist of history. The term "photon" first appeared in the title of a paper written in 1926.[40] The title: "The conservation of photons." The author: the distinguished physical chemist Gilbert Newton Lewis (1875–1946) from Berkeley. The subject: a speculation that light consists of "a new kind of atom . . . uncreatable and indestructible [for which] I . . . propose the name photon." This idea was soon forgotten, but the new name almost immediately became part of the language. In October 1927 the fifth Solvay conference was held; its subject was "Electrons et Photons."

When Einstein introduced light quanta in 1905, these were *energy* quanta satisfying Eq. (14.1). There was no mention in that paper of Eqs. (14.2) and (14.3). In other words, the full-fledged particle concept embodied in the term photon was not there all at once. For this reason I make the distinction between light quantum ("$E = h\nu$ only") and photon in this section. The dissymmetry between energy and momentum in the 1905 paper is of course intimately connected with the origins of the light quantum postulate in equilibrium statistical mechanics. In the statistical mechanics of equilibrium systems one derives important relations between the overall energy and other macroscopic variables. The overall momentum plays a trivial and subsidiary role. These distinctions between energy and momentum are much less pronounced when one considers fluctuations around the equilibrium state. It was via the analysis of statistical fluctuations of blackbody radiation that Einstein eventually came to associate a definite momentum with a light quantum. This happened in 1917. Before I describe what he did I should again draw the attention of the reader to the remarkable fact that it took the father of special relativity theory twelve years to write down the relation $p = h\nu/c$ side by side with $E = h\nu$. I shall have more to say about this in Sec. 8.3.

Now to the fluctuations. In 1909 Einstein derived two fluctuation formulas, both of which will play a role in what is to follow. The first one concerns energy fluctuations. Consider a subvolume v of a cavity filled with radiation in thermal equilibrium at a temperature T. The mean energy $\bar{\varepsilon}$ in the frequency interval between ν and $\nu + d\nu$ is given by $\rho(\nu, T)vd\nu$, where ρ is the spectral density of the blackbody radiation. Let $\overline{\varepsilon^2}$ denote the mean energy squared in this interval and $\langle\varepsilon^2\rangle$ the mean square energy fluctuation. Einstein derived the formula

$$< \varepsilon^2 > = \overline{\varepsilon^2} - \bar{\varepsilon}^{2} = kT^2 vd\nu \cdot \frac{\partial \rho}{\partial T}, \qquad (14.4)$$

which holds regardless of the explicit form of $\rho(\nu, T)$. According to Planck

$$\rho(\nu, T) = \frac{8\pi\nu^2}{c^3} \cdot \frac{h\nu}{e^{h\nu/kT} - 1}. \qquad (14.5)$$

From (14.4) and (14.5)

$$< \varepsilon^2 > = \left[\rho h\nu + \frac{c^3\rho^2}{8\pi\nu^2}\right] vd\nu, \qquad (14.6)$$

the first of the Einstein fluctuation formulas of 1909. I shall return to Eq. (14.6) in Sec. 5.2.

The second formula deals with momentum fluctuations and is pertinent to the question of photon momentum. Einstein considered the case of a plane mirror with mass m and area f placed inside the cavity. The mirror moves perpendicular to its own plane and has a velocity v at time t. During a small time interval from t to $t + \tau$, its momentum changes from mv to $mv - Pv\tau + \Delta$. The second term describes the drag force due to the radiation pressure (P is the corresponding friction constant). This force would eventually bring the mirror to rest were it not for the momentum fluctuation term Δ induced by the fluctuations of the radiation pressure. In thermal equilibrium the mean square momentum $m^2 <v^2>$ should remain unchanged over the interval τ. Hence[42] $<\Delta^2> = 2mP\tau <v^2>$. The equipartition law applied to the kinetic energy of the mirror implies that $m < v^2 > = kT$. Hence,

$$< \Delta^2 > = 2P\, \tau kT\,. \tag{14.7}$$

Einstein computed P in terms of ρ for the case in which the mirror is fully transparent for all frequencies except those between ν and $\nu + d\nu$, which it reflects perfectly. Using Eq. (14.5) for ρ he found that

$$< \Delta^2 > = \frac{1}{c}\left[\rho h\nu + \frac{c^3\rho^2}{8\pi\nu^2}\right] f\tau d\nu\,. \tag{14.8}$$

The parallels between Eqs. (14.6) and (14.8) are striking. The respective first terms dominate if $h\nu / kT >> 1$, the regime in which Eq. (14.5) is approximated by Wien's exponential law. The first term in Eq. (14.6) extends Einstein's energy quantum postulate of 1905: "Monochromatic radiation of low density [i.e.] within the domain of the Wien radiation formula) behaves in thermodynamic respect as if it consists of mutually independent energy quanta of magnitude $h\nu$"[43] to energy fluctuations. One might expect that the first term in Eq. (14.8) would have led Einstein to state, in 1909, the "momentum quantum postulate": Monochromatic radiation of low density behaves in regard to pressure fluctuations as if it consists of mutually independent momentum quanta of magnitude $h\nu/c$. It is unthinkable to me that Einstein did not think so. But he did not quite say so.

This is what he did say: "If the radiation were to consist of very few extended complexes with energy $h\nu$ which move independently through space and which are independently reflected — a picture which represents the roughest visualization of the light quantum hypothesis — then as a consequence of fluctuations in the radiation pressure such momenta would act on our plate as are represented by the first term only of our formula [(14.8)]."[44] He did not refer explicitly to momentum quanta nor to the relativistic connection between $E = h\nu$ and $p = h\nu/c$. Yet a particle concept (the photon) was clearly on his mind, since he went on to conjecture that "the electromagnetic fields of light are linked to singular points similar to the occurrence of electrostatic fields in the theory of electrons."[45] It seems fair to paraphrase this statement as follows: Light quanta may well be particles in the same sense that electrons are particles. The association between the particle concept and a high degree of spatial localization is typical for that period. It is, of course, not correct in general.

The photon momentum made its explicit appearance in that same year, 1909. Johannes Stark (1874–1957) had attended the Salzburg meeting at which Einstein had discussed the radiative fluctuations.[46] A few months later Stark stated that according to the light quantum hypothesis, "the total electromagnetic momentum emitted by an accelerated electron is different from zero and . . . in absolute magnitude is given by $h\nu/c$."[47] As an example he mentioned bremsstrahlung, for which he wrote down the equation

$$m_1\mathbf{v}_1 + m_2\mathbf{v}_2 = m_1\mathbf{v}_1{}' + m_2\mathbf{v}_2{}' + \frac{h\nu}{c^2}\,\mathbf{c}\,, \tag{14.9}$$

the first occasion on record in which the photon enters explicitly into the law of momentum conservation for an elementary process.

Einstein himself did not touch the blackbody radiation problem again for the next seven years. During the intervening period he was in the main otherwise engaged! Then came his last three papers on electromagnetic radiation, two in 1916 and one early in 1917.[48]

All three papers deal with a joint equilibrium system consisting of a molecular gas in interaction with radiation. Einstein introduced the following "kinetic Stosszahlansatz" for this interaction. Consider two energy levels E_m, E_n of a molecule, $E_m > E_n$. Then it is assumed that the total number dW of transitions in the molecular gas per time interval dt is given by

$$dW_{mn} = N_m(\rho B_{mn} + A_{mn})dt \qquad \text{for } m \rightarrow n, \tag{14.10}$$

$$dW_{nm} = N_n \rho B_{nm}\, dt \qquad \text{for } n \rightarrow m\ . \tag{14.11}$$

ρ is the spectral density, here a function of $E_m - E_n$ and T. N_m, N_n are the Maxwell distribution occupations of the levels E_m and E_n, respectively. $B_{mn}(B_{nm})$ is the coefficient of induced emission (absorption); A_{mn} is the coefficient of spontaneous emission. The best-remembered result of these papers is Einstein's derivation of the Planck formula. As is discussed by Klein in the preceding chapter, this derivation demands that a few simple additional requirements be satisfied. I draw attention to one of these: namely,

$$E_m - E_n = h\nu\ . \tag{14.12}$$

The content of this equation is far more profound than a definition of the symbol ν. It is a compatibility condition. Its physical content is this: in order for Eqs. (14.10) and (14.11) to lead to Planck's law it is necessary that the transition $m \rightarrow n$ be accompanied by a single *monochromatic* radiation quantum. By this remarkable reasoning Einstein therefore established a bridge between blackbody radiation and Bohr's theory of spectra.

The 1917 paper contained another result, which Einstein himself considered far more important: photons carry momentum $h\nu/c$. Again the reasoning is centered on a compatibility condition, but applied this time to momentum fluctuations. Einstein raised the following problem. In equilibrium the molecules have a Maxwell distribution for the translational velocities. How is this distribution maintained in time considering the fact that the molecules are subject to the influence of radiation pressure? In other words, what is the Brownian motion of molecules in the presence of radiation?

Technically the following issue arises. If a molecule emits or absorbs an amount ε of radiative energy all of which moves in the same direction, then it experiences a recoil of magnitude ε/c. There is no recoil if the radiation is not directed at all, as for a spherical wave. Question: What can one say about the degree of directedness of the emitted or absorbed radiation for the system under consideration? Einstein began the discussion of this question in the same way he had treated the mirror problem in 1909. Instead of the mirror, he now considered molecules that all move in the same direction. Then there is again a drag force $-P v \tau$ and a fluctuation term Δ. Equipartition gives again $m \langle v^2 \rangle = kT$ and one arrives once more at Eq. (14.7). Next comes the issue of compatibility. With the help of Eqs. (14.10) and (14.11) Einstein could compute separately expressions for $\langle \Delta^2 \rangle$ as well as for P in terms of the A's, B's, and ρ, where ρ is now given by Planck's law. I shall not reproduce the details of these calculations but do note the crux of the matter. In order to obtain the same answer for the quantities on the left-hand side and the right-hand side of Eq. (14.7), Einstein had to invoke a condition of directedness: "If a bundle of radiation causes a molecule to emit or absorb an energy amount $h\nu$ then a momentum $h\nu/c$ is transferred to the molecule, directed along the bundle for absorption and oppositely to the bundle for [induced] emission."[49] (The question of spontaneous emission is discussed below.) Thus Einstein found that consistency with the Planck distribution [and Eqs. (14.10), (14.11)] requires that the radiation be fully directed. (This was often called *Nadelstrahlung*.) And so with the help of his trusted and beloved fluctuation methods Einstein once again produced a major insight, the association of momentum quanta with energy quanta. Indeed if we set aside the question of spin we may say that *Einstein abstracted not only the light quantum but also the more general photon concept entirely from statistical mechanical considerations*. (Some comments on the photon spin will be found in Sec. 4.)

2.2 *Earliest Unbehagen about Chance*

Einstein prefaced his statement about photon momentum that I just quoted with the remark that this conclusion can be considered "als ziemlich sicher erwiesen" ("as fairly certainly proven"). If he had some lingering reservations, this was mainly because he had derived some of his equations on the basis of "the quantum theory [which is] incompatible with the Maxwell theory of the electromagnetic field."[50] Moreover, his momentum condition was a sufficient, not a necessary, condition, as was emphasized by Wolfgang Pauli (1900–58) in a review article completed in 1924: "From Einstein's considerations it could . . . not be seen with complete certainty that his assumptions were the only ones which guarantee thermodynamic-statistical equilibrium."[51] Nevertheless his 1917 results led Einstein to drop his caution and reticence about light quanta. They had become real to him. In a letter to Besso about the needle rays he wrote: "Damit sind die Lichquanten so gut wie gesichert" ("With that [the existence of] the light quanta is practically certain").[52] And, in a phrase contained in another letter about two years later, "I do not doubt anymore the reality of radiation quanta, although I still stand quite alone in this conviction," he underlined the word *Realität*.[53]

On the other hand, at about the same time that Einstein lost any remaining doubts about the existence of light quanta we also encounter the first expressions of his Unbehagen, his discomfort with the theoretical implications of the new quantum concepts in regard to *Zufall*, chance.

This earliest unease stemmed from the conclusion concerning spontaneous emission that Einstein had been forced to draw from his consistency condition (14.7): The needle ray picture applies not only to induced processes (as was mentioned before) but also to spontaneous emission. That is, in a spontaneous radiative transition the molecule suffers a recoil $h\nu/c$. However, the recoil direction cannot be predicted! Einstein stressed (quite correctly of course) that it is "a weakness of the theory . . . that it leaves time and direction of elementary processes to 'chance.'"[54] What decides when the photon is spontaneously emitted? What decides in which direction it shall go?

These questions were not new. They also apply to another class of emission processes the spontaneity of which had puzzled physicists since the turn of the century: radioactive transformations. A spontaneous emission coefficient was in fact first introduced by Rutherford in 1900 when he wrote down the equation $dN = -\lambda N dt$ for the decrease of the number N of radioactive thorium emanation atoms in the time interval dt.[55] Einstein himself drew attention to this similarity: "It speaks in favor of the theory that the statistical law assumed for [spontaneous] emission is nothing but the Rutherford law of radioactive decay."[56] I have written elsewhere about the ways physicists responded to this baffling lifetime problem.[57] I should now add that Einstein was the first to realize that the probability for spontaneous emission is a nonclassical quantity. No one before Einstein in 1917 saw as clearly the depth of the conceptual crisis generated by the occurrence of spontaneous processes with a well-defined (partial) lifetime. He expressed this in prophetic terms: "The properties of elementary processes required by [Eq. (14.7)] make it seem almost inevitable to formulate a truly quantized theory of radiation."[58]

Immediately following his comment on "chance," to which I have just referred, Einstein continued: "Nevertheless I have full confidence in the route which has been taken."[59] If Einstein was confident at that time about the route, he also felt strongly that this route would be a long one. The chance character of spontaneous processes meant that something was amiss with classical causality. That would forever deeply trouble him. As early as March 1917 Einstein had written on this subject to Besso: "I feel that the real joke which the eternal inventor of enigmas has presented us with has absolutely not been understood as yet."[60] It is believed by nearly all of us that the joke was understood soon after 1925, when it became possible to calculate Einstein's A_{mn} from first principles. As I shall discuss, Einstein eventually accepted these principles, but he never considered them to be *first* principles. Einstein's attitude throughout the rest of his life was: The joke has not been understood as yet. One further example may show how from 1917 on Einstein could not make his peace with the quantum theory. In 1920 he wrote as follows to Born: "That business about causality causes me a lot of trouble too. Can the quantum absorption and emission of light ever be understood in the sense of the complete causality requirement, or would a statistical residue remain? I must admit that there I lack the courage of a conviction. But I would be very unhappy to renounce complete causality."[61]

2.3 *The Compton Effect*

We now come to the *dénouement* of the photon story.

Since, after 1917, Einstein firmly believed that light quanta were here to stay, it is not surprising that he would look for new ways in which the existence of photons might lead to observable deviations from the classical picture. In this he did not succeed. At one point, in 1921, he thought he had found a new quantum criterion,[62a] but it soon turned out to be a false lead.[62b] In fact, after 1917 nothing particularly memorable happened in regard to light quanta until capital progress was achieved when Arthur Compton (1892–1962) and Peter Jozef Debye (1884–1966) independently wrote down the relativistic kinematics for the scattering of a photon on an electron at rest[63]:

$$h\mathbf{k} = \mathbf{p} + h\mathbf{k}' , \tag{14.13}$$

$$hc\,|\,\mathbf{k}\,| + mc^2 = hc|\,\mathbf{k}'\,| + \sqrt{c^2p^2 + m^2c^4}\,. \tag{14.14}$$

Why were these elementary equations not published five or even ten years earlier, as well they could have been? Even those opposed to quantized radiation might have found these relations to their liking since (independent of any quantum dynamics) they yield at once significant differences from the classical theories of the scattering of light by matter[64] and, therefore, provide simple tests of the photon idea.

I have no entirely satisfactory answer to this question. In particular it is not clear to me why Einstein himself did not consider these relations. However, there are two obvious contributing factors. First, because photons were rejected out-of-hand by the vast majority of physicists, few may have felt compelled to ask for tests of an idea in which they did not believe to begin with. Secondly, it was only in about 1922 that strong evidence became available for deviations from the classical picture. This last circumstance impelled both Compton and Debye to pursue the quantum alternative.[65] Debye, incidentally, mentioned his indebtedness to Einstein's work on needle radiation. Compton in his paper does not mention Einstein at all.[66]

The same paper in which Compton discusses Eqs. (14.13) and (14.14) also contains the result of a crucial experiment. These equations imply that the wavelength difference $\Delta\lambda$ between the final and the initial proton is given by

$$\Delta\lambda = \frac{h}{mc}(1 - \cos\theta) \,, \tag{14.15}$$

where θ is the photon scattering angle. Compton found this relation to be satisfied within the error.[67] The quality of the experiment is well demonstrated by the value he obtained for the Compton wavelength: $h/mc \simeq 0.0242$ Å, which is within less than 1 percent of the modern value.[68] Compton concluded: "The experimental support of the theory indicates very convincingly that a radiation quantum carries with it directed momentum as well as energy."[69]

This discovery "created a sensation among the physicists of that time."[70] There were the inevitable controversies surrounding a discovery of such major proportions. Nevertheless the photon idea was rapidly accepted. Sommerfeld incorporated the Compton effect in his new edition of *Atombau und Spektrallinien* in 1924 with the comment: "It is probably the most important discovery which could have been made in the current state of physics."[71]

What about Einstein's own response? A year after Compton's experiments he wrote a popular article for the *Berliner Tageblatt* that ends as follows: "The positive result of the Compton experiment proves that radiation behaves as if it consisted of discrete energy projectiles, not only in regard to energy transfer but also in regard to *Stosswirkung* (momentum transfer)."[72] Here then, in projectile (that is, particle) language, is the "energy-momentum quantum postulate," phrased in close analogy to the language of 1905. In both cases we encounter the phraseology: "Radiation . . . behaves . . . as if it consists of . . . ."

Still, Einstein was not (and would never be) satisfied. There was as yet no real theory. In the same article he also wrote: "There are therefore now two theories of light, both indispensable, and — as one must admit today in spite of twenty years of tremendous effort on the part of theoretical physicists — without any logical connection."

The period 1923–24 marks the end of the first phase of Einstein's apartness in relation to the quantum theory. Yet there remained one important bastion of resistance to the photon, centering around Niels Bohr.

3. Interlude: the BKS Proposal

Sie haben sich heiss und innig geliebt.

Helen Dukas

In January 1924 Niels Bohr, Hendrik Anton Kramers (1894–1952), and John Clarke Slater (1900–76) submitted to the *Philosophical Magazine* an article that contained drastic theoretical proposals concerning the interaction of light with matter.[73] It was written after Compton's discovery, yet it rejected the photon. It was written also after Einstein and Bohr had met. This discussion of their proposal (hereafter called the BKS proposal) is intended to serve a twofold purpose: it is a postscript to the story of the photon and a prelude to the Bohr–Einstein dialogue, which will occupy us more fully in later sections.

I have already mentioned that Einstein was immediately and strongly impressed by Bohr's work of 1913. They did not yet know each other at that time. A number of years was to pass before their first encounter; meanwhile they followed each others' published work. Also, Ehrenfest kept Einstein informed of the progress in Bohr's thinking. "Ehrenfest tells me many details from Niels Bohr's 'Gedankenküche' [thought kitchen]; he must be a very first rate mind, extremely critical and far seeing, which never loses track of the grand design."[74] Einstein remained forever deeply respectful of Bohr's pioneering work. When he was nearly seventy he wrote: "That this insecure and contradictory foundation [of physics in the teens] was sufficient to enable a man of Bohr's unique instinct and tact to discover the major laws of the spectral lines and of the electron shells of the atoms together with their significance for chemistry appeared to me like a miracle — and appears to me as a miracle even today. This is the highest form of musicality in the sphere of thought."[75]

Einstein and Bohr finally met in the spring of 1920, in Berlin. At that time both had already been widely recognized as men of destiny who would leave their indelible marks on the physics of the twentieth century. The impact of their encounter was intense and went well beyond a meeting of minds. Shortly after his visit, Einstein wrote to Bohr: "Not often in life has a human being caused me such joy by his mere presence as you did."[76] Bohr replied: "To meet you and to talk with you was one of the greatest experiences I have ever had."[77] Some years later Einstein began a letter to Bohr as follows: "Lieber oder vielmehr geliebter Bohr" ("Dear or rather beloved Bohr").[78] Once when I talked with Helen Dukas, Einstein's devoted secretary (in 1978, fifty years after she had started to work for Einstein) about the strong tie between these two men she made the comment that is quoted at the head of this section: "They loved each other warmly and dearly."

All who have known Bohr will be struck by this characterization that Einstein gave of him much later: "He utters his opinions like one perpetually groping and never like one who believes to be in the possession of definite truth."[79] Bohr's style of writing makes clear for all to see how he groped and struggled. "Never express yourself more clearly than you think," he used to admonish himself and others. Bohr's articles are sometimes dense. Having myself assisted him on a number of occasions when he was attempting to put his thoughts on paper, I know to what enormous lengths he went to find the most appropriate turn of phrase. I have no such first-hand information about the way Einstein wrote. But, again for all to see, there are his papers, translucent. The early Einstein papers are brief; their content is simple; their language is sparse. They exude finality even when they deal with a subject in flux. For example, no statement made in the 1905 paper on light quanta needs to be revised in the light of later developments. Whether he published in German or English, he initially wrote in German. He had a delicate musical sense of language and a keen insight into people, as his description of Bohr illustrates.

Their meeting, in 1920 took place some years before they found themselves at scientific odds on profound questions of principle in physics. They did not meet very often in later times. They did correspond, but not voluminously. I was together a few times with both of them some thirty years after their first encounter when their respective views on the foundations of quantum mechanics had long since become irreconcilable. More about that later. Let me note here only that neither the years nor later events ever diminished the mutual esteem and affection in which they held one another.

Let us now turn to the BKS proposal.

As I have already stressed in Sec. 1.1, it was the position of most theoretical physicists during the first decades of the quantum era that the conventional continuous description of the free radiation field should be protected at all cost and that the quantum puzzles concerning radiation should eventually be resolved by a revision of the properties of interaction between radiation and matter. The BKS proposal represents the extreme example of this position. Its authors suggested that radiative processes have highly unconventional properties "the cause of [which] we shall not seek in any departure from the electrodynamic theory of light as regards the laws of propagation in free space, but in the peculiarities of the interaction between the virtual field of radiation and the illuminated atoms."[80] Before describing these properties I should point out that the BKS paper represents a program rather than a detailed research report. It contains no formalism whatsoever.[81] This program was not to be the right way out of the difficulties of the old quantum theory, yet the paper had a lasting impact in that (as we shall see) it stimulated important experimental developments. Let us discuss next the two main paradoxes which BKS addressed.

3.1 *The First Paradox.*

Consider an atom that emits radiation in a transition from a higher to a lower state. Bohr, Kramers, and Slater assume that in this process "energy [is] of two kinds, the continuously changing energy of the field and the discontinuously changing atomic energy."[82] But how can there be conservation of an energy that consists of two parts, one changing discontinuously, the other continuously? The BKS answer: "As regards the occurrence of transitions, which is the essential feature of the quantum theory, we abandon . . . a direct application of the principles of conservation of energy and momentum."[83] Energy and momentum conservation, they suggested, does not hold true for individual elementary processes but should only hold statistically, as an average over many such processes.

The idea of energy nonconservation had already been on Bohr's mind a few years prior to the time of the BKS proposal.[84] However, it was not Bohr but Einstein who had first raised — and rejected — this possibility. In 1910 Einstein wrote to a friend: "At present I have high hopes for solving the radiation problem and that without light quanta. I am enormously curious how it will work out. One must renounce the energy principle in its present form."[85] A few days later he was disenchanted: "Once again the solution of the radiation problem is getting nowhere. The devil has played a rotten trick on me."[86] He raised the issue one more time at the 1911 Solvay meeting, noting that his formula for the energy fluctuations of blackbody radiation could be interpreted in two ways: "One can choose between the [quantum] structure of radiation and the negation of an absolute validity of the energy conservation law." He rejected the second alternative: "Who would have the courage to make a decision of this kind? . . . We will agree that the energy principle should be retained."[87] But others were apparently not as convinced. In 1916 the suggestion of statistical energy conservation was taken up by Nernst.[88] Not later than January 1922 Sommerfeld remarked that the "mildest cure" for reconciling the wave theory of light with quantum phenomena would be to relinquish energy conservation.[89] Thus the BKS proposal must be regarded as an attempt to face the consequences of an idea that had been debated for quite some time.

In order to understand Bohr's position in 1924 it is above all important to realize that the correspondence principle was to him the principal reliable bridge between classical and quantum physics. But the correspondence principle is of course no help in understanding light quanta; the controversial issue of photons versus waves lies beyond this principle. To repeat, the photon-wave duality was the earliest known instance of what was later to be called a complementary situation. The BKS theory, with its rejection of photons and its insistence on the continuous picture of light at the price of nonconservation, historically represents the last stand of the old quantum theory. For very good reasons this proposal was characterized some years later by one of the principal architects of the quantum mechanics as representing the height of the crisis in the old quantum theory.[90] Nor was nonconservation of energy and momentum in individual processes the only radical proposal made by BKS.

3.2 *The Second Paradox*

This paradox concerns a question that had troubled Einstein since 1917 (as we have seen): How does an electron know *when* to emit radiation in making a spontaneous transition?

In its general form the BKS answer to this question was: There is no *truly* spontaneous emission. They associated to an atom in a given state a "virtual radiation field" that contains all the possible transition frequencies to other stationary states and assumed that "the transitions which in [the Einstein theory of 1917] are designated as spontaneous are, on our view, *induced* [my italics] by the virtual field." According to BKS, the spontaneous transition to a specific final state is connected with the virtual field mechanism "by probability laws which are analogous to those which in Einstein's theory hold for induced transitions." In this way "the atom is under no necessity of knowing what transition it is going to make ahead of time."[91] Thus spontaneous emission is ascribed to the action of the virtual field, but this action is noncausal. I shall not discuss details of the BKS picture of induced emission and absorption and other radiative processes. Suffice it to say that all of these are supposed to be due to virtual fields and that in all of these causality is abandoned. In a paper completed later in 1924 Slater noted that the theory "has unattractive features . . . [but] it is difficult at the present stage to see how [these are] to be avoided."[92]

But what about the Compton effect? The successfully verified Eq. (14.15) rests on the conservation laws (14.13) and (14.14). However (BKS argued), these equations do hold in the average, and the experiment on $\Delta\lambda$ refers only to the average change of the wavelength. In fact at the time of the BKS proposal *there did not exist any direct experimental proof of energy-momentum conservation nor of causality in any individual elementary process.* This is one of the reasons that the objections to BKS (held by many, "perhaps the majority" of physicists[93]) were initially expressed in a somewhat muted fashion. Thus, Pauli wrote to Bohr that he did not believe in his theory but that "one cannot prove anything logically and also the available data are not sufficient to decide for or against your view."[94] All this was to change soon.

There was a second reason, I believe, for the subdued character of comments by others. The physics community was witness to a rare occurrence. Here were the two leading authorities of the day locked in conflict. (The term conflict was used by Einstein himself.[95]) To take sides meant to choose between the two most revered physicists. Ideally, personal considerations of this kind ought to play no role in matters scientific. But this ideal is not always fully realized. Pauli reflected on this in a letter concerning the BKS issue: "Even if it were psychologically possible for me to form a scientific opinion on the grounds of some sort of belief in authority (which is not the case, however, as you know), this would be logically impossible (at least in this case) since here the opinions of two authorities are so very contradictory."[96]

Even the interaction between the two protagonists was circumspect during that period. They did not correspond on the BKS issue.[97] Nor (as best I know) were there personal meetings between them in those days even though Bohr had told Pauli repeatedly how much he would like to know Einstein's opinion.[98] Werner Heisenberg (1902–76) had written to Pauli that he had met Einstein in Göttingen and that the latter had "a hundred objections."[99] Sometime later Pauli also met Einstein, whereupon he sent Bohr a detailed list of Einstein's criticisms.[100]

Einstein of course never cared for BKS. He had given a colloquium on this paper at which he had raised objections. The idea, he wrote Ehrenfest, "is an old acquaintance of mine, which I do not hold to be an honest fellow however." ("den ich aber für keinen reellen Kerl halte").[101] At about that time he drew up a list of nine objections, which I shall not reproduce here in detail but shall sample: "what should condition the virtual field which corresponds to the return of a previously free electron to a Bohr orbit? (very questionable). . . . Abandonment of causality *as a matter of principle* should only be permitted in the most extreme emergency."[102] The causality issue, which had plagued him for seven years by then, was clearly the one to which he took exception most strongly. He confided to Born that the thought was unbearable to him that an electron could choose freely the moment and direction in which to move.[103] The causality question would continue to nag him long after experiment revealed that the BKS answers to both paradoxes were incorrect.

3.3 *The Experimental Verdict on Causality*

The BKS ideas stimulated Walther Bothe (1891–1957) and Hans Geiger (1882–1945) to develop counter coincidence techniques[104] for the purpose of measuring whether (as causality demands) the secondary photon and the knocked-on electron are produced simultaneously in the Compton effect. Their result[105]: these two particles are both created in a time inverval $<10^{-3}$ sec. Within the limits of accuracy causality had been established and the randomness (demanded by BKS) of the relative creation times disproved. Since then this time interval has been narrowed down experimentally to $<10^{-11}$ sec.[106]

3.4 *The Experimental Verdict on Energy-Momentum Conservation*

The validity of these conservation laws in individual elementary processes was established for the Compton effect by Compton and A. W. Simon. From cloud chamber observations on photoelectrons and knock-on electrons they could verify[107] the validity of the relation

$$\tan \phi = -\left\{\left(1 + \frac{h\nu}{mc^2}\right)\tan\frac{\theta}{2}\right\}^{-1} \qquad (14.16)$$

in individual events, where ϕ, θ are the scattering angles of the electron and photon, respectively, and ν is the incident frequency.

And so the last resistance to the photon came to an end. Einstein's views had been fully vindicated. The experimental news was received generally with great relief.[108] Bohr took the outcome in good grace and proposed "to give our revolutionary efforts as honourable a funeral as possible."[109] He was now prepared for an even more drastic resolution of the quantum paradoxes. In July 1925 he wrote: "One must be prepared for the fact that the required generalization of the classical electrodynamical theory demands a profound revolution in the concepts on which the description of nature has until now been founded."[110]

These remarks by Bohr end with references to de Broglie's thesis and also to Einstein's work on the quantum gas (the subject of the next section): the profound revolution had begun.

4. A Loss of Identity: The Birth of Quantum Statistics

4.1 *Boltzmann's Axiom*

This episode begins with a letter[111] dated June 1924 written by a young Bengali physicist from the University of Dacca, now in Bangladesh. His name was Satyendra Nath Bose (1894–1974). The five papers he had published by then were of no particular distinction. The subject of his letter was his sixth paper. He had sent it to the *Philosophical Magazine.* A referee had rejected it.[112] Bose's letter was addressed to Einstein, then forty-five years old, and already recognized as a world figure by his colleagues and by the public at large. In this section, I shall describe what happened in the scientific lives of these two men during the half year following Einstein's receipt of Bose's letter. For Bose the consequences were momentus. From virtually an unknown he became a physicist whose name will always be remembered. For Einstein this period was only an interlude.[113] Already he was deeply engrossed in a project that he was not to complete: his search for a unified field theory. Such is the scope of Einstein's *oeuvre* that his discoveries in that half year do not even rank among his five main contributions, yet they alone would have sufficed as well for Einstein to be remembered forever.

Bose's sixth paper deals with a new derivation of Planck's law.[114] Along with his letter he had sent Einstein a copy of his manuscript, written in English, and asked him to arrange for publication in the *Zeitschrift für Physik*, if he were to think the work of sufficient merit. Einstein acceded to Bose's request. He personally translated the paper into German and submitted it to the *Zeitschrift für Physik*. He also added this translator's note: "In my opinion Bose's derivation of the Planck formula constitutes an important advance. The method used here also yields the quantum theory of the ideal gas, as I shall discuss elsewhere in more detail."

It is not the purpose of this section to discuss the history of quantum statistics, but rather to describe Einstein's contribution to the subject. Nevertheless, I include a brief outline of Bose's work (Sec. 4.2). There are numerous reasons for doing so: 1) It will give us some insight into what made Einstein diverge temporarily from his main pursuits. 2) It will facilitate the account of Einstein's own research on the molecular gas. This work is discussed in Sec. 4.3, with the exception of one major point that is reserved for section 5: Einstein's last encounter with fluctuation questions. 3) It will be of help to explain Einstein's ambivalence to Bose's work. In a letter to Ehrenfest, written in July, Einstein did not withdraw, but rather qualified, his praise of Bose's paper: Bose's "derivation is elegant but the essence remains obscure."[115] 4) It will help to make clear how novel the photon concept still was at that time, and it will throw an interesting sidelight on the question of photon spin.

Bose recalled, many years later, that he had not been aware of the extent to which his paper defied classical logic: "I had no idea that what I had done was really novel . . . . I was not a statistician to the extent of really knowing that I was doing something which was really different from what Boltzmann would have done, from Boltzmann statistics. Instead of thinking of the light quantum just as a particle, I talked about these states. Somehow this was the same question which Einstein asked when I met him [in October or November 1925]: How had I arrived at this method of deriving Planck's formula?"[116] In order to answer Einstein's question it is necessary (though it may not be sufficient) to understand what gave Bose the idea that he was doing what Ludwig Boltzmann (1844–1906) would have done. This in turn demands a brief digression on classical statistics.

Suppose I show someone two identical balls lying on a table. Next I ask him to close his eyes and a few moments later to open them again. I then ask him whether or not I have meanwhile exchanged the two balls. He cannot tell, since the balls are identical. Yet I do know the answer. If I have exchanged the balls then I have been able to follow the *continuous* motion that brought the balls from the initial to the final configuration. This simple example illustrates Boltzmann's first axiom of classical mechanics, which says, in essence, that identical particles that cannot come infinitely close to each other can be distinguished by their initial conditions and by the continuity of their motion. This assumption, Boltzmann stressed, "gives us the sole possibility of recognizing the same material point at different times."[117] As Schroedinger has emphasized, "Nobody before Boltzmann held it necessary to define what one means by [the term] the same material point."[118]

Thus we may speak classically of a gas with energy E consisting of N identical distinguishable molecules. Let there be n_1 particles with energy ε_1, . . . n_i particles with energy ε_i . . so that

$$N = \Sigma n_i \quad , \qquad E = \Sigma n_i \varepsilon_i \tag{14.17}$$

The number of states corresponding to this partition is given by

$$w = N! \, (\Pi n_i!)^{-1} \, , \tag{14.18}$$

where the factors $n_i!$ reflect the application of Boltzmann's axiom. These states are the microstates of the system and the description in terms of them is the so-called fine-grained description. For the limited purpose of analyzing the macroscopic properties of the system, one contracts this description to the so-called coarse-grained description.

Divide the one particle phase space into cells ω_1, ω_2, . . . such that a particle in ω_A has the *mean* energy E_A. Partition the N particles such that there are N_A particles in ω_A,

$$N = \Sigma N_A \, , \tag{14.19}$$

$$E = \Sigma N_A E_A \; . \tag{14.20}$$

The set (N_A, E_A) defines a coarse-grained state. For the special case of an ideal gas the relative probability of this state is given by

$$W = N! \prod_A \frac{\omega_A^{N_A}}{N_A!} \; . \tag{14.21}$$

Equation (14.21) rests on two independent ingredients: 1) Boltzmann's distinguishability axiom and 2) statistical independence; that is, for an ideal classical gas the individual molecules have no a priori preference for any particular region in phase space. Boltzmann's principle states that in thermodynamic equilibrium the entropy is given by

$$S = k \ln W_{,\max} + C \, , \tag{14.22}$$

where C is an arbitrary additive constant; that is, C does not depend on the n_i. W_{max} is the maximum of W as a function of the N_A subject to the constraints (14.19), (14.20). The classical Boltzmann distribution then follows from the extremal conditions

$$\Sigma\delta N_A\,(\ln\omega_A - \ln N_A + \lambda + \beta^{-1}E_A) = 0 \qquad (14.23)$$

and $\beta = kT$ follows from $\partial S/\partial E = T^{-1}$. All these elementary facts have been recalled in order to accentuate the differences between classical and quantum statistics.

Both logically and historically, classical statistics developed via the sequence:

Fine-grained counting → coarse-grained counting.

This is of course the logic of quantum statistics as well, but its historical development went the reverse way, from coarse-grained to fine-grained. For the oldest quantum statistics, the Bose-Einstein (BE) statistics, the historical order of events was as follows:

1924–5: A new coarse-grained counting is introduced first by Bose, then Einstein.

1925–6: Nonrelativistic quantum mechanics is discovered. It is not at once obvious[119] how one should supplement the new [120] theory with a fine-grained counting principle that leads to BE statistics.

1926. This principle[121] is discovered by Paul Adrien Maurice Dirac (C. 1902); Eq. (14.18) is to be replaced by

$$w = 1,\ \text{independent of the } n_i, \qquad (14.24)$$

only the single microstate is allowed that is symmetric in the N particles. Dirac notes that (14.24) leads to Planck's law. After more than a quarter century, the search for the foundations of this law comes to an end.

Equation (14.24) was of course not known when Bose and Einstein embarked on their explorations of a new statistics. Let us next turn to Bose's contribution.

4.2 *Bose*

The paper by Bose[122] is the fourth and last of the revolutionary papers of the old quantum theory [the other three are, respectively, by Planck (December 1900), Einstein (1905), and Bohr (1913)]. Bose's arguments divest Planck's law of all supererogatory elements of electromagnetic theory and base its derivation on the bare essentials. It is the thermal equilibrium law for particles with the following properties: they are massless; they have two states of polarization; the number of particles is not conserved; and the particles obey a new statistics. In Bose's paper two new ideas enter physics almost stealthily. One, the concept of a *particle* with two states of polarization, mildly puzzled Bose. The other is the nonconservation of photons. I doubt whether Bose even noticed this fact. It is not explicitly mentioned in his paper.

Bose's letter to Einstein begins as follows: "Respected Sir, I have ventured to send you the accompanying article for your perusal. I am anxious to know what you will think of it. You will see that I have ventured to deduce the coefficient $8\pi\nu^2/c^3$ in Planck's law independent of the classical electrodynamics. . . . "[123] Einstein's 1924 letter to Ehrenfest contains the phrase: "The Indian Bose has given a beautiful derivation of Planck's law including the constant [i.e., $8\pi\nu^2/c^3$]." Neither letter mentions the other parts of Planck's formula. Why this emphasis on $8\pi\nu^2/c^3$?

In deriving Planck's law one needs to know the number of states Z^s in the frequency interval between ν^s and $\nu^s + d\nu^s$. It was customary to compute Z^s by counting the number of standing *waves* in a cavity with volume V, which gives

$$Z^s = 8\pi(\nu^s)^2 \, V d\nu^s/c^3 \, . \tag{14.25}$$

Bose was so pleased because he had found a new derivation of this expression for Z^s that enabled him to give a new meaning to this quantity in terms of *particle* language. His derivation rests on the replacement: counting wave frequencies → counting cells in one-particle phase space. He proceeded as follows: Integrate the one-particle phase space element $d\mathbf{x}d\mathbf{p}$ over V and over all momenta between p^s and $p^s + dp^s$. Supply a further factor 2 to count polarizations. This produces the quantity $8\pi V(p^s)^2 \, dp^s$, which equals $h^3 Z^s$ by virtue of the relation $p^s = h\nu^s/c$. Hence, Z^s is the number of cells of size h^3 that is contained in the particle phase space region considered. How innocent is looks, yet how new it was. Recall that the kinematics of the Compton effect had only been written down about a year and a half earlier. Here was a new application of $p = h\nu/c$!

Before I turn to the rest of Bose's derivation I should like to digress briefly on the subject of photon spin. In his paper, when Bose introduced his polarization factor of 2, he noted that "it seems required" to do so. This slight hesitation is understandable. Who in 1924 had ever heard of a particle with two states of polarization? For some time this remained a rather obscure issue. After the discovery of the electron spin, Ehrenfest asked Einstein "to tell me how the analogous hypothesis is to be stated for light-corpuscles, in a relativistically correct way."[124] As is well known, this is a delicate problem since there exists, of course, no rest frame definition of spin in this instance. Moreover, gauge invariance renders ambiguous the separation into orbital and intrinsic angular momentum.[125] It is not surprising that in 1926 the question of photon spin seemed quite confusing to Einstein. In fact, he went as far as to say that he was "inclined to doubt whether the angular momentum law can be maintained in the quantum theory. At any rate its significance is much less deep than the momentum law."[126] I believe this is an interesting comment on the state of the art some fifty years ago and that otherwise not too much should be made of it.

Let us return to Bose. His new interpretation of Z^s was in terms of "number of cells," not "number of particles." This must have led him to follow Boltzmann's counting but *to replace everywhere "particles" by "cells,"* a procedure that he neither did, nor could, justify — but one that gave the right answer. He partitioned Z^s into numbers p_r^s, where p_r^s is defined as the number of cells that contains r quanta with frequency ν^s. Let there be N^s photons in all with this frequency and let E be the total energy. Then

$$Z^s = \sum_r p_r^s \, , \tag{14.26}$$

$$N^s = \sum_r r p_r^s \, , \tag{14.27}$$

$$E = \sum_s N^s h\nu^s \tag{14.28}$$

while

$$N = \sum_s N^s \,, \tag{14.29}$$

is the total number of photons. Next Bose introduced a new coarse-grained counting:

$$W = \prod_s \frac{Z^s!}{p^s_0!\, p^s_1! \ldots}. \tag{14.30}$$

He then maximized W as a function of the p_r^s holding Z^s and E fixed so that

$$\sum_{s,r} \delta p_r^s \{1 + \ln p_r^s + \lambda^s + \frac{1}{\beta} rh\nu^s\} = 0 \,, \tag{14.31}$$

then derived Planck's law for $E(\nu, T)$ by standard manipulations — and therewith concluded his paper without further comments.

Bose considered his *Ansatz* [Eq. (14.30)] to be "evident."[127] I venture to guess that to him the cell counting (14.30) was the perfect analog of Boltzmann's particle counting (14.21) and that his cell constraint, hold Z^s fixed, was similarly the analog of Boltzmann's constraint, hold N fixed. Likewise the two Lagrange parameters in (14.31) are his analogs of the parameters in (14.23). Bose's replacement of fixed N by fixed Z^s already implies that N is not conserved. The final irony is that the constraint of fixed Z^s is irrelevant: If one drops this constraint then one must drop λ^s in Eq (14.31). Even so, it is easily checked that one still finds Planck's law! This is in accordance with the now familiar fact that Planck's law follows from Bose statistics with E held fixed as the only constraint. In summary, Bose's derivation introduces three new features:

1. Photon number is not conserved.
2. Bose's cell partition numbers p_r^s are defined by asking how many particles are in a cell, not which particles are in a cell. Boltzmann's axiom is gone.
3. The *Ansatz* Eq. (14.30) implies statistical independence of cells. Statistical independence of particles is gone.

The astounding fact is that Bose was correct on all three counts. (In his paper he commented on none of them.) I believe there has been no such successful shot in the dark since Planck introduced the quantum.

4.3 *Einstein*

As long as Einstein lived he never ceased to struggle with quantum physics. Insofar as his constructive contributions to this subject are concerned, these came to an end with three papers, the first published in September 1924, the last two in early 1925. In the true Einstein style, their conclusions are once again reached by statistical methods, as was the case for all his important earlier contributions to the quantum theory. The best known result is his derivation of the Bose–Einstein (BE) condensation phenomenon. I shall discuss this topic next and shall leave for the subsequent section another result contained in these papers, one that is perhaps not as widely remembered although it is even more profound.

First, a postscript to Einstein's light quantum paper of 1905. Its logic can be schematically represented in the following way.

Einstein 1905: Wien's law
} → light quanta.
gas analogy

One last question needs to be considered before this logic is complete. We know that BE is the correct statistics when radiation is treated as a photon gas. Then how could Einstein correctly have conjectured the existence of light quanta using Boltzmann statistics? The answer is that according to BE statistics, the most probable value $< n_i >$ of the n_i for photons is given by $< n_i > = \{\exp(h\nu_i/kT) - 1\}^{-1}$. This implies $< n_i > << 1$ in the Wien regime $h\nu_i >> kT$. Therefore up to an irrelevant factor N! Eqs. (14.18) and (14.24) coincide in the Wien limit.[128] This asymptotic relation in the Wien region fully justifies, ex post facto, Einstein's extraordinary step forward in 1905!

Bose's reasoning in 1924 went as follows:

Bose 1924: Photons
} → Planck's law
Quantum statistics

and in 1924–5 Einstein came full circle:

Einstein 1924–5: Bose statistics
} → the quantum gas.
Photon analogy

It was inevitable, one might say, that he would do so. "If it is justified to conceive of radiation as a quantum gas, then the analogy between the quantum gas and a molecular gas must be a complete one."[129]

In his 1924 paper[130] Einstein adopted Bose's counting formula [Eq. (14.30)], but with two modifications. He needed of course the Z^s appropriate for nonrelativistic particles with mass m:

$$h^3 Z^s = 2\pi V (2m)^{3/2} (E^s)^{1/2}\, dE^s,\ 2mE^s = (p^s)^2\,. \tag{14.32}$$

Secondly (and unlike Bose!) he needed the constraint that N is held fixed. This is done by adding a term,

$$-r \ln A \tag{14.33}$$

inside the curly brackets of Eq. (14.31).[131] One of the consequences of the modified (14.31) is that the Lagrange multiplier ($-\ln A$) is determined by

$$N = \sum_s N^s = \sum_s \left[\frac{1}{A} \exp \frac{E^s}{kT} - 1 \right]^{-1} . \tag{14.34}$$

Hence, Einstein noted, the "degeneracy parameter" A^{-1} must satisfy

$$A \leqslant 1 . \tag{14.35}$$

In his first paper[132] Einstein discussed the regime in which A does not reach the critical value 1. He proceeds to the continuous limit in which the sum in Eq. (14.34) is replaced by an integral over phase space and finds[133]:

$$\text{if } A < 1 : \frac{1}{v} = \frac{\phi_{3/2}(A)}{3} , \quad \frac{p}{kT} = \frac{\phi_{5/2}(A)}{3} , \tag{14.36}$$

$$\phi_n (A) = \sum_{m=1}^{\infty} m^{-n} A^m ,$$

with $\lambda^2 = h^2/2\pi mkT$ and $v = V/N$. He then discusses the region $A << 1$ where the equation of state [obtained by eliminating A between the two equations (14.36)] shows perturbative deviations from the classical ideal gas. All this is good physics, though unusually straightforward for a man like Einstein.

In his second paper,[134] the most important one of the three, Einstein begins with the v - T relation at $A = 1$:

$$kT_0 = \frac{h^2}{2\,m[v_0\phi_{3/2}(1)]^{3/2}} , \tag{14.37}$$

and asks what happens if T drops below T_0 (for given v_0). His answer: "I maintain that in this case a number of molecules steadily growing with increasing density goes over in the first quantum state (which has zero kinetic energy) while the remaining molecules distribute themselves according to the parameter value $A = 1$ . . . a separation is effected; one part "condenses," the rest remains a "saturated ideal gas."[135] Einstein had come upon the first purely statistically derived example of a phase transition (a term he himself did not use at that time), which now is called Bose–Einstein condensation.

Until 1938 this phenomenon had "the reputation of having only a purely imaginary character."[136] I shall discuss elsewhere in some detail the reasons for this[137] and confine myself here to two brief comments on this subject. 1) In 1926 George Uhlenbeck raised the following objection[138]: The quantity N^0 in Eq. (14.34) $\rightarrow \infty$ as $A \rightarrow 1$ (for fixed T), hence also $N \rightarrow \infty$. Thus if $A = 1$ it is impossible to implement the constraint that N is a fixed *finite* number. Therefore $A = 1$ can only be reached asymptotically and there is no two phase régime. Ehrenfest communicated this objection to Einstein, who agreed (G. E. Uhlenbeck, private communication). Uhlenbeck and Einstein were both right, however. The point is that a sharp phase transition (in the BE or in any other case of a gas) can only occur in the so-called thermodynamic limit $N \rightarrow \infty$, $V \rightarrow \infty$. v fixed. 2) The second reason for this "imaginary character" was lack of experimental evidence for the condensation phenomenon. Recall that the He I – He II phase transition in liquid He^4 ws not discovered until 1928[139] and that the interpretation of this transition as a BE condensation was not proposed until 1938, by Fritz London (1900–54).[140] I shall not pursue these developments further, however, but will, rather, turn to other important facets of the three Einstein papers.

1. *Einstein on statistical dependence.* After the papers by Bose[141] and the first one by Einstein[142] had come out, Ehrenfest and others objected (so we read in Einstein's second paper[143]) that "the quanta and molecules respectively are not treated as statistically independent, a fact which is not particularly emphasized in our papers" (that is, that of Bose and Einstein's first paper). Einstein replied to them: "This [objection] is entirely correct,"[144] and went on to stress that the differences between the Boltzmann and the BE counting "express indirectly a certain hypothesis on a mutual influence of the molecules which for the time being is of a quite mysterious nature." With this remark Einstein came to the very threshold of the quantum mechanics of identical particle systems. The mysterious influence is of course the correlation induced by the requirement of totally symmetric wave functions.

2. *Einstein on indistinguishability.* In order to illustrate further the differences between the new and the old counting of macrostates Einstein cast W in a form alternative to Eq. (14.30).[145] He counted the number of ways in which N^s indistinguishable particles in the dE^s-interval can be partitioned over the Z^s cells. This yields

$$W = \prod_s \frac{(N^s + Z^s - 1)!}{N^s!(Z^s - 1)!} . \tag{14.38}$$

Einstein's equation [(14.38)] rather than Bose's equation [(14.30)] is the one now used in all textbooks.

3. *Einstein on the third law of thermodynamics.* Einstein was the first to observe[146] that the BE gas does satisfy this law. Since all particles go into the zero state as $T \to 0$, we have in this limit $N^0 = N$, all other $N^s = 0$. Hence $W \to 1$ and $S \to 0$ as $T \to 0$. A Boltzmann gas does not satisfy the third law, Einstein further noted. It was as important to him that a molecular BE gas yield Nernst's law as that a BE photon gas yield Planck's law.

4. *Einstein and nonconservation of photons.* After 1917 Einstein ceased to write scientific articles on questions related to radiation.[147] The only mention of radiation in the 1924–5 papers is that "the statistical method of Herr Bose and myself is by no means beyond doubt, but seems only a posteriori justified by its success for the case of radiation."[148]

There can be no doubt that he must have noted the nonconservation of photons. In his language this is implemented by putting $A = 1$ in Eq. (14.33). Yet I have not found any reference to nonconservation, either in his scientific writings or in the correspondence I have seen. I cannot state with certainty why he chose to be silent on this and all further issues regarding photons. However, I do believe it is a fair guess that Einstein felt he would have nothing fundamental to say about photons until such time as he could find his own way of dealing with the lack of causality he had noted in 1917.

Such times never came.

5. Einstein as a Transitional Figure: The Birth of Wave Mechanics

We now leave the period of the old quantum theory and turn to the time of transition during which matter waves were being discussed by a tiny group of physicists at a time when matter wave mechanics had not yet been discovered. This period begins in September 1923 with two brief communications by Louis de Broglie (1892–) to the French Academy of Sciences.[149] It ends in January 1926 with the first paper by Erwin Schroedinger (1887–1961) on wave mechanics.[150] The main purpose of this section is to stress Einstein's key role in these developments, his influence on de Broglie, de Broglie's subsequent influence on him, and finally, the influence of both on Schroedinger.

Neither directly nor indirectly did Einstein contribute to an equally fundamental development that preceded Schroedinger's discovery of wave mechanics: the discovery of matrix mechanics by Heisenberg.[151] Therefore I shall have no occasion in this article to comment in any detail on Heisenberg's major achievements.

5.1 *From Einstein to de Broglie*

During the period that began with Einstein's work on needle rays (1917) and ended with Debye's and Compton's papers on the Compton effect (1923) there were a few other theoreticians who did research on photon questions. Among those, the only one whose contribution lasted was de Broglie.[152]

De Broglie had finished his studies before the First World War. In 1919, after a long tour of duty with the French forces, he joined the physics laboratory headed by his brother Maurice (1875–1960), where X-ray photo effects and X-ray spectroscopy were main topics of study. Thus he was much exposed to questions concerning the nature of electromagnetic radiation, a subject on which he published several papers. In one of these de Broglie evaluated independently of Bose (and published before him) the density of radiation states in terms of particle (photon) language[153] (see Sec. 4.2). That was in October 1923 — one month after his enunciation of the epochal new principle that particle-wave duality should apply not only to radiation but also to matter. "After long reflection in solitude and meditation," he later wrote, "I suddenly had the idea, during the year 1923, that the discovery made by Einstein in 1905 should be generalized in extending it to all material particles and notably to electrons."[154]

He made the leap in his September 10, 1923 paper:[155] $E = h\nu$ shall hold not only for photons but also for electrons to which he assigns a "fictitious associated wave." In his September 24 paper he indicated the direction in which one "should seek experimental confirmations of our ideas": A stream of electrons traversing an aperture whose dimensions are small compared with the wavelength of the electron waves "should show diffraction phenomena."[156]

Other important aspects of de Broglie's work are beyond the scope of this paper. The articles mentioned were extended to form his doctor's thesis, which he defended on November 25, 1924.[157] Einstein received a copy of this thesis from one of de Broglie's examiners, Paul Langevin (1872–1946). A letter he wrote to Lorentz (on December 15, 1924) shows that Einstein was impressed and also that he had found a new application of de Broglie's ideas: "A younger brother of . . . de Broglie has undertaken a very interesting attempt to interpret the Bohr–Sommerfeld quantum rules (Paris Dissertation 1924). I believe it is a first feeble ray of light on this worst of our physics enigmas. I too have found something which speaks for his construction."

5.2 *From de Broglie to Einstein*

In 1909 and again in 1917 Einstein had drawn major conclusions about radiation from the study of fluctuations around thermal equilibrium. It goes without saying that he would again examine fluctuations when, in 1924, he turned his attention to the molecular quantum gas.

In order to appreciate what he did this time it is helpful to repeat Eq. (14.6) given in Sec. 2, the formula for the mean square energy of electromagnetic radiation:

$$< \varepsilon^2 > = \left(\rho h\nu + \frac{c^3\rho^2}{8\pi\nu^2} \right) V d\nu \; .$$

Put $V\rho d\nu = n(\nu)h\nu$ and $<\varepsilon^2> = \Delta(\nu)^2(h\nu)^2$. $n(\nu)$ can be intepreted as the average number of quanta in the energy interval $d\nu$ and $\Delta(\nu)^2$ as the mean square fluctuation of this number. One can now write Eq. (14.6) as

$$\Delta(\nu)^2 = n(\nu) + \frac{n(\nu)^2}{Z(\nu)} \, , \tag{14.39}$$

where $Z(\nu)$ is the number of states per interval $d\nu$ given in Eq. (14.25).* In his paper submitted on January 8, 1925,[125] Einstein showed that Eq. (14.39) holds equally well for his quantum gas, as long as one defines ν in the latter case by $E = h\nu = p^2/2m$ and uses Eq. (14.32) instead of Eq. (14.25) for the number of states.

When discussing radiation, in 1909, Einstein recognized the second term of Eq. (14.6) as the familiar wave term and the first one as the unfamiliar particle term. When, in 1925, he revisited the fluctuation problem for the case of the quantum gas, he noted a reversal of roles. The first term, at one time unfamiliar for radiation, was now the old fluctuation term for a Poisson distribution of (distinguishable) particles. What to do with the second term (which incorporates indistinguishability effects of particles) for the gas case? Since this term was associated with waves in the case of radiation, Einstein was led to "interpret it in a corresponding way for the gas, by associating with the gas a radiative phenomenon." He added: "I pursue this interpretation further since I believe that here we have to do with more than a mere analogy."[159]

But what were the waves?

At that point Einstein turned to de Broglie's thesis,[160] "a very notable publication." He suggested that a de Broglie type wave field should be associated with the gas and pointed out that this assumption enabled him to interpret the second term in Eq. (14.39). Just as de Broglie had done, he also noted that a molecular beam should show diffraction phenomena, but he added that the effect should be extremely small for manageable apertures. Einstein also remarked that this wave field had to be a scalar. (The polarization factor equals 2 for Eq. (14.25), as noted above, but it equals 1 for Eq. (14.32)!)

It is another Einstein feat that he would be led to state the necessity of the existence of matter waves from the analysis of fluctuations. One may wonder what the history of twentieth-century physics would have been like had Einstein pushed the analogy still further. However, with the achievement of an independent argument for the particle-wave duality of matter, the twenty-year period of highest scientific creativity in Einstein's life, at a level probably never equaled, came to an end.

5.2.1 *Postscript, summer of 1978.* In the course of preparing this article I noticed a recollection by Pauli[161] of a statement made by Einstein during a physics meeting held in Innsbruck in 1924.

*In Eq. (14.39) I drop the index s occurring in Eq. (14.25)

According to Pauli, Einstein proposed in the course of that meeting "to search for interference and diffraction phenomena with molecular beams." On checking the dates of that meeting I found them to be September 21–27. This intrigued me. Einstein came to the particle-wave duality of matter via a route independent of the one taken by de Broglie. The latter defended his thesis in November. If Pauli's memory is correct, then Einstein made his remark about two months prior to that time. Could Einstein have come upon the wave properties of matter independently of de Broglie? After all, Einstein had been thinking about the molecular gas since July. The questions arise: When did Einstein become aware of de Broglie's work? In particular, when did Einstein receive de Broglie's thesis from Langevin? Clearly it would be most interesting to know what Professor de Broglie might have to say about these questions. Accordingly I wrote to him. He was kind enough to reply. With his permission I quote from his answers.

De Broglie does not believe that Einstein was aware of his three short publications written in 1923.[162] "Nevertheless, since Einstein would receive the Comptes Rendus and since he knew French very well, he might have noticed my articles" (private communication, August 9, 1978). De Broglie noted further that he had given Langevin the first typed copy early in 1924. "I am certain that Einstein knew of my Thèse since the spring of 1924" (private communication, September 26, 1978). This is what happened: "When in 1923 I had written the text of the Thèse de Doctorat which I wanted to present in order to obtain the Doctorat ès Sciences, I had three typed copies made. I handed one of these to M. Langevin so that he might decide whether this text could be accepted as a Thèse. M. Langevin, probably rather astonished by the novelty of my ideas ("probablement un peu etónné par la nouveauté de mes idées"), asked me to furnish him with a second typed copy of my Thèse for transmittal to Einstein. It was then that Einstein declared, after having read my work, that my ideas seemed quite interesting to him. This made Langevin decide to accept my work." (Private communication, August 9, 1978.)

Thus Einstein was not only one of the three fathers of the quantum theory, but also the sole godfather of wave mechanics.

5.3 *From de Broglie and Einstein to Schroedinger*

Late in 1925 Schroedinger completed an article entitled "On Einstein's gas theory."[163] It was his last paper prior to his discovery of wave mechanics. Its content is crucial to an understanding of the genesis of that discovery.

In order to follow Schroedinger's reasoning it is necessary to recall first a derivation of Planck's formula given by Debye in 1910.[164] Consider a cavity filled with radiation oscillators in thermal equilibrium. The spectral density is $8\pi\nu^2\varepsilon(\nu, T)/c^3$, where ε is the equilibrium energy of a radiation field oscillator with frequency ν. Debye introduced the quantum prescription that the only admissible energies of the oscillator shall be $nh\nu$, $n = 0,1,2, \ldots$. In equilibrium the nth energy level is weighed with its Boltzmann factor. Hence $\varepsilon = \Sigma nh\nu y^n/\Sigma y^n$, $y = \exp(-h\nu/kT)$. This yields Planck's law.[165]

Now back to Schroedinger.[166] By his own admission he was not much taken with the new BE statistics. Instead, he suggested, why not evade the new statistics by treating Einstein's molecular gas according to the Debye method? That is, *why not start from a wave picture of the gas* and superimpose on that a quantization condition à la Debye? Now comes the key sentence in the article: "That means nothing else but taking seriously the de Broglie–Einstein wave theory of moving particles." And that is just what Schroedinger did. It is not necessary to discuss further details of this article, which was received by the publisher on December 25, 1925.

Schroedinger's next paper[167] was received on January 27, 1926. It contains his equation for the hydrogen atom. Wave mechanics was born. In this new paper Schroedinger acknowledged his debt to de Broglie and Einstein: "I have recently shown[168] that the Einstein gas theory can be founded on the consideration of standing waves which obey the dispersion law of de Broglie. . . . The above considerations about the atom could have been presented as a generalization of these considerations."[169] In April 1926 Schroedinger again acknowledged the influence of de Broglie and "brief but infinitely far seeing remarks by Einstein."[170]

6. Einstein's Response to the New Dynamics

Everyone familiar with modern physics knows that Einstein's attitude regarding quantum mechanics was one of skepticism. No biography of him fails to mention his saying that God does not throw dice. He was indeed given to such utterances (as I know from experience), and stronger ones such as "It seems hard to look in Gods' cards. But I cannot for a moment believe that he plays dice and makes use of "telepathic" means (as the current quantum theory alleges he does)."[171] However, remarks such as these should not create the impression that Einstein had abandoned active interest in quantum problems in favor of his quest for a unified field theory. Far from it. In fact even in the search for a unified theory the quantum riddles were very much on his mind, as I shall discuss later on in Sec. 8. In the present section, I shall attempt to describe how Einstein's position concerning quantum mechanics evolved in the course of time. To some extent this is reflected in his later scientific papers. It becomes evident more fully in several of his more autobiographical writings and in his correspondence. My own understanding of his views has been helped much by discussions with him.

To begin with I shall turn to the period 1925–31, during which he was much concerned with the question: Is quantum mechanics consistent?

6.1 *1925–31: The Debate Begins*

As I mentioned earlier, the three papers of 1924–5 on the quantum gas were only a temporary digression from Einstein's program begun several years earlier to unify gravitation with electromagnetism. During the very early days of quantum mechanics[172] we find him "working strenuously on the further development of a theory on the connection between gravitation and electricity."[173] Yet the great importance of the new developments in quantum theory was not lost on him. Bose, who visited Berlin in November 1925, has recalled that "Einstein was very excited about the new quantum mechanics. He wanted me to try to see what the statistics of light quanta and the transition probabilities of radiation would look like in the new theory."[174]

Einstein's deep interest in quantum mechanics must have led him to write to Heisenberg rather soon after the latter's paper[175] had been published. All the letters of Einstein to Heisenberg have been lost.[176] However, a number of letters by Heisenberg to Einstein are extant.[177] One of these (dated November 30, 1925) is clearly in response to an earlier lost letter by Einstein to Heisenberg in which Einstein appears to have commented on the new quantum mechanics.[178] One remark by Heisenberg in the letter is of particular interest: "You are probably right that our formulation of the quantum mechanics is more adapted to the Bohr–Kramers–Slater attitude; but this [BKS theory] constitutes in fact one aspect of the radiation phenomena. The other is your light quantum theory and we have the hope that the validity of the energy and momentum laws in our quantum mechanics will one day make possible the connection with your theory." I find it remarkable that Einstein apparently sensed that there was some connection between the BKS theory and the quantum mechanics. No such connection exists of course. Nevertheless the BKS proposal contains statistical features,[179] as we have seen. Could Einstein have surmised as early as 1925 that *some* statistical element is inherent in the quantum mechanical description?!

During the following months, Einstein vacillated in his reaction to the Heisenberg theory. In December 1925 he expressed misgivings.[180] But in March 1926 he wrote to the Borns: "The Heisenberg–Born concepts leave us all breathless, and have made a deep impression on all theoretically oriented people. Instead of a dull resignation, there is now a singular tension in us sluggish people."[181] The next month he expressed again his conviction that the Heisenberg–Born approach was off the track. That was in a letter in which he congratulated Schroedinger on his new advance.[182] In view of the scientific links between Einstein's and Schroedinger's work, it is not surprising that Einstein would express real enthusiasm about wave mechanics: "Schroedinger has come out with a pair of wonderful papers on the quantum rules."[183] It was the last time he would write approvingly about quantum mechanics.

There came a parting of ways.

Nearly a year passed after Heisenberg's paper before there was a first clarification of the conceptual basis of quantum mechanics. It began with the observation[184] by Max Born (1882–1970) in June 1926 that the absolute square of a Schroedinger wave function is to be interpreted as a probability density. Born's brief and fundamental paper goes to the heart of the problem of determinism. Regarding atomic collisions he wrote: "One does not get an answer to the question "what is the state after collision" but only to the question "how probable is a given effect of the collision. . . . From the standpoint of our quantum mechanics there is no observable [Grösze] which causally fixes the effect of a collision in an individual event. Should we hope to discover such properties later . . . and determine [them] in individual events? . . . I myself am inclined to renounce determinism in the atomic world. But that is a philosophical question for which physical arguments alone do not set standards."

Born's paper had a mixed initial reception. Several leading physicists found it hard if not impossible to swallow the abandonment of causality in the classical sense; among them was Schroedinger. More than once Bohr mentioned to me that Schroedinger told him he might not have published his papers had he been able to foresee what consequences they would unleash.[185] Einstein's position in the years to follow can be summarized succinctly by saying that he took exception to every single statement contained in the lines I have quoted from Born. His earliest expressions of discomfort I know of date from late 1926, when he wrote Born: "Quantum mechanics is certainly imposing. But an inner voice tells me that it is not yet the real thing. The theory says a lot, but does not really bring us closer to the secret of the 'old one.'"[186]

"Einstein's verdict . . . came as a hard blow" to Born.[187] There are other instances as well in which Einstein's reactions were experienced with a sense of loss, of being abandoned by a venerated leader in battle. Thus Samuel Goudsmit (1902–78) told me (private communication, January 16, 1978) of a conversation that had taken place in mid-1927 (to the best of his recollection) between Ehrenfest and himself. In tears, Ehrenfest had said that he had to make a choice between Bohr's and Einstein's position and that he could not but agree with Bohr. Needless to say Einstein's reactions affected the older generation more intensely than the younger.

Of the many important events in 1927, three are particularly significant for the present account.

6.1.1 *March 1927.* Heisenberg states the uncertainty principle.[188] In June Heisenberg writes a letter to Einstein that begins as follows: "Many cordial thanks for your kind letter; although I really do not know anything new, I would nevertheless like to write once more why I believe that indeterminism, that is, the nonvalidity of rigorous causality, is *necessary* [his italics] and not just consistently possible."[189] This letter is apparently in response to another, lost, letter by Einstein triggered, most probably, by Heisenberg's work in March. I shall return to Heisenberg's important letter in Sec. 8. I only mention its existence at this point to emphasize that once again Einstein did not react to these new developments as a passive bystander. In fact, at just about that time, he was doing his own research on quantum mechanics (his first, I believe). "Does Schroedinger's wave mechanics determine the motion of a system completely or only in the statistical sense?"[190] he asked. Heisenberg had heard indirectly that Einstein "had written a paper in which you . . . advocate the view that it should be possible after all to know the orbits of particles more precisely than I would wish." He asked for more information "especially because I myself have thought so much about these questions and only came to believe in the uncertainty relations after many pangs of conscience, though now I am entirely convinced."[191] Einstein eventually withdrew his paper.

6.1.2 *September 16, 1927.* At the Volta meeting in Como Bohr enunciates for the first time the principle of complementarity: "The very nature of the quantum theory . . . forces us to regard the space-time coordination and the claim of causality, the union of which characterizes the classical theories, as complementary but exclusive features of the description, symbolizing the idealization of observation and definition respectively."[192]

6.1.3 *October 1927.* The fifth Solvay conference convenes. All the founders of the quantum theory were there, from Planck, Einstein, and Bohr to de Broglie, Heisenberg, Schroedinger, and Dirac. During the sessions "Einstein said hardly anything beyond presenting a very simple objection to the probability interpretation . . . Then he fell back into silence."[193]

However, the formal meetings were not the only place of discussion. All participants were housed in the same hotel and there, in the dining room, Einstein was much livelier. Otto Stern has given this first-hand account: "Einstein came down to breakfast and expressed his misgivings about the new quantum theory, every time [he] had invented some beautiful experiment from which one saw that it did not work . . . . Pauli and Heisenberg who were there did not react to these matters, 'ach was, das stimmt schon, das stimmt schon' ['ah well, it will be allright, it will be allright']. Bohr on the other hand reflected on it with care and in the evening, at dinner, we were all together and he had cleared up the matter in detail."[194]

Thus began the great debate between Bohr and Einstein. Both men refined and sharpened their positions in the course of time. No agreement between them was ever reached. Between 1925 and 1931 the only objection by Einstein which appeared in print in the scientific literature is the one at the 1927 Solvay Conference.[195] However, there exists a masterful account of the Bohr–Einstein dialogue during these years, published by Bohr in 1949.[196] I have written elsewhere about the profound role the discussions with Einstein played in Bohr's life.[197]

The record of the Solvay meeting contains only minor reactions to Einstein's comments. Bohr's later article analyzed them in detail.[198] Let us consider next the substance of Einstein's remarks.

Einstein's opening phrase tells more about him than does many a book: "Je dois m'excuser de n'avoir pas approfondi la mécanique des quanta" ("I must apologize for not having examined quantum mechanics in depth").[199]

He then discusses an experiment in which a beam of electrons hits a (fixed) screen with an aperture in it. The transmitted electrons form a diffraction pattern that is observed on a second screen. Question: Does quantum mechanics give a complete description of the individual electron events in this experiment? His answer: This cannot be. For let *A* and *B* be two distinct spots on the second screen. If I know that an individual electron arrives at *A* then I know instantaneously that it did not arrive at *B*. But this implies a peculiar instantaneous action at a distance between *A* and *B* contrary to the relativity postulate. Yet (Einstein notes) in the Geiger–Bothe experiment on the Compton effect[200] there is no limitation of principle to the accuracy with which one can observe coincidences in individual processes, and that without appeal to action at a distance. This circumstance adds to the sense of incompleteness of the description for diffraction.

Quantum mechanics provides the following answer to Einstein's query: It does apply to individual processes, but the uncertainty principle defines and delimits the optimal amount of information that is obtainable *in a given experimental arrangement.* This delimitation differs incomparably from the restrictions on information inherent in the coarse-grained description of events in classical statistical mechanics. There the restrictions are wisely self-imposed in order to obtain a useful approximation to a description in terms of an ideally knowable complete specification of momenta and positions of individual particles. In quantum mechanics the delimitations mentioned earlier are not self-imposed but are renunciations of first principle (on the fine-grained level, one might say). It is true that one would need action at a distance if one were to insist on a fully causal description involving the localization of the electron at every stage of the experiment in hand. Quantum mechanics denies that such a description is called for and asserts that *in this experiment* the final position of an individual electron cannot be predicted with certainty. Quantum mechanics nevertheless makes a prediction in this case concerning the probability for an electron to arrive at a given spot on the second screen. The verification of this prediction demands, of course, that the "one electron experiment" be repeated as often as is necessary to obtain this probability distribution with a desired accuracy.

Nor is there a conflict with Geiger–Bothe, since now one refers to *another experimental arrangement* in which localization in space-time is achieved, but this time at the price of renouncing information on sharp energy-momentum properties of the particles observed in coincidence. From the point of view of quantum mechanics these renunciations are expressions of laws of nature. They are also applications of the saying "Il faut reculer pour mieux sauter" ("It is necessary to take a step back in order to jump better"). As we shall see, what was and is an accepted renunciation to others was an intolerable abdication in Einstein's eyes. On this score he was never prepared to give up anything.

I have dwelt at some length on this simple problem since it contains the germs of Einstein's position, which he stated more explicitly in later years. Meanwhile the debate in the corridors between Bohr and Einstein continued during the sixth Solvay conference (on magnetism) in 1930. This time Einstein thought he had found a counterexample to the uncertainty principle. The argument was ingenious. Consider a box with a hole in its walls that can be opened or closed by a shutter controlled by a clock inside the box. The box is filled with radiation. Weigh the box. Set the shutter to open for a brief interval during which a single photon escapes. Weigh the box again some time later. Then (in principle) one has found to arbitrary accuracy both the photon energy and its time of passage, in conflict with the energy-time uncertainty principle.

"It was quite a shock for Bohr . . . he did not see the solution at once. During the whole evening he was extremely unhappy, going from one to the other and trying to persuade them that it couldn't be true, that it would be the end of physics if Einstein were right; but he couldn't produce any refutation. I shall never forget the vision of the two antagonists leaving the club [of the Fondation Universitaire]: Einstein a tall majestic figure, walking quietly, with a somewhat ironical smile, and Bohr trotting near him, very excited. . . . The next morning came Bohr's triumph."[201] I shall not reproduce here at length Bohr's splendid refutation, found in detail in his 1949 article.[202] The point is that the displacement of the box in the gravitational field used for the weighing affects the frequency of the clock governing the photon emission precisely in the amount needed to satisfy the energy-time uncertainty relation.

By 1931 Einstein's position on quantum mechanics had undergone a marked change.

First of all, his next paper on quantum mechanics, submitted in February 1931, shows that he had accepted Bohr's criticism.[203] It deals with a new variant of the clock-in-the-box experiment. Experimental information about one particle is used to make predictions about a second article. This paper, a forerunner of the Einstein–Podolsky–Rosen article to be discussed subsequently, need not be remembered for its conclusions.[204]

A far more important expression of Einstein's opinions is found in a letter he wrote the following September. In this letter, addressed to the Nobel committee in Stockholm, Einstein nominates Heisenberg and Schroedinger for the Nobel Prize.[205] In his motivation he says about quantum mechanics: "Diese Lehre enthält nach meiner Überzeugung ohne Zweifel ein Stück endgültiger Wahrheit" ("It is my conviction that this tenet contains without doubt a part of the ultimate truth"). Einstein himself was never greatly stirred by honors and distinctions. Even so, his nominations do not only reveal extraordinarily clearly what his thoughts were but are also deeply moving as an expression of his freedom of spirit and generosity of mind. They show that Einstein had come to accept that quantum mechanics was not an aberration but rather a truly professional contribution to physics.

It is not that from then on he desisted from criticizing quantum mechanics. He had recognized it to be part of the truth but was and forever remained deeply convinced that it was not the whole truth. From 1931 on, the issue for him was no longer the consistency of quantum mechanics but rather its completeness.

During the last twenty-five years of his life Einstein maintained that quantum mechanics is incomplete. He no longer believed that quantum mechanics was wrong, but that the common view of the physics community was wrong in ascribing to the postulates of quantum mechanics a degree of finality that he held to be naive and unjustified. The content and shape of his dissent will gradually unfold in what is to follow.

In November 1931 Einstein gave a colloquium in Berlin "On the uncertainty relation." The report of this talk[206] does not state that Einstein objected to Heisenberg's relations. Rather, it conveys a sense of his discomfort about the freedom of choice to measure precisely either the color of a light ray or its time of arrival. H. B. G. Casimir has written to me (private communication, December 31, 1977) about a colloquium that Einstein gave in Leiden, with Ehrenfest in the chair. To the best of Casimir's recollection this took place in the winter of 1931–2. In his talk Einstein discussed several aspects of the clock-in-the-box experiments. In the subsequent discussion it was mentioned that no conflict with quantum mechanics exists. Einstein reacted to this statement as follows: "Ich weiss es, widerspruchsfrei ist die Sache schon, aber sie enthält meines Erachtens doch eine gewisse Härte" ("I know, this business is free of contradictions, yet in my view it contains a certain unreasonableness").

By 1933 Einstein stated explicitly his conviction that quantum mechanics does not contain logical contradictions. In his Spencer lecture he said of the Schroedinger wave functions: "These functions are only supposed to determine in a mathematical way the probabilities of encountering those objects in a particular place or in a particular state of motion, if we make a measurement. This conception is logically unexceptionable and has led to important successes."[207]

It was in 1935 that Einstein stated for the first time his own desiderata in a precise form. This is the criterion of objective reality. He continued to subscribe to this for the rest of his life.

6.2 *Einstein on Objective Reality*

In his Como address Bohr had remarked that quantum mechanics, like relativity theory, demands refinements of our everyday perceptions of inanimate natural phenomena: "We find ourselves here on the very path taken by Einstein of adapting our modes of perception borrowed from the sensations to the gradually deepening knowledge of the laws of Nature."[208] Already then, in 1927, he emphasized that we have to treat with extreme care our use of language in recording the results of observations that involve quantum effects. "The hindrances met with on this path originate above all in the fact that, so to say, every word in the language refers to our ordinary perception," he observed. Bohr's deep concern with the role of language in the appropriate interpretation of quantum mechanics never ceased. In 1948 he put it as follows: "Phrases often found in the physical literature, as 'disturbance of phenomena by observation' or 'creation of physical attributes of objects by measurements' represent use of words like 'phenomena' and 'observation' as well as 'attribute' and 'measurement' which is hardly compatible with common usage and practical definition and, therefore, is apt to cause confusion. As a more appropriate way of expression, one may strongly advocate limitation of the use of the word *phenomenon* to refer exclusively to observations obtained under specified circumstances, including an account of the whole experiment."[209] This usage of "phenomenon," if not generally accepted, is the one to which nearly all physicists now adhere.

In contrast to the view that the concept of phenomenon *irrevocably* includes the specifics of the experimental conditions of observation, Einstein held that one should seek for a deeper-lying theoretical framework that permits the description of phenomena independently of these conditions. That is what he meant by the term "objective reality." After 1933 it was his almost solitary position that quantum mechanics is logically consistent but that it is an incomplete manifestation of an underlying theory in which an objectively real description is possible.

In an article written in 1935 together with Boris Podolsky and Nathan Rosen,[210] Einstein gave reasons for his position by discussing an example, simple as always. This paper "created a stir among physicists and has played a large role in philosophical discussions."[211] It contains the following definition: "If without in any way disturbing a system, we can predict with certainty (i.e., with a probability equal to unity) the value of a physical quantity, then there exists an element of physical reality corresponding to this physical quantity." The authors then consider the following problem: Two particles with respective momentum and position variables (p_1, q_1) and (p_2, q_2) are in a state with definite total momentum $P = p_1 + p_2$ and relative distance $q = q_1 - q_2$. This of course is possible since P and q commute. The particles are allowed to interact. Observations are made on particle 1 long after the interaction has taken place. Measure p_1 and one knows p_2 without having disturbed particle 2. Therefore (in their language) p_2 is an element of reality. Then measure q_1 and one knows q_2 again without having disturbed particle 2. Therefore q_2 is also an element of reality. Therefore both p_2 and q_2 are elements of reality. But quantum mechanics tells us that p_2 and q_2 cannot be simultaneously elements of reality because of the noncommutativity of the momentum and position operators of a given particle. Therefore quantum mechanics is incomplete.

The authors stess that they "would not arrive at our conclusion if one insisted that two . . . physical quantities can be regarded as simultaneous elements of reality *only when they can be simultaneously measured or predicted*" (their italics). Then follows a remark that is the key to Einstein's philosophy and that I have italicized in part.

"This [simultaneous predictability] makes the reality of p_2 and q_2 depend upon the process of measurement carried out on the first system which does not disturb the second system in any way. *No reasonable definition of reality could be expected to permit this.*" The only part of this article that will ultimately survive, I believe, is this last phrase, which so poignantly summarizes Einstein's views on quantum mechanics in his later years. The content of this paper has been referred to on occasion as the Einstein–Podolsky–Rosen paradox. It should be stressed that this paper contains neither a paradox nor any flaw of logic. It simply concludes that objective reality is incompatible with the assumption that quantum mechanics is complete. This conclusion has not affected subsequent developments in physics and it is doubtful that it ever will.

"It is only the mutual exclusion of any two experimental procedures, permitting the unambiguous definition of complementary physical quantities which provides room for new physical laws," Bohr wrote in his rebuttal.[212] He did not believe that the Einstein–Podolsky–Rosen paper called for any change in the interpretation of quantum mechanics. Most physicists (including myself) agree with this opinion. I shall reserve for the next section a further comment on the completeness of quantum theory.

This concludes an account of Einstein's position. He returned to his criterion for objective reality in a number of later papers[213] in which he repeated the Einstein–Podolsky–Rosen argument on several occasions. These papers add nothing substantially new. In one of them[214] he discussed a further example (I omit the details) stimulated by the question whether the quantum mechanical notion of phenomenon should also apply to bodies of everyday size. The answer is, of course, in the affirmative.

Bohr was, of course, not the only one to express opposition to objective reality; nor was Einstein the only one critical of the complementarity interpretation.[215] I have chosen to confine myself to the exchanges between Einstein and Bohr because I believe that Einstein's views come out most clearly in juxtaposing them with Bohr's. Moreover I am well acquainted with their thoughts on these issues because of discussions with each of them. Bohr was in Princeton when he put the finishing touches to his 1949 article[216] and we did discuss these matters often at that time. (It was during one of these discussions that Einstein sneaked in to steal some tobacco.[217]) However, it needs to be stressed that other theoretical physicists and mathematicians have made important contributions to this area of problems. Experimentalists have actively participated as well. A number of experimental tests of quantum mechanics in general and also of the predictions of specific alternative schemes have been made.[218] This has not led to surprises.

The foregoing was a brief sketch of the substance of Einstein's arguments. There is another question that at least to me is far more fascinating, the one of motivation. What drove Einstein to use methods he himself called "quite bizarre as seen from the outside?"[219] Why would he continue "to sing [his] solitary old little song"[220] for the rest of his life? As I shall discuss in Sec. 8 the answer has to do with a grand design for a synthetical physical theory that Einstein conceived early (before the discovery of quantum mechanics). He failed to reach this synthesis.

So to date have we all.

The phenomena to be explained by a synthetical theory of particles and fields have become enormously richer since the days when Einstein embarked on his unified theory. Theoretical progress has been very impressive, but an all embracing theory does not exist. The need for a new synthesis is felt ever more keenly as the phenomena grow in complexity.

Therefore, in the year of Einstein's centennial, any assessment of Einstein's visions can only be made from a vantage point that is necessarily tentative. It may be useful to record ever so briefly what this vantage point appears to be to at least one physicist. This will be done next, with good wishes for the second centennial.

7. A Time Capsule

Einstein's life ended . . . with a demand on us for synthesis.

W. Pauli[221]

When Einstein and others embarked on programs of unification, three particles (in the modern sense) were known to exist: the electron, the proton, and the photon, and there were two fundamental interactions, electromagnetism and gravitation.[222] At present the number of particles runs into the hundreds. A further reduction to more fundamental units appears inevitable. The number of fundamental interactions is now believed to be at least four. The grand unification of all four types of forces — gravitational, electromagnetic, weak, and strong interactions — is an active topic of current exploration. It has not been achieved as yet.

Relativistic quantum field theories (in the sense of special relativity) are the principal tools for these explorations. Our confidence in the general field theoretic approach rests first and foremost on the tremendous success of quantum electrodynamics (Q.E.D.). One number, the g-factor of the electron, may illustrate the level of predictability that this theory has reached:

$$\frac{1}{2}(g-2) = \begin{cases} 1\,159\,652\,375(261) \times 10^{-12}, & \text{predicted by pure Q.E.D.,}^{223} \\ 1\,159\,652\,410(200) \times 10^{-12}, & \text{observed.} \end{cases}$$

It is nevertheless indicated that this branch of field theory has to merge with the theory of other fields.

"If we could have presented Einstein with a synthesis of his general relativity and the quantum theory, then the discussion with him would have been considerably easier."[224] To date, this synthesis is beset with conceptual and technical difficulties. The existence of singularities associated with gravitational collapse is considered by some as an indication for the incompleteness of the general relativistic equations. It is not known whether or not these singularities are smoothed out by quantum effects.

The ultimate unification of weak and electromagnetic interactions has not been achieved, but a solid beachhead appears to have been established in terms of local non-Abelian gauge theories with spontaneous symmetry breakdown. As a result it is now widely believed that weak interactions are mediated by massive vector mesons. Current expectations are that such mesons will be observed within another decade.

It is widely believed that strong interactions are also mediated by local non-Abelian gauge fields. Their symmetry is supposed to be unbroken, so that the corresponding vector mesons are massless. The dynamics of these "non-Abelian photons" is supposed to prohibit their creation as single free particles. The technical exploration of this theory is only in its very early stages.

Since electromagnetism and gravitation are also associated with local gauge fields it is commonly held that the grand unification will eventually be achieved in terms of a multicomponent field of this kind. This may be said to represent a program of geometrization that bears resemblance to Einstein's attempts, although the manifold subject to geometrization is larger than he anticipated.

In the search for the correct field theory, model theories have been examined that reveal quite novel possibilities for the existence of extended structures (solitons, instantons, monopoles). In the course of these investigations topological methods have entered this area of physics. More generally, it has become clear in the past decade that quantum field theory is much richer in structure than was appreciated earlier. The renormalizability of non-Abelian gauge fields with spontaneous symmetry breakdown, asymptotic freedom, and supersymmetry are cases in point.

The proliferation of new particles has led to attempts at a somewhat simplified underlying description. According to the current picture the basic constituents of matter are: two families of spin 1/2 particles, the leptons and the quarks; a variety of spin 1 gauge bosons, some massless, some massive; and (more tentatively) some fundamental spin 0 particles. The only gauge boson observed so far is the photon. To date, three kinds of charged leptons have been detected. The quarks are hypothetical constituents of the observed hadrons. To date, at least five species of quarks are needed. The dynamics of the strong interactions is supposed to prohibit the creation of quarks as single free particles. This prohibition, confinement, has not as yet been implemented theoretically in a convincing way. No criterion is known that would enable one to state how many species of leptons and of quarks should exist.

Weak, electromagnetic, and strong interactions have distinct intrinsic symmetry properties, but this hierarchy of symmetries is not well understood theoretically. Perhaps the most puzzling are the noninvariance under space reflexion at the weak level and the noninvariance under time reversal at an even weaker level. It adds to the puzzlement that the latter phenomenon has been observed so far in only a single instance, namely in the K^0-$\bar{K}^0$ system. (These phenomena were first observed after Einstein's death. I have often wondered what might have been his reactions to these discoveries, given his "conviction that pure mathematical construction enables us to discover the concepts and the laws connecting them."[225]

It is not known why electric charge is quantized, but it is plausible that this will be explicable in the framework of a future gauge theory.

In summary, at the time of the centenary of the death of James Clerk Maxwell (June 13, 1831–November 5, 1879) and the birth of Albert Einstein (March 14, 1879–April 18, 1955) the evidence is overwhelming that the theory of particles and fields is incomplete. Einstein's earlier complaint remains valid to this day: "The theories which have gradually been connected with what has been observed have led to an unbearable accumulation of independent assumptions."[226] At the same time no experimental evidence or internal contradiction exists to indicate that the postulates of general relativity, of special relativity, or of quantum mechanics are in mutual conflict or in need of revision or refinement. We are therefore in no position to affirm or deny that these postulates will forever remain unmodified.

8. Particles, Fields, and the Quantum Theory: Einstein's Vision

Apart. . . . 6. Away from common use for a special purpose.

Oxford English Dictionary

8.1 *Some Reminiscences*

The rest of this chapter is based in part on what I learned from discussions with Einstein. I should like to mention first some reminiscences of my encounters with him.

I knew Einstein from 1946 until the time of his death.[227] I would visit with him in his office or accompany him (often together with Kurt Gödel) at lunchtime on his walk home to 112 Mercer Street. Less often I would visit him there. In all I saw him about once every few weeks. We always talked German, the language best suited to grasp the nuances of what he had in mind as well as the flavor of his personality. Only once did he visit my apartment. The occasion was a meeting of the Institute for Advanced Study faculty for the purpose of drafting a statement of our position in the 1954 Oppenheimer affair. I shall not go into Einstein's outspoken opinions on world affairs and public policy.

Einstein's company was comfortable and comforting to those who knew him. Of course, he knew well that he was a legendary figure in the eyes of the world. He accepted this as a fact of his life. There was nothing in his personality to promote such attitudes. Nor did he relish them. Privately he would express annoyance if he felt that his position was misused. I recall the case of Professor X who had been quoted by the newspapers as having found solutions to Einstein's generalized equations of gravitation. Einstein said to me "Der Mann ist ein Narr" and added that in his opinion X could calculate but could not think. X had visited Einstein to discuss this work and Einstein, always courteous, said to him that if his results were true they would be important. Einstein was chagrined that he had been quoted in the papers without this proviso. He said that he would keep silent on the matter but would not receive X again. According to Einstein the whole thing started because, in his enthusiasm, X had told some colleagues who saw the value of it as publicity for their university.

To those physicists who could follow his scientific thought and who knew him personally, the legendary aspect was never in the foreground — yet it was never wholly absent. I recall an occasion, in 1947, when I gave a talk at the Institute about the newly discovered π- and μ-mesons. Einstein walked in just after I had begun. I remember that I was speechless for the brief moment necessary to overcome a sense of the unreal. I recall a similar moment during a symposium held in Frick Chemical laboratory in Princeton on March 19, 1949.[228] The occasion was Einstein's seventieth birthday. Most of us were in our seats when Einstein entered the hall. Again there was this brief hush before we stood to greet him.

Nor do I believe that such reactions were typical only of those who were much younger than he. There were a few occasions when Pauli and I were both together with him. Pauli, not known for an excess of awe, was just slightly different in Einstein's company. One could perceive his sense of reverence. I have also seen Bohr and Einstein together. Bohr too was affected in a somewhat similar way, differences in scientific outlook notwithstanding.

Whenever I met Einstein, our conversations might range far and wide, but invariably the discussion would turn to physics. Such discussions would touch only occasionally on matters having to do with the period before 1925 and then they would mainly concern relativity. I recall asking Einstein once what influence Poincaré's work had had on him. Einstein replied that he never had read Poincaré. On the other hand, Einstein held no one in higher esteem than Lorentz. He once told me that without Lorentz he would never have been able to take "den Schritt" ("the step"). (Einstein always talked about relativity in an impersonal way.) Lorentz was to Einstein the most well-rounded and harmonious personality he had met in his entire life. He had also a great veneration for Planck.

Our discussions, however, centered first and foremost on the present and the future. When relativity was the issue he would often talk of his efforts to unify gravitation and electromagnetism and of his hopes for the next steps. His faith rarely wavered in the path he had chosen. Only once did he express a reservation to me when he said, in essence: I am not sure that differential geometry is the framework for further progress, but if it is, then I believe I am on the right track. (This remark must have been made sometime during his last few years.)

The main topic of discussion was quantum physics, however. Einstein never ceased to ponder the meaning of the quantum theory. Time and time again the argument would turn to quantum mechanics and its interpretation. He was explicit in his opinion that the most commonly held views on this subject could not be the last word, but he had also more subtle ways of expressing his dissent. For example, he would never refer to a wave function as "die Wellenfunktion" but would always use a mathematical terminology: "die Psifunktion." We often discussed his notions on objective reality. I recall that during one walk Einstein suddenly stopped, turned to me, and asked whether I really believed that the moon exists only when I look at it. The rest of this walk was devoted to a discussion of what a physicist should mean by the term *to exist*.

I was never able to arouse much interest in Einstein about the new particles. It was apparent that he felt that the time was not ripe to worry about such things and that these particles would eventually appear as solutions of the equations of a unified theory. In some sense he may well prove to be right.

It was even more difficult to discuss quantum field theory with him. He was willing to admit that quantum mechanics was successful on the nonrelativistic level. However, he did not believe that this theory provided a secure enough basis for relativistic generalizations.[229] Relativistic quantum field theory was repugnant to him.[230] Valentine Bargmann has told me that Einstein once asked him to give a private survey of quantum field theory, beginning with second quantization. Bargmann did so for about a month. Thereafter Einstein's interest waned.

An unconcern with the past is a privilege of youth. In all the years I knew Einstein I never read a single one of his papers from the 1905–25 period on the quantum theory. It is now clear to me that I might have asked him some interesting questions, had I been less blessed with ignorance. I might then have known some interesting facts by now, but at a price. My discussions with Einstein were not historical interviews, they concerned live physics. I am glad it never was otherwise.

8.2 *Einstein, Newton, and Success*

''It seems to be clear . . . that the Born statistical interpretation of the quantum theory is the only possible one,'' Einstein wrote in 1936.[231] He also called the statistical quantum theory ''the most successful physical theory of our period.''[232] Then why was he never convinced by it? I believe Einstein himself answered this indirectly in his 1933 Spencer lecture — perhaps the clearest and most revealing expression of his mode of thinking.[233] The key is to be found in his remarks on Newton and classical mechanics.

In this lecture Einstien notes that ''Newton felt by no means comfortable about the concept of absolute space, . . . of absolute rest . . . [and] about the introduction of action at a distance.'' Einstein then goes on to refer to the success of Newton's theory in these words: ''The enormous practical success of his theory may well have prevented him and the physicists of the eighteenth and nineteenth centuries from recognizing the fictitious character of the principles of his system.'' It is important to note that by ''fictitious'' Einstein means free inventions of the human mind. Whereupon he compares Newton's mechanics with his own work on general relativity: ''The fictitious character of the principles is made quite obvious by the fact that it is possible to exhibit two essentially different bases each of which in its consequences leads to a large measure of agreement with experience.''

Elsewhere Einstein has addressed Newton as follows: ''Newton forgive me: you found the only way which, in your age, was just about possible for a man with the highest power of thought and creativity.''[234] Only one man in history could have possibly written that line.

In the Spencer lecture Einstein mentioned the success not only of classical mechanics but also of the statistical interpretation of quantum theory: ''This conception is logically unexceptionable and has led to important successes.'' But, he added,''I still believe in the possibility of giving a model of reality which shall represent events themselves and not merely the probability of their occurrence.''

From this lecture as well as from discussions with him on the foundations of quantum physics I have gained the following impression: Einstein tended to compare the successes of classical mechanics with those of quantum mechanics. In his view both were on a par, being successful but incomplete. For more than a decade Einstein had pondered the single question of how to extend to general motions the invariance under uniform translations. His resulting theory, general relativity, had led to only small deviations from Newton's theory (instances where these deviations are large were discussed only much later). He was likewise prepared to undertake his own search for objective reality, fearless of how long it would take. He was also prepared for the survival of the practical successes of quantum mechanics, with perhaps only small modifications. It is quite plausible that the very success of his highest achievement, general relativity, was an added spur to Einstein's apartness. Yet it should not be forgotten that this trait characterized his entire oeuvre and style.

Einstein was not oblivious to others' reactions to his own position. ''I have become an obstinate heretic in the eyes of my colleagues,'' he wrote to one friend,[235] and ''in Princeton they consider me an old fool,'' he said to another.[236] He knew, and on occasion would even say, that his road was a lonely one,[237] yet he held fast. ''Momentary success carries more power of conviction for most people than reflections on principle.''[238]

Einstein was neither saintly nor humorless in defending his solitary position on the quantum theory. On occasion he could be acerbic. At one time he said that Bohr thought very clearly, wrote obscurely, and thought of himself as a prophet.[239] Another time he referred to Bohr as a mystic.[240] On the other hand, in a letter to Bohr, Einstein referred to his own position by quoting an old rhyme: "Über die Reden des Kandidaten Jobses/Allgemeines Schütteln des Kopses" (roughly: "There was a general shaking of heads concerning the words of candidate Jobs)." There were moments of loneliness: "I feel sure that you do not understand how I came by my lonely ways. . . ."[241] Einstein may not have expressed all his feelings on these matters. But that was his way. "The essential of the being of a man of my type lies precisely in *what* he thinks and *how* he thinks, not in what he does or suffers."[242]

The crux of Einstein's thinking on the quantum theory was not his negative position in regard to what others had done but, rather, his deep faith in his own distinct approach to the quantum problems. His beliefs may be summarized as follows:

1. Quantum mechanics represents a major advance, yet it is only a limiting case of a theory that remains to be discovered. "There is no doubt that quantum mechanics has seized hold of a beautiful element of truth, and that it will be a test stone for a future theoretical basis, in that it must be deducible as a limiting case from that basis, just as electrostatics is deducible from the Maxwell equations of the electromagnetic field or as thermodynamics is deducible from statistical mechanics."[243]
2. One should not try to find the new theory by beginning with quantum mechanics and trying to refine it or reinterpret it. "I do not believe that quantum mechanics will be the starting point in the search for this basis just as one cannot arrive at the foundations of mechanics from thermodynamics or statistical mechanics."[244]
3. Instead — and this was Einstein's main point — one should start all over again, as it were, and endeavor to obtain the quantum theory as a by-product of a general relativistic field theory. As an introduction to a further discussion of this last issue it is useful to comment first on the profound differences between Einstein's attitude to relativity and to the quantum theory.

8.3 *Relativity Theory and Quantum Theory*

Einstein's paper on light quanta was submitted in March 1905, his first two papers on relativity in June and September of that year, respectively.[245] In a letter to a friend written early in 1905 he promised him a copy of his March paper "about radiation and the energy properties of light [which] is very revolutionary."[246] In the same letter he also mentioned that a draft of the June paper was ready and added that "the purely kinematic part of this work will surely interest you." It is significant that Einstein would refer to his light quantum paper, but not to his relativity paper, as a revolutionary step.

If a revolutionary act consists in overthrowing an existing order, then to describe the light quantum hypothesis in those terms is altogether accurate. It is likewise fitting not to apply these terms to relativity theory, since it did not overthrow an existing order but, rather, brought immediate order to new domains.

Einstein was one of the freest spirits there ever was. But he was not a revolutionary, as the overthrow of existing order was never his prime motivation. It was his genius that made him ask the right question. It was his faith in himself that made him persevere until he had the answer. If he had a god it was the god of Spinoza. He had to follow his own reasoning regardless of where it would lead him. He had a deep respect for the traditions of physics. But if his own reasoning indicated answers that lay outside the conventional patterns, he accepted these answers, not for the sake of contradiction, but because it had to be. He had been free to ask the question; he had no choice but to accept the answer. This deep sense of destiny led him farther than anyone before him — but not as far as finding his own answer to the quantum theory.

It is striking how, from the very beginning, Einstein kept his scientific writing on relativity theory separate from that on quantum theory. This was evident already in 1905. In his first relativity paper Einstein noted: "It is remarkable that the energy and frequency of a light complex vary with the state of motion of the observer according to the same law."[247] Here was an obvious opportunity to refer to the relation $E = h\nu$ of his paper on light quanta, finished a few months earlier. But Einstein did not do that. Also in the September paper,[248] he referred to radiation but not to light quanta. In his 1909 address at Salzburg,[249] Einstein discussed his ideas both on relativity theory and on quantum theory but kept these two areas well separated. As we have seen, in his 1917 paper[250] Einstein ascribed to light quanta an energy $E = h\nu$ and a momentum $p = h\nu/c$. This paper concludes with the following remark: "Energy and momentum are most intimately related; therefore a theory can only then be considered as justified if it has been shown that according to it the momentum transferred by radiation to matter leads to motions as required by thermodynamics." Why is only thermodynamics mentioned, why not also relativity?

I believe that the reason Einstein kept the quantum theory apart from relativity theory is that he considered the former to be provisional (as he had said already in 1911[251]), whereas relativity to him was the revealed truth. Einstein's destiny reminds me in more than one way of the destiny of Moses.

The road to general relativity theory had been local classical field theory. The same road, he hoped, would also lead to the implementation of objective reality.

8.4 *Einstein's Vision*

In 1923 Einstein published an article entitled "Does field theory offer possibilities for the solution of the quantum problem?"[252] It begins with a reminder of the successes achieved in electrodynamics and general relativity theory in regard to a causal description: events are causally determined by differential equations combined with initial conditions on a spacelike surface. However, Einstein continued, this method cannot be applied to quantum problems without further ado. As he put it, the discreteness of the Bohr orbits indicates that initial conditions cannot be chosen freely. Then he asked: Can one nevertheless implement these quantum constraints in a (causal) theory based on partial differential equations? His answer: "Quite certainly: we must only "overdetermine" the field variables by [appropriate] equations." Next he states his program, based on three requirements: 1) The theory should be generally covariant. 2) The desired equations should at least be in accordance with the gravitational and the Maxwell theory. 3) The desired system of equations that overdetermines the fields should have static spherically symmetric solutions that describe the electron and the proton. If this overdetermination can be achieved, then "we may hope that these equations co-determine the mechanical behavior of the singular points (electrons) in such a way that also the initial conditions of the field and the singular points are subject to restrictive conditions." He goes on to discuss a tentative example and concludes as follows: "To me the main point of this communication is the idea of overdetermination."

This paper contains all the essential ingredients of the vision on particles, fields, and the quantum theory that Einstein was to pursue for the rest of his life.

To Einstein the concept of a unified field theory meant something different from what it meant and means to anyone else. He demanded that the theory shall be strictly causal, that it unify gravitation and electromagnetism, that the particles of physics shall emerge as special solutions of the general field equations, and that the quantum postulates *shall be a consequence* of the general field equations. Einstein had all these criteria in mind when he wrote, in 1949: "Our problem is that of finding the field equations of the total field."[253]

Already in 1923 he had been brooding on these ideas for a number of years. In 1920 he had written to Born: "I do not seem able to give tangible form to my pet idea ["meine Lieblingsidee"], which is to understand the structure of the quanta by redundancy in determination, using differential equations."[254] It is the earliest reference to Einstein's strategy that I am aware of. It would seem likely that ideas of this kind began to stir in Einstein soon after 1917, when he had not only completed the general theory of relativity but also had confronted the lack of causality in spontaneous emission.[255] The early response of others to these attempts by Einstein has been recorded by Born: "In those days [early in 1925] we all thought that his objective . . . was attainable and also very important."[256] Einstein himself felt that he had no choice: "The road may be quite wrong but it must be tried."[257]

As I already intimated in the introduction, it is essential for the understanding of Einstein's thinking to realize that there were two sides to his attitude concerning quantum physics. There was Einstein the critic, never yielding in his dissent from complementarity, and there was Einstein the visionary, forever trying to realize the program just outlined, which went well beyond a mere reinterpretation of quantum mechanics.[258] His vision predates quantum mechanics, it was certainly with him in 1920 and probably even a few years earlier.

A detailed description of his efforts in this direction belongs to a history of unified field theory, a topic that cannot be dealt with here. In concluding this paper I shall confine myself to a few brief observations concerning Einstein's own attitude toward his program.

Einstein believed that the field equations would generate particles with nonzero spin as particlelike solutions that are not spherically symmetrical (V. Bargmann, private communication). Presumably he hoped that his idea of overdetermination would lead to discrete spin values.[259] He also hoped that the future theory would contain solutions that are not absolutely localized and that would carry quantized electric charge.[260] (In 1925 Einstein noted[261] that if the combined gravitational/electromagnetic field equations have particlelike solutions with charge e and mass m, then there should also be solutions[262] with $(-e,m)$! This led him to doubt for some time that the unification of gravitation and electromagnetism was consistently possible.)

Einstein's correspondence shows that the unified field theory and the quantum problems were very often simultaneously on his mind. Here are but a few examples. In 1925, while he was at work on a theory with a nonsymmetric metric, he wrote to a friend: "Now the question is whether this field theory is compatible with the existence of atoms and quanta."[263] He discussed the same generalized theory in a letter written in 1942: "What I am doing now may seem a bit crazy to you. One must note, however, that the wave-particle duality demands something unheard of."[264] In 1949: "I am convinced that the . . . statistical [quantum] theory . . . is superficial and that one must be backed by the principle of general relativity."[265] And in 1954: "I must seem like an ostrich who forever buries his head in the relativistic sand in order not to face the evil quanta."[266]

A similar concurrent preoccupation with general relativity and quantum theory is evident in a paper by Einstein and Rosen completed in 1935, two months after the Einstein–Podolsky–Rosen article. It deals with singularity-free solutions of the gravitational-electromagnetic field equations. One phrase in this paper, "One does not see a priori whether the theory contains the quantum phenomena,"[267] illustrates once again the scope of the program that was on Einstein's mind.

Simplicity was the guide in Einstein's quest: "In my opinion there is *the* correct path and . . . it is in our power to find it. Our experience up to date justifies us in feeling sure that in Nature is actualized the ideal of mathematical simplicity."[268] Already in 1927 Heisenberg stressed, in a letter to Einstein mentioned earlier,[269] that Einstein's concept of simplicity and the simplicity inherent in quantum mechanics cannot both be upheld. "If I have understood correctly your point of view then you would gladly sacrifice the simplicity [of quantum mechanics] to the principle of [classical] causality. Perhaps we could comfort ourselves [with the idea that] the dear Lord could go beyond [quantum mechanics] and maintain causality. I do not really find it beautiful, however, to demand physically more than a description of the connection between experiments."[270]

As Einstein's life drew to a close, doubts about his vision arose in his mind.

He wrote to Born,[271] probably in 1949: "Our respective hobby horses have irretrievably run off in different directions . . . even I cannot adhere to [mine] with absolute confidence." I have mentioned before the reservations that Einstein expressed to me in the early fifties. (V. Bargmann informs me that Einstein made similar remarks to him in the late 1930s.) To his dear friend Besso he wrote, in 1954: "I consider it quite possible that physics cannot be based on the field concept, i.e. on continuous structures. In that case *nothing* remains of my entire castle in the air, gravitation theory included, [and of] the rest of modern physics."[272] It is to be doubted whether any physicist can be found who would not respectfully and gratefully submit that this judgment is unreasonably harsh.

Otto Stern has recalled a statement that Einstein once made to him: "I have thought a hundred times as much about the quantum problems as I have about general relativity theory" (R. Jost, private communication, August 17, 1977). He kept thinkng about the quantum till the very end. Einstein wrote his last autobiographical sketch[273] in Princeton, in March 1955, about a month before his death. Its final sentences deal with the quantum theory: "It appears dubious whether a field theory can account for the atomistic structure of matter and radiation as well as of quantum phenomena. Most physicists will reply with a convinced 'No,' since they believe that the quantum problem has been solved in principle by other means. However that may be, Lessing's comforting word stays with us: the aspiration to truth is more precious than its assured possession."

9. Epilogue

I saw Einstein for the last time in December 1954.

As he had not been well, he had for some weeks been absent from the Institute, where he normally spent a few hours each morning. Since I was about to take a term's leave from Princeton, I called Helen Dukas and asked her to be kind enough to give my best wishes to Professor Einstein. She suggested I might come to the house for a brief visit and a cup of tea. I was, of course, glad to accept. After I arrived, I went upstairs and knocked at the door of Einstein's study. There was his gentle "Come." As I entered he was seated in his armchair, a blanket over his knees, a pad on the blanket. He was working. He put his pad aside at once and greeted me. We spent a pleasant half hour or so; I do not recall what was discussed. Then I told him I should not stay any longer. We shook hands, and I said good-bye. I walked to the door of the study, not more than four or five steps away. I turned around as I opened the door. I saw him in his chair, his pad on his lap, a pencil in his hand, oblivious to his surroundings.

He was back at work.

Acknowledgements

No one alive is more familiar with the circumstances of Einstein's life and with his collected correspondence than Helen Dukas. I am deeply grateful for her friendship and generosity through the years, which have helped me most importantly in gathering key information contained in this article. I am also much indebted to Res Jost, Martin Klein, and George Uhlenbeck for advice and guidance. Numerous discussions with Sam Treiman about the contents of this article have been invaluable. I also wish to thank the Princeton Physics Department for its hospitality and the staff of Fine Hall Library for continued help. Finally I want to thank the executors of the Einstein estate and of the Pauli estate for permission to quote from unpublished documents.

This work was supported in part by the U. S. Department of Energy under Contract Grant No-Ey-C-02-2232 B* 000.

NOTES

1. Rev. Mod. Phys. **21** (3) (July 1949).
2. A. Pais, letter dated December 9, 1948.
3. It was decided later that also L. de Broglie, M. von Laue, and P. Frank should do so.
4. R. A. Millikan, Rev. Mod. Phys. **21**, 343 (1949).
5. R. A. Millikan, Phys. Rev. **7**, 18 (1916).
6. R. A. Millikan, Phys. Rev. **7**, 355 (1916).
7. No detailed references to the literature will be given, in keeping with the brevity of my comments on this subject.
8. J. J. Thomson, *Recollections and Reflections* (G. Bell and Sons, London, 1936), p. 341.
9. It is often said, and not without grounds, that the neutron was actually anticipated. In fact, twelve years before its discovery, in one of his Bakerian lectures, Ernest Rutherford (1871–1937) spoke of "the idea of the possible existence of an atom of mass one which has zero nuclear charge" [Proc. Roy. Soc. **A97**, 374 (1920)]. Nor is there any doubt that the neutron being in the air at the Cavendish was of profound importance to its discoverer James Chadwick (1891–1974) [see Proceedings of the Tenth International Congress of Historians of Science, Ithaca, N.Y. Vol. 1, (1962), p. 159]. Even so, not even a Rutherford could have guessed that his 1920 neutron (then conjectured to be a tightly bound proton-electron system) was so essentially different from the particle that would eventually go by that name.
10. It is too early to include the graviton.
11. P. A. M. Dirac, *Proc. Roy. Soc.* **A133**, 60 (1931).
12. H. Weyl, *The Theory of Groups and Quantum Mechanics*, 2nd ed. (Dover, New York, original text published in 1930), pp. 263–4 and preface.
13. "All these 50 years of pondering have not brought me closer to answering the question 'What are light quanta?'"[letter by A. Einstein to M. Besso, 12 December 1951, in *Albert Einstein, Michele Besso Correspondence 1903–1955* (Hermann, Paris, 1972), p. 453].
14. A. Einstein, Ann. d. Phys. **17**, 132 (1905).
15. In modern notation, Wien's law states that $\rho(\nu,T)$, the spectral density per unit volume, is given by $8\pi h\nu^3/c^3 \cdot \exp(-h\nu/kT)$.
16. *La théorie du rayonnement et les quanta*, edited by P. Langevin and M. de Broglie (Gauthier-Villars, Paris 1912), p. 443.
17. A. Sommerfeld, Verh. d. Ges. D. Naturf. und Ärzte, **83**, 31 (1912).

18. R. A. Millikan, Science **37**, 119 (1913).

19. R. A. Millikan, Phys. Rev. **7**, 355 (1916).

20. M. Planck, Verh. d. Deutsch. Phys. Ges. **2**, 237 (1900).

21. Planck considered a material linear oscillator with proper frequency ν. Let $U(\nu, T)$ be its average energy when in thermal equilibrium with radiation at temperature T. He derived the important classical relation $c^3\rho(\nu, T) = 8\pi^2 U(\nu, T)$. From this equation he obtained his blackbody radiation law by imposing quantum conditions on U that in turn *imply* the quantum theoretical form of ρ.

22. M. Planck, letter to A. Einstein, 6 July 1907.

23. M. Planck, Phys. Z. **10**, 825 (1909).

24. H. A. Lorentz, letter to W. Wien, April 12, 1909, quoted by A. Hermann in *Frühgeschichte der Quantumtheorie* (Mosbach, Baden, 1969), p. 68.

25. N. Bohr, Phil Mag. **26**, 1 (1913).

26. N. Bohr, Nature **92**, 231 (1913).

27. G. de Hevesy, letter to E. Rutherford, 14 October 1913, quoted in A. S. Eve, *Rutherford* (Cambridge University, Press, Cambridge, England, 1939), p. 226.

28. Cf. R. H. Stuewer, *The Compton Effect* (Science History, New York, 1975), Chap. 2.

29. E. von Schweidler, Jahrb. der Radioakt. u. Elektr. **1**, 358 (1904).

30. R. A. Millikan, Phys. Rev. **7**, 18 (1916).

31. W. Duane and F. Hunt, Phys. Rev. **6**, 166 (1915).

32. Phys. Rev. **7**, 18 (1916).

33. E. Rutherford, J. Röntgen Soc. **14**, 81 (1918).

34. S. Arrhenius in *Nobel Lectures in Physics*, Vol. **1** (Elsevier, New York, 1965), pp. 478–81. Einstein could not attend the festivities since he was in Japan at that time. He showed his indebtedness one year later by going to Göteborg and giving an address on relativity theory.

35. G. Kirchhoff and R. Bunsen, Ann. d. Phys. und Chemie **110**, 160 (1860).

36. J. J. Balmer, Ann. d. Phys. **25**, 80 (1885).

37. G. Kirchhoff, Ann. d. Phys. und Chemie **109**, 275 (1860).

38. There have been occasional speculations that the photon might have a tiny nonzero mass. Direct experimental information on the photon mass is therefore a matter of interest. The best determinations of this mass come from astronomical observations. The present upper bound is $8 \cdot 10^{-49}$ grm [L. Davis, A. S. Goldhaber, and M. M. Nieto, Phys. Rev. Letters **35**, 1402 (1975)]. In what follows the photon mass is taken to be strictly zero.

39. E. P. Wigner, Ann. Math **40**, 149 (1939).

40. G. N. Lewis, Nature **118**, 874 (1926). For ease I shall often use the term "photon" also in referring to times prior to 1926.

41. A. Einstein, Phys. Z. **10**, 185 (1909), and Phys. Z. **10**, 817 (1909).

42. Terms $0(\tau^2)$ are dropped. $\langle v\Delta \rangle = 0$ since v and Δ are uncorrelated.

43. A. Einstein, Ann. d. Phys. **17**, 132 (1905).

44. A. Einstein, Phys. Z. **10**, 817 (1909).

45. Ibid.

46. Ibid.

47. J. Stark, Phys. Z. **10**, 902 (1909). Stark wrote n instead of ν.

48. A. Einstein, Verh. D. Phys. Ges. **18**, 318 (1916), Mitt. Ph. Ges. Zurich **16**, 47 (1916), and Phys. Z. **18**, 121 (1917).

49. Phys. Z. **18**, 121 (1917).

50. Ibid.

51. W. Pauli, *Collected Scientific Papers , Vol. 1* (Interscience, New York, 1964), p. 630.
52. A. Einstein, letter to M. Besso, 6 September 1916, in *Einstein, Besso Correspondence,* op. cit. in n. 13, p. 82.
53. A. Einstein, letter to M. Besso, 29 July 1918, in *Einstein, Besso Correspondence*, op. cit. in n. 13, p. 130.
54. A. Einstein, Phys. Z. **18**, 121 (1917).
55. E. Rutherford, Phil. Mag. **49**, 1 (1900). Here a development began which, two years later, culminated in the transformation theory for radioactive substances published by E. Rutherford and F. Soddy in Phil. Mag. **4**, 370, 569 (1902).
56. A. Einstein, Verh. D. Phy. Ges. **18**, 318 (1916).
57. A. Pais, Rev. Mod. Phys. **49**, 925 (1977).
58. Phys. Z. **18**, 121 (1917).
59. Ibid.
60. A. Einstein, letter to M. Besso, 9 March 1917, in *Einstein, Besso Correspondence*, op. cit. in n. 13, p. 103.
61. A. Einstein, letter to M. Born, 27 January 1920, in *The Born-Einstein Letters* (Walker, New York, 1971), p. 23.
62a. A. Einstein, S. B. Preuss, Akad. W. (1921), p. 882.
62b. A. Einstein, S. B. Preuss., Akad. W. (1922), p. 18. For further details see M. Klein, Hist. St. Phys. Sci. **2**, 1 (1970).
63. A. H. Compton, Phys. Rev. **21**, 483 (1923); P. Debye, Phys. Z. **24**, 161 (1923). Einstein attached great importance to an advance in another direction that took place in the intervening years: the effect discovered by Otto Stern (1888–1969) and Walther Gerlach (1889–1979) (see his letter to Besso, 24 May 1924, *Einstein, Besso Correspondence*, op. cit. in n. 13, p. 201). Together with Ehrenfest he made a premature attempt at its interpretation [A. Einstein and P. Ehrenfest, Z. f. Phys. **11**, 31 (1922)].
64. For details on these classical theories see R. H. Stuewer's fine monograph on the Compton effect, op. cit. in n. 28.
65. Nor is it an accident that these two men came forth with the photon kinematics at about the same time. In his paper [Phys. Z. **24**, 161 (1923)] Debye acknowledges a 1922 report by Compton in which the evidence against the classical theory was reviewed. A complete chronology of these developments in 1922 and 1923 is found in Stuewer, op. cit. in n. 28, p. 235.
66. For a detailed account of the evolution of Compton's thinking see Stuewer, ibid., ch. 6.
67. *K* line X rays from a molybdenum anticathode were scattered off graphite. Compton stressed that one should only use light elements as scatterers so that the electrons will indeed be quasi-free. Scattered X rays at 45°, 90°, and 135° were analyzed.
68. For the current status of the subject see *Compton Scattering*, edited by B. Williams (McGraw-Hill, New York, 1977).
69. The work of Compton and Debye led Pauli to extend Einstein's work of 1917 to the case of radiation in equilibrium with free electrons [Z. f. Phys. **18**, 272 (1923)]. Einstein and Ehrenfest subsequently discussed the connection between Pauli's and Einstein's *Stosszahlansatz* [Z. f. Phys. **19**, 301 (1923)].
70. S. K. Allison, Biogr. Mem. NAS **38**, 81 (1965).
71. A. Sommerfeld, *Atombau und Spektrallinien*, 4th ed. (Vieweg, Braunschweig, 1924), p. viii.
72. A. Einstein, "Das Komptonsche Experiment," Berliner Tageblatt (20 April 1924).
73. N. Bohr, H. A. Kramers, and J. C. Slater, Phil. Mag. **47**, 785 (1924).
74. A. Einstein, postcard to M. Planck, 23 October 1919, quoted by C. Seelig in *Albert Einstein* (Europa Verlag, Zürich, 1954), p. 193.
75. A. Einstein, "Autobiographical Notes," in *Albert Einstein: Philosopher-Scientist*, edited by P. A. Schilpp, (Library of Living Philosophers, Evanston, Ill., 1949), p. 3.
76. A. Einstein, letter to N. Bohr, 2 May 1920.
77. N. Bohr, letter to A. Einstein, 24 June 1920.
78. A. Einstein, letter to N. Bohr, 11 January 1923.
79. A. Einstein, letter to B. Becker, 20 March 1954.
80. Bohr, Kramers, and Slater, op. cit. in n. 73.

81. The same is true for a sequel to this paper which Bohr wrote in 1925 (Z. f. Phys. **34**, 142). Schroedinger [Naturw. **36**, 720 (1924)] and especially Slater [Phys. Rev. **25**, 395 (1925)] did make attempts to put the BKS ideas on a more formal footing. See also Slater's own recollection of that period [Int. J. Qu. Chem. **15**, 1 (1967)].
82. Slater, Phys. Rev. **25**, 395 (1925).
83. Bohr, Kramers, and Slater, op. cit. in n. 73.
84. N. Bohr, Z. f. Phys. **13**, 117 (1923), esp. Sec. 4. A letter from Ehrenfest to Einstein shows that Bohr's thoughts had gone in that direction at least as early as 17 January 1922.
85. In Seelig, op. cit. in n. 74, p. 137.
86. Ibid.
87. Langevin and de Broglie, eds., op. cit. in n. 16, pp. 429, 436.
88. The title of Nernst's paper is (in translation): "On an attempt to revert from quantum mechanical considerations to the assumption of continuous energy changes." W. Nernst, Verh. d. Deutsch. Phys. Ges. **18**, 83 (1916).
89. A. Sommerfeld, *Atombau und Spektrallinien*, 3rd ed. (Vieweg, Braunschweig, 1922), p. 311. See M. Klein, Hist. St. Phys. Sci. **2**, 1 (1970) for similar speculations by other physicists.
90. W. Heisenberg, Naturw. **17**, 490 (1929).
91. Slater, Phys. Rev. **25**, 395 (1925).
92. Ibid.
93. W. Pauli, letter to N. Bohr, 2 October 1924. Born, Schroedinger, and R. Ladenburg were among the physicists who initially believed that BKS might be a step in the right direction.
94. Ibid.
95. On October 25, 1924, the Danish newspaper *Politiken* carried a news item on the Bohr-Einstein controversy. This led the editor of a German newspaper to send a query to Einstein (K. Joel, Letter to A. Einstein, 28 October 1924). Einstein sent a brief reply (A. Einstein, letter to K. Joel, 3 November 1924) acknowledging that a conflict existed, but also stating that no written exchanges between himself and Bohr had ensued.
96. W. Pauli, letter to N. Bohr, 2 October 1924.
97. A. Einstein, letter to K. Joel, 3 November 1924.
98. W. Pauli, letter to N. Bohr, 2 October 1924.
99. W. Heisenberg, letter to W. Pauli, 8 June 1924.
100. W. Pauli, letter to N. Bohr, 2 October 1924.
101. A. Einstein, letter to P. Ehrenfest, 31 May 1924.
102. A. Einstein, undated document in the Einstein Archives, obviously written in 1924.
103. A. Einstein, letter to M. Born, 29 April 1924, in the Born-Einstein Letters, op. cit. in n. 61.
104. W. Bothe and H. Geiger, Z. f. Phys. **26**, 44 (1924).
105. W. Bothe and H. Geiger, Naturw. **13**, 440 (1925), and Z. f. Phys. **32**, 639 (1925).
106. Z. Bay, V. P. Henri, and F. McLennon, Phys. Rev. **97**, 1710 (1955).
107. A. H. Compton and A. W. Simon, Phys. Rev. **26**, 889 (1925).
108. See e.g. W. Pauli, letter to H. A. Kramers, 27 July 1925. Pauli's description of BKS, written early in 1925, is found in Pauli, op. cit., in n. 51, pp. 83–86.
109. N. Bohr, letter to R. H. Fowler, 21 April 1925, quoted in Steuwer, op. cit. in n. 28, p. 301.
110. N. Bohr, Z. f. Phys. **34**, 142, Appendix (1925).
111. S. N. Bose, letter to A. Einstein, 4 June 1924.
112. W. A. Blanpied, Am. J. Phys. **40**, 1212 (1972).
113. In 1925 Einstein said of his work on quantum statistics: "That's only by the way" [quoted by E. Salaman, Encounter, p. 19 (April 1979)].
114. S. N. Bose, Z. f. Phys. **26**, 178 (1924).

115. A. Einstein, letter to P. Ehrenfest, 12 July 1924.
116. J. Mehra, "Satyendra Nath Bose," Biogr. Mem. FRS **21**, 117 (1975). Such a lack of awareness is not uncommon in times of transition. But it is not a general rule. Einstein's light quantum paper of 1905 is an example of a brilliant exception.
117. L. Boltzmann, *Vorlesungen über die Principe der Mechanik*, Vol. **1**, (Leipzig, Barth, 1897), p. 9; reprinted (Wiss. Buchges. Darmstadt, 1974).
118. Letter by E. Schroedinger to E. Broda, reproduced in E. Broda, *Ludwig Boltzmann* (Deuticke, Vienna, 1955), p. 65.
119. Cf. W. Heisenberg, Z. f. Phys. **38**, 411 (1926).
120. P. A. M. Dirac, Proc. Roy. Soc. **A 112**, 661 (1926).
121. I shall leave aside all reference to Fermi–Dirac statistics.
122. Bose, op. cit. in n. 114.
123. 4 June 1924.
124. P. Ehrenfest, letter to A. Einstein, 7 April 1926.
125. See e.g. J. M. Jauch and F. Rohrlich, *The Theory of Photons and Electrons* (Addison-Wesley Advanced Book Program, Reading, Mass., 1955), p. 40.
126. A. Einstein, letter to P. Ehrenfest, 12 April 1926.
127. Bose, op. cit. in n. 114.
128. The $N!$ is irrelevant since it only affects C in Eq. (14.22). The constant C is interesting nevertheless. For example, its value bears on the possibility of defining S in such a way that it becomes an extensive thermodynamic variable. The interesting history of these normalization questions has been discussed in detail by M. Klein [Proc. Kon. Ak. W. Amsterdam **62**, 41, 51 (1958)].
129. A. Einstein, S. B. Preuss. Akad. Berlin, p. 3 (1925).
130. A. Einstein, S. B. Preuss, Akad. Berlin, p. 261 (1924).
131. $A^{-1} = \exp(-\mu/kT)$, where μ is the chemical potential. Einstein, of course, never introduced the superfluous λ^s in these curly brackets. In Eqs. (14.33)–(14.38) I deviate from Einstein's notations.
132. Op. cit. in n. 130.
133. All technical steps are now found in standard textbooks.
134. Op. cit. in n. 129.
135. Ibid.
136. F. London, Nature **141**, 643 (1938).
137. In an article "Einstein and the Quantum Theory," Rev. Mod. Phys. **51**, 863 (October 1979).
138. G. E. Uhlenbeck, *Over statistische methoden in de theorie der quanta*, Ph.D. thesis (Nyhoff, the Hague, 1927).
139. Cf. W. H. Keesom, *Helium* (Elsevier, New York, 1942).
140. Op. cit. in n. 136.
141. Op. cit. in n. 114.
142. Op. cit. in n. 130.
143. Op. cit. in n. 129.
144. Ibid.
145. Ibid.
146. Ibid.
147. Except for a brief refutation of an objection to his work on needle radiation [Z. f. Phys. **31**, 784 (1925)]. I found a notice [S. B. Preuss, Akad. Berlin, p. 543 (1930)] by Einstein in 1930 announcing a new paper on radiation fluctuations. This paper was never published, however.
148. A. Einstein, S. B. Preuss, Akad. Berlin, p. 18 (1925).
149. L. de Broglie, Comp. Rend. Paris **177**, 507 (1923), and Comp. Rend. Paris. **177**, 548 (1923).

150. E. Schroedinger, Ann. d. Phys. **79**, 361 (1926).

151. W. Heisenberg, Z. f. Phys. **33**, 879 (1926).

152. The other ones I know of are L. Brillowin [J. de Phys. **2**, 142 (1921)], M. Wolfke [Phys. Z. **22**, 315 (1921)], W. Bothe [Z. f. Phys. **20**, 145 (1923)], H. Bateman [Phil. Mag. **46**, 977 (1923)] and L. S. Ornstein and F. Zernike, Proc. Kon Wet. Amsterdam **28**, 280 (1919).

153. L. de Broglie, Comp. Rend. Paris **177**, 630 (1923).

154. L. de Broglie, preface to his reedited 1924 Ph.D. thesis, *Recherches sur la Théorie des Quanta* (Masson et Cie, Paris, 1963), p. 4.

155. Op. cit. in n. 149, p. 507.

156. Op. cit. in n. 149, p. 548.

157. Op. cit. in n. 154.

158. Op. cit. in n. 129.

159. Ibid.

160. Op. cit. in n. 154.

161. W. Pauli in *Albert Einstein: Philosopher-Scientist*, op. cit. in n. 75, p. 156.

162. Op. cit. in n. 149, pp. 507, 548, 630.

163. E. Schroedinger, Phys. Z. **27**, 95 (1926).

164. P. Debye, Ann. d. Phys. **33**, 1427 (1910).

165. This derivation differs from Planck's in that the latter quantized material rather than radiation oscillators. It differs from the photon gas derivation in that the energy $nh\nu$ is interpreted as the nth state of a single oscillator, not as a state of n particles each with energy $h\nu$.

166. Op. cit. in n. 163.

167. E. Schroedinger, Ann. d. Phys. **79**, 361 (1926).

168. Op. cit. in n. 167, p. 373.

169. Op. cit. in n. 167.

170. E. Schroedinger, Ann. d. Phys. **79**, 734 (1926), fn. on p. 735.

171. A. Einstein, letter to C. Lanczos, 21 March 1942.

172. Recall that Heisenberg's first paper on this subject [Z. f. Phys. **33**, 879 (1926)] was completed in July 1925, Schroedinger's in January 1926 [Ann. d. Phys. **79**, 361 (1926)].

173. A. Einstein, letter to R. A. Millikan, 1 September 1925.

174. Quoted in J. Mehra, op. cit. in n. 116. It was not for Bose but for Dirac to answer this question. In his 1927 paper, which founded quantum electrodynamics, Dirac gave the dynamical derivation of expressions for the Einstein A and B coefficients [Proc. Roy. Soc. **A114**, 243 (1927)].

175. Op. cit. in n. 151.

176. According to Helen Dukas.

177. The letters by Heisenberg to Einstein referred to are now in the Einstein archives in Princeton.

178. Heisenberg begins his 30 November 1925 letter by thanking Einstein for his letter and then proceeds with a rather lengthy discussion of the role of the zero point energy in the new theory. This seems to be in response to a point raised by Einstein. It is not clear to me from Heisenberg's letter what Einstein had in mind.

179. Heisenberg remarked much later that "the attempt at interpretation by Bohr, Kramers and Slater nevertheless contained some very important features of the later, correct, interpretation [of quantum mechanics]" [W. Heisenberg in *Niels Bohr and the Development of Physics,* edited by W. Pauli (McGraw-Hill, New York, 1955), p. 12]. I do not share this view but shall not argue this issue beyond what has been said in Sec. 3.

180. A. Einstein, letter to M. Besso, 25 December 1925, in *Einstein, Besso Correspondence*, op. cit. in n. 13., p. 215.

181. A. Einstein, letter to Mrs. Born, 7 March 1926, in *The Born-Einstein Letters*, op. cit. in n. 61, p. 88.

182. A. Einstein, letter to E. Schroedinger, 26 April 1926, reproduced in *Letters on Wave Mechanics,* edited by M. Klein (Philosophical Library, New York, 1967).

183. A. Einstein, letter to M. Besso, 1 May 1926, in *Einstein, Besso Correspondence*, op. cit. in n. 13, p. 224.
184. M. Born, Z. f. Phys. **37**, 863 (1926).
185. Schroedinger retained reservations on the interpretation of quantum mechanics for the rest of his life. Cf. W. T. Scott, *Erwin Schroedinger* (U. of Massachusetts Press, Amherst, Mass., 1967).
186. A. Einstein, letter to M. Born, 4 December 1926 in *The Born-Einstein Letters*, op. cit. in n. 61, p. 90.
187. See *The Born-Einstein Letters*, op. cit. in n. 61, p. 91.
188. W. Heisenberg, Z. f. Phys. **43**, 172 (1927).
189. W. Heisenberg, letter to A. Einstein, 10 June 1927.
190. This is the title of a paper which Einstein submitted at the May 5, 1927 meeting of the Prussian Academy in Berlin. The records show that this paper was in print when Einstein requested by telephone that it be withdrawn. The unpublished manuscript is in the Einstein Archives. I would like to thank Dr. John Stachel for bringing this material to my attention.
191. W. Heisenberg, letter to A. Einstein, 19 May 1928.
192. N. Bohr, Nature **121**, 580 (1928). Einstein had been invited to this meeting but did not attend.
193. L. de Broglie, *New Perspectives in Physics* (Basic Books, New York, 1962), p. 150.
194. In a discussion with R. Jost, taped 2 December 1961. I am very grateful to R. Jost for making available to me a transcript of part of this discussion.
195. A. Einstein, in *Proceedings of the Fifth Solvay Conference* (Gauthier-Villars, Paris, 1928), p. 253.
196. In *Albert Einstein, Philosopher-Scientist*, op. cit. in n. 75, p. 199.
197. A. Pais, in *Niels Bohr*, edited by S. Rozenthal (North Holland, Amsterdam, 1967), p. 215.
198. In *Albert Einstein, Philosopher-Scientist*, op. cit. in n. 75, p. 199.
199. Op. cit in n. 195.
200. Op. cit. in n. 105.
201. L. Rosenfeld, in *Proceedings of the Fourteenth Solvay Conference* (Interscience, New York, 1968), p. 232.
202. In *Albert Einstein, Philosopher-Scientist*, op. cit. in n. 75.
203. A. Einstein, R. Tolman, and B. Podolsky, Phys. Rev. **37**, 780 (1931). The *Gedanken* experiment in this paper involves a time measurement. The authors take care to arrange things so that "the rate of the clock . . . is not disturbed by the gravitational effects involved in weighing the box."
204. The authors are "forced to conclude that there can be no method for measuring the momentum of a particle without changing its value." This statement is of course unacceptable.
205. A. Einstein, letter to Nobel Committee in Stockholm, 30 September 1931.
206. A. Einstein, Z. f. angew. Chemie **45**, 23 (1932).
207. A. Einstein, *On the Method of Theoretical Physics* (Oxford University Press, New York, 1933); reprinted in Phil. of Sci. **1**, 162, (1934).
208. Op. cit. in n. 192.
209. N. Bohr, Dialectica **2**, 312 (1948).
210. A. Einstein, B. Podolsky, and N. Rosen, Phys. Rev. **47**, 777 (1935).
211. N. Bohr, in *Albert Einstein, Philosopher-Scientist*, op. cit. in n. 75, p. 232. This stir reached the press. On 4 May 1935 the *New York Times* carried an article under the heading "Einstein attacks quantum theory," which also includes an interview with a physicist. Its 7 May issue contains a statement by Einstein in which he deprecated this release, which did not have his authorization.
212. N. Bohr, Phys. Rev. **48**, 696 (1935).
213. A. Einstein, Dialectica **2**, 320 (1948); in *Albert Einstein, Philosopher-Scientist*, op. cit. in n. 75, p. 1; in *Scientific Papers Presented to Max Born* (Hafner, New York, 1951), p. 33; in *Louis de Broglie, Physician et Penseur* (A. Michel, Paris, 1953), p. 5.
214. In *Scientific Papers Presented to Max Born,* op. cit. in n. 213.

215. In a letter written 22 December 1950 to Schroedinger (reproduced in *Letters on Wave Mechanics*), Einstein mentioned Schroedinger and von Laue as the only ones who shared his views. There were, of course, many others who at that time (and later) had doubts about the complementarity interpretation, but their views and Einstein's did not necessarily coincide or overlap. Cf. A. Einstein, letter to M. Born, 12 May 1952, in *The Born–Einstein Letters*, op. cit. in n. 61, p. 192. Note also that the term "hidden variable" does not occur in any of Einstein's papers or letters, as far as I know.

216. In *Albert Einstein, Philosopher-Scientist*, op. cit. in n. 75, p. 192.

217. A. Pais, in *Niels Bohr,* op. cit. in n. 197.

218. Typically, these tests deal with variants of the EPR arrangement, such as long-range correlations between spins or polarizations. I must admit to be insufficiently familiar with the extensive theoretical and experimental literature on these topics. My main guides have been a book by M. Jammer [*The Philosophy of Quantum Mechanics* (Wiley, New York, 1974)] and a review article by F. M. Pipkin [Adv. At. Mol. Phys. **14**, 281 (1978]. Both contain extensive references to other literature.

219. A. Einstein, letter to L. de Broglie, 8 February 1954.

220. A. Einstein, letter to N. Bohr, 4 April 1949.

221. W. Pauli, Neue Zürcher Zeitung, (January 12, 1958); reprinted in *W. Pauli's Scientific Papers*, Vol. 2 (Interscience, New York, 1964), p. 1362.

222. This section is meant to provide a brief record without any attempt at further explanation or reference to literature. It can be skipped without loss of continuity.

223. In this prediction (which does not include small contributions from muons and hadrons) the best value of the fine structure constant α has been used as an input: $\alpha^{-1} = 137.035\ 987(29)$. The principal source of uncertainty [T. Kinoshita, *New Frontiers in High Energy Physics* (Plenum Press, New York, 1978), p. 127] in the predicted value of $g - 2$ stems from the experimental uncertainties of α.

224. Pauli, op. cit. in n. 221.

225. A. Einstein, op. cit. in n. 207.

226. A. Einstein, *Lettres à Maurice Solovine* (Gauthier-Villars, Paris, 1956), p. 130.

227. In 1941 I received my Ph.D. in Utrecht with L. Rosenfeld. Some time later I went into hiding in Amsterdam. Eventually I was caught and sent to the Gestapo prison there. Those who were not executed were released a few days before V. E. day. Immediately after the war, I applied for a postdoctoral fellowship at Niels Bohr's institute as well as at the Institute for Advanced Study in Princeton, where I hoped to work with Pauli. I received a letter from Pauli saying he would support my application. I was accepted at both places and went first for one year to Copenhagen. When I finally arrived at Princeton in September 1946 I found that Pauli had in the meantime gone to Zürich. Bohr also came to Princeton that same month. Both of us attended the Princeton Bicentennial Conference (where P. A. Schilpp approached Bohr for a contribution to the Einstein biography). Shortly thereafter Bohr introduced me to Einstein.

My stay at the Institute had lost much of its attraction because Pauli was no longer there. As I was contemplating returning to Copenhagen the next year, Oppenheimer contacted me to inform me that he had been approached for the directorship of the Institute and to ask me to join him in building up physics there. I accepted. A year later I was appointed to a five-year membership at the Institute and in 1950 to professor. I remained at the Institute until 1963.

228. The speakers were J. R. Oppenheimer, I. I. Rabi, E. P. Wigner, H. P. Robertson, S. M. Clemence, and H. Weyl.

229. Cf. also A. Einstein, letter to M. Born, 22 March 1934, in *The Born–Einstein Letters*, op. cit., p. 121, and letter to A. Sommerfeld, 14 December 1946, in *Albert Einstein/Arnold Sommerfeld Briefwechsel* edited by A. Hermann (Schwabe Verlag, Basel), p. 121.

230. Cf. also M. Born, letter to A. Einstein, 10 October 1944, in *The Born–Einstein Letters*, op. cit. in n. 61, p. 155. W. Thirring has written to me (29 November 1977) of conversations with Einstein in which "His objections became even stronger when it concerned quantum field theory and he did not believe in any of its consequences."

231. A. Einstein, J. Franklin Institute **221**, 313 (1936).

232. In *Albert Einstein, Philosopher-Scientist*, op. cit. in n. 75, p. 1.

233. Op. cit. in n. 207.

234. In *Albert Einstein, Philosopher-Scientist*, op. cit. in n. 75, p. 1.

235. A. Einstein, letter to M. Besso, 8 August 1949, in *Einstein, Besso Correspondence,* op. cit in n. 13, p. 407.

236. *The Born–Einstein Letters*, op. cit. in n. 61, p. 131.

237. A. Einstein, letter to M. Born, 18 March 1948, in *The Born–Einstein Letters,* op. cit. in n. 61, p. 163.

238. A. Einstein, letter to M. Besso, 24 July 1949, in *Einstein, Besso Correspondence*, op. cit. in n. 13, p. 402.

239. R. S. Shankland, Am. J. Phys. **31**, 47 (1963).

240. A. Einstein, letter to E. Schroedinger, 9 August 1939, reproduced in *Letters on Wave Mechanics,* op. cit. in n. 182, p. 32.

241. A. Einstein, letter to M. Born, 18 March 1948 in *The Born–Einstein Letters,* op. cit. in n. 61, p. 162.

242. Einstein, in *Einstein, Philosopher-Scientist*, op. cit. in n. 75.

243. Einstein, op. cit. in n. 231.

244. Ibid.

245. A. Einstein, Ann. d. Phys. **17**, 891 (1905), and Ann. d. Phys. **18**, 639 (1905).

246. A. Einstein, letter to Conrad Habicht, undated but probably written in March 1905, reproduced in C. Seelig, op. cit in n. 74.

247. Ann. d. Phys. **17**, 891 (1905).

248. Ann. d. Phys. **18**, 639 (1905).

249. Phys. Z. **10**, 817 (1909).

250. Phys. Z. **18**, 121 (1917).

251. Op. cit. in n. 16.

252. A. Einstein, S. B. Preuss. Akad Berlin, p. 359 (1923).

253. In *Albert Einstein, Philosopher-Scientist* op. cit. in n. 75.

254. A. Einstein, letter to M. Born, 3 March 1920, in *The Born–Einstein Letters*, op. cit. in n. 61, p. 26.

255. Phys. Z. **18**, 121 (1917).

256. *The Born–Einstein Letters*, op. cit. in n. 61, p. 88.

257. A. Einstein, letter to M. Besso, 5 January 1924, in *Einstein, Besso Correspondence*, op. cit. in n. 13, p. 197.

258. A. Einstein, letter to M. Born, 12 May 1952, in *The Born–Einstein Letters,* op. cit. in n. 61.

259. I note in passing that in 1925 Einstein gave a helping hand to Uhlenbeck and Goudsmit in the explanation of the origins of the spin-orbit coupling of electrons in atoms. [G. E. Uhlenbeck, Physics Today **29**, 43 (June 1976)].

260. Einstein, op. cit. in n. 207.

261. A. Einstein, Physica **5**, 330 (1925).

262. The proof involves the application of time reversal to the combined equations. In related context the existence of the $(\pm e, m)$ solutions was first noted by Pauli in 1919 [Phys. Z. **20**, 457].

263. A. Einstein, letter to M. Besso, 28 July 1925, in *Einstein, Besso Correspondence*, op. cit. in n. 13, p. 209.

264. A. Einstein, letter to M. Besso, August 1942, in *Einstein, Besso Correspondence*, op. cit., p. 366.

265. A. Einstein, letter to M. Besso, 16 August 1949, in *Einstein, Besso Correspondence*, op. cit., p. 409.

266. A. Einstein, letter to L. de Broglie, 8 February 1954.

267. A. Einstein and N. Rosen, Phys. Rev. **48**, 73 (1935).

268. A. Einstein, *On the Method of Theoretical Physics*, op. cit.

269. W. Heisenberg, letter to A. Einstein, 10 June 1927.

270. Ibid.

271. A. Einstein, letter to M. Born, undated, in *The Born–Einstein Letters*, op. cit. in n. 61, p. 180

272. A. Einstein, letter to M. Besso, 10 August 1954, in *Einstein, Besso Correspondence*, op. cit. in n. 13, p. 525.

273. A. Einstein, in *Helle Zeit—Dunkle Zeit* edited by C. Seelig (Europa Verlag, Zürich, 1956).

15. COMMENT ON "EINSTEIN ON PARTICLES, FIELDS AND QUANTUM THEORY"

Res Jost
Translated by Sonja Bargmann

1. Einstein's Historical Greatness

> It simply wasn't possible. The only thing that was really known about quantum theory was Planck's formula. Period!
>
> — Otto Stern, Reminiscences from Zurich, 1913

> I won't conceal either that an American colleague maintained that Berlin was altogether deteriorating. . . . The same colleague also contended the situation would have improved if I hadn't refused the call to Berlin.
>
> — Ludwig Boltzmann, 1905

The heading is a desperate attempt to put into German the English word "apartness." It has become a cliche. Actually, the concept "historical greatness" must be understood here in the sense of Jacob Burckhardt. As a justification, let me adduce his statement: "Not an explanation, but a further paraphrase of greatness can be obtained from this point on by the words: *uniqueness, irreplaceability.*"

Unique and *irreplaceable* were indeed Einstein's endeavors also in his papers about the quantum theory, more specifically about light quanta; to be more explicit: without his paper, "On a Heuristic Point of View Concerning the Creation and Transformation of Light," of March 17, 1905, the development of physics in our century is unthinkable.[1]

This statement ought to be rigorously proven, which I am unable to do. I can only illustrate it. First of all, it is not meant to and cannot detract from Planck's achievement. Anyone who takes the trouble to trace in some detail Planck's struggle for the correct radiation formula can only revere this man. And Planck comprehended quite clearly the epochal significance of his achievement, in particular of the discovery of his constants k and h: he presented them in powerful language. His words remained unheard. A glance at the literature confirms that. The discovery of

Harry Woolf (ed.), Some Strangeness in the Proportion: A Centennial Symposium to Celebrate the Achievements of Albert Einstein
ISBN 0-201-09924-1

the quantum of action in December 1900, although noticed, did not produce a sensation. Even Boltzmann, who in the beginning had still been polemically involved in Planck's endeavors, had lost interest and had retained his bad opinion of Planck, as can be seen from the quotation at the head of this section.[2]

If two decades later the scene has completely changed, this is only due to Einstein. Let us hear Walter Nernst in 1919:

> That Planck's radiation formula was a tremendous step forward, fertilizing by its consequences the theoretical physics of the last two decades in an almost unique way, can no longer be doubted by anyone; it is another question, however, whether it is the last word on the radiation laws of black bodies.
>
> For the further development of our experimental methods . . . as well as our theoretical interpretations (in particular of the quantum hypothesis) the question whether Planck's formula is a strict law of nature is of the deepest interest.
>
> . . . but experimental physics has hardly achieved anything in this area, while theoretical physics, on the other hand, has filled our literature with a great many papers which, strictly speaking, would not make any sense if Planck's formula were not an exact law of nature.[3]

Each reader will have his own ideas about Nernst's attempt to improve on Planck's formula. It is not altogether wrong in principle because, indeed, the photon gas is not, strictly speaking, an ideal quantum gas. We, however, will keep in mind that Planck achieved his well-deserved fame only through Einstein's work.

It is true, Einstein's revolutionary conclusions from Planck's formula met mostly with opposition, and this went on for more than two decades. But it is the mark of a great individual that everyone has to come to terms with his achievements, that indifference to him is impossible. One may argue whether it was the success of his special theory of relativity that produced the astonished interest in the paper on light quanta: what is singularly important is that *both* are due to Einstein.

But the resistance against the light quanta may need an explanation; for the experimental evidence in favor of this hypothesis, though of a qualitative nature, was overwhelming. In addition to Einstein's own arguments here is a statement by J. J. Thomson:

> These results [on the ionization of gases by ultra-violet light] can however be reconciled by the view stated above that a wave of light is not a continuous structure, but that its energy is concentrated in units (the places where the lines of force are disturbed) and that the energy in each of these units does not diminish as it travels along its line of force. . . . We should know at once the coarseness of the structure corresponding to light of any intensity if we knew the amount of energy in each unit of light. . . . The greater the frequency of the light the greater is the energy in each unit. . . . The coarseness of structure of light even of feeble intensity is probably almost as nothing in comparison with that of the γ rays. . . . Thus the units in the γ rays will, unless the intensity is exceedingly great, be very widely separated; as these units possess momentum as well as energy they will have all the properties of material particles, except that they cannot move at any other speed than that of light.[4]

These are highly interesting statements, which can be traced back to Thomson's Silliman lectures at Yale University in 1902, but comparing them with Einstein's precise and inexorable conclusions from Wien's limiting law one will be dissatisfied also in this case with E. T. Whittaker's presentation[5] of Einstein's role in the development of physics. Indeed, Einstein's analysis of this one equation of Planck's for the spectral distribution of blackbody radiation, which extends over twenty years and in which he finally derives *all* properties of the photon (apart from the rather trivial spin), has as far as I know no equal in the history of the sciences. The basic principles of this analysis — which often may have taken place in a mood close to despair, as the quotation of Otto Stern at the beginning of this section indicates — are expressed with all desirable clearness and brevity in the Proceedings of the Solvay Congress in 1911 in Einstein's contribution to the discussion:

> Einstein: I take it we all agree that the so-called quantum theory in its present state, though a useful tool, is not a theory in the proper sense of the word, . . .
>
> The question arises which general laws of physics we can still hope to find valid in the field with which we are concerned. To start with, we'll all be agreed that the energy principle must be retained.
>
> Another principle whose validity must be retained under all circumstances, it seems to me, is Boltzmann's probability definition of entropy. The faint glimmer of theoretical light which is spread today over the statistical equilibrium states in processes of an oscillatory character we owe to this principle.[6]

But we are evading the question of why the light quantum hypothesis met with such general rejection. One reason could be that Johannes Stark was the first to believe that he could profit by availing himself of the foreign theory. We shall not concern ourselves with this truly dreadful person any more than is absolutely necessary and refer to the literature.[7] That much is clear: a man of this kind causes damage to anything and endangers anybody utilized by him for his purposes. Let me quote as an example the concluding sentence of a polemical article of Arnold Sommerfeld's against Stark: "At the same time, this reinforces in my opinion our confidence in the validity of the electromagnetic theory also for the elementary processes of the electric field, which seemed to have become shaken to some extent by the latest speculations about light quanta."[8]

But basically such uproar is again only a paraphrase of historical greatness, which consists in turning the development in the face of very great opposition (and danger) into new directions. *He,* the outsider, with full insight into all human weaknesses and with forebearance for them as long as they did not originate from sheer evil, performed this work carefully and *without sacrificing an iota of humanity.* Here we find a quality extending beyond historical greatness — so often an attribute of brutal tyrants: He knew that he was serving truth. The analysis of Planck's formula is a selfless search for truth.

Before I continue, let me put things into perspective. Einstein was *continuously* concerned with the quantum riddle; therefore, his written utterances about it can only represent elucidations of momentary situations. The continuum of the development is hidden from us; it does not leave a trace for posterity. The passage quoted by Pais from the letter to Paul Ehrenfest of May 31, 1924, illumines this situation: Einstein says there that the basic idea of the theory of Bohr, Kramers, and Slater has long been an old acquaintance of his, "whom he doesn't consider, however, to be the genuine article."

2. The Momentum of the Photon

> Another thing was that Einstein always wracked his brain about the law of radioactive decay. He constructed such models . . . .
>
> — Otto Stern, Reminiscences from Prague, 1912

I follow the terminology of my friend Pais, using the term "photon" as an abbreviation for "light quantum with energy $h\nu$ and momentum $h\nu/c$ in the direction of radiation."

In general, it is justified to consider Einstein's papers about the quantum theory of radiation from the years 1916–17 "as among the most important and most profound to influence the development of modern physics."[9] Nevertheless, none of the thoughts expressed there entered into Einstein's summing up of his life in 1949.[10] In those "Autobiographical Notes" Einstein remembered with obvious relish an earlier argument in favor of the momentum of the photon: the Brownian motion of a selectively reflecting mirror in the field of radiation.[11] Was he held back by an unease about the time and direction of the elementary process (of spontaneous emission of light) being left to "chance," to the uncanny, which had already been a preoccupation of his in Prague in 1912?

The decisive reversal in favor of the photon is subsequently brought about by the important paper by A. H. Compton[12] on the scattering of X rays by (almost free) electrons, whose significance I see most of all in the experimental sphere. Compared to the supreme mastery exhibited in the experiments, the kinematic computations of the collision between electron and photon is elementary, and it is not improbable that it has been performed apart from Debye also by others, so presumably by H. A. Kramers in Copenhagen (according to a remark by W. Pauli to me). Clarifying as Compton's paper was with respect to the kinematics of the photon, it produced at the same time some confusion with respect to another question. It contains a formula for the differential cross-section, which, though derived theoretically from contradictory assumptions, seems to be brilliantly confirmed by the experiment. A. H. Compton remarks that there is no frequency shift in the center of mass system. From this he would like to conclude on the basis of the correspondence principle that Thomson's formula for the cross-section holds. The scattering thus produced a spherical wave originating in the center of mass.[13] See Fig. 15.1.

One can now interpret in a natural way the theory of Bohr, Kramers, and Slater, extensively discussed by A. Pais, as an attempt to introduce a logical connection with Compton's strange method of computation, which mixes classical and quantum-theoretical points of view.

Leafing through Bohr's papers, I am seized by a feeling of depression with no way out, which is relieved only by the *postscript* of the second paper quoted above written after the experiments of W. Bothe and H. Geiger.[14] The decision had been made, against Bohr, Kramers, and Slater and in favor of Einstein. Niels Bohr writes:

> Giving up space-time models is characteristic for the formal treatment of problems in radiation theory as well as in mechanical thermodynamics, which is being attempted in the latest papers of de Broglie's and Einstein's. In particular, in view of the perspective which these papers are opening up, I have . . . decided to publish my paper unchanged, although the ideas on which it is based may now well seem hopeless.[15]

Einstein's papers on the theory of ideal quantum gases will be the subject of Sec. 4.

3. Einstein Facing the Riddle of the Quantum

> You can hardly imagine how hard I tried to find a satisfactory mathematical treatment of quantum theory. So far without success.
>
> — Albert Einstein to J. Stark on July 31, 1908

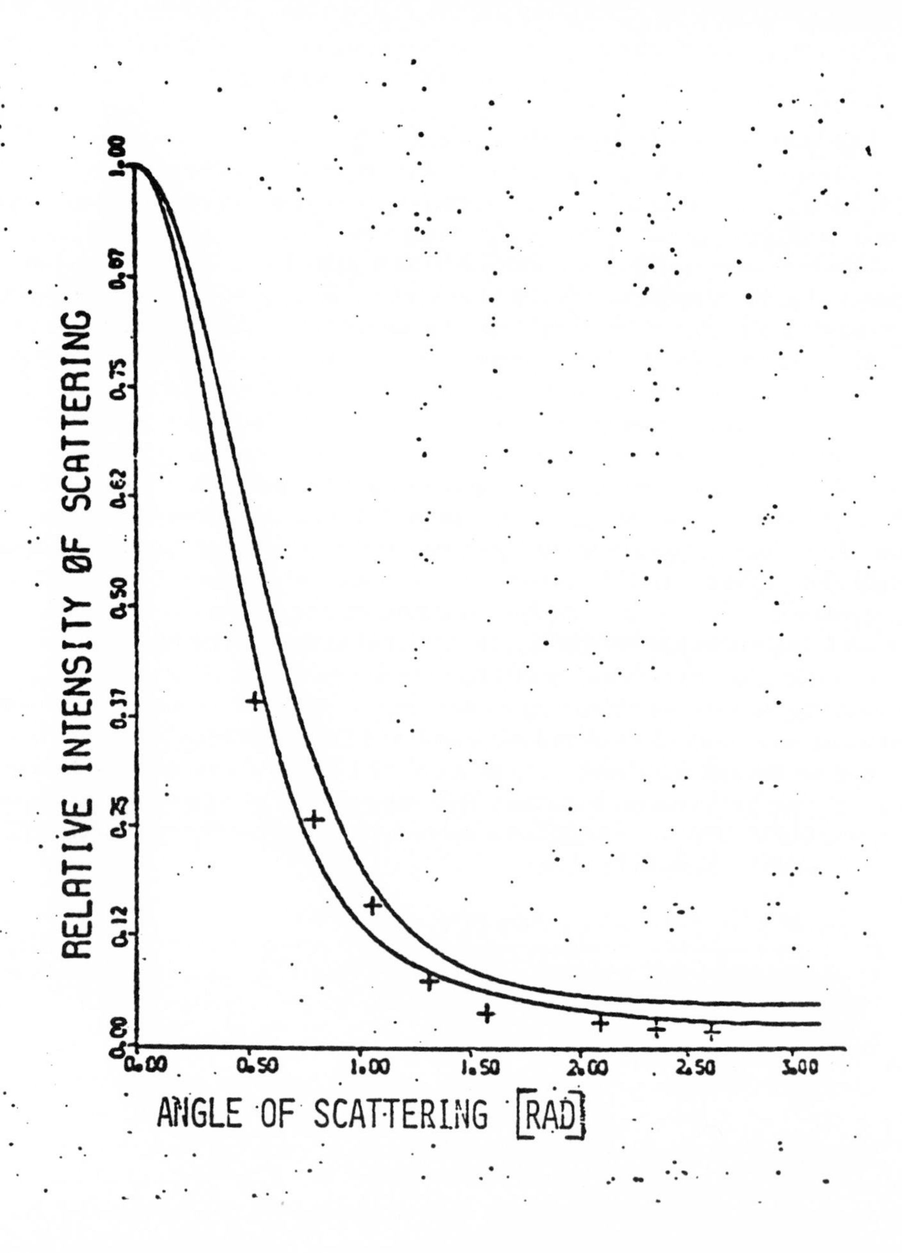

Fig. 15.1. *Lower curve:* Compton's cross-section. *Upper curve:* Cross-section according to Klein and Nishina. The cross-sections are normalized to 1 at $\theta = 0$. The crosses are Compton's measured values. It is assumed that $\alpha = 1.19$.

Einstein's lecture at the Eighty-first Congress of the German Natural Scientists and Physicians in Salzburg in the fall of 1909 concludes with the following sentences:

> Still, for the time being it seems to me most natural to assume that the occurrence of the electromagnetic fields of light is just as much linked to singular points as is, according to the electron theory, the occurrence of electrostatic fields. It is conceivable that in such a theory the total energy of the electromagnetic field could be considered localized in these singularities, just as in the old theory of action at a distance. I imagine each such singular point to be surrounded by a field of force, which essentially has the character of a plane wave and whose amplitude decreases with the distance from the singularity. If many of these singularities exist in intervals small compared to the dimensions of the field of force of a singularity, the fields of force will be superimposed upon each other and in their totality will produce an undulatory field of force which may hardly differ from an undulating field of the present-day electromagnetic theory of light. That such a model need not be taken seriously, as long as it does not lead to an exact theory, hardly requires any special mention.[16]

Although this picture cannot really be considered scientifically valid, still it exerts a strange fascination. Its effect, it seems to me, can still be found many years later in Louis de Broglie's ideas. At the Solvay congress in 1927 he says with respect to matter waves: "First, the author of this report always assumed that a material point occupies a well-defined position in space. Hence, the amplitude *f* would have to include a singularity or, at least, exhibit abnormally high values in a very small area."[17] Einstein confirms him in the general discussion: "but that at the same time the particle is localized during its propagation. I think M. de Broglie is right to search in this direction."[18]

But Einstein must also play godfather to Max Born's decisive breakthrough to the probability interpretation of Schroedinger's wave function. There, though without reference, the following opinion or statement is attributed to him:

> I would like to try to give here a third interpretation and to test its usefulness in collision processes. I am referring here to a remark of Einstein's concerning the relationship of wave field and light quanta; he said essentially that the waves exist only in order to show the way to the corpuscular light quanta, and in this sense he spoke of a "ghost field." This determines the probability for a light quantum, the carrier of energy and momentum, to follow a certain path: the field proper, however, possesses neither energy nor momentum.[19]

Whether and where Einstein ever said this, is not known to me,[20] and as the further development shows, he may very well not have accepted the descent of Born's theory from his own. But at that time everybody liked to refer to the unattainable model.

Einstein seems to have broken away from the aforementioned picture more than any other physicist. The reason for it is evidently his unprecedented and finally successful struggle for the general theory of relativity. Somewhere hidden in the background Einstein's preoccupation with Bohr's ideas[21] also plays a part.

In 1923, there appeared in the Berlin Proceedings a paper of Einstein's putting the rhetorical question: "Does the Field Theory Offer Possibilities for the Solution of the Quantum Problem?"[22]

It contains "Einstein's vision" to which we shall now turn.

The paper consists of two sections. The first, headed "General Remarks," presents the framework within which a solution to the problem of the quantum can be expected. The second describes a particular model within this framework.

The first section attempts nothing less than to reverse a main tendency of the development of the modern exact sciences. There we read:

> The essential element in the theoretical development up to now is the fact that it employes differential equations which in a four-dimensional space-time continuum uniquely determine the course of events if it is known for a space-like section. The unique determination of the continuation of the process in time by partial differential equations is the method by which we do justice to the law of causality. . . . According to the existing theories, the initial state of a system can be freely chosen; differential equations will then provide the continuation in time.[23]

We are not changing the content of the statement, we are only shifting the emphasis, if we are inclined to perceive "in the development up to now" (that is, since Newton) a noteworthy renunciation with respect to the predictive power of the mathematical description of nature. Side by side with laws, employed in the description of the time development, we find arbitrariness in the form of freely chosen initial conditions. Our theories can never explain reality; they only reduce it to the initial conditions on a spacelike section. As an example of an evidently relevant situation, which is, however, ignored by classical celestial mechanics, I mention the Titius–Bode rule[24] concerning the planetary distances whose predictive power led to the discovery of the planetoids. (I am quite aware that this example is open to criticism in light of the *historical* age of the planetary system and the modern knowledge and expectations from *qualitative* celestial mechanics.) I do not know to what extent the renunciation had been felt at the time. Presumably, any hesitations were swept away by the pleasure in computing (that is, by opportunism); for as a recompense for giving up explaining reality a beautifully simple, though abstract, mathematics was obtained compared with the previous state of affairs. Moreover, the free choice of the initial conditions reflects the penchant for experimental research.

With this schema, venerable by virtue of age and success, Einstein now interferes in a highly original manner by trying to restrict the possible initial conditions by differential equations. He explicitly refers to Bohr's theory of stationary states distinguished by quantum conditions. Thus Einstein is searching for a theory in which the role of chance (due to restrictions on the initial conditions) is being curtailed. It is well known that present-day quantum mechanics, to the extent to which it may at all be used as a comparison, took exactly the opposite path.

Of course, Einstein considers only realistic models, which include gravitational theory. This leads to intractable mathematical problems, and for the first time we hear the complaint and the hope in the following sentences:

> As is always the case in the theory of general relativity, in this case, too, it is difficult to obtain from the equations information about their solutions which can be compared with the conclusive results obtained from experiments, in this case specifically with those of quantum theory. . . . My exposition has served its purpose if it induces mathematicians to collaborate in this endeavor and convinces them that the road taken here is practicable and that it is absolutely necessary to think it through to its end.[25]

Einstein never deviated from this conviction to any noticeable extent.

4. An Erratic Block: The Papers on the Ideal Quantum Gas

> That this decisive step forward [of Kepler's] did not leave any traces in Galileo's lifework is a grotesque illustration of the fact that creative individuals are often not receptive to the ideas of others.
>
> — Albert Einstein, Foreword to Galileo

The two papers on the quantum theory of the mono-atomic ideal gas from the years 1924 and 1925 occupy a separate position in Einstein's work.[26] Belonging as they do to analysis rather than to synthesis and speculation, they nevertheless are closest to quantum mechanics, as accepted today. Indeed, they reach it completely for a special, not at all trivial, model. In addition, they originated directly from an outside stimulus, which is quite a rare event for Einstein. It is true, an infallible sense for what is correct and essential was required in order to recognize the significance of S. N. Bose's manuscript. The papers contain the wave-corpuscle duality for (material) atoms and the reference to the thesis of Louis de Broglie.

Without these papers, in particular without the "infinitely far-seeing comments" on pages 9–11 of Einstein's second paper, Schroedinger's wave mechanics would not have come about. The content of those pages proves however that Einstein (to quote Schroedinger again) "has taken the de Broglie–Einstein undulation theory [more] seriously"[27] than is immediately evident from the text; to wit, he computed the interference fluctuations for the quantum gas for which he obviously had to use the wave equation for a nonrelativistic force-free mass point, or at least its solutions.

Here the main question arises: Why didn't Einstein himself discover wave mechanics? To try to answer this question is hardly within the competence of a nonexpert. He is amazed at the apparent ease with which the principal and the technical problems are set up and solved — and is disappointed when the author, after apparently bringing his goal within reach, suddenly stops in order to consolidate what has been achieved with general considerations left for a third paper, "On the Quantum Theory of the Ideal Gas."[28] Einstein's letter to M. Besso of June 5, 1925, contains the sentences: "The quantum problem seems to me to require something like a particular scalar and I found a plausible way to introduce it. I also have developed a quantum theory of the ideal gas."[29] These sentences may indicate that the desire for a fusion of quantum theory and general relativity theory was an obstacle, but it is not very likely. As is well known, the bridge from classical physics to wave mechanics connects nonrelativistic mechanics with nonrelativistic quantum mechanics. A connection between relativistic field theory and relativistic quantum mechanics does not exist in a proper sense: the foundations are lacking.

5. The Separation

> If God kept hidden in his right all truth and in his left only the always active urge for truth, though with the proviso that I would always be mistaken, and said to me: choose! I would humbly fall into his left.
>
> — Gotthold Ephraim Lessing, 1778

In the same year 1925 Werner Heisenberg, Max Born, and Pascual Jordan conceived matrix mechanics. The most candid judgment of Einstein's of this development is probably to be found in the letter to M. Besso of December 25, 1925, and reads as follows: "The most interesting result that the theory has lately produced is the theory of quantum states of Heisenberg–Born–Jordan. It is veritable black magic, in which infinite determinants (matrices) take the place of the Cartesian coordinates. Highly ingenious, and due to its great complexity, sufficiently protected against proof of incorrectness."[30] The last sentence is easier to understand if one knows that in January of the same year Einstein refuted with ease, on two pages, a rebuttal by Pascual Jordan of his famous paper of 1916–17 "On the Quantum Theory of Radiation."[31]

Although it was still possible to be skeptical about the correctness of matrix mechanics because of its lack of transparency, Schroedinger's wave mechanics could only be doubted if one had a false recollection of the Schroedinger equation. This is exactly what happened to Einstein, as is shown by his letters to Erwin Schroedinger of April 16 and 22, 1926.[32] That Einstein guessed the *correct* Schroedinger equation on the basis of the requirement of energy additivity for the composition of uncoupled systems is the most impressive testimony to his insight into the structure and consistency of Schroedinger's theory. When Erwin Schroedinger in March of the same year had finally proven the equivalence of his quantum mechanics and that of Heisenberg–Born–Jordan, no further objection to the latter was possible on logical grounds. And with the accumulation of empirical confirmations of nonrelativistic quantum mechanics (the only one considered here so far) it was no longer possible to doubt the practical success of the theory.

Nevertheless, Einstein, de Broglie, Planck, von Laue, Schroedinger on the one hand, and Bohr, Born, and almost all of the then young generation on the other, went their separate ways.

The renunciation of predictive power, which characterized the new dynamics, was for Einstein much too high a price to pay for the unquestionable successes of the modish-modern theory. The epistomological speculations and philosophisms, stimulated by quantum mechanics, which occasionally overshot their goal, seemed to him unjustified. The paradoxes of the new mechanics worried him, and not only him.[33]

For the opposition, the rejection was according to the person in question a cause for regret, even for sadness, up to a hardly concealed triumph (e.g., with respect to Schroedinger). For Bohr the victory over Einstein after the "big debate" (about which, however, we possess only one-sided information) represented a high honor, which he remembers with pleasure in his later years.

Einstein remained true to himself, faithful to his search; in his last years he may have become resigned. His personal relationship to Bohr seems to have been rather that of perplexed incomprehension. To Hermann Weyl he said, "It is strange with Bohr. If you can hear him, the mind can't follow; if you think you can understand him, one's hearing fails." (I owe Einstein's remark to an oral communication from Hermann Weyl.) But still, in his advanced years he recognized his own image in Bohr. Anyone with any understanding is well aware that the words, "That this insecure and contradictory basis [of the old quantum mechanics] was sufficient to enable a person of Bohr's unique instinct and sensibility to discover the major laws of the spectral lines and the electron shells of atoms, as well as their significance for chemistry, appeared like a miracle to me — and still does so today. This is the highest musicality in the sphere of thought"[34] mutatis, mutandis characterize Einstein himself.

It can hardly be a question at this point of my describing once more the historical events in *my* way based on the same source material. I have nothing more to add to the excellent presentation of my friend A. Pais. I remember fondly my few encounters with Einstein, but for the public they are of no interest; the perusal of a comprehensive exposition of the philosophy of quantum mechanics confuses me — I have a hard time remembering even the names of the various quantum-mechanical churches and conventicles. Thus I shall restrict myself to some remarks that are loosely connected with Einstein's concern.

5.1 *On Nonrelativistic Quantum Mechanics*

Some paradoxes of quantum mechanics are due to the application of the quantum laws to macroscopic systems. Examples are Schroedinger's cat experiment (abbreviated by Einstein to: "Radioactive atom + Geiger counter + amplifier + powder charge + cat in a box, with a ψ-function of the system which describes the cat alive as well as decomposed into its parts."[35]), discussion of the orbits of celestial bodies, the theory of the measuring process. Here it seems to me that frequently it is disregarded in an impermissible manner that macroscopic bodies, for any purpose outside of mathematics, must be considered as composed of infinitely many atoms. Without this limit of Boltzmann's[36] for infinitely many degrees of freedom no clarity will ever be reached. It is true, as far as I know, that this limiting process of Boltzmann's has been rigorously carried out for only *one* realistic (and classical) system, by Oscar Lanford.[37] A discussion of the measuring process from this point of view is contained in Klaus Hepp.[38]

Notwithstanding all learned qualifying statements, it can be said with certainty that a coherent superposition of a state of the living cat and a state of the dead cat cannot be distinguished in any reasonable way from the corresponding (incoherent) mixture. This does away with the paradox that a curious glance into the box will "reduce the wave packet" and liberate the cat into reality from being suspended between life and death. The thermodynamic limit might correspond reasonably exactly to the "amplifier" in Einstein's abbreviation.

Next, a comment concerning the question whether quantum mechanics can be embedded into a complete theory without considerable changes in its assertions. I follow here a fine paper by S. Kochen and E. P. Specker.[39]

Years ago I learned from Helen Dukas that P. A. M. Dirac's book, *Quantum Mechanics,* was Einstein's constant companion, even on vacation. Dirac's conceptual system of observables and states has become exemplary. I follow him here. If A is an observable and α a state, then a unique probability distribution $w(A, \alpha; \Delta)$ will result from the theory, which gives for every interval Δ the probability to find for a measurement of A in the state α a value in Δ. One distinction between classical physics and quantum physics consists in the fact that in the former, there are states α_0 in which all *observables* assume exact values, hence where all probability distributions $w(A, \alpha_0, \cdot)$ are concentrated in *one* point $\alpha_0(A)$. From these *classical-extremal* states, *all* states can be generated by formation of *centers of gravity*, just as the points of a triangle can be represented as the centers of gravity of weights in the corners: the proportion of the weights is uniquely determined by the triangle point. In quantum mechanics, there is not a single state in which all observables assume exact values. The space of the states, moreover, resembles a circular disk rather than a triangle: although each point can be interpreted as the center of gravity of weights in the boundary, the proportion of the weights is not uniquely determined.

The question of the completion of quantum mechanics amounts, roughly speaking, to increasing the number of the states of quantum mechanics by adding classical-extremal states such that the classical situation of the "triangle" is obtained. That is always possible to achieve, but only by a completely unacceptable sacrifice (if the dimension of the "Hilbert space" exceeds three), which I now would like to explain. I shall start with an example: Assuming that I know the probability distribution of A in the state α, I certainly also know the probability distribution of A^2 in the state α, since the probability of finding a value for A^2, say, in the interval [0,4] must be equal to the probability of finding a value for A in the interval [−2,2]. *In general:* it must be possible (and so it is in quantum mechanics) to speak of the function $f(A)$ of an observable and to *define* it by

$$w(f(A), \alpha; \Delta) := w(A, \alpha, f^{-1}(\Delta))$$

(valid for all Δ and α).

This definition leads to a contradiction if one tries to add classical-extremal states. Extending the system of the quantum-mechanical states by a classical-extremal state α_0 one cannot demand that $f(A)$ assume the exact value $f(\alpha_0(A))$ when A has the value $\alpha_0(A)$. No exotic functions f are necessary: quadratic functions will do. Nor is it necessary to test the displayed formula for very many observables *A*, thirty-nine (and probably fewer) will suffice.[40]

Thus the desire for a classically complete theory will not easily be satisfied on the level of nonrelativistic quantum mechanics.

The experience with quantum theory has led most physicists of an intermediate generation to the more or less conscious conviction that an extension of the mathematical description of nature to a qualitative new area forces us to give up in principle some of the predictive power of the older theory. This conviction is coupled with the hope for a simpler, though more abstract, mathematical structure. And indeed, it is much easier to solve the qualitative problems of nonrelativistic quantum mechanics than the corresponding problems of classical mechanics.

Whether or not there exist hidden parameters determining, for example, the exact point in time of a radioactive decay can be asked from nature itself, as we know. The exclusion principles are only applicable to truly indistinguishable systems, which again was first recognized by Albert Einstein.[41] But the band spectrum of the tritium molecule T_2 shows that the radioactive nuclei obey the exclusion principle,[42] hence not even nature can predict the exact time of decay of a radioactive nucleus from hidden qualities.

There seems to be significant agreement that nonrelativistic quantum mechanics within its framework is a theory hardly to be improved upon. The question whether a discussion of its foundations is meaningful cannot be answered unequivocably. Einstein would probably have answered: no. The tremendous literature on this subject does not exactly gladden the heart of the nonspecialist. There always comes to my mind ''Mach's little horse,'' about which Einstein writes: ''But it is incapable of bearing a foal, all it can do is to eradicate vermin.''[43]

5.2 *Relativistic Quantum Mechanics*

It seems that Einstein in particular rejected relativistic quantum mechanics. We still have no counterargument. It is true, as is demonstrated by the figures in Sec. 6 of Pais's paper, that quantum electrodynamics is within a clearly defined range of validity a *computational method,* which can hardly be improved upon. I cannot call it a theory at this point, because it is still not known whether this computational method ever computes anything rigorously. It is still uncertain whether the unification of relativity and quantum theory within electrodynamics is free from contradictions. ''If a complete system of theoretical physics consists of concepts, fundamental laws . . . and of conclusions to be reached by logical deductions,''[44] the consistency of the fundamental laws must certainly be assured. The situation does not lack a certain bitter irony. Einstein's attempts during the second half of his life frequently ran up against the *qualitative problem of the existence* of solutions (free of singularities) of partial differential equations. Our present renormalizable quantum field theories run up against the same qualitative problem of existence.

As far as the gauge theories themselves are concerned, it is known well enough that they have developed from Hermann Weyl's (unsuccessful) attempt to geometrize gravitation and the electromagnetic field in a unified structure. About this theory Weyl writes to Carl Seelig on May 19, 1952: ''I myself gave up this theory long ago, since its correct core, gauge invariance, has been saved as a principle and become part of quantum theory . . . tying the wave field of the electron to the electromagnetic field.''

Saving the correct core was achieved in the paper, ''Electron and Gravitation,''[45] which, if historical development did not proceed in a meandering way, could have formed the immediate point of departure for the modern non-Abelian gauge theories.

In answer to the congratulations to his seventieth birthday, Einstein wrote to Maurice Solovine on March 28, 1949:

> You seem to imagine that I am looking back with deep satisfaction on the work of a lifetime. But from close by it looks quite different. There isn't a single concept of which I am convinced that it will endure, and I feel uncertain that I am altogether on the right road. What concerns my contemporaries, they see in me at the same time a heretic and a reactionary, who kind of outlived himself. That, it is true, has to do with fashion and short-sightedness, but the feeling of inadequacy originates within. Well — it hardly can be otherwise if one is critical and honest, and a sense of humor and perspective keeps me in a state of equilibrium, notwithstanding the surrounding influences.[46]

Those are brave and humane words of the great seeker after truth. They also are an admonition to the physicist to remain critical and honest and active.

Never in the past fifty years has differential geometry entered into as close a link with the theory of elementary interactions as during the last years. The development, one may say, has taken place, though not away from quantum theory, still, toward Einstein. But let us not forget at this point Einstein's admonition, a complement to the quotation from Lessing's Theological Disputations, which is contained in a letter to Hermann Weyl of May 26, 1923:

> BUT ABOVE STANDS THE MARBLE SMILE OF INEXORABLE NATURE, WHO HAS ENDOWED US WITH MORE LONGING THAN MIND.

The present comment was finished in November 1978, based on a draft of the contribution of my good friend A. Pais. Unavoidable circumstances made it impossible for me to rewrite this paper, for which I am asking the reader's indulgence.

Acknowledgment

It is a great pleasure for me to express my warm thanks to Sonja Bargmann for her thoughtful and painstaking translation of this paper.

NOTES

1. A. Einstein, "Über einen die Erzeugung und Verwandlung des Lichtes betreffenden heuristischen Gesichtspunkt," Ann. d. Phys. (4) **17**, 132–48 (1905).
2. The quotation is from p. 408 of "Aus Reise eines deutschen Professors ins Eldorado," in Ludwig Boltzmann, Populäre Schriften (Leipzig, 1905), p. 404ff.
3. W. Nernst and T. Wulf, "Über eine Modifikation der Planckschen Strahlungsformel auf experimenteller Grundlage," Deutsch. Phys. Ges. Verh. **21**, 294–337 (1919). Einleitung von Walter Nernst.
4. J. J. Thomson, "On the Ionization of Gases by Ultra-Violet Light and on the Evidence as to the Structure of Light Afforded by Its Electrical Effects," Proc. Cambridge Phys. Soc. **14**, 417–24 (1907).
5. E. T. Whittaker, *A History of the Theories of Aether and Electricity,* Bd. II (London, 1953), 78ff.
6. Einstein, "Theorie der Strahlung und der Quanten," Abhandlungen der Deutschen Bunsen Gesellschaft, No. 7 (Halle a.S., 1914), p. 353. The quotations from Otto Stern heading the first and second sections are taken from a tape deposited at the ETHZ of November 25 and December 2, 1961.
7. Armin Hermann, "Albert Einstein und Johannes Stark," Sudhoffs Archiv **50**, 267–85 (1966), and "Die frühe Diskussion zwischen Stark und Sommerfeld uber die Quantenhypothese (1)," Centaurus **12**, 38–59 (1968).
8. A. Sommerfeld, "Über die Verteilung der Intensität bei der Emission von Röntgenstrahlen," Phys. Z. **10**, 969–76 (1909).
9. The quotation is from P. Jordan, "Zur Theorie der Quantenstrahlung," Z. f. Phys. **33**, 297–319 (1924). See A. Einstein, "Zur Quantentheorie der Strahlung," Phys. Z. **18**, 121–28 (1917).

10. A. Einstein, "Autobiographical Notes," pp. 2–95 in *Albert Einstein, Philosopher-Scientist,* edited by P. A. Schilpp (Library of Living Philosophers, Evanston, Ill., 1949), esp. p. 48ff.

11. A. Einstein, "Über die Entwicklung unserer Anschauungen über das Wesen und die Konstitution der Strahlung," Phys. Z. **10,** 817–25 (1909), p. 823ff.

12. A. H. Compton, "A Quantum Theory of Scattering of X Rays by Light Elements," Phys. Rev. **21,** 483–502 (1923).

13. Ibid., p. 491ff esp. p. 493 formula (27). See also W. Pauli, *Collected Scientific Papers,* Vol. I (Interscience Publishers, New York, 1964), p. 288ff, esp. p. 291f; and N. Bohr, H. A. Kramers and J. C. Slater, "Über die Quantentheorie der Strahlung," Z. f. Phys. **24,** 69–87 (1924).

14. See op. cit. in n. 13; and Niels Bohr, "Über die Wirkung von Atomen bei Stössen," Z. f. Phys. **34,** 142–57 (1925). W. Bothe and H. Geiger, "Uber das Wesen des Comptoneffekts, ein experimenteller Beitrag zur Theorie der Strahlung," Z. f. Phys. **32,** 639–63 (1925).

15. Bohr, op. cit. in n. 14.

16. Einstein, op. cit. in n. 11, p. 824f.

17. Louis de Broglie, "La nouvelle dynamique des quanta," in *Electrons et Photons, 5ème Conseil de Physique Solvay* (Gauthiers-Villars, Paris, 1928), pp. 105–32, esp. p. 108f.

18. Einstein, in *Electrons et Photons,* op. cit., p. 256.

19. Max Born, "Quantenmechanik der Stossvorgänge," Z. f. Phys. **38,** 803–27 (1926), p. 803f.

20. See however the most interesting contribution of Eugene Wigner to this volume, in particular his section on "The Extension of Einstein's Interest in Physics."

21. A. Einstein, "Zum Quantensatz von Sommerfeld und Epstein," Deutsch. Phys. Ges. Verh. **19,** 82–92 (1917), where the name Niels Bohr does not yet appear.

22. A. Einstein, "Bildet die Feldtheorie Möglichkeiten für die Lösung des Quantenproblems?" Berl. Ber., pp. 359–64 (1923).

23. Ibid., pp. 359, 360.

24. See Otto Struve, *Astronomie,* (Walter de Gruyter, Berlin, 1963), p. 174ff.

25. Op. cit. in n. 22, p. 361.

26. A. Einstein, "Quantentheorie des einatomigen idealen Gases," Berl. Ber., pp. 261–67 (1924); Berl. Ber., 3–14 (1925).

27. E. Schroedinger, "Zu Einsteins Gastheorie," Phys. Z. **27,** 95–101 (1926), sec. 1.

28. A. Einstein, "Quantentheorie des idealen Gases," Berl. Ber., pp. 18–25 (1925).

29. In *Albert Einstein, Michele Besso, Correspondence 1903–1955,* edited by Pierre Speziali (Herman, Paris, 1972), p. 204.

30. Ibid., p. 215.

31. A. Einstein, "Bemerkungen zu P. Jordans Abhandlung Zur Theorie der Quantenstrahlung," Z. f. Phys. **31,** 784–85 (1925). This is a reply to Jordan, op. cit. in n. 9.

32. The letters are printed in *Briefe zur Wellenmechanik,* edited by K. Przibram (Vienna, Springer, 1963), p. 21ff.

33. Also Paul Ehrenfest, see his "Inquiries Concerning Quantum Mechanics," where he refers directly to Einstein. ["Einige die Quantenmechanik betreffende Erkundigungsfragen," Z. f. Phys. **78,** 555–59 (1932).] The first section expresses vividly the "ideological pressure" of the neophytes and prophets of the new teaching — and may be regrets that Einstein no longer was to be found *above* the parties but *within* one of them. S. Goudsmit's recollections (quoted by A. Pais in the preceding chapter of this volume) have the ring of truth.

34. Op. cit. in n. 10, p. 44ff.

35. Letter from Einstein to Schroedinger of 22 December 1950, in *Briefe zur Wellenmechanik,* op. cit. in n. 32, p. 36.

36. "The question whether matter has an atomistic or a continuous constitution is thus reduced to the question whether those properties, assuming an extremely large *finite* number of particles or their *limit* of a constantly increasing number of particles, represent . . . the observed properties of matter . . ." [Ludwig Boltzmann, "On Statistical Mechanics," lecture given in St. Louis in 1904, in Boltzmann, *Populare Schriften,* op. cit. in n. 2, p. 345ff, esp. p. 358ff].

37. Oscar E. Lanford III, "Time Evolution of Large Classical Systems," in *Dynamical Systems, Theory and Applications,* edited by J. Moser, Lecture Notes in Physics **38,** 1–111 (1975).

38. Klaus Hepp, "Quantum Theory of Measurement and Macroscopic Observables," Helv. Phys. Acta **45,** 237–47 (1972).

39. S. Kochen and E. P. Specker, "The Problem of Hidden Variables in Quantum Mechanics," J. Math. Mech. **17,** 59–68 (1967).

40. For this enumeration, R. Jost, "Measures on the Finite Dimensional Subspaces of a Hilbert Space: Remarks to a Theorem of A. M. Gleason," in Studies in Mathematical Physics, Essays in Honor of Valentine Bargmann," edited by E. H. Lieb, B. Simon, and A. S. Wightman (Princeton University Press, Princeton, N. J., 1976).

41. See the last paragraph of the 1924 paper cited in n. 26 for the paradox and p. 10 of the 1925 cited ibid. for the solution of the paradox.

42. G. H. Dieke and F. S. Tomkins, "The Molecular Spectrum of Hydrogen," Phys. Rev. **76,** 283–89 (1949). It is very much to be regretted that the analysis of the T_2 spectrum was not continued (personal communication from K. Dressler).

43. A. Einstein, letter to M. Besso, 13 May 1917, in *Einstein, Besso Correspondence,* op. cit. in n. 29, p. 114. About the literature on the philosophy of quantum mechanics, see Max Jammer, *The Philosophy of Quantum Mechanics* (John Wiley, New York, 1974).

44. A. Einstein, Spencer lecture; German original in A. Einstein, *Mein Weltbild,* edited by Carl Seelig (Europa Verlag, Ulm, 1972), p. 113ff.

45. Hermann Weyl, "Elektron und Gravitation," Z. f. Phys. **56,** 330–52 (1929).

46. In *A. Einstein, Lettres a Maurice Solovine,* edited by M. Solovine (Gauthier-Villars, Paris, 1956).

Open Discussion

Following Papers by A. PAIS and R. JOST

Chairman, JULIAN SCHWINGER

J. Stachel (Boston U.): I wanted to make two additional or supplementary comments on Professor Pais's very beautiful paper, which I think has left us all speechless, but I will try to overcome that small difficulty. First of all, I would like to underline the point that he made about the problem that Einstein found in the Einstein–Podolsky–Rosen paper, which I think comes out even more clearly in the Einstein–Born correspondence. It seems to me that the thing that worried Einstein most about quantum mechanics, if it were to be accepted as an ultimate theory, was the inseparability even more than the indeterminism: that once two quantum mechanical systems have interacted, they are forever inextricably intertwined, no matter how far apart they get in space-time. And this to him was even more paradoxical as an ultimate characteristic of reality in his sense than the indeterminism. One could speculate that if the theory had not been inseparable, he might have been less inclined to reject the theory. At any rate, I think that particularly comes out in the Born–Einstein correspondence — that was the real sticking point with him. Secondly, I think what Einstein said about quantum mechanics was that for a theory that started out with the concepts with which it did, it went about as far as it could. I think what he meant was that it was based on mechanical concepts, position, momentum, and so forth. Of course what Bohr showed so masterfully was that, although one could not apply these concepts simultaneously to any individual situation, complementarity indicated just what the limits of applicability of any particular subset of these concepts were in a given situation; and that was the way to analyze it. But for Einstein, it was starting with mechanical concepts that led to the ultimate limitations of the theory. For him the relativity principle, the principle of general covariance, was so fundamental that any theory that was to advance further would have to start by incorporating general covariance, which did not seem possible in a theory that began with mechanical concepts; and therefore it seemed to him that field theory was the only line along which he could advance. But even though he admitted one might have to abandon the continuum entirely and that an algebraical, combinatorial theory might be the ultimate answer, he felt that such a theory also could not be based on mechanical concepts. That too was a major sticking point in his approach to quantum theory as an ultimate theory.

Pais: I have no particular comment to make except to say that there is one point you have touched on which is certainly a very delicate point — it not only was a delicate point for Einstein but remains a very delicate point for us, and that is of course the technical implementation of the principles of quantum mechanics in the context of a general relativistic theory. We will hear more about that later, and that is a matter on which the books are certainly not closed today.

B. Hoffmann (Queens College, City U. of New York): I just want to add a footnote to history; when Infeld and I were working with Einstein on one occasion we were discussing the Einstein–Podolsky–Rosen (EPR) paradox and Einstein remarked somewhat humorously, though truly, that he had received various pieces of correspondence from people — physicists — telling why the EPR paradox was not a paradox, was wrong and so on, and he said they were all offering different reasons. Incidentally, he did take Bohr seriously. I don't think he took any of the others seriously.

Y. Ne'eman (Tel-Aviv U.): Just a small story that I have from Murray Gell-Mann, and I've not checked it with David Bohm. Murray told me that David Bohm, who was an active physicist at the time in a variety of fields, had had his doubts about the foundations of quantum mechanics. Then, he had written his book, which was an orthodox book — he had written the book in order to get rid of all his doubts — and felt extremely happy after he had written the book, and was sort of ready to start a new life and felt he had reached the correct conclusion. And then a phone call came that Einstein wanted to speak to him and he went over to Einstein and he came out all shaken, and said Einstein showed him that it was all wrong, and he has not recovered to this day with respect to this point.

Pais: I would like to make one brief remark about this. I have said that Einstein was so apart in his assessment of quantum mechanics. You must understand the following: there were others who had their doubts of one kind or another. There are people today who have their doubts, and I do not wish to talk about it in an all-embracing fashion. There is a literature about that. In regard to Einstein himself, I know the following. That in regard to Bohm, he thought the work was very interesting, but it was not what he had in mind, it was not what he really wanted. To the best of my knowledge, the only two people with whom Einstein felt he shared certain concerns were Schroedinger and von Laue.

Anonymous: I want to recount a very small incident that perhaps throws another light on Einstein's efforts and aims. It was in the forties, early fifties — I don't remember when — when I spoke to him about the new discoveries in mesons and all that and I asked him, "Does this make any important difference to you in your point of view?" He said, "Well, I already know that the charge of the electrons is quantized. I already know that its mass is quantized. How many more things should I put into a theory?" In other words he was looking at something really comprehensive.

Pais: Of course, I had often talked with Einstein about the question of where his dissent or dissatisfaction lay. We never got really all that far in questions of relativistic field theory. It started on what we would call simple nonrelativistic applications of the quantum theory. Really he simply would sort of not listen to questions of what we call, in the sense of special relativity, relativistic field theories. It was very hard to draw him out on that.

A. Hermann (U. of Stuttgart): Professor Pais, if I have understood you correctly, you stated that at first when the "Drei-Männer-Arbeit," or the Göttingen quantum mechanics, came out, Einstein was quite in favor of it and only when the Born paper in 1926 came out did he change his mind. If I have understood you correctly, then I want to say that this cannot be true. I remember a letter to Ehrenfest, and I think it is from Einstein to Ehrenfest in September or October of 1925, in which he stated, "Heisenberg hat ein grosses Quanten-Ei gelegt. In Göttingen glauben sie daran, ich nicht" ["Heisenberg has laid a huge quantum-egg. In Göttingen they believe it, I don't"]. This is important since Pauli and especially Heisenberg always had the impression that their Göttingen quantum mechanics was based on the same philosophical principle as Einstein had utilized in his paper on special theory of relativity. However, as Heisenberg recalled, when he discussed the matter with Einstein, Einstein had changed his mind and Einstein said, "erst die Theorie entscheidet darüber was man beobachten kann" ["The theory decides (or determines) what one can observe"].

Pais: I do know the letter to which you refer and that it is exactly from the time which you indicate. What I tried to say was this: that there was a time, what I said about Einstein, in November of 1925 or Fall of 1925, when he was very excited about the Heisenberg theory. If you follow the letters, there are such letters as you mention. There are also others. I have tried to convey that there was a time of vacillation. He said to Born, you know, there is a great sense of excitement in the air. He wrote the letter that you correctly mentioned. It is that vacillating period that comes to an end in my opinion, as I have traced it with a little care, at the time of the Born paper. From then on, so to speak, you do not find a good word anymore. That is my reading, which does not conflict with what you say.

VI. Relativity and Its Ramifications

16. GENERAL RELATIVITY AND DIFFERENTIAL GEOMETRY

Shiing-shen Chern

The title for this paper originally suggested by the organizing committee was "The Interaction of General Relativity and Differential Geometry." It is a strange feeling to speak on a topic of which I do not know half of the title. I will speak as a differential geometer looking at the impressive structure of general relativity. As I understand it, general relativity is physics; its basis is physical experimentation. The aim of geometry should be the study of spaces. But this is a tautological statement. It is perhaps better to say that geometrical investigations are guided by tradition and continuity. Their criterion is mathematical originality, simplicity, and depth, and a good combination and compromise of the latter. Geometry has thus more freedom and can indulge a little bit more in ideal topics. It has been, however, a historical fact that it could be rudely awakened to find its abstract objects close to reality. Such an example is furnished by the relation between differential geometry and general relativity.

General relativity was born in 1915. Differential geometry at its early stage was identical with the infinitesimal calculus, the derivative and the tangent line and the integral and the area being synonymous objects. As an independent discipline differential geometry was born in 1827. It was the year that Gauss published his "General Investigations of Curved Surfaces (Disquisitiones circa superficies curvas)," in which he laid the foundations of a local two-dimensional geometry based on a quadratic differential form. Even Gauss could not have foreseen that its four-dimensional generalization would be the basis of gravitation.

1. Differential Geometry before Einstein

In a historical paper in 1854, "Über die Hypothesen, welche der Geometrie zugrunde liegen," Riemann generalized Gauss's work to high dimensions and laid the foundations of Riemannian geometry. In that paper he first introduced the idea of an n-dimensional manifold whose points are described by n real numbers as coordinates. This was a giant step from Gauss, whose curved surfaces lie in the three-dimensional Euclidean space and are not intrinsic. Being mathematically a purist, Einstein was reluctant to accept such an idea. It took him seven years to pass from his special relativity in 1908 to his general relativity in 1915. He gave the following reason: "Why were another seven years required for the construction of the general theory of relativity? The main

Harry Woolf (ed.), Some Strangeness in the Proportion: A Centennial Symposium to Celebrate the Achievements of Albert Einstein
ISBN 0-201-09924-1

reason lies in the fact that it is not so easy to free oneself from the idea that coordinates must have an immediate metrical meaning."[1]

The fundamental problem in Riemannian geometry is the form problem. Given two quadratic differential forms

$$ds^2 = \sum_{i,k} g_{ik}\, dx^i\, dx^k \tag{16.1}$$

$$ds'^2 = \sum_{i,k} g'_{ik}\, dx'^i\, dx'^k, \quad 1 \leqslant i,k \leqslant n, \tag{16.2}$$

in two different systems of coordinates $x^1, \ldots, x^n$ and $x'^1, \ldots, x'^n$, find the conditions that there is a coordinate transformation

$$x'^i = x'^i(x^1, \ldots, x^n), \quad 1 \leqslant i \leqslant n, \tag{16.3}$$

carrying one into the other. This problem was solved by E. B. Christoffel and R. Lipschitz in 1869. Christoffel's solution involves the symbols associated with his name and the notion of covariant differentiation. Based on it, Ricci developed in 1887–96 the tensor analysis that plays a fundamental role in general relativity. An account of the Ricci calculus was given in a historical memoir, "Méthodes de calcul differentiel absolu et leurs applications," published in *Mathematische Annalen* in 1901, by him and his student, T. Levi-Civita. Christoffel taught at the Eidgenossische Technische Hochschule in Zürich (where Einstein was later a student) and thereby had an influence on Italian geometers. It may be of interest to note that this year is the one-hundred-fiftieth anniversary of his birth.

Important as these developments were, the main activities in differential geometry at the turn of the century centered on the geometry in Euclidean space, following the tradition of Euler and Monge. A representative work was Darboux's four-volumed "Théorie des surfaces," which was and still is a classic. It was difficult for geometers to free themselves from an absolute ambient space, which is usually the Euclidean space.

At about the same time as the Christoffel–Lipschitz solution of the form problem, Felix Klein formulated in 1872 his *Erlangen* program. This is to define geometry as the study of a space with a continuous group of automorphisms, such as the Euclidean space with the group of rigid motions, the projective space with the group of projective collineations, and so on. The *Erlangen* program unified geometry under group theory and was the guiding principle of geometry for about a half century. In practice it can be used, as a consequence of the isomorphism of groups, to derive from a given geometrical result new and seemingly unrelated results. A famous example is the line-sphere transformation of Sophus Lie. A more elementary example is based on Study's dual numbers. From the theorem in elementary geometry that the three heights of a triangle meet in a point one can derive the following theorem of Morley–Petersen: Given a simple skew hexagon in space whose adjacent sides are perpendicular, there is a line that meets perpendicularly the three common perpendiculars of the three pairs of opposite sides.[2]

Klein's *Erlangen* program fits perfectly with the special theory of relativity, one of whose principles is the invariance of the field equations under the Lorentz group. As a result Klein, the most influential German mathematician at the turn of the century, was one of the earliest supporters of the special theory of relativity. The Lorentzian structure plays a fundamental role in relativity. It has also geometrical interpretations. Indeed this happens when we study the geometry of spheres in space. All the contact transformations in space carrying spheres to spheres form a 15-parameter Lie group, and those that carry planes to planes a 10-parameter subgroup. The latter is isomorphic to the Lorentzian group in four variables. The resulting geometry is known as the Laguerre sphere geometry.[3]

The great success of Klein's *Erlangen* program naturally led to the study of differential geometry in a Klein space, or a homogeneous space as it is now called. In particular, projective differential geometry was initiated in Halphen's thesis in 1878 and later developed by the American school under E. J. Wilczynski beginning in 1906 and the Italian school under G. Fubini beginning in 1916.

At the beginning of the twentieth century, global differential geometry was at its infancy. The four-vertex theorem was formulated by Mukhopadhyaya in 1909. The topological conclusion drawn from the Gauss–Bonnet formula that the integral of the Gaussian curvature of a closed orientable surface is equal to $2\pi\chi$, where χ is the Euler characteristic of the surface, was deduced by von Dyck in 1888. With typical foresight Hilbert wrote a paper in 1901 on surfaces of constant Gaussian curvature, in which he gave a new proof of Liebmann's theorem that a closed surface of constant Gaussian curvature is a sphere and the theorem (Hilbert's theorem) that a complete surface of constant negative curvature cannot be regular everywhere. Under Hilbert's supervision Zoll found in 1903 closed surfaces of revolution, not the sphere, all of whose geodesics are closed. Motivated by dynamics, Poincaré and G. D. Birkhoff proved the existence of closed geodesics on convex surfaces.

The final aim of differential geometry is global results. However, local differential geometry should not be minimized because every global result must have a local basis. To have a systematic development of global differential geometry its foundations must be laid. This should come from topology. General relativity provided the impetus.

2. The Impact of General Relativity

When Einstein founded general relativity, the mathematical tools were available in the form of Riemannian geometry treated by the Ricci calculus. Einstein introduced the useful summation convention. The effect on differential geometry was electrifying; Riemannian geometry became the central topic. It may be of interest to note that almost all the standard books on Riemannian geometry, by Schouten, Levi-Civita, Elie Cartan, and Eisenhart, appeared in the period 1924–26.

These developments immediately led to generalizations. It soon became clear that in the applications of Riemannian geometry to relativity, the Levi-Civita parallelism, and not the Riemannian metric itself, plays the crucial role. In his famous book *Raum, Zeit, Materie*, published in 1918, Hermann Weyl introduced the notion of an affine connection. It is a structure, not necessarily Riemannian, where parallelism and covariant differentiation are defined, with the desired properties. Weyl's affine connections are symmetric or without torsion.

A definitive treatment of affine connections, together with a generalization to connections with torsion, was given by Elie Cartan in his fundamental paper, "Sur les variétés a connexion affine et la théorie de la relativité généralisé," published in 1923–24. The paper did not receive the attention it deserved for the simple reason it was ahead of its time. For it is more than a theory of affine connections. Its ideas can be easily generalized to give a theory of connections in a fiber bundle with any Lie group, for whose treatment the Ricci calculus is no longer adequate. The paper shows, among other things, how Einstein's theory is a direct generalization of Newton's theory when the latter is expressed in an intrinsic form, that is, in general coordinates. Specifically the following contributions could be singled out: 1) the introduction of the equations of structure and the interpretation of the so-called Bianchi identities as the result of exterior differentiation of the equations of structure; 2) the recognition of curvature as a tensor-valued exterior quadratic differential form.

Geometrically an affine connection is a family of affine spaces, the fibers, parametrized by a space, the base space, such that the family is locally trivial and that there is a law of "development" of the fibers along the curves of the base space, preserving the linear properties. In the same vein we can take Klein spaces as fibers and replace the general linear group by a Lie group acting on the Klein space, with a corresponding law of development. Such a structure was called by Cartan a generalized space ("espace généralisé"). In general, the connection is nonholonomic; that is, the development depends on the curve in the base space. In other words, the space does not return to its original position after being developed along a closed curve; the measure of the deviation is given by the curvature of the connection. Clearly, a Klein space is itself a generalized space, with curvature identically zero.

When Klein formulated his *Erlangen* program, it was observed that Riemannian geometry was not included, because a generic Riemannian space admits no isometry other than the identity. From Cartan's viewpoint it is a generalized space with Euclidean spaces as fibers and provided with the Levi–Civita connection. This settles a foundational issue in differential geometry, as we have now a notion that includes Klein spaces and Riemannian spaces, and generalizations of both.

In practice, geometrical structures are given in a nonintuitive form. Frequently it is either a metric defined by an integral or a family of submanifolds defined by a system of differential equations. The two most familiar examples are the Riemannian metric and the paths defined by a system of ordinary differential equations of the second order. The association of a connection is not an easy problem. In fact, even the definition of the Levi–Civita connection of a Riemannian metric is, from the purely algebraic viewpoint, fairly nontrivial. As to be expected, the geometry of paths (E. Cartan, O. Veblen, T. Y. Thomas) involves a projective connection.

These developments in what is commonly described as non-Riemannian geometry were accompanied by parallel developments in general relativity. With special relativity for the electromagnetic field and general relativity for the gravitational field there was clearly a need for a unified field theory where the two could be combined. The first significant step was taken by Hermann Weyl in 1918 in his gauge theory. Weyl used a generalized space having as group the group of homothetic transformations. It was found to be physically untenable. It is now understood that his gauge group should not be the noncompact group of homothetic transformations but the compact circle group. Recent developments on the gauge theory will be discussed in Sec. 4.

Following Weyl other unified field theories were proposed, among which were those of Kaluza–Klein, Einstein–Mayer (1931), and Veblen (the projective theory of relativity, 1933). A common feature is the introduction of a fifth dimension, to account for electricity and magnetism. (Veblen's theory is four-dimensional, but the tangent projective space has five homogeneous coordinates.) Veblen's projective theory is geometrically simple, in taking as a starting point the paths in the space, which are to be the trajectories of the charged particles.

Einstein himself pursued his search for a unified field theory throughout his last years, often with collaborators. In this respect I wish to insert a personal note. I came to the Institute for Advanced Study in 1943 from Kunming in Southwest China. It was at the height of the Second World War, and Einstein greeted me with great warmth and sympathy. It was my great fortune to be able to discuss with him from time to time different topics, general relativity included. I soon saw the extreme difficulty of his problem and the difference between mathematics and physics. The famous problems in mathematics are usually well formulated, but in physics the formulation is part of the problem.

Einstein, with a strict standard for the final answer, was not content with the aforementioned proposals, and in fact with many others. He tested the geometrical structures that might possibly underlie a unified field theory. Among them are the following:

1. A nonsymmetric g_{ij}
2. Four-dimensional complex space with an Hermitian structure
3. Metric spaces more general than Riemannian

The geometry of general metric spaces was founded and investigated by Karl Menger, and Einstein's friend Kurt Gödel made important contributions to it. Structure 1, the g_{ij}, splits uniquely into a symmetric and an antisymmetric part. If the former is nondegenerate, the structure is equivalent to a pseudo-Riemannian structure with an exterior quadratic differential form. The pseudo-Riemannian structure is Riemannian or Lorentzian if the signature of the symmetric part of g_{ij} is $+ + + +$ or $+ + + -$. Structure 2 is closely related to the geometry of complex algebraic varieties and to functions of several complex variables, which are areas of mathematics with spectacular developments in the last decades.

3. Positive-Mass Conjecture, Minimal Surfaces, and Manifolds of Positive Scalar Curvature

In the post-Einstein era general relativity made great progress in its emphasis on a global theory (or "large-scale space-time"). The origin was cosmology, in which Einstein himself took an active part, but the influence of the developments in global differential geometry is unmistakable. The universe is identified as a four-dimensional connected Lorentzian manifold, and physics and geometry are intertwined more than ever. However, the purely geometrical problems are usually simpler. Two of the reasons are that geometry is Pythagoriasian or Riemannian and the geometers can idealize their spaces by assuming them to be compact.

It is natural to record the data at a given instant by a data-set, which is a hypersurface Σ having everywhere a timelike normal (so that the induced metric is Riemannian). In this way the theory of hypersurfaces in a four-dimensional manifold, an immediate generalization of classical surface theory, plays a role in general relativity. The local invariants of Σ are given by two quadratic differential forms, its first and second fundamental forms. The trace of the second fundamental form is called the mean curvature. Its vanishing characterizes the maximal hypersurfaces. On the other hand, the induced Riemannian metric on Σ has a scalar curvature. All these quantities are connected by the Gauss–Codazzi equations. The mass density μ and the momentum density J^a are, as a consequence of Einstein's field equations, combinations of the coefficients of the first and second fundamental forms and their covariant derivatives. Since momentum density should not exceed the mass density, we have

$$\mu \geqslant \left| \sum_{1 \leqslant a \leqslant 3} J^a J_a \right|^{1/2} \geqslant 0 . \tag{16.4}$$

On a maximal hypersurface the scalar curvature is non-negative.

A data-set is called asymptotically flat if for some compact C, $\Sigma - C$ consists of a finite number of components N_i (to be called the ends), such that each N is diffeomorphic to the complement of a compact set in R^3 and has asymptotically the metric

$$ds^2 = \left(1 + \frac{M_i}{2r}\right)^4 \left(\sum_{1 \leqslant a \leqslant 3} (dx^a)^2\right), \tag{16.5}$$

r being the distance from the origin. The number M_i coincides with the Schwarzschild mass in the case of the Schwarzschild metric and is hence called the total mass of N_i. The positive-mass conjecture states that for an asymptotically flat data set, each end has total mass $M_i \geqslant 0$, and that if one $M_i = 0$, the data set is flat in the sense that the induced Riemannian metric is flat and the second fundamental form is zero. This conjecture is of fundamental importance in general relativity. For physical reasons Einstein assumed it to be true.

Under the assumption that Σ is a maximal hypersurface this was proved in its full generality in 1978 by R. Schoen and S. T. Yau.[6] The complete story of this work is a perfect example of the fruit of contact and collaboration between relativists and differential geometers. At the Differential Geometry Summer Institute of the American Mathematical Society at Stanford University in 1973, Robert Geroch was invited to give a series of lectures on general relativity. The positive-mass conjecture is clearly one of the open problems. To simplify its statement Geroch formulated some pilot conjectures, one of which says the following: "On the three-dimensional number space R^3 consider a Riemannian metric which is flat outside a compact set. If its scalar curvature is $\geqslant 0$, then the metric is flat." By enclosing the compact in a big box and identifying the opposite faces, J. Kazdan and F. Warner reformulated the conjecture as follows: "A Riemannian metric on the three-dimensional torus with scalar curvature $\geqslant 0$ is flat." Geroch remarked: "It is widely felt that proofs of several of these special cases could be generalized to a proof of the full conjecture."

This Geroch conjecture falls in the realm of differential geometry; Schoen and Yau proved it first. The idea of the proof makes use of closed minimal surfaces. In fact, from the formula for the second variation of area it is seen that in a three-dimensional compact orientable Riemannian manifold of positive scalar curvature a closed minimal surface of positive genus is not stable; that is, it will have a smaller area by a perturbation. On the other hand, the three-dimensional torus has a large fundamental group (isomorphic to $Z \oplus Z \oplus Z$), and has second Betti number equal to 3. These topological properties should imply the existence of a closed regular surface of smallest area in a nonzero homotopy class of closed surfaces. Such a result is a generalization of the fact that on a compact Riemannian manifold there is in every nonzero homotopy class of closed curves a shortest closed smooth geodesic. The proof of a corresponding result for minimal surfaces is of course much more subtle. Schoen and Yau went on to a proof of the positive-mass conjecture for a maximal hypersurface. Later in 1978 the result was generalized to higher dimensions.

These developments touch on topics that are dear to differential geometers. They are: minimal surfaces and manifolds of positive scalar curvature.

Early investigations on minimal surfaces were concentrated to the plateau problem: Given a closed curve in R^3, find a surface of smallest area bounded by it. Only in recent years has attention been directed to the study of closed or complete minimal surfaces in a given manifold, such as the n-dimensional Euclidean space R^n *or* the n-dimensional unit sphere S^n. The study generalizes that of closed geodesics, which have been found to play an important role in the geometry and topology of Riemannian manifolds.[8] Closed or complete minimal surfaces, particularly the regular ones, are destined to be a richer and even more interesting object. It is natural to take as a data-set a maximal hypersurface. Recently, J. Sachs and K. Uhlenbeck proved that a compact simply connected Riemannian manifold always has a minimally immersed 2-sphere.

The basic question on manifolds of positive scalar curvature is: What compact manifolds can be given a Riemannian metric of positive scalar curvature? The interest in this question is enhanced by the fact that every compact manifold of dimension $\geqslant 3$ can be given a Riemannian metric of negative scalar curvature. In fact, it can be given a metric whose total scalar curvature (that is, the integral of the scalar curvature) is $\leqslant 0$. The latter can then be conformally deformed into one with constant negative scalar curvature. On the other hand, by studying harmonic spinors, A. Lichnerowicz proved in 1963 that if a compact spin manifold has a Riemannian metric of positive scalar curvature, its $\hat{A}$-genus is zero. Yau and Schoen showed that the three-dimensional torus cannot have a Riemannian metric of positive scalar curvature; the same has since been proved true for the n-dimensional torus (M. Gromov, B. Lawson, R. Schoen, S. T. Yau). Motivated by general relativity, these authors are close to a complete topological description of all compact manifolds that can carry a Riemannian metric of positive scalar curvature.

The same questions can be asked for the Ricci curvature or the sectional curvature. A classical theorem of S. Myers says that a complete Riemannian manifold of positive Ricci curvature must be compact and hence must have a finite fundamental group. The condition for a Riemannian manifold to have positive sectional curvature is even stronger, and such manifolds are expected to be few. The compact symmetric spaces of rank 1 have this property, but there are others of a sporadic nature. A complete topological description of compact Riemannian manifolds of positive sectional curvature seems to be difficult.

4. Gauge Field Theory

Gauge field theory was introduced by Hermann Weyl in his paper "Gravitation und Elektrizität" in 1918. The idea is to use a pair of quadratic and linear differential forms

$$ds^2 = \sum_{i,k} g_{ik}\, dx_i\, dx_k\,, \tag{16.6}$$

$$\phi = \sum_i \phi_i\, dx_i\,, \quad 1 \leqslant i,k \leqslant 4, \tag{16.7}$$

defined up to the gauge transformation

$$\begin{aligned} ds^2 &\rightarrow \lambda ds^2\,, \\ \phi &\rightarrow \phi + d \log \lambda\,. \end{aligned} \tag{16.8}$$

With ϕ as the electromagnetic potential and its exterior derivative $d\phi$ as the electromagnetic strength or Faraday, this was the first attempt of a unified field theory. Einstein objected to the indeterminacy of ds^2 but expressed admiration of the depth and boldness of Weyl's proposal.

All the objections, then and later, disappear if we interpret Weyl's theory as based on the geometry of a circle bundle over a Lorentzian manifold.[9] Then the form ϕ, subject to the gauge transformation, can be interpreted as defining a connection in the circle bundle and ds^2 remains unaltered, eliminating Einstein's objection.

The mathematical basis of gauge field theory lies in vector bundles and the connections in them. The notion of a fiber bundle or fiber space, being global in character, arose in topology. At first it was an attempt to find new examples of manifolds.[10] Fiber spaces are locally, but not globally, product spaces. The presence of such a distinction is a sophisticated mathematical fact. The development of fiber spaces has to wait until invariants are found to distinguish the fiberings or even to show that globally there are nontrivial ones. The first such invariants are the characteristic classes introduced by H. Whitney and by E. Stiefel in 1935.[11] Topology, however, forgets the algebraic structure, and in applications vector bundles, with the linear structure intact, are more useful. A vector bundle $\pi : E \rightarrow M$ over a manifold M is, roughly speaking, a family of vector spaces parametrized by M such that it is locally a product. The vector space $E_x = \pi^{-1}(x)$ corresponding to $x \in M$ is called the fiber at x. Examples are the tangent bundle of M and all tensor bundles associated to it. A more trivial bundle is the product bundle $M \times V$, where V is a fixed vector space and (x, V), $x \in M$, is the fiber at x. A vector bundle is called real or complex according as the fiber is a real or complex vector space. Its dimension is the dimension of the fibers.

It is important that the linear structure on the fibers has a meaning so that the general linear group plays a fundamental role in matching the fibers; it is called the structure group. A real (respectively complex) vector bundle is called Riemannian (respectively Hermitian) if the fibers are provided with inner products. In this case the structure group is reduced to $O(n)$ (respectively $U(n)$), n being the dimension of the fibers; the bundle is then called an $O(n)$ (respectively $U(n)$)-bundle. Similarly, we have the notion of an $SU(n)$-bundle.

A section of the bundle E is an attachment, in a continuous and smooth manner, to every point $x \in M$, a point of the fiber E_x. In other words, it is a continuous mapping $s: M \rightarrow E$ such that the composition $\pi \circ s$ is the identity. The notion is a natural generalization of a vector-valued function and of a tangent vector field. In order to differentiate s we need a so-called connection in E. The latter allows the definition of the covariant derivative $D_X s$ (X being a vector field in M), which is a new section of E. Covariant differentiation is generally not commutative; that is, $D_X \circ D_Y \neq D_Y \circ D_X$ for two vector fields X, Y in M. The measure of the noncommutativity gives the curvature of the connection; this is an analytic version of the geometric concept of nonholonomy described in Sec. 2. Following Elie Cartan it is important to regard the curvature as a matrix-valued exterior quadratic differential form. Its trace is a closed 2-form. More generally, the sum of all its principal minors of order k is a closed $2k$-form. It is called a characteristic form (Pontrjagin form or Chern form, according as the bundle is real or complex). By the de Rham theory the characteristic form of degree $2k$ determines a cohomology class of dimension $2k$, to be called a characteristic class. Whereas the characteristic forms depend on the connection, the characteristic classes depend only on the bundle. They are the simplest global invariants of the bundle. It must be an act of nature that the nontriviality of a vector bundle is recognized through the need of a covariant differentiation and that its noncommutativity accounts for the first global invariants. This introduction of the characteristic classes gives emphasis on its local character, and the characteristic forms contain more information than the classes. When M is a compact oriented manifold, a characteristic class of the top dimension (that is, of dimension equal to that of M) gives by integration a characteristic number. When it is an integer, it is called a topological quantum number.

These differential-geometric notions have been found to be the likely mathematical basis of a unified field theory. Weyl's gauge theory deals with a circle bundle or a $U(1)$-bundle, that is, a complex Hermitian bundle of dimension one.

In studying the isotopic spin Yang–Mills used what is essentially a connection in an $SU(2)$-bundle. It is the first instance of a non-Abelian gauge theory. From the connection the "action" can be defined. A connection in an $SU(2)$-bundle at which the action takes the minimum is called an instanton. Its curvature has a simple expression and is called self-dual. An instanton is thus a self-dual solution of the Yang–Mills equation. When the space R^4 is compactified into the four-dimensional sphere S^4, the $SU(2)$-bundles are determined up to an isomorphism by a topological quantum number k, which is an integer. Atiyah, Hitchin, and Singer[12] proved that over S^4 the moduli (or parameter space) for the set of connections with self-dual curvature on the $SU(2)$-bundle with given $k > 0$ is a smooth manifold of dimension $8k - 3$. In physical terms this is the dimension of the space of instantons with topological quantum number $k > 0$.

Atiyah and Ward observed that the self-dual Yang–Mills fields fit well into the Penrose Twistor program. They were able to translate the problem of finding all self-dual solutions into a problem of algebraic geometry: classifying holomorphic vector bundles on the complex projective 3-space. This problem had been looked at closely by K. Barth, G. Horrocks, R. Hartshone, and others. Using their results, one can finally find all self-dual solutions,[13] and in fact, coming back to physics, translate the mathematical results into explicit formulas that physicists find satisfactory.[14]

Instantons can claim a relation to Einstein through the following result. The group $SO(4)$ is locally isomorphic to $SU(2) \times SU(2)$, so that a Riemannian metric on a four-dimensional manifold M gives rise through projection to connections in the $SU(2)$-bundles. M is an Einstein manifold if and only if these connections are self-dual or anti-self-dual.[15]

It remains to be seen whether a satisfactory unified field theory, including even weak and strong interactions, will be achieved through a nonabelian gauge theory. It suffices to say that the geometric notion of bundle and connection, as explained here, is exceedingly simple. I believe Einstein would have liked it.

5. Concluding Remarks

My own limitation restricts the scope of this account, which is conspicuously incomplete. The most obvious omission is the important topic of singularities as culminated in the work of Penrose and Hawking.

Finally, as an amateur I wish to express the hope that general relativity not be restricted to the gravitational field. A global unified field theory, whatever it is, must be closer to Einstein's grand design. More mathematical concepts and tools are now available.

Acknowledgement

This material is based on work supported by the National Science Foundation under Grant MCS 77-23579.

NOTES

1. A. Einstein, "Autobiographical Notes," in *Albert Einstein: Philosopher-Scientist* edited by P. A. Schilpp (Open Court, La Salle, Ill., 1949) Vol. 1, p. 67.
2. F. Klein, *Höhere Geometrie* (Springer, 1926), p. 314.
3. W. Blaschke, *Differentialgeometrie*, Vol. III (Springer, 1929).
4. See Appendix II of *The Meaning of Relativity*, 5th ed. (1955).
5. Y. Choquet-Bruhat, A. E. Fischer, and J. E. Marsden, "Maximal Hypersurfaces and Positivity of Mass," *Isolated Gravitating Systems in General Relativity*, Ed. J. Ehlers, Italian Physical Society, 1979, 396–456.

6. R. Schoen and S. T. Yau, "Incompressible Minimal Surfaces, Three-dimensional Manifolds with Nonnegative Scalar Curvature, and the Positive Mass Conjecture in General Relativity," Proc. NAS, USA, **75**, 2567 (1978); "On the Proof of the Positive Mass Conjecture in General Relativity," Comm. Math. Phys., **65**, 45–76 (1979) "Complete Manifolds with Nonnegative Scalar Curvature and the Positive Action Conjecture in General Relativity," Proc. NAS, USA **76**, 1024–25 (1979).
7. Robert Geroch, "General Relativity," in *Proceedings of the Symposium in Pure Math*, 27, Pt 2, (1975), pp. 401–14.
8. W. Klingenberg, *Lectures on Closed Geodesics* (Springer, 1978).
9. S. Chern, "Circle Bundles," Geometry and Topology, III Latin American School of Mathematics, Springer Lecture Notes, No. 597 (1977) pp. 114–31.
10. H. Hotelling, "Three Dimensional Manifolds of States of Motion," Trans. Amer. Math. Soc. **27**, 329–44 (1929), and **28**, 479–90 (1929); H. Seifert, "Topology 3 dimensionaler gefaserter Raume," Acta Math. **60**, 137–238 (1932).
11. E. Stiefel, "Richtungsfelder und Fernpurallelismus in Mannigfaltigkeiten," Comm. Math. Helv. **8**, 3–51 (1936); H. Whitney, "Sphere Spaces," Proc. Math. Acad. Sci. USA **21**, 462–68 (1935).
12. M. F. Atiyah, N. J. Hitchin, and I. M. Singer, "Deformations of Instantons," Proc. NAS, USA, **74**, 2662–63 (1977); "Self-duality in Four-dimensional Riemannian Geometry," Proc. R. Soc. Lond. **A362**, 425–61 (1978).
13. M. F. Atiyah, et. al. "Construction of Instantons," Phys. Lett. **65A**, 185 (1978).
14. N. H. Christ, E. J. Weinberg, and N. K. Stanton, "General Self-Dual Yang-Mills Solutions," Phys. Rev. **D 18**, 2013–25 (1978).
15. Atiyah, Hitchin, and Singer, op. cit. in n. 12.

17. MATHEMATICS AND PHYSICS

Tullio Regge

It is hard to match Professor Chern's masterly and lucid lecture on the general relationship between general relativity and differential geometry. Perhaps I am better off in offering a few comments on ideas he purposely omitted that are related to contemporary developments toward a unified theory. I would like also to emphasize some basic differences in the general philosophy of physicists and mathematicians on the matter of geometry and physics.

One general observation that springs to mind in hearing and reading papers on relativity is the persistent use of old-fashioned tensor calculus and the almost total neglect of the language of forms. This reflects more than a mere choice of mathematical tool, on a purely technical level; rather, it demonstrates a lack of appreciation of the role of gravity as a gauge theory.

This role was discovered long ago by Cartan, but its importance was not understood by physicists until much later when gauge theories became fashionable. The importance of Cartan's work became apparent to physicists essentially through the work of Sciama, Kibble, and Utiyama, who introduced into relativity the concepts of tetrad and connection. The metric tensor thus became an auxiliary quantity, Christoffel symbols disappeared from the scene and it became quite easy to deal with spinors. Also gravity began to look like a real Yang–Mills theory. A few formulas may help to make the point clearer. As usual, we write η_{ab} for special relativity metric tensor ($\eta_{00} = +1$) so that we have:

$$g_{\mu\nu} = e_{\mu}{}^{a} \otimes e_{\nu}{}^{b} \eta_{ab} \ , \tag{17.1}$$

where the tetrad $e_{\mu}{}^{a}$ now defines a form $e^{a} = e_{\mu}{}^{a}\, dx^{\mu}$. The connection $\omega_{\mu}{}^{ab}$ is now defined by the equations:

$$T^{a} = de^{a} + \omega^{a}{}_{b} \wedge e^{b} = 0 \ . \tag{17.2}$$

Finally the curvature is defined by:

$$R^{ab} = d\omega^{ab} + \omega^{al} \wedge \omega_{l}{}^{b} \ . \tag{17.3}$$

Harry Woolf (ed.), Some Strangeness in the Proportion: A Centennial Symposium to Celebrate the Achievements of Albert Einstein
ISBN 0-201-09924-1

It is a standard result that the components of the curvature calculated from (17.3) coincide essentially with those of the Riemann tensor as calculated from the metric tensor. This equivalence rests on the assumption that the torsion T^a vanishes, as stated by (17.2). This condition and the equation of motion are better stated in a first-order formalism. Einstein's action is given by:

$$A = \text{const} \cdot \int_{\mathcal{R}^4} R^{ab} \wedge e^c \wedge e^d \varepsilon_{abcd} \ . \tag{17.4}$$

Variation of this action in the e^a and ω^{ab} independently implies (17.2) and

$$R^{ab} \wedge e^c \, \varepsilon_{abcd} = 0 \ . \tag{17.5}$$

which turn out to be the standard field equations for gravity. This brings us to a further point. The set e^a, ω^{ab} seems to be beautifully arranged as a set of Yang–Mills potentials for a gauge theory with the Poincaré group as a structural group; indeed, torsion now would appear as the translational part of the curvature, and ordinary curvature and the rotational one. We may now write instead of e^a, ω^{ab}, the unique set of ω^A where A runs over the Lie algebra of the Poincaré group P and, instead of T^a, R^{ab}, the unique set R^A, and cast both (17.2) and (17.3) into the form:

$$R^A = d\omega^A + \frac{1}{2} C_{BD}{}^A \, \omega^B \wedge \omega^D \ , \tag{17.6}$$

where the $C_{BD}{}^A$ are now the structure constants of the Poincaré group. Gravity, therefore, seems to be perfectly cast for the role of a gauge theory with P as a structural group. We must not forget, however, that physics, so to speak, is geometry plus an action principle. Strangely enough, the action principle (17.4) is not invariant under the general set of Poincaré gauge transformations, but only under the Lorentz ones. Notice that general coordinate transformations now are a quite trivial consequence of the language of forms. It is instructive to see how lack of translational invariance works on (17.4). An infinitesimal translation has the form (λ^a 0-form):

$$\begin{aligned} \delta e^a &= d\lambda^a + \omega^a{}_b \, \lambda^b \quad \delta\omega^{ab} = 0, \\ \delta T^a &= R^a{}_b \lambda^b \quad \delta R^{ab} = 0 \ , \end{aligned} \tag{17.7}$$

and plainly does not leave (17.4) invariant. It is hard to give up invariance, so we find often in the trade what I may call an "amended" gauge transformation. Here one introduces a curvature-dependent correction in the connection, as follows:

$$\delta\omega^{ab} = - \, e^c \, \lambda^d \, R_{cd}{}^{ab} \ . \tag{17.8}$$

This correcting term is chosen in order to force the torsion T^a to stay zero, as natural in a second-order formalism. Miraculously the action (17.4) also is invariant. Unfortunately (17.8) is nothing but a disguised coordinate transformation, and it does not reflect any special invariance property of the action. In fact, any action that can be written in terms of forms admits a formula of the type (17.8). Therefore, I could say that (17.8) is very useful, but I feel that it is definitely misleading at this stage to call it a gauge transformation. But this brings me to the main point of my discussion, to face the rather embarrassing notion that (17.4) does not admit the whole Poincaré invariance that we expected from the grand geometrical formula (17.6). A similar and related lack of invariance appears in the interesting paper of MacDowell and Mansouri, who work their way through a de Sitter formalism. And I could raise a similar point for supergravity and for many of the proposed generalizations of supergravity.

I am not questioning either the usefulness or the validity of conventional supergravity, the so-called gauge transformations, but rather, I am emphasizing a formal point. It appears that nature in setting up gravity has prepared a beautiful geometrical setting but has decided at the last moment to forego invariance under the whole group P. This leaves us with the problem of finding an alternative constructive principle, not based on invariance, but equally restrictive. We could, for instance, give up the interpretation of e as a translational Yang–Mills potential. I believe that this way out is unsatisfactory, in that the general trend in unified theories is to interpret matter fields as objects whose geometrical interpretation is not yet known and which will turn out to be some forgotten component of an unnamed Yang–Mills potential.

But whatever the details of future unified theory, I feel that we can all agree that gauge theories are here to stay and that geometry and physics are forever beautifully linked. Perhaps it is enough to cite the instanton idea and the related work of Atiyah and his collaborators to note that we physicists have still a lot to learn and to profit from the infusion of new mathematical ideas into physics.

Open Discussion

Following Papers by S. S. CHERN and T. REGGE

Chairman, E. AMALDI

R. Penrose (Oxford U.): I would like to make a remark in connection with Professor Chern's talk — in connection with the very beautiful work by Schoen and Yau. Lest people think that the subject is now closed, their result does depend, in fact, upon the assumption of the existence of a maximal hypersurface, which is perhaps a less physically obvious thing than the result they prove. It is certainly important to remove this restriction. (A recent new result by Schoen and Yau, based on work by Jang, now appears to have done just this!) Also, in Stephen Hawking's lecture, he mentioned the "null" version of the energy conjecture, which refers to the Bondi mass and it would be very nice also to have a theorem there in order to know whether that mass is necessarily positive.

Chern: I think this Schoen–Yau result assumes a maximal hypersurface, and eventually the hope is that this can be removed. So, the problem is by no means closed. It is very open. (In a recent work of Schoen and Yau this assumption has been removed.)

Y. Ne'eman (Tel-Aviv U.): A question to Professor Chern of a more history-of-fashions type. We learned from listening to your very interesting talk now, that the *Erlangen* program was based on applying group theory to geometry. So group theory was, in a very obvious way, the guide to much of the world of that period. How come when we read what Hermann Weyl writes, he talks of the Group-Pest as if there was some kind of backlash against group theory. There seems to be an apology in the approach employing group theory, and the general impression is that group theory is not there to last, that it will just go away, that it's a temporary thing! Could you say something about that? Was there a period in which there was a reaction against that?

Chern: Group theory was accepted more readily by the mathematicians. It was a new concept, and was difficult to grasp.

V. Bargmann (Princeton U.): The word Group-Pest originated with the physicists and it was an obvious response. They had to learn something that they had never learned before. Nobody likes to do that, and if somebody can convince them that this is unnecessary, they can get the same results without learning group theory, they will be very grateful. So, first of all, this word of course is due to the physicists who didn't like it. It was a particular generation — the present-day generation has changed and the whole thing has disappeared. But then it was a very strong reaction.

Chern: I would like to say that the main emphasis in pure mathematics in the nineteenth century was analysis, that is, function theory, integration, and so on. Group theory became popular mainly because of Felix Klein. [Somebody says, "Jordan."] But Jordan dealt with "finite groups," which is an easier concept than continuous groups.

C. N. Yang (State U. of New York, Stony Brook): Could I make two very short historical remarks about Professor Chern's talk? The first is: I hope everybody realizes that the name gauge field is a misnomer. Gauge transformations and gauge fields were invented by Hermann Weyl in 1918, as we heard, and at that time he was gauging things because he wanted to have a scale invariance. He called it at first "Mass stab invariance." But later on we knew the "gauge" got turned 90° in the complex plane and there's an *i* added to it. The *i* removed all the objections to his ideas, but of course when the scale is changed by 90° in the complex plane, it becomes a phase. So it is not gauge transformations we are talking about, it is phase transformations, and it is not gauge fields we are talking about but phase fields. But, of course, it is impossible to change names now. A second remark I would like to make, a very simple one, is that we at Stony Brook learned really what is meant by a fiber bundle in 1975 when we invited Jim Simons, who was chairman of the mathematics department, to tell us what fiber bundles are all about. He spent considerable time very patiently telling us about De Rham theorems and so on and so forth, and we finally caught glimpses of the very beautiful theory of characteristic classes. Then, in turn, I believe it was Dan Freedman and I, who showed him the beautiful paper written by Dirac in 1931 that discussed the magnetic monopole and the resultant quantization of the magnetic and electric charges because of the marriage of Maxwell equations plus the magnetic monopole with quantum mechanics. After Jim read that, he said of course the Dirac quantization is the classification according to the $U(1)$ Chern class and he added, so Dirac, in fact, discovered the Chern–Weil theorem first.

E. Wigner (Princeton U.): I would like to ask the mathematicians a question that I have asked very often but never got an answer. I hope this time I do. It would be natural in physics to characterize a point in the four-dimensional space not by arbitrary coordinates but by four invariants. If we did that, how many equations are needed? How many more invariants have to be introduced so that, if these other invariants are the same functions of the original invariants, then the spaces are identical?

Chern: Four dimension is rather large. Now you have the same question in two dimensions. In two dimensions instead of the Riemann–Christoffel curvature tensor, you have the Gaussian curvature. And what you have in mind is to get other invariants, through covariant differentiation or differential parameters. And that has been classified and it depends on how many more independent functions you get. If your Riemann metric is very homogeneous, you do not get — I mean, for example if the Gaussian curvature is constant, then you do not get any more invariants. So it has to be generic, I suppose.

Wigner: The number of invariants is limited?

Chern: Yes, limited. If the space is homogeneous like the sphere, then you cannot get more scalar invariants and don't need them. For your purpose you need the condition that the metric is generic. The program can be carried out in the generic case.

Wigner: You know this has been a big and important question because the coordinates that we use are arbitrary and have no physical meaning. Therefore, for example, if we wanted to give a truly meaningful general relativistic description of the motion of a particle, we would have to specify the combination of the values of the four invariants that it encounters during its life. More specifically, three invariants would be given as functions of one, just as, in nonrelativistic theory, the three space coordinates are functions of the time coordinate. But in general relativity, the usually introduced coordinates are entirely arbitrary and have no real meaning. My question is, therefore, whether for instance such a formulation of the laws of motion of a particle can be given, whether it should be given, whether we know what invariants would have to be used.

P. Bergmann (Syracuse U.): I think this brings us right back to an elementary problem that you stated as the first problem, one of the first in differential geometry. You said that, in the beginning, the problem of the equivalence of quadratic differential forms was solved by Christoffel and Lipschitz. Now, I remember that Professor Wigner and several of us relativists had heated discussions at summer schools about how to go about this question in general relativity and, if I remember it correctly according to my studies of the subject from the literature, particularly from Cartan, the answer that is offered to this equivalence problem is of such a kind that one reduces the question of the equivalence to whether a certain system of exterior differential equations is solvable or not, and this is almost as difficult, in a way, as the original question that one asks. What one would much prefer in physics is an answer that intrinsically first characterizes each of the two spaces separately. And, as far as this has been attempted in general relativity, it seems there is, so to speak, no simple, constructive, explicit answer. One has to distinguish between a lot of different cases and there is a complicated classification. So my question is: Is there an elegant general solution of this equivalence problem also for indefinite metrics?

Chern: First of all, there is no difference between the definite and indefinite metric, provided the indefinite metric is nondegenerate, because it is a local problem. And secondly, the problem was solved by Christoffel and Lipschitz in the following sense. Originally you wanted to know whether two elements of arc are equivalent under the group of all point transformations — where the x''s are arbitrary functions of the x''s. Now, by constructing the Riemann–Christoffel tensors and the successive covariant derivatives, you reduce this problem to a problem relative to the general linear group. These tensors have to be transformed by transformations of the general linear group according to tensor algebra. So the answer is that there is a solution of the original problem if, and only if, the system of algebraic equations in the tensors have a solution. In that sense, by reducing a problem on differential equations to an algebraic problem, the problem was solved. Professor Wigner wanted to reduce the problem further by considering, instead of tensor, the actual functions. This is possible, and actually for two dimensions was carried out in detail in Darboux's book that I mentioned. In the four-dimensional case there was the work of Petrov. In the general case you need some generality assumptions in order that the problem can be controlled. Otherwise there are just too many cases. Incidentally, I think that the solution of Christoffel–Lipschitz I mentioned can be found in Veblen's book on invariants of quadratic differential forms. There it is presented in a much more elementary way.

Finally, let me answer Professor Wigner's last question. The insight gained by mathematicians through the works of Christoffel, Lipschitz, and Ricci leads to the realization that it is desirable to stay with general coordinates; and this is the foundation of tensor analysis. The existence of special coordinates, as proposed by Professor Wigner, will give deeper understanding of the problems, both physically and mathematically. It is only possible for special problems. My precise answer to Professor Wigner's question is: Such a formulation of the laws of motion *can* be given, but should *not* be given (because it will be complicated), and *I* do *not* know what invariants to be used.

Stachel: I think this discussion may be an example of one of Einstein's cynical comments about mathematics. He said, quoting loosely: "Mathematicians know many interesting things but never quite the ones we physicists want to know."

VII. The Universe

18. THE SIZE AND SHAPE OF THE UNIVERSE

Martin J. Rees

Einstein's first paper on cosmology appeared in 1917, twelve years before Hubble (building on earlier work by Slipher, Wirtz, and Lundmark) discovered the universal expansion, and even before the "spiral nebulae" were agreed to be extragalactic. Einstein deployed the concepts of general relativity to construct — for the first time — a self-consistent model for a homogeneous unbounded static universe. He had felt forced to introduce the "cosmological constant" Λ in order to obtain his static solution; but in 1922 Friedmann showed that, even for $\Lambda = 0$, Einstein's equations had solutions that represented expanding universes. Until the discovery of compact objects in the 1960s, there seemed no observational context other than cosmology where general relativity played any major role. During the fifty years that have passed since the pioneering researches of Einstein, de Sitter, Friedmann, and Lemaitre, much theoretical attention has been focused on devising other more complex and more general cosmological models: models with rotation, shear or gross inhomogeneity. But the present evidence allows us to trace the history of our universe back to a hot dense phase within one second of the "big bang." Over this entire stretch of time, its evolution has proceeded in amazingly precise accordance with a Friedmann model — the most straightforward and mathematically tractable of all.

In this report, I shall summarize the evidence that our universe is kinematically described by a Robertson–Walker metric and then assess the extent to which the dynamics of the universe — the relationships between expansion rate, curvature, and density — conform to Einstein's equations. Finally I shall discuss the origin of structures — galaxies, clusters, and so forth — and the validity and limitations of the homogeneity hypothesis.

1. Evidence for an Isotropically Expanding "Robertson–Walker" Model

1.1 *Hubble's Constant: The Present Scale of the Observable Universe*

The tight relation between the red shift and the apparent magnitude of the brightest elliptical galaxies in clusters tells us two things: first, these galaxies are indeed reasonable "standard candles"; and, second, their motions accord with those expected in a Robertson–Walker space-time for a set of co-moving objects whose distances are smaller than the overall radius of curvature

Harry Woolf (ed.), Some Strangeness in the Proportion: A Centennial Symposium to Celebrate the Achievements of Albert Einstein
ISBN 0-201-09924-1

of the hypersurfaces of homogeneity. The constant of proportionality between recession velocity and distance — the Hubble constant — is thus the key parameter determining the scale of the observable universe and the time scale over which its properties can change.

The preferred values of H_0 have dropped drastically since Hubble's original estimate of 530 kilometers per second per megaparsec (which corresponded to an embarassingly short "Hubble time" of $H_0^{-1} \simeq 2 \times 10^9$ years). Different experts currently quote values spanning the general range 50–80 km/sec/Mpc, so substantial uncertainties remain. At a recent conference in Paris attended by most of the astronomers specializing in the cosmic distance scale, no strong objections were voiced to a value $H_0 \simeq 70$ km/sec/Mpc.

The extragalactic distance scale is built up via a complex set of interrelated procedures. The main steps are these. The distances of nearby galaxies are determined by measuring the apparent brightness of Cepheids, R. R. Lyrae stars, novae, or supergiants in them. The absolute magnitudes of these objects — the so-called primary indicators — can be calibrated by observing them in our own galaxy. The goal is to extend the procedures for direct distance determination out to some galaxies far enough away that the "Hubble flow" is not significantly influenced by local pecularities in the velocity field; but no class of "primary indicator" is intrinsically bright enough to be seen in any sufficiently remote galaxy. One must thus have recourse to "secondary indicators." These are objects (for example, globular clusters or H-II regions) whose absolute properties can be calibrated in galaxies $\leqslant 5$ Mpc away (and thus latched onto the scale established from primary indicators), but which are detectable out to greater distances; another class of secondary indicators involve morphological criteria, for example the prominence of spiral structures, that appear to be correlated with a galaxy's absolute magnitude. I shall delve no further into the subject now but hope that Professor de Vaucouleurs will have an opportunity to offer an authoritative appraisal later in this volume.

Three types of development should gradually improve the precision with which H_0 is known:

1. Astronomers are discovering and calibrating an increasing variety of distance indicators. An attack on the problem that combines several methods should dilute systematic errors and "spread the risks," thereby increasing our confidence in the combined procedure.
2. The NASA/ESA Space Telescope, due for launch in 1983, should extend by a factor ~ 5 the range out to which Cepheids, globular clusters, and bright stars can be detected and used as distance indicators.
3. There are well-founded hopes of developing some feasible technique that can directly establish the metric distance to a remote extragalactic object, thereby bypassing the hierarchy of intermediate steps hitherto involved in establishing the distance scale.

   a. Application of the so-called Baade–Wesselink method to distant supernovae can in principle determine their distance if their rate of brightening, photospheric expansion velocity, and photospheric temperature are known.
   b. If the kinematics of "superluminal" variations in compact radio sources were understood, then metric distances could be inferred from (very-long-baseline) interferometer data.
   c. The hot gas in clusters of galaxies emits bremsstrahlung X rays, and scatters microwave radiation to produce a characteristic depression in the measured background temperature. For a cluster whose red shift, angular size, and temperature

are known, the X ray and microwave effects depend on $n_e^2 H_0^{-1}$ and $n_e H_0^{-1}$, respectively (n_e being the electron density); so the combined observations yield n_e *and* H_0.

d. If an intervening galaxy (or other massive object) produced a gravitational lens image of a variable quasar, then the structure and variability of the lens image essentially yields a metric distance by simple geometrical optics. It is not unduly improbable that astronomers may discover instances of the lens effect.

1.2 *The Isotropic Microwave Background*

The only subsequent cosmological discovery that has matched Hubble's work in importance was the detection by Penzias and Wilson in 1965 of the cosmic microwave background radiation. In the same issue of *Astrophysical Journal* in which this discovery was announced, Dicke and his Princeton colleagues interpreted the microwave background as a relic of a "primordial fireball" phase when the universe was hot, dense, and opaque. This concept had been developed many years earlier by Gamow, Alpher, and Herman.

The characteristics of this radiation — its isotropy, its apparently thermal spectrum, and its high energy density ($\sim$0.25 ev/cm^3 or $\gtrsim$ 10 times the energy density of intergalactic starlight) — were quickly recognized to be irreconcilable with most conventional astrophysical interpretations: A consensus thus emerged very quickly, even among the "conservative element" in astronomy, that this radiation was a relic of very early epochs indeed. Observations at various wavelengths have confirmed a broadly thermal spectrum. Searches for anisotropies have revealed no effect exceeding $\sim$0.1% on any angular scale down to $\sim$1 are min. (Being relative measurements these are much more precise than the intensity measurements.)

According to the "hot big bang" model the material in the early universe would constitute a hot plasma, strongly coupled to a thermal radiation field. The radiation temperature would decrease as $\rho_m^{1/3}$, where ρ_m is the particle density. When adiabatic expansion had cooled the material to approximately 4,000 °K, the plasma would recombine and the radiation would no longer undergo substantial scattering or absorption. The microwave photons that are now observed have not been scattered since an epoch when the mean density of matter was $\sim 10^9$ times higher than it is now — long before any galaxies came into existence. The effective source of the background is thus a "cosmic photosphere" at a red shift $z \simeq 1{,}000$. The measured isotropy then implies that the observable part of the universe can be described by a Robertson–Walker metric with a precision of $\lesssim$ 1 part in 10^3.

The only positive anisotropy so far detected, at the $\sim$0.1 percent level, can be attributed *either* to a $\sim$500 km/sec velocity of our galaxy (and its neighbors) relative to the mean velocity of distant matter on the cosmic photosphere, *or* to a $\sim$0.1 percent deviation from Robertson–Walker. I shall mention in Sec. 3 how we might distinguish between these two options. For the moment, the noteworthy inference is that the gross kinematics of the universe are exceedingly symmetrical. The microwave background establishes this with greater precision than the data on which the Hubble law is based. Unless we adopt an anti-Copernican viewpoint, this forces us to adopt a Robertson--Walker metric. In this metric the expansion is described by a single scale factor $R(t)$; the only other parameter is a curvature constant k, such that the spacelike 3-surfaces of homogeneity at time t have a curvature $\propto k(R(t))^{-2}$. The present Hubble constant is then $H_0 = \dot{R}(t_0) / R(t_0)$, and any observed source has a red shift z such that $(1 + z)$ equals the factor by which $R(t)$ has increased between emission and reception of its radiation. Of course, it is only because the universe possesses this gross simplicity that observational cosmology is feasible at all. Our empirical evidence is essentially limited to observations along our past light cone, and to inferences about the history of the region near our world line. Only because of the homogeneity can we infer any resemblance between distant events along our past light cone and events in our own galaxy's past.

Everything I have discussed so far is essentially kinematical: no dynamics. But one must introduce field equations in order to derive the form of $R(t)$ and to relate this function and the Robertson–Walker curvature to the content of the Universe. General relativity tells us that, for $\Lambda = 0$, a homogeneous universe is infinite and ever-expanding (negative curvature of 3-surfaces), or finite and destined to recollapse (positive curvature of 3-surfaces), according as $\rho \lessgtr \rho_{crit}$, where $\rho_{crit} = (3H_0^2/8\pi G)$. It has become conventional to define a density parameter $\Omega = (\rho/\rho_{crit})$.

To test whether we have a Friedmann universe governed by Einstein equations, we must relate the density, deceleration, and dynamics.

2. Testing the Validity of Friedmann-type Models

2.1 *Determining q_0 From the Magnitude-Red Shift Relation*

The best known "classical" cosmological test involves measuring the deviations from linearity at large red shifts in the Hubble law for the brightest galaxies in clusters, and thereby determining the deceleration parameter $q_0 = -(R_0\ddot{R}_0/\dot{R}_0^2)$. In Friedmann models where the pressure is $\ll \rho c^2$ (and with $\Lambda = 0$) q_0 is related to the present density parameter Ω_0 by $\Omega_0 = 2q_0$. The results remain disappointingly inconclusive, despite the intensive efforts of observers over many years.

The most obvious difficulty with the classical program for determining q_0 is that the galaxies become almost invisibly faint — even with the best telescopes and detectors — before one can probe deep enough into space for the effects of deceleration to really show up. This makes it hard to measure magnitudes accurately for relevant galaxies. But there are several further problems, among which are the following:

1. No class of galaxies constitutes a "standard candle" with a precision better than $\sim$30 percent; this introduces an inevitable scatter into the magnitude-red shift relation.
2. Galaxies are not point sources, nor do they have sharp edges; their surface brightness falls with increasing radius and eventually merges into the background. Thus the measured magnitude must be corrected to allow for the fraction of the galaxy's total light contained within the aperture. This correction is z-dependent, but what is worse is that it depends also on q_0, (because the angular size corresponding to a given linear dimension and red shift is a function of q_0).
3. A systematic error may stem from the inclusion in the chosen sample of progressively more exceptional and intrinsically luminous galaxies at larger z (the "catchment volume" rises roughly as z^3).
4. Galaxies with large red shifts are being seen at a younger stage in their evolution; this may make the luminosities depend on z. There are two competing evolutionary effects.
   a. The evolution of the stellar population probably causes elliptical galaxies to fade with age, as progressively lower-mass stars evolve off the main sequence and onto the giant branch (the importance of this effect depends on the mass function of the stars). *But:*

b. Massive galaxies tend to gain mass, and therefore luminosity, by capturing and swallowing smaller galaxies in the same cluster. The importance of these "mergers" and "galactic cannibalism" has only recently been appreciated.

There are additional uncertainties that will need to be assessed before we can determine q_0 precisely. But at the moment the corrections due to evolution (item 4 in the preceding list) are so uncertain that they would change q_0 by an amount $\pm$ 1.

The situation may improve when better data are available on large samples of galaxies with $z \simeq 0.5$. Pushing the Hubble diagram to still larger red shifts will not in itself help, because the evolutionary corrections then become bigger and still more uncertain. The Space Telescope, which should be able to resolve the profiles of individual galaxies out to $z \simeq 0.5$, will elucidate the evolutionary effects. The astrophysical and geometrical problems are so intertwined that there is little hope of determining q until we understand galaxies better. Extragalactic research must be pursued on a broad front, in the hope that solutions to cosmological and astrophysical problems will swim into focus simultaneously.

Formally, the effect of the deceleration on the magnitude-red shift relation is, to lowest order, a purely kinematic one. In a power series expansion in z the effects due to focusing by matter along the light cone are of higher order. In practice, however, the test is feasible only if observations extend out to $z \simeq 0.5$; it is therefore best to fit to a specific model rather than use a series expansion. In Friedmann models, the angular sizes start to *increase* at large red shifts. Since the energy flux received from a distant object is proportional to (surface brightness) $\times$ (solid angle subtended) $\times$ $(1 + z)^{-4}$, magnitude-red shift tests and angular-diameter tests are essentially equivalent. Sadly, a well defined and nonevolving "rigid rod" has proved even more elusive than a suitable "standard candle." (In a high-density universe where the matter is "clumped," the expected magnitudes and angular diameters of a sample of distant objects are different from in a strictly homogeneous Friedmann model with the same Ω, owing to the effects of shear and multiple weak gravitational lenses along the line of sight).

2.2 *Quasars and Radio Galaxies*

Quasars and radio galaxies, even if we do not understand their precise nature, might seem potentially important for observational cosmology, because they are hyperluminous beacons allowing us to probe much deeper into space (that is, further back into the past) than is possible by means of normal galaxies. There is, however, a huge spread in their intrinsic properties, and no subclass can be distinguished that serves as a good "standard candle." One must therefore rely on statistical correlations between magnitude, radio flux density, and red shift.

There is now substantial evidence that the density of quasars and radio galaxies depends steeply on cosmic epoch. The density of powerful sources, both radio galaxies and quasars, apparently increases by a factor 10^2 to 10^3 *per co-moving volume* — that is, over and above the $(1 + z)^3$ compression factor) between the present epoch and the epoch corresponding to $z = (2–3)$. This result is rather insensitive to the cosmological model that is adopted, and one certainly does not know enough about the relevant astrophysics to be able to deduce the evolution, as one might hope for normal galaxies, and thereby disentangle the geometrical effects and estimate q.

It is perhaps disappointing that the discovery of objects with very large red shifts has not led to any progress in "classical" or "geometrical" cosmology, and that it is still studies of normal bright galaxies (despite the smaller red shift involved) that are likely to yield the most significant limits on the deceleration parameter.

The most dramatic inference from these analyses is that the density of powerful sources would have been about 1,000 times higher at early epochs than at the present time. (The inference was particularly important in the early 1960s when it was originally made by the Cambridge radio astronomers since it provided the first firm evidence against the steady-state theory.) The most distant observed known sources — quasars with red shifts ~ 3.5 — are so far away that they emitted the radiation that we now detect when the universe was <20 percent of its present age. By observing such objects we are thus in effect probing 80 percent of cosmic history, and maybe it should not surprise us when we infer such dramatic differences between conditions at these ancient epochs and at the present day. A hypothetical astronomer observing only 2 billion years after the initial singularity would perceive a vastly more active and dramatic environment: whereas our nearest bright quasar, 3C 273, is about 800 Mpc distant, he would be likely to find a similar object only 15 Mpc away, or 50 times closer, and appearing as bright as a fourth-magnitude star. This evidence is obviously crucially important to our understanding of galaxy formation and evolution. For some reason, galaxies had a much greater propensity to develop explosive nuclei when they were young than they do today.

(Galactic nuclei do not feature elsewhere in the program of this conference, so it may be appropriate to recall the growing consensus among astrophysicists that the central energy source in quasars and radio galaxies involves a massive ($\gtrsim 10^8 M_\odot$) black hole. Active galactic nuclei mark places where the space of our universe has been punctured by the accumulation and collapse of large masses. It would be exciting indeed to relativists if there were some crucial diagnostic whereby, for instance, the radiation emitted by gas swirling towards a black hole gave quasars some observable characteristic that directly indicated the metric, and thus led to an observational test of "strong field" general relativity. It is perhaps these massive holes (rather than the stellar-mass black holes in X-ray sources) that hold out the best prospects for a "clean" test of gravitation theory. This is because a black hole of stellar mass develops only after collapse to nuclear densities, with all the physical uncertainties entailed by high-density physics; a black hole as massive as the one postulated in the elliptical galaxy M87 would, on the other hand, have formed before the mean density of its constitutent material exceeded that of air. An experimenter falling into such an object would still have several hours for leisured observation before being discomfited by tidal forces or imminent incorporation in the singularity.)

2.3 *Age Estimates*

Given the substantial uncertainties, there seem to be no embarassing conflicts between the astrophysically inferred ages of the oldest galactic objects and the estimated time since the big bang. The age of a (decelerating) Friedmann model is bounded by H_0^{-1}; for $\Omega_0 << 1$ it is $\sim H_0^{-1}$; for $\Omega_0 \gtrsim 1$ it is $\leqslant \frac{2}{3} H_0^{-1}$. A serious "age problem" would arise only if H_0 were close to the upper end of the currently quoted range of values (~ 80 km/sec/Mpc) and Ω_0 were $\gtrsim 1$; improved age estimates might set tighter constraints on permissible models.

The concordance, at least within a factor ~ 2, between H_0^{-1} and the age of the oldest stars is a natural consequence of a big bang model, but would be a remarkable and unaccountable coincidence in some "unconventional" theories (for example, tired light, steady state, or chronometric cosmology).

2.4 *The Expansion Time Scale at Very Early Epochs*

The initial dense "fireball" phase is very brief compared to the later stages. The thermal radiation field (and not the baryons) then provides the dominant mass-energy, so that $\rho \simeq aT^4 f(T)/c^2$, where $f(T)$ is a factor (increasing with T) that allows for the other species, apart from photons, that are present in equilibrium radiation at temperature T. The Friedmann models yield a unique relation between the expansion rate and density at early times: $(R/\dot{R}) = (3/8\pi G\rho)^{1/2}$. This translates into a unique temperature-time relation. The amount of helium synthesized in the big bang is controlled by the expansion rate prevailing when $T \simeq 10^{10}$ °K (and $t \simeq 2$ sec); the present helium abundance can thus set constraints on the dynamics of the universe at epochs vastly earlier than those accessible to any direct scrutiny.

The standard hot big bang model predicts that the primordial material should emerge with approximately 25 percent (by mass) of helium. There are no astronomical objects whose helium abundance is definitely below this value, and this high percentage of helium could not have been synthesized by the kind of nucleosynthetic processes occurring in stars without the concurrent production of an excessive abundance of heavier elements. It is thus tempting to attribute the production of most cosmic helium to the big bang itself. Helium is synthesized during the first 100 seconds of the expansion (temperatures $\sim 10^9$ °K). The amount produced depends on the neutron–proton ratio, which "freezes out" when the universe cools to the stage when weak interactions become too slow to maintain thermal equilibrium. This ratio is very sensitive to the expansion rate: if this rate differed by even 50 percent from that in the standard model, the helium abundance would disagree with that observed.

There are some cosmological models (involving anisotropic expansion, scalar fields, or a variable gravitational constant) that are consistent with all other data but predict expansion rates at the helium formation epoch that are up to 10^6 times faster than in the standard model. Thus the measured helium abundance in the universe strongly supports the simplest version of the hot big bang theory. It also suggests that we are perhaps justified in extrapolating the laws of microphysics back to these epochs. Cosmic deuterium may have a primordial origin also.

Many theorists have discussed processes occurring at still earlier phases and higher temperatures — even the first microsecond. Here one has even less confidence in the applicability or completeness of known physics. The matter density is well below nuclear densities at all times $\gtrsim 10^{-5}$ sec. At times $< 10^{-5}$ sec, however, the temperature is so high that nucleon-antinucleon pairs would be as abundant as photons, the densities would exceed nuclear densities, and the equation of state is unknown. Nevertheless, the explanation for the observed isotropy, and maybe other key features of the universe as well, may lie in processes associated with the extreme conditions prevailing at the earliest times, when exotic processes in particle physics, and even quantum gravitational effects, hold sway. These will be discussed by Professors Sciama and Weinberg in this volume.

2.5 *The Mean Density of Matter*

The critical density ρ_{crit} is 4.7×10^{-30} $(H_0/50$ km/sec/Mpc$)^2$ g/cm^3. The mean density of matter in the luminous visible part of galaxies falls short of this by a factor of about 30. Even if we include the extra "dark mass" whose presence in galactic halos and clusters is inferred from dynamical considerations, the best estimate of Ω is only ~ 0.2. This estimate is uncertain by a factor ~ 3; it depends on (the square of) the often poorly known velocity dispersion in gravitationally bound systems. However, it seems unlikely that the universe has a density $> \rho_{\text{crit}}$ unless there is some "missing mass" that is more smoothly distributed than the mass we see, and not concentrated in the potential wells of galaxies and clusters. Professor Field will be discussing the possibility of a very hot intergalactic gas, and there are other more exotic possibilities. Einstein was well aware of the "missing mass" problem. In the appendix on cosmology in later editions of *The Meaning of Relativity* he stressed that no estimate of the mean density should be considered as anything other than a lower limit; he also argued that the proportion of matter in the "dark" form should be larger outside galaxies than within.

If the observed interstellar deuterium is primordial, the standard hot big bang model implies that the value of ρ_m must be $\lesssim 5 \times 10^{-31}$ g/cm^{-3}. Deuterium production is an intermediate stage in helium synthesis, and only if the baryon density is low does a significant amount of D get left behind. During the last few years, theorists have tried hard to think of synthesizing deuterium by current astrophysical processes (for example, by spallation of helium in shock waves from supernovae). So far those efforts have been unsuccessful. This may merely mean that the theorists have not been ingenious enough; on the other hand, it may be a genuine argument in favor of a low-density universe. As almost always happens in astrophysics, no conclusion can be stated without inserting provisos and "escape clauses." Even if D *is* made in the big bang, Ω could be high if the missing mass were in some form such as small black holes or leptons of nonzero rest mass: such entities could make the main contribution to ρ now, but at the nucleosynthesis epoch they would neither affect the expansion rate (which is controlled by the radiation) nor contribute to the baryon density.

2.6 *Cosmological Tests of Einstein's Equations*

A crucial and direct cosmological test of Einstein's field equations will be feasible when (and not until) we know both q_0 and Ω_0; the Friedmann models predict $\Omega_0 = 2q_0$ (provided that pressure is now dynamically negligible, which should be the case unless a big contribution to Ω comes from some undiscovered type of relativistic particle). This test also demands sufficient uniformity in the matter distribution (cf. Sec. 3) to ensure that the measured value of Ω prevails out to the distances of the objects from which q_0 is determined.

The Friedmann models yield two contrasting scenarios for the future of the universe. Which one is realized depends on whether the density parameter Ω is $\leqslant 1$ or > 1. In the former case, the expansion will continue indefinitely: the galaxies will disperse and fade, the endpoint — after time scales quite immense compared with the present Hubble time — being the incorporation of material into black holes, which then evaporate via quantum radiance effects. If $\Omega > 1$ the expansion will eventually reverse, galaxies and stars being destroyed and finally engulfed in a "big crunch" — a "fireball" differing from that with which our universe began in having higher entropy and being less smooth (owing to cumulative entropy production and gravitational clustering during the expansion and recollapse phases).

If it turns out that $\Omega_0 \neq 2q_0$, Einstein's theory could be salvaged by reviving the Λ-term as an extra parameter; such a model makes specific predictions about the magnitude-z relation and could be checked if this relation were extended to greater distances (or to the next order in z in a power series expansion).

Einstein himself would have been disappointed by such an eventuality. He introduced the cosmical constant, or Λ-term, in order that his field equations should permit a static homogeneous universe. But he abandoned the Λ-term (as his "greatest blunder") when Friedmann showed that there were nonstatic models with $\Lambda = 0$; and when Hubble's great work showed that static solutions were irrelevant. Since the 1930s the Λ-term has been generally regarded as an ugly appendage to general relativity, or as an unphysical field "acting on everything but acted on by nothing"; most cosmologists feel that the theory is well rid of it. But a minority have considered, contrariwise, that it is a legitimate parameter whose value must be determined empirically; observations of galaxies and clusters set upper limits to Λ, and it can have no significant manifestations except of the cosmological scale. About a decade ago, there seemed to be evidence for a pile-up of quasar red shifts at a particular value. This triggered a flurry of interest in Eddington–Lemaitre models (with $\Lambda \neq 0$); in these models the universe pauses in its expansion and can experience a long "coasting phase" where gravitation is balanced by the cosmical repulsion. Later data failed to corroborate the strong pile-up in quasar red shifts, and the Eddington–Lemaitre model has been put back on the shelf. Recently there has been renewed discussion of the Λ-term in a more modern guise by identifying it with a small negative energy density associated with the quantum vacuum.

If general relativity proved inapplicable to the cosmos, after having been vindicated by "local" and astronomical tests, the status of cosmology as a genuine physical science would suffer a drastic setback. We should then be forced to conclude that the dynamics and large-scale structure of our universe were governed by laws and processes that have no other observable manifestation, and so cannot be investigated in any more local or controllable context. Cosmology would then drop to the scientific level of "Just-So" stories.

3. Homogeneity

3.1 *The Observed Large-Scale Distribution of Luminous Matter*

There is strong evidence, as noted previously, that the universe, on large scales, is homogeneous and isotropic in its kinematic properties. If one accepts general relativity, this must imply some degree of uniformity in the matter distribution. Such limits are, however, rather weak if Ω is low; the gravitational effect even of an "overdense" region is then weak and cannot cause much deviation from uniform undecelerated Hubble flow. Moreover, if one regards the universe as an arena for *testing* general relativity, independent evidence on the matter distribution is essential.

The distribution of matter is evidently by no means uniform on the scale of individual galaxies — nor, of course, on any smaller scale. What, then, is the independent evidence that the matter distribution is, even in a "broad brush" sense, homogeneous?

Galaxies are aggregated in groups and clusters; the statistics of the clustering have been studied by many astronomers over the years, culminating in the elaborate analyses by Peebles and his associates. The main message of these investigations is that there is no preferred scale for a "cluster of galaxies"; the galaxies are grouped in such a fashion that $\xi(r) \propto r^{-1.7}$, the probability of a galaxy lying in a volume element δV a distance r from a given galaxy being proportional to $(1 + \xi(r))\,\delta V$. The characteristic radius r_0 for which $\xi(r_0) = 1$ is

$$\sim 10\,(H_0/50 \text{ km/sec/Mpc})^{-1} \text{ Mpc} .$$

On scales $r \gtrsim r_0$ the inhomogeneities in the density of galaxies generally have amplitudes less than unity. There is no firm evidence that $\xi(r)$ follows the same straight power law when $r > r_0$ (and reasons for suspecting that it does not); there is good evidence for some clumping on scale of 50

Mpc. But the universe does seem to be more uniform in scales > 100 Mpc than on any scales < 10 Mpc. The evidence for this uniformity comes from counts of faint galaxies in different directions, and to different depths in the same direction. Analyses of the sources in radio surveys (many of which are quasar-type objects with $z \simeq 1$) fail to reveal any anisotropy. The X-ray background, which is largely due to discrete sources with $z \simeq 1$, or to diffuse gas, is isotropic to a precision of 1 to 2 percent. Even if one sets aside all the kinematical arguments in Sec. 1, there is good radio, optical, and X-ray evidence that the distribution on the sky of all kinds of luminous objects becomes smoother as we look deeper. This suggests, given a "Copernican" assumption, that the spatial distribution becomes smoother as we average over cells running from ~10 Mpc to ~3,000 Mpc in size (and is inconsistent with any Charlier-style hierarchy).

The X-ray background is likely to prove a sensitive probe for the large scale distribution of matter. Any very large-scale (~1,000 Mpc) region that is slightly "overdense" will perturb the X-ray background in two ways: 1) the enhanced density of sources in the region will show up as an excess X-ray flux in the corresponding direction; and 2) the overdense region will induce a slight shear in the motions of all objects influenced by its gravitational field, and will give our galaxy a peculiar velocity. For a lump of given size and location, effect (1) depends on $\delta\rho/\rho$, but effect (2) depends on $\Omega(\delta\rho/\rho)$. The X-ray data may soon be sufficiently precise to permit interesting investigations along these lines. By combining such data with information on anisotropies in the microwave background, it should be possible to test whether the latter are due to our peculiar velocity or to nonuniformities on the "cosmic photosphere" at $z \simeq 1,000$.

3.2 *The Origin of Galaxies and Clusters*

The observed inhomogeneity of the universe indicates that, on co-moving scales of 1–10 Mpc, "curvature fluctuations" of amplitude 10^{-4} were imprinted into the initial conditions. If these are "given," recent work suggests that gravitational clustering and dissipative processes can evolve them into galaxy and clusters. These fluctuations may be part of a smooth power-law spectrum extending over a much larger range of scales. The only irregularities that are seen now — and that need ever have existed in the past — are *weak* perturbations of the metric, in the sense that, on length scales r, $(G\delta\rho r^2) << c^2$. The microwave background isotropy guarantees that the universe is extensive and homogeneous enough to satisfy the "trapped surface" condition of the singularity theorems, implying that it had a singular origin. The universe is thus smooth and uniformly curved only in the same sense as is a golf ball or an orange: there is a well-defined overall curvature, but superposed on this are localized small-amplitude deviations from the mean. There is no explanation for the number $\varepsilon \simeq 10^{-4}$ that characterizes the "roughness" of the spacelike hypersurfaces. Cosmologists divide into two camps. Some (adducing arguments from "simplicity," thermodynamic equilibrium, and so forth) regard a strictly homogeneous cosmos as "natural," and are therefore surprised that ε is so large compared to the very tiny value expected from, for instance, $1/\sqrt{N}$ fluctuations in the baryon distribution. Other cosmologists regard a "chaotic" universe, with more macroscopic degrees of freedom open to it, as more likely than one described by the Robertson–Walker metric. They are consequently surprised that ε is as small as 10^{-4} — that the initial state did not more fully avail itself of these degrees of freedom — and regard the universe's large-scale smoothness, rather than its small-scale roughness, as the greater enigma. To "explain" a specific amplitude like 10^{-4} must entail exploring nonlinear processes, perhaps at some very early epoch.

3.3 *Homogeneity of the Whole "Hypersphere"?*

Einstein attached preeminence to the hypothesis of global homogeneity; he felt "urged towards the conviction that there exists an average density of matter in space which differs from zero." A finite ever-expanding "island universe" would be unacceptable on Machian grounds. This homogeneity, implicit in Friedmann models, is broadly vindicated by all observations in the part of the universe accessible to us. But our direct evidence is restricted, even in principle, to the domain within our present particle horizon, which does not (except in some models with $\Lambda \neq 0$) encompass the entire content of the universe; one may therefore query the plausibility of extrapolating into the unknown region. (I shall not explore the possibility of "multiply connected" topology, which would in principle permit an ever-expanding universe where we see the same finite volume over and over again in a lattice structure.)

In a finite closed universe the fraction of the matter content that we can (in principle) already see at time t_0 is $\gtrsim t_0/t_{cyc}$, where t_{cyc} is the time for the entire cycle. Unless Ω exceeds unity by only a very tiny amount, we have thus already seen a significant fraction of "everything there is," so only a limited extrapolation is involved in inferring genuine global homogeneity. If we have already seen halfway around the universe, it is not too implausible to conjecture that the other half is similar, and meshes smoothly onto the part we see.

On the other hand, an ever-expanding Friedmann model with $\Omega < 1$ contains an *infinite* amount of matter. The amount within our horizon is, of course, finite, but grows with t (the growth being proportional to t^2 in the limit $t \rightarrow \infty$ for a $k = -1$ ($\Omega < 1$) model).

The extrapolation to global homogeneity is thus a far greater one for an open universe, and less evidently warranted. Even though we know the deviations from homogeneity are small on the scale of our present horizon (and can infer *something* about even larger scales, which would manifest themselves as a shear within our horizon) there is no prohibition on adopting a finite model whose edge may be seen in some future era. (We may recall Einstein's preference for a closed, bounded universe, in which this prospect would seem excluded.)

4. Conclusions

The most remarkable outcome of fifty years of observational cosmology has been the realization that the universe is more isotropic and uniform than the pioneer theorists of the 1920s would ever have suspected. The microwave background attests rather unambiguously to a singular origin; its discovery has brought the physics of the very early universe within the framework of serious scientific discussion. The hot big bang cosmology should not become a dogma, but it has more than fashion and beauty to commend it, and is justifiably being adopted by most cosmologists as a working framework until some glaring contradiction emerges. There can, however, be no confirmation that the dynamics of the expansion currently proceed in accordance with Einstein's equations until the density of matter and the deceleration are better determined; in fact we have stronger evidence (via the cosmic helium abundance) that the Friedmann equations applied when the universe was 1 second old than we have for their present-day applicability.

Two goals lie before cosmologists in the next decade:

1. To investigate and understand the physical processes whereby the primordial plasma gradually condensed into galaxies, and the internal evolution of the individual galaxies.
2. To address the fundamental question of *why* our universe actually has the entropy, scale, structure, and overall symmetry that is a prerequisite for scientific progress in cosmology.

19. COMMENT ON "THE SIZE AND SHAPE OF THE UNIVERSE"

Phillip J. E. Peebles

Martin Rees has given a clear and fair account of our present understanding of the large-scale structure of the universe: a nearly homogeneous mass distribution with no visible edge and no preferred center. It is interesting to ask how Einstein hit upon this picture of the universe and how it came to be accepted. Following are some of the main points I could find in the literature.

In this volume several authors remark that Einstein was not one to be deterred by experimental details, particularly when the phenomena were complex and poorly explored, and that seems to be the case here. A good picture of the observational situation in 1916 is found in Eddington's book, *Stellar Movements and the Structure of the Universe.*[1] The universe in Eddington's book is the Milky Way galaxy of stars. The count of stars as a function of brightness and direction in the sky indicated that our galaxy is a finite system shaped like a disk. This is still the accepted interpretation although we now know the radius of the disk was substantially underestimated because the effect of obscuration by interstellar dust was not fully appreciated. It was speculated that the spiral nebulae are similar island universes of stars, and in support of this Eddington noted that the dark lanes seen in edge-on spirals resemble the rifts in the Milky Way caused by obscuring interstellar clouds. A problem with this idea was that the spirals are not seen in the directions of the Milky Way. If the spirals are star systems comparable to our own, one would not expect them to prefer the poles of our galaxy. Again, the problem was that the effect of obscuration by dust in the disc was not fully appreciated.

De Sitter's second paper on relativity and Einstein's 1917 paper on cosmology give us some impression of Einstein's ideas of what the universe might be like in the large.[2] He argued that, in the conventional Newtonian theory, the universe of matter could not be homogeneous and unbounded, for that would make the gravitational potential diverge, giving the stars arbitrarily large peculiar velocities. He advanced two arguments against the idea that the stars are concentrated in an island in otherwise empty asymptotically flat space. First, the distribution is secularly unstable: thermal relaxation causes some stars to evaporate from the system while others become concentrated in a tight cusp at the center. That is, an island universe could not be a permanent arrangement. I doubt that Einstein considered this very seriously, for the same problem is to be found on smaller scales. Given sufficient time, the planets in the solar system must escape or fall into the sun, and, if energy is conserved, the stars eventually must stop shining. Einstein's second objection

Harry Woolf (ed.), Some Strangeness in the Proportion: A Centennial Symposium to Celebrate the Achievements of Albert Einstein
ISBN 0-201-09924-1

to the island universe is that a star that escapes from it moves arbitrarily far from all other matter but yet retains all its inertial properties. This violates a principle Einstein took very seriously, that "there can be no inertia relative to 'space' but only an inertia of masses relative to one another."[3]

One way to avoid the problem is to imagine that the gravitational potential diverges outside the island of matter, so nothing can escape. In 1916 Einstein was considering the idea that in the absence of matter the components of the metric tensor might naturally degenerate to

$$g_{ij} = \begin{matrix} 0 & 0 & 0 & \infty \\ 0 & 0 & 0 & \infty \\ 0 & 0 & 0 & \infty \\ \infty & \infty & \infty & \infty \end{matrix} . \tag{19.1}$$

An island of matter would force g_{ij} to deviate from this natural form to one that, with a suitable choice of coordinates, is close to the Minkowski form:

$$g_{ij} = \begin{matrix} -1 & 0 & 0 & 0 \\ 0 & -1 & 0 & 0 \\ 0 & 0 & -1 & 0 \\ 0 & 0 & 0 & 1 \end{matrix} . \tag{19.2}$$

Thus in accordance with Mach's principle matter would be the source of space, and outside the realm of matter g_{ij} would degenerate to a natural form where inertia is meaningless.

At the boundary of the realm of matter there would have to be masses that cause space to curl up into the singular form of Eq. (19.1). De Sitter[4] was willing to agree that such masses may exist, but he felt that little is gained because hypothetical absolute space has only been replaced with "hypothetical masses." Since all measured spectra of stars and nebulae show spectral lines quite close to the wavelengths measured in the laboratory, all these objects must be within the region where Eq. (19.2) is a good approximation, and so they could not be the hypothetical masses. Thus, although Einstein's scheme is philosophically attractive as an explanation for inertia, de Sitter objected that otherwise undetected matter is invoked ad hoc to account for the one phenomenon of inertia.

Einstein soon abandoned the scheme and hit upon another that proved much more popular.[5] He assumed that the mass density is homogeneous in the large-scale average, with space curved so it closes in on itself. Thus the universe is finite but has no hypothetical masses at the boundaries, and indeed has no boundaries. Einstein's argument from Mach's principle seemed to be met in a way that avoided de Sitter's objections.

De Sitter had mentioned conversations with Einstein at Leiden.[6] This should have exposed Einstein to the observational situation on the large-scale structure of the universe. I have found no clues to how much attention Einstein gave to the observations or what he made of them. De Sitter quickly took up the homogeneity assumption and in his third relativity paper he was willing to speculate that the spiral nebulae are islands of stars like the Milky Way and that these nebulae are uniformly distributed so that Einstein's homogeneous universe is the realm of the nebulae.[7] It is somewhat surprising that de Sitter did not discuss possible observational checks on the homogeneity assumption although a crude test was at hand. For every resolved nebula there are many faint nebulae that could be spirals at much greater distances.[8] That is what one would expect if the nebulae were uniformly distributed, because there is more room for nebulae at greater distances.

Just as the variation of star counts with brightness had been used to show that the Milky Way is bounded, nebula counts could be used to explore how the nebulae are distributed. It appears that Hubble in 1926 was the first to apply this test and to show that the counts do behave about as expected for a homogeneous distribution in approximately flat space.[9] The expected variation of nebula counts with apparent magnitude m,

$$\frac{dN}{dm} \propto 10^{0.6m} , \tag{19.3}$$

might well be counted as another Hubble's law as significant as his relation between apparent recession velocities and distances of galaxies,

$$v = Hr , \tag{19.4}$$

where H now is called Hubble's constant.

The data Hubble had available for this test were quite limited, and the results thus were only a preliminary indication, as Hubble well realized; much of his work in the next decade was devoted to improving the test. The lively discussions of cosmologies that followed Hubble's first tentative arguments for Eqs. (19.3) and (19.4) almost entirely concentrated on homogeneous world models. It is clearer in hindsight than it was at the time that this was a considerable extrapolation from crude observational evidence.

Modern galaxy counts reach apparent magnitude $J \sim 24$, where the galaxy red shifts are on the order of unity, and typical galaxy distances are $r \sim cH^{-1} \sim 1 \times 10^{10}$ light years. The trend of the counts with apparent magnitude agrees with Hubble's conclusion: there is no indication of an edge to the realm of the nebulae. However, it is amusing to note that the observed counts dN/dm closely approximate Eq. (19.3) in only a very limited range of galaxy brightnesses, from perhaps twelfth to sixteenth magnitude. At the bright end, the scope of dN/dm is shallow because of the effect of the local concentration of galaxies in and around the Virgo cluster of galaxies, of which we are an outlying number. From sixteenth to twenty-fourth magnitude the power-law index in Eq. (19.3) is about 0.43 rather than 0.6. This is at least in part the result of the cosmological red shift, which moves the spectrum of a distant galaxy toward the red and out of the fixed band of observation in the visible. Unfortunately the size of the effect on the apparent magnitudes is not yet known very well because we do not know enough about the ultraviolet spectra of galaxies. Thus, although the deep counts do not contradict the homogeneity assumption, they do not rule out an appreciable radial gradient in density. I would guess that the mean co-moving number density at $r \sim cH^{-1}$ could differ from the "local" mean density at $r \sim 0.01\ cH^{-1}$ by a factor of 3.

The isotropy of the deep counts can be tested to much better precision. For example the counts in Kron's two well-separated fields differ by no more than 20 percent over the magnitude range $J = 20$ to 23.[10] It may prove difficult to make the limit much stronger than this because of the problem with variable obscuration in our galaxy.

The isotropy of the galaxy counts is considered good evidence of a like degree of large-scale homogeneity, for the alternative is a universe spherically symmetric about a point near us, which would mean that the view of the universe from most galaxies that seem to be equally good homes for observers would be quite different from the view from our own galaxy, and that seems unreasonable. This is the Copernican argument Rees mentioned.

The most precise measure of the large-scale isotropy of the universe comes from the microwave background, which is isotropic to better than $\delta T/T \sim 10^{-3}$ on angular scales $\gtrsim 10'$. This says that the red shift to the horizon is isotropic to like accuracy. Therefore, save for special initial conditions, mass density fluctuations on the scale of the Hubble length cH^{-1} could not be greater than $\delta\rho/\rho \lesssim 10^{-3}$, for otherwise the perturbation to the radiation temperature that arises before the density fluctuation came within the horizon would have exceeded the bound on $\delta T/T$.

The recent observations yield fairly convincing evidence that the universe is accurately homogeneous in the average over scales comparable to cH^{-1}. This must be counted as a remarkable success for Einstein's physical intuition. The great enigma is whether we can make more of it than that, whether Mach's principle really forces the universe to be accurately homogeneous on the scale of the horizon.

NOTES

1. A. S. Eddington, *Stellar Movements and the Structure of the Universe* (London: Macmillan, 1914).
2. W. de Sitter, Monthly Notices Roy. Astron. Soc. **77**, 181 (1916); A. Einstein, S. B. Preuss. Akad. Wiss., p. 142 (1917).
3. Einstein, op. cit. in n. 2.
4. de Sitter, op. cit. in n. 2.
5. Einstein, op. cit. in n. 2.
6. de Sitter, op. cit. in n. 2.
7. W. de Sitter, Monthly Notices Roy. Astron. Soc. **78**, 3 (1917).
8. R. F. Sanford, Bull. Lick Obs. **9**, 80 (1917).
9. E. Hubble, Astrophys. J. **64**, 321 (1926).
10. R. G. Kron, Ph.D. thesis, University of California, Berkeley, 1978.

Open Discussion

Following Papers by M. J. REES and P. J. E. PEEBLES

Chairman, D. SCIAMA

Ehlers (Munich): If one accepts the value of 100 kilometers per second per megaparsec, isn't one then again in some trouble with stellar evolution ages — the oldest ones? This cannot be used as an argument against the measurements as such; but isn't that then already a somewhat serious problem?

Y. Ne'eman (Tel-Aviv U.): To Professor Rees. You said that it is clear that after the microwave screen, we are in a Friedmann universe. However, you also drew some evidence from homogeneity before the screen. To what extent do you feel that this evidence could not be explained by having a nonhomogeneous start with mixmaster or any kind of dissipative arrangement afterwards — I mean something that would turn the nonhomogeneity into a homogeneous system afterwards.

Rees: The only direct evidence that the universe was Friedmannian long before the last scattering is the evidence from the helium. I do not rate that as convincing at more than, say, the 60 percent level: I think one should be open-minded about the possibility that helium is made when the first stars form and that most of the entropy is produced there. So, I do not think one should be too convinced by the evidence for the Friedmannian time scale at the epoch of nucleogenesis. Now the other issue is whether a universe that was initially chaotic could have isotropized itself. Probably Professor Misner will be addressing this question. My impression is that the people who have tried to discuss this type of model have not found it possible to erase the inhomogeneity, and so it does seem very difficult to explain the present large-scale structure unless one does somehow imprint it at the threshold of classical cosmology.

G. B. Field (Harvard U.): Concerning the point as to whether the helium that we observe is, in fact, produced in the early big bang, I think one piece of evidence is quite relevant. And that has to do with the fact that in observing H-II regions in external galaxies, it is found that there is a definite gradient in the abundances of the elements as one moves from the center of the galaxy to the outer parts. This is due largely to the work of Piembert and his collaborators. The nature of this change is that both helium and the other heavy elements are positively correlated, and furthermore they are correlated in such a way that one would predict this correlation from the theory of nucleosynthesis in stars. However, when one extrapolates to zero metal abundance, typically in the outer parts of these galaxies, one finds that one is left with a very large residual helium, which I think is properly interpreted as cosmological.

Rees: That is evidence that the helium is mainly pregalactic, but that does not mean, I would argue, that it has to be made when the universe was less than 1 minute old. It could be made in a pregalactic generation of massive stars.

In response to another question, from the floor, about uniformity in nature, I would emphasize that the homogeneity and overall smoothness of the universe is, so far as I know, a complete mystery in the context of all present-day theory. And one must not forget that the fact that we observe a homogeneous universe and that the Friedmann models were already known to theorists, does not mean that there was any reason for expecting the universe to possess this simplicity. It is still a mystery.

Sciama: That mystery will be dispelled this afternoon!

20. GALAXIES AND INTERGALACTIC MATTER

George B. Field

1. The Role of Intergalactic Matter in Astronomy

In the last twenty years intergalactic matter has been playing an increasing role in astronomy. Such matter interacts strongly with galaxies in clusters, and it may even be a major contributor to the mass density of the universe.

1.1 *Appearances Can Be Deceptive*

Deep-space photographs reveal galaxies, great groups of up to 10^{12} stars held together by their mutual gravitation (Fig. 20.1); the space between the galaxies appears to be empty, unlike the space between the stars within our Milky Way galaxy (Fig. 20.2). There we see extended emission composed of hot ($T \sim 10^4$ °K) ionized hydrogen (H-II), as well as dark clouds containing interstellar dust particles. The dust clouds also contain large amounts of atomic hydrogen (H-I) and molecular hydrogen (H_2), as demonstrated by studies using radio emission in the 21-cm hyperfine line of H-I and absorption of starlight in the far-ultraviolet Lyman band of H_2.[1] One wonders why dust and/or H-II do not appear between the galaxies even in very long exposures of regions like those in Fig. 20.1.

The absence of dust is easily explained. Interstellar dust is composed of silicate, graphite, and other materials containing elements heavier than hydrogen and helium; such elements are made within galaxies and are therefore presumably very scarce in intergalactic space. The absence of H-II emission is more subtle. Such emission in our galaxy is due to recombinations of electrons and protons, a process depending on the square of the gas density. We shall see later that a reasonable estimate for the mean density of intergalactic gas is $\sim 10^{-6}$ atoms/cm^3, some 10^6 times smaller than in a typical galactic H-II region. For this reason, the emissivity of hot intergalactic gas is only $\sim 10^{-12}$ that of galactic gas, and even though intergalactic path lengths may be 10^7 times larger than those in the galaxy, the integrated emission of intergalactic gas is therefore expected to be $\sim 10^{-5}$ that in the galaxy, explaining its lack of detection.

What about H-I and H_2? Early searches for a broad 21-cm emission line characteristic of intergalactic H-I expanding with the universe failed, as have later attempts.[2] With the discovery of

Harry Woolf (ed.), Some Strangeness in the Proportion: A Centennial Symposium to Celebrate the Achievements of Albert Einstein
ISBN 0-201-09924-1

Fig. 20.1 In this deep-sky photograph dozens of galaxies can be seen. Distinguished from foreground stars by their fuzzy outlines, the galaxies in this group include spirals, ellipticals, and lenticulars. *Source:* Hale Observatories.

Fig. 20.2 A densely populated part of the Milky Way, containing stars (bright spots), dark clouds of dust and molecular hydrogen (H_2), and emission regions composed of ionized hydrogen (H-II). The dark clouds, to the right, obscure the more distant stars and H-II over an area shaped like the east Atlantic and Caribbean, while the H-II emission is shaped like North America, giving this region its name, the North American Nebula. *Source:* Hale Observatories.

a quasar having a red shift of about 2, it became possible to search for the absorption at Lyman α due to intergalactic H-I between us and the quasar, receding at nearly as great a velocity as does the quasar. Gunn and Peterson showed in this way that n (H-I) $< 10^{-11}/cm^3$ at that red shift.[3]

1.2 *Galaxy Formation and Closure of the Universe*

How much intergalactic gas does one expect? According to the standard big-bang model of the universe,[4] galaxies may originate in low-density fluctuations present in the early universe. When the expanding mixture of plasma and radiation cools to about 4,000 °K, the plasma recombines to form H-I, which interacts weakly with radiation. Henceforward, radiation pressure can no longer act on the matter to counter the mutual attraction within regions of somewhat higher density, and galaxies form by gravitational instability.[5] In this process some of the gas will be left behind, so that as a crude estimate for its density we can take the smoothed-out density of matter within galaxies, estimated to be about 3×10^{-31} g/cm^3, equivalent to about 1.5×10^{-7} atoms/cm^3.[6] Of course the actual density could easily be larger or smaller than this value.

Particular interest centers on whether the mean density of intergalactic gas equals or exceeds the cosmologically critical value of 4.7×10^{-30} g/cm^3, which, as discussed by M. J. Rees in this volume, separates closed from open Friedmann models of the universe if Hubble's constant is H_0 = 50 km/sec/Mpc. This critical density corresponds to 3×10^{-6} atoms/cm^3, about 20 times the amount of matter within galaxies. In terms of the parameter of $\Omega = \rho/\rho_c$ defined by Dr. Rees, Ω (galaxies) ~ 0.05, where $\Omega \geqslant 1$ is needed to close the universe. From the observed intensity of the cosmic microwave background radiation one calculates that its contribution to Ω is 10^{-4}. Other possible contributors, such as very faint stars, black holes, and massive neutrinos, will be very difficult to detect. But there is some hope of detecting gas if it is there.

We stated earlier that n(H-I) $<10^{-11}/cm^3$ at $z = 2$, corresponding to Ω(H-I) $<3 \times 10^{-7}$. It is hard to believe that galaxy formation is efficient enough to reduce Ω for the intergalactic medium as a whole to such a low value. Rather, it seems more reasonable to suppose that whatever gas is present is highly ionized (and therefore hard to detect, as explained previously). If we adopt n = $1.5 \times 10^{-7}/cm^3$, equivalent to the matter in galaxies, the degree of ionization would have to be so high that very high temperatures, -10^6 °K or greater, would be required to explain it.[7] This suggests a novel method of detection based on the emission of X rays by the hot gas. In fact, the first rocket flight that definitely detected X rays from outside the solar system also showed the existence of a diffuse background of 2–6 keV X rays, which could conceivably be due to intergalactic gas at temperatures above about 10^7 °K.[8] I will return in Sec. 4 to this interpretation, but first let us consider the X-ray emission by intergalactic gas within clusters of galaxies.

2. Intergalactic Matter in the Coma Cluster of Galaxies

2.1 *Characteristics of the Coma Cluster*

This cluster (Fig. 20.3) is one of the several thousand groups of galaxies easily recognized on the Palomar Survey and catalogued by Abell.[9] Its some 10^3 galaxies are distributed in a regular manner in and around a core whose radius is 0.25 Mpc[10]; its distance is 140 Mpc if H_0 = 50. A small percentage of all galaxies is believed to be in Abell clusters; the rest, although outside clusters rich enough to be recognized by Abell, are to a large extent associated in groups of a few to a few dozen members. Abell clusters generally fall into two broad classes. Those like Coma are relatively compact, circular in shape, and composed dominantly of elliptical (E) and lenticular (SO) galaxies that contain little or no interstellar gas. Others, like the great cluster in Virgo, are more diffuse, irregular in shape, and composed of both ellipticals and spiral (S) galaxies, which contain interstellar gas. Some of these differences are related to the presence of intergalactic matter.

Fig. 20.3 The center of the Coma cluster of galaxies. The two brightest galaxies are giant ellipticals; other ellipticals and lenticulars are grouped in their vicinity. *Source:* Kitt Peak National Observatories, operated by Associated Universities, Inc.

The Coma Cluster was discovered to be an X-ray source by Gursky et al. in 1971.[11] Its spectrum has now been studied over the entire range from 2 to 60 keV by an experiment on board the OSO-8 spacecraft by Mushotzky et al.[12] As can be seen in Fig. 20.4, the continuous spectrum is fit well by a thermal bremsstrahlung spectrum of $T = 1.0 \times 10^8$ °K. At this temperature one would expect to observe the complex of Fe-XXV and Fe-XXVI lines at 6.8 keV, and indeed, that is the case, as shown in the figure. The analysis of the iron line indicates an iron abundance of 1.4×10^{-5} by number, about one-third that in the sun.

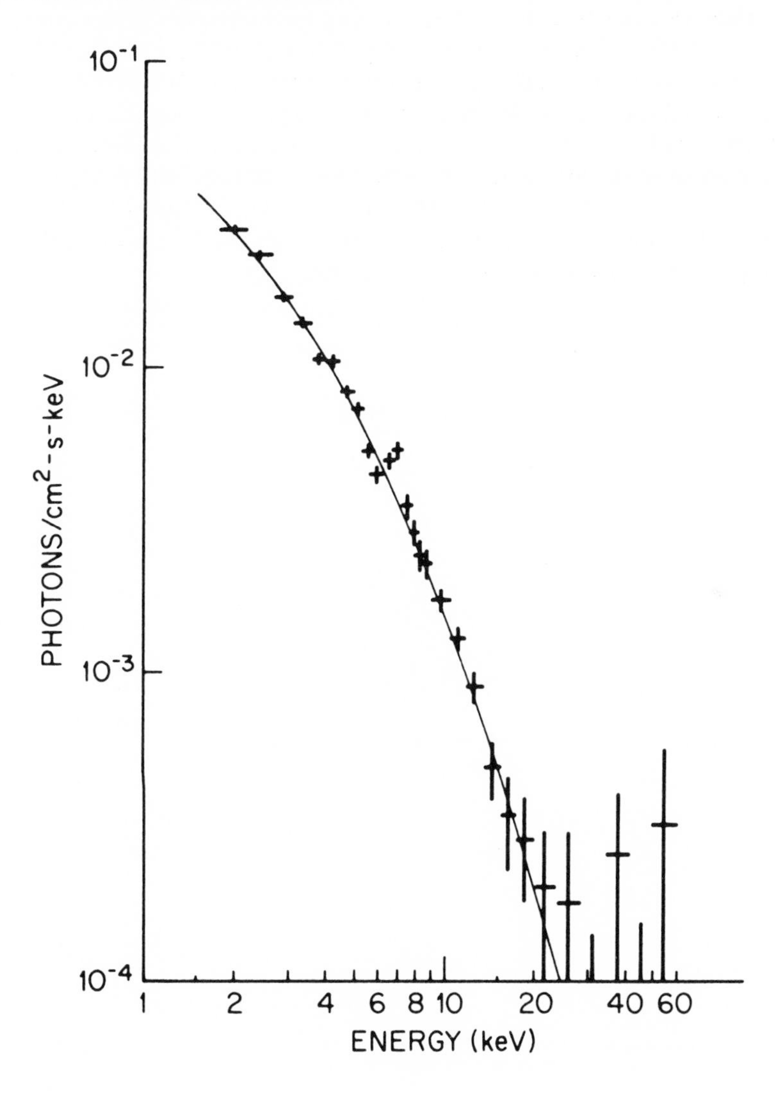

Fig. 20.4 The X-ray spectrum of the Coma cluster from 2 to 60 keV; note the iron line at 6.8 keV. The curve is a thermal bremsstrahlung spectrum corresponding to $T = 10^8$ °K. *Source:* R. F. Mushotzky et al., Astrophys. J. **225**, 21 (1978).

The rms velocity along the line of sight of the galaxies in Coma has been determined from the Doppler shifts in their lines to be about 1,000 km/sec. The temperature of a hydrogen-helium plasma having the same energy per unit mass is close to the temperature deduced from the X-ray spectrum. Hence it was immediately suggested that the hot gas, like the galaxies, is supported against the gravitational potential of the cluster by its internal kinetic energy; this would predict that the gas should be smoothly distributed like the galaxies. That prediction has been recently tested by observing the X rays from Coma with an imaging proportional counter (see Fig. 20.5).[13] A detailed model has been constructed by Cavaliere and Fusco-Femiano,[14] according to which the amount of gas within 5 Mpc of the center is 4×10^{14} solar masses ($M_\odot$), or about 20 percent of the "virial mass" determined from the inferred gravitational potential, $2 \times 10^{15}\, M_\odot$. The "luminous mass," estimated by multiplying the observed visual luminosity of the galaxies ($1.3 \times 10^{13}\, L_\odot$) by the mass-to-light ratio of 30 $M_\odot/L_\odot$ believed to be valid for elliptical and lenticular galaxies, is $4.0 \times 10^{14}\, M_\odot$, about equal to that of the gas. Note that the virial mass is about 2.5 times the sum of the gas mass and luminous mass: $1.2 \times 10^{15}\, M_\odot$ remains to be accounted for. Quite possibly it is due to faint stars in the halos of individual galaxies or to stars residing in intergalactic space after having been stripped from such halos by tidal interactions.

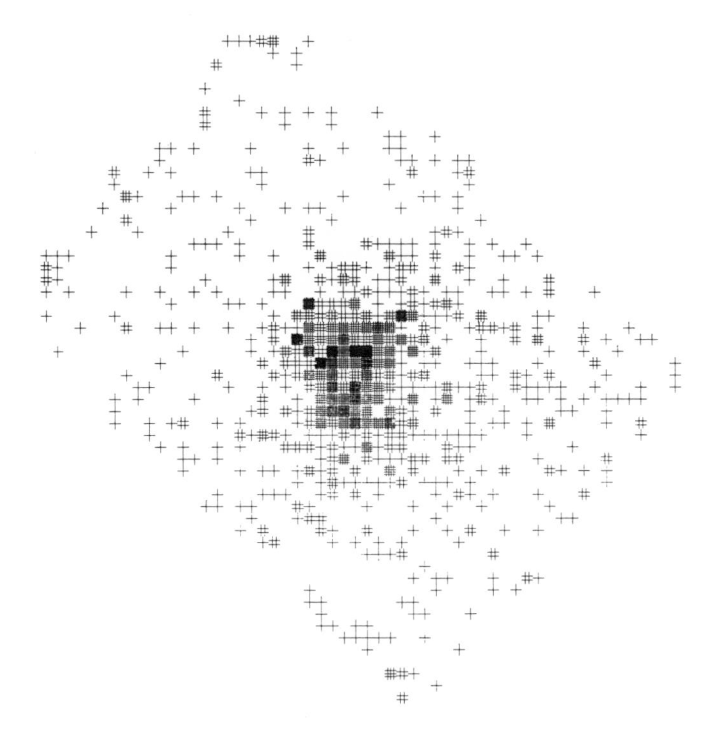

Fig. 20.5 An X-ray plot of the Coma cluster made by a rocket-borne imaging proportional counter. The resolution is about 1 arc minute. *Sources:* D. G. Fabricant, Ph.D. Thesis, Harvard University, 1978, and P. Gorenstein et al., Astrophys. J. (1979).

The presence of iron in the gas strongly suggests that at least some of it originated in the galaxies of the cluster.[15] Some of the gas, however, could have been left over when the galaxies formed, or it could have fallen into the cluster later from intercluster space.[16] The heat stored in the gas is at least partially due to the kinetic energy associated with the orbital motion of the parent galaxies (for that component originating in the galaxies[17]), or with the orbital motion of clouds of leftover gas, or with the shock heating of infalling intercluster gas.[18] However, supernova explosions within the cluster galaxies also contribute.[19]

One scenario for the formation of the intergalactic gas in Coma is that of Mathews and Baker.[20] Even if there was little remaining after the formation of the galaxies, the elliptical galaxies present would lose mass to intergalactic space in "galactic winds." Such winds occur because the mass lost in the form of planetary nebulae by intermediate-mass stars would be heated above the escape temperature by supernova explosions occurring in the same galaxy. The winds, expanding around each elliptical at ~300 km/sec, would collide because of the orbital motions of the parent galaxies at ~2,000 km/sec and shock heat to over 10^8 °K.

Subsequently, even those galaxies that did not for some reason or other develop winds would be stripped of their interstellar gas by the ram pressure of the first-generation intergalactic gas, due to the orbital motion of the parent galaxy through it.[21] This gas would also be heated to high temperature as its orbital kinetic energy thermalized (Fig. 20.6).

The consequences of this model have been worked out by DeYoung.[22] When corrected by a small factor[23] DeYoung's results indicate that the mass of iron in Coma is predicted to be about half of that actually observed, while the total mass of gas (mostly hydrogen and helium) is predicted to be only about 8 percent of that observed. He also predicts that supernovae should deliver about twice the energy now stored in the gas as thermal energy. The agreement of theory and observation as regards the iron is adequate, but as regards the hydrogen, it is not; to me, this

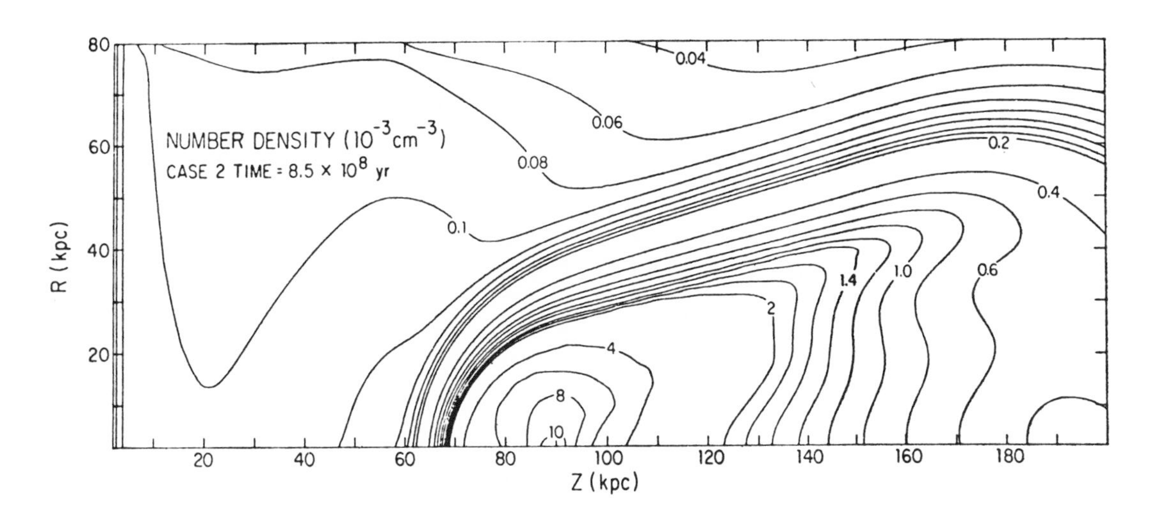

Fig. 20.6 The calculated distribution of gas around a galaxy moving to the left through intergalactic gas at a velocity near 1,000 km/sec. The gas in the "tail" is lost to the galaxy, and becomes mixed into the intergalactic medium. *Source:* S. M. Lea and D. S. DeYoung, Astrophys. J. **210,** 647 (1976).

suggests that a substantial fraction of the gas was either left over from galaxy formation or else fell in later. It is probably too early to draw strong conclusions about the energy contributed by supernovae, as the difference between models incorporating it and those based solely on orbital kinetic energy is only a factor of 3.

Based on the Mathews–Baker and DeYoung model, one can make simple predictions for other clusters:

1. All clusters should be X-ray sources. In fact, forty-eight have been identified so far.[24]
2. Their spectra should be thermal bremsstrahlung. In fact, in most cases this gives a better fit, according to Mushotzky et al.[25]
3. The iron line should be present. Again, eight out of twenty clusters studied by Mushotzky et al. manifest the iron line, and in all twenty cases, the data are consistent with Fe/H = 1.4×10^{-5}.
4. If the high temperature T is due to the gravitational potential (either through orbital motion or infall), it should correlate with the mean-square velocity of the cluster galaxies $(\Delta v)^2$. In fact, according to Mushotzky et al.[26] there is a rough correlation (Fig. 20.7), with a regression line indicating that $T \simeq \mu (\Delta v)^2/\mathcal{R}$, where $\mu = 0.61$ is the molecular weight of ionized H + 10 percent He and $\mathcal{R}$ is the gas constant. This suggests that supernova heating is often not dominant.

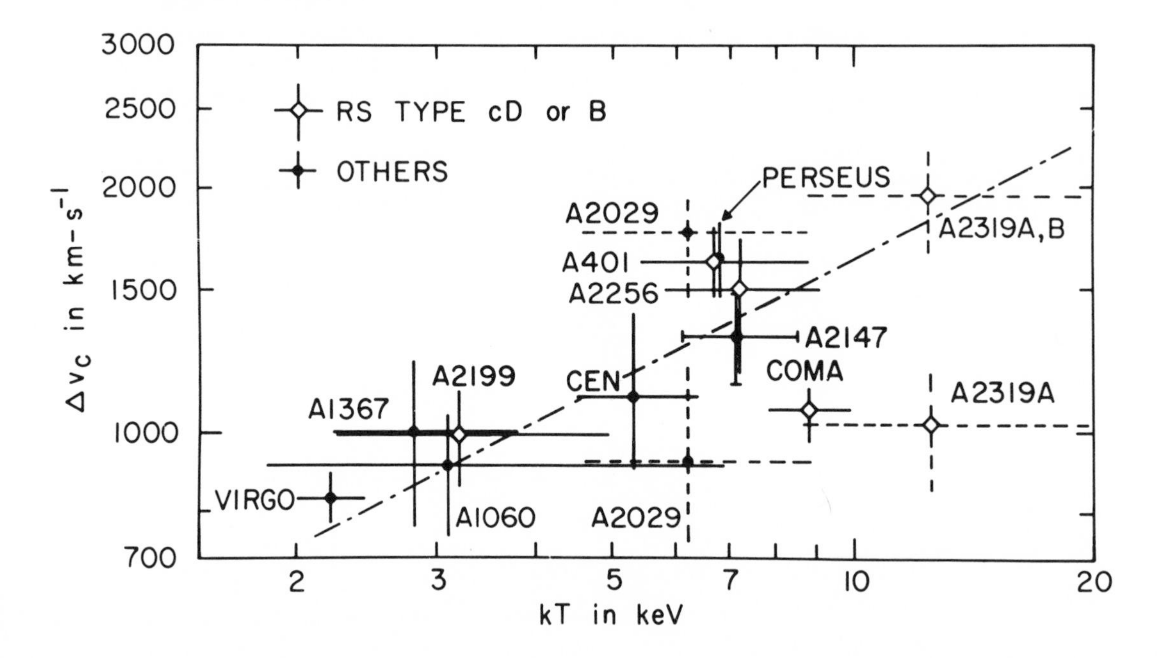

Fig. 20.7 The correlation between velocity dispersion Δv_c of the cluster galaxies and the temperature T deduced from the X-ray spectrum. The regression line corresponds to T being proportional to $(\Delta v_c)^2$. *Source:* R. F. Mushotzky et al., Astrophys. J. **225**, 21 (1978).

5. One should be able in some cases to observe the dimming of the microwave background radiation due to its Compton scattering to slightly higher frequencies by the hot electrons.[27] In fact, this has been accomplished convincingly in one case, Abell 2218, even though the dimming is only 1.0×10^{-3} °K (by Birkinshaw et al., as shown in Figure. 20.8).

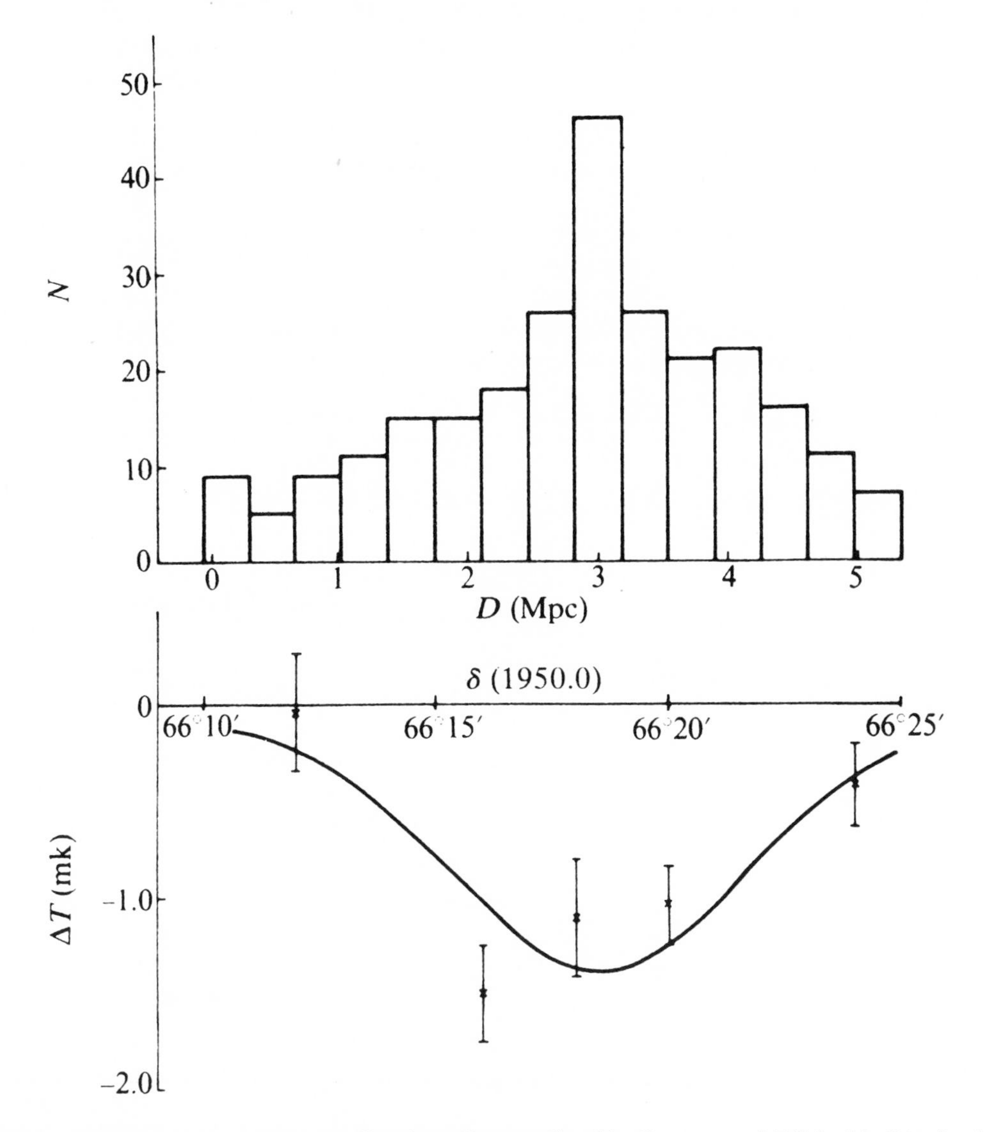

Fig. 20.8 Detection of the microwave dimming effect predicted by Sunyaev and Zel'dovich,[28] in the cluster Abell 2218. The upper panel indicates the number of galaxies in strips at various distances from the center of the cluster; the lower one indicates the amplitude of the dimming effect in units of 10^{-3} °K. The correlation indicates that the hot gas is distributed like the galaxies. *Sources:* M. Birkinshaw et al. Monthly Notices, Roy. Astron. Soc. **185,** 245 (1978), and Nature **275,** 40 (1978).

6. The X-ray luminosity in a fixed energy range L_x should correlate with the number of galaxies N_0 within 0.5 Mpc. If the fraction of mass going into gas is constant, higher N_0 implies higher emissivity not only because of the higher density, but also because of the higher temperature implied by the virial theorem. Such a correlation has been observed and explained quantitatively by Bahcall (see Fig. 20.9).
7. Clusters of higher N_0 and L_x (by Fig. 20.9) should have fewer gas-rich spiral galaxies, since the higher density of intergalactic gas would create a higher ram pressure, capable of stripping the gas from such systems and converting them to lenticular galaxies. Again, such an effect has been observed and quantitatively interpreted by Bahcall (Fig. 20.10).
8. Finally, if there were some way to detect the interstellar gas being stripped from galaxies, it should show up as a tail stretching back along the orbit of the galaxy. In fact, such tails are observed in the radio spectrum; 3C 129 is a good example (Fig. 20.11).

I conclude that the Mathews–Baker and DeYoung model meets the tests to which it has been subjected. However, the relative importance of leftover or infalling gas versus galactric gas, and of supernova energy versus orbital or gravitational infall energy, is still unclear.

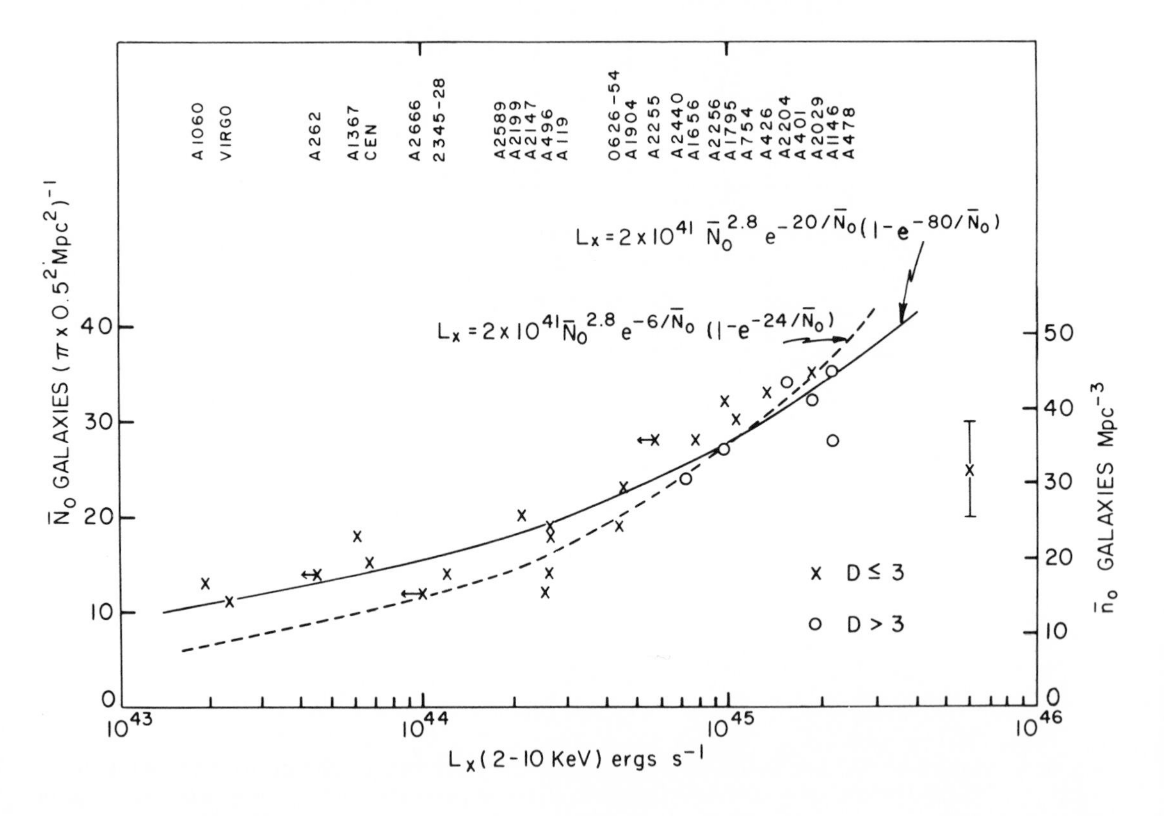

Fig. 20.9 The observed correlation between the X-ray luminosity L_x of a cluster and the number of galaxies N_0 within 0.5 Mpc of its center. The two lines are theoretical fits. *Source:* N. Bahcall, Astrophys. J. **217**, L77 (1977).

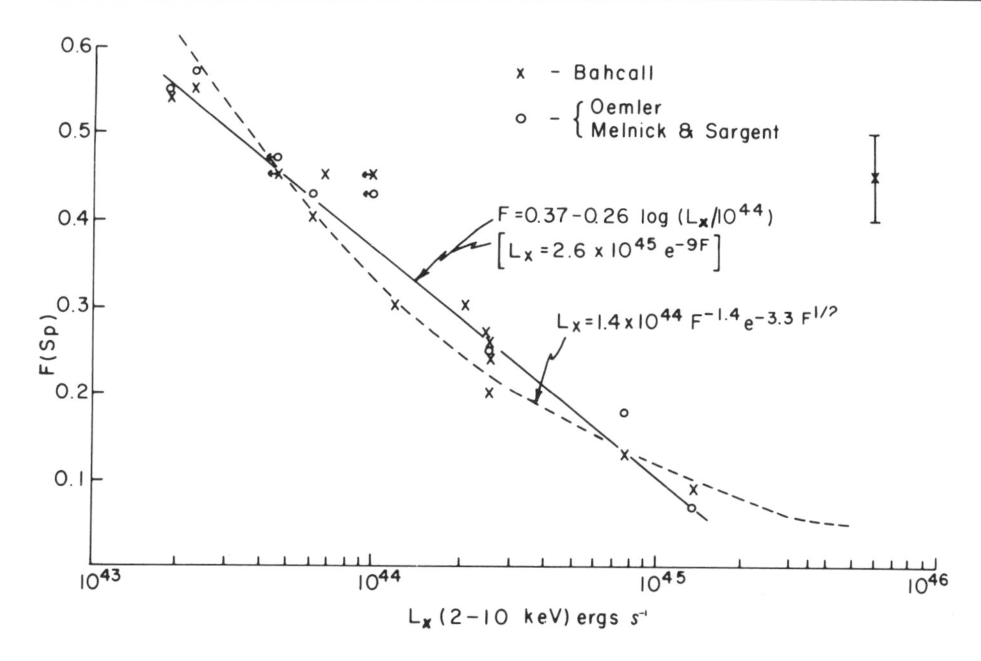

Fig. 20.10 The correlation between the X-ray luminosity L_x cluster and the fraction of galaxies $F(Sp)$ in that cluster that is spirals. The lines are theoretical fits. *Source:* N. Bahcall, Astrophys. J. **218,** L93 (1977).

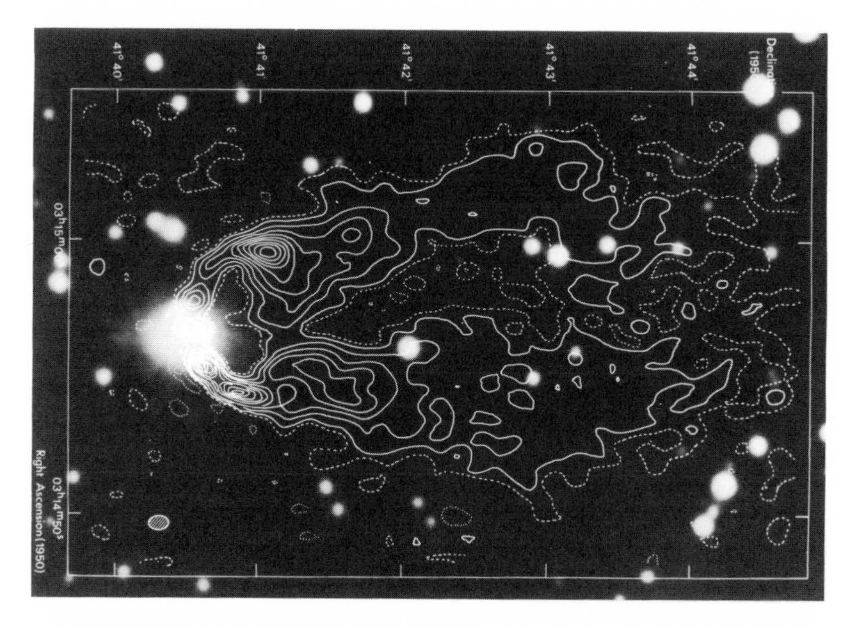

Fig. 20.11 The radio-tail galaxy 3C 129; radio isophotes are superimposed on a photograph of the galaxy (*bottom center*).

4. Intergalactic Matter Outside of Clusters

From what has been said, it would not be surprising if there were significant amounts of gas outside of clusters, as well as in clusters. Since the amount of gas in Coma and other clusters is roughly equivalent to the luminous mass in galaxies, we might expect that the same amount of gas would accompany galaxies outside of clusters. Indeed, there is evidence for such gas, but the amounts appear to be surprisingly large.

4.1 *Cluster X-ray Halos*

The first line of evidence is the discovery by Forman et al.[29] that cluster X-ray sources appear to be more intense when observed with a large (5° × 5°) collimator than with a small (5° × ½°) one. Their plot of the results for twenty-three clusters is reproduced in Fig. 20.12. In no case does the large collimator yield a significantly smaller flux, but in eight cases it yields a significantly larger one. Forman et al. attribute this effect to a halo of hot gas extending to very large distances from the parent cluster, and they show that even in the cases where the effect is not apparent in Fig. 20.12, the results are consistent with the presence of such gaseous halos.

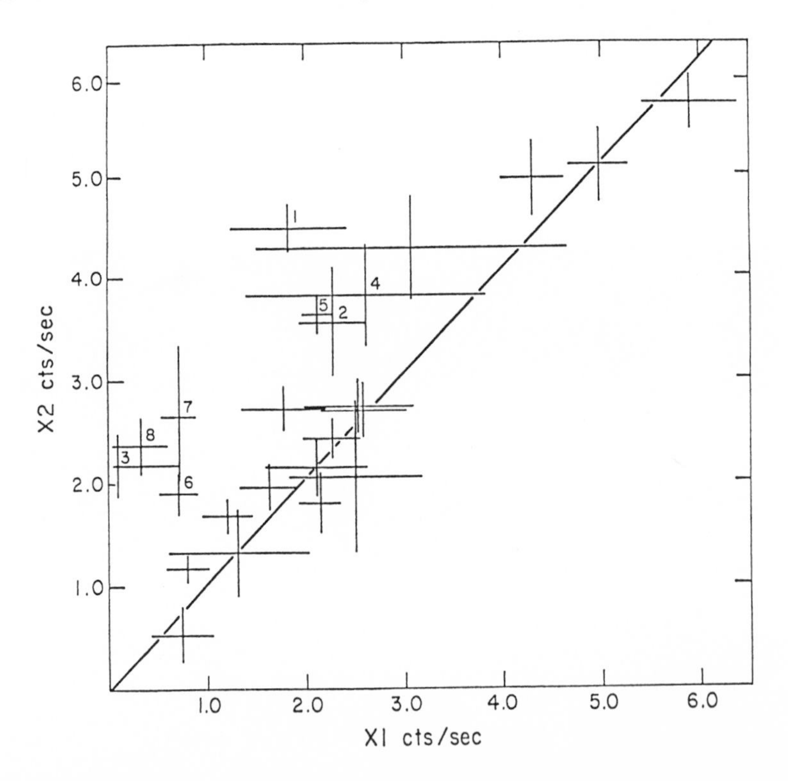

Fig. 20.12 Fluxes of X-ray clusters detected with two different collimators (X1 and X2) aboard the UHURU spacecraft. With few exceptions, the flux obtained with the larger collimator (X2) exceeds that obtained with the smaller one (X1), indicating that not all the flux is entering X1, and that therefore there must be a halo around each cluster. *Source:* W. Forman et al., Astrophys. J. **225,** L1 (1978).

Since the spectra of the halos are not detectably different from those observed with the narrow collimator alone, Forman et al. attribute the same temperature to each component and calculate halo radii and masses on this basis. The radii vary from 3 to 12 Mpc, with a mean of 8 Mpc (some 20 times a typical X-ray core radius), and the amounts of gas vary from $2 \times 10^{15} M_{\odot}$ within 8 Mpc to $1 \times 10^{16} M_{\odot}$ within 30 Mpc. These masses are astonishingly large; if correct, they imply that there is 5 to 25 times as much mass in gas as in the luminous mass of the cluster galaxies. If this result applied to galaxies in general, I estimate that the corresponding Ω would be 0.05 to 0.25.

4.2 *The Diffuse X-ray Background*

The other line of evidence comes from the diffuse X-ray background discussed in Sec. 1.2. The origin of this background is still not known. Most investigators believe that it is due to large numbers of unresolved sources, such as active galaxies, but there are enough difficulties with that interpretation that it is reasonable to consider the alternative that it is thermal bremsstrahlung emission by very hot intergalactic gas.[30]

Its spectrum is now very well known over the range 1 to 100 keV (Fig. 20.13), and it is known to be nearly, but not quite, isotropic, demonstrating its extragalactic orgin; Schwartz obtains deviations from isotropy of 2.6 percent (rms).[31]

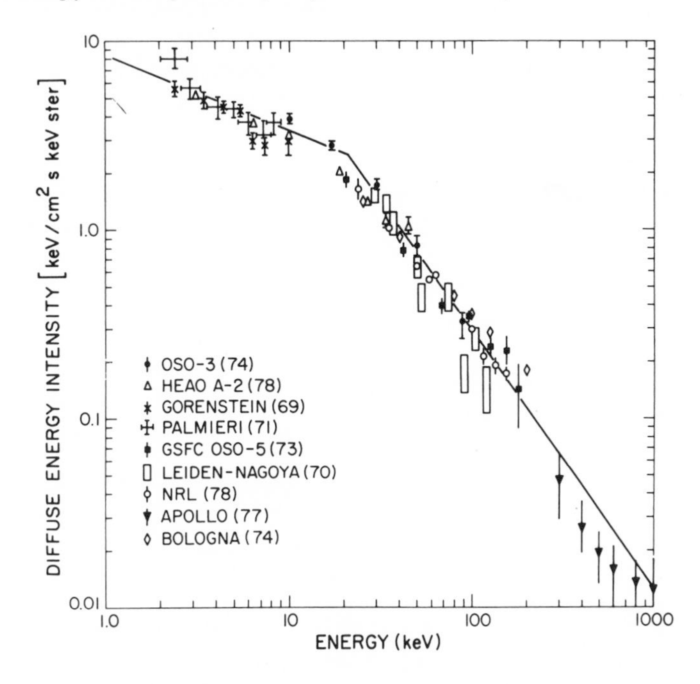

Fig. 20.13 The spectrum of the diffuse X-ray background over the range 1–100 keV, as compiled by Schwartz. The straight lines representing two power laws joining at 20 keV have no particular significance but serve to show that a single power law does not represent the data. *Source:* D. A. Schwartz, in *Proceedings of COSPAR — International Astronomical Union Symposium on X-ray Astronomy, Innsbruck, 1977,* edited by L. Peterson and W. Baity (Pergamon Press, New York, 1979).

There are three problems to be solved if the background is due to unresolved sources. First, the sources must have spectra compatible with that of the background. So far no class of extragalactic sources has been identified that meets this requirement; the spectra of clusters are too soft, whereas those of active galaxies are too hard.[32] Second, the sources must be bright enough to account for the intensity of the background. Clusters fail this test by at least a factor of 3 at a few keV, and by a larger amount at energies above 10 keV, while active galaxies fail by even larger factors. Here one might hope for help from evolutionary effects, with larger numbers of brighter sources at higher red shifts. Theoretical work by Perrenod shows that the opposite is the case for clusters.[33] The situation for active galaxies is unknown. Finally, there must be a sufficient number of sources that the observed isotropy is not violated. Clusters fail on this score, but again, active galaxies might do. In summary, clusters are unlikely on a number of counts to account for the diffuse X-ray background, but active galaxies are a viable option if a new class is found with appropriate spectra and numbers increasing rapidly with red shift.

Turning to the interpretation in terms of hot gas, it has been shown by Boldt that the spectrum fits a single-temperature thermal bremsstrahlung spectrum rather well between 2 and 60 keV (but not above 100 keV according to Field and Perrenod);[34] the deduced temperature, 5×10^8 °K, confirms that the gas responsible for the emission (if such is the case) cannot be within normal clusters of galaxies. Cowsik and Kobetich suggested that the X-ray background could be accounted for by truly diffuse hot intergalactic gas; this proposal was further analyzed by Field and Perrenod, who considered that to heat the gas to such high temperatures would require energy in the form of fast particles (cosmic rays) generated by galaxies, presumably in active phases.[35] The local population of active galaxies is inadequate to do the job, but because the number of one class of active galaxies, quasars, is known to be much larger in the past[36] the bulk of the heating could well take place at red shifts of the order of 2 to 3. Figure 20.14 shows the thermal history of the gas calculated by Field and Perrenod[38]; the temperature is normalized to its present value, 4.4×10^8 °K in this model. The fit of the model to the 1–100 keV data (Fig. 20.15) yields for the mean density of intergalactic gas, $\Omega = 0.5 \pm$ a factor of 2. If this were in fact the correct interpretation of the X-ray background, intergalactic gas is one of the main components of the universe. Within the errors, one could not be certain that the universe is not closed by such gas.

4.3 *Difficulties with Supposing the X-ray Background is Due to Hot Gas*

I hasten to point out several serious objections to such a model:

1. The energy in fast particles required from each galaxy averaged over all of its active phases is 2×10^{10} times the rest energy of the sun. Although unusual galaxies such as M87 may have massive black holes in their nuclei (see W. L. W. Sargent's chapter, which follows this one), there is no evidence that all galaxies do; yet I can think of no other sources for such a large amount of energy in fast particles.
2. The amount of energy required from each galaxy, $2 \times 10^{10} M_\odot c^2$, is comparable to the amount of thermal energy now observed in Abell clusters like Coma. If even a tiny fraction of the galaxies in such clusters in fact delivered such an enormous amount of energy to the surrounding intergalactic gas, this gas would be blown out of the cluster, contrary to observation. The only way out would be to suppose that galaxies in clusters are dramatically different from those outside. Although reasons for supposing that some dif-

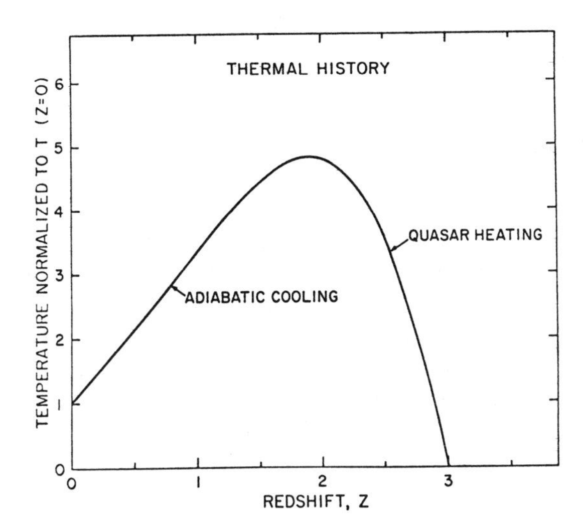

Fig. 20.14 The variation of temperature with red shift z for diffuse intergalactic gas, normalized to its local value $T(z = 0)$ according to a model of Field and Perrenod. The sharp increase starting at $z = 3$ is due to heating by the large numbers of quasars at that period; the decrease at smaller z is due to adiabatic cooling as the universe expands. To obtain a fit to the observed X-ray background (Fig. 20.15), the present temperature must be about 4.4×10^8K, and the maximum value, about 2×10^9K. *Source:* G. B. Field and S. C. Perrenod, Astrophys. J. **215,** 717 (1977).

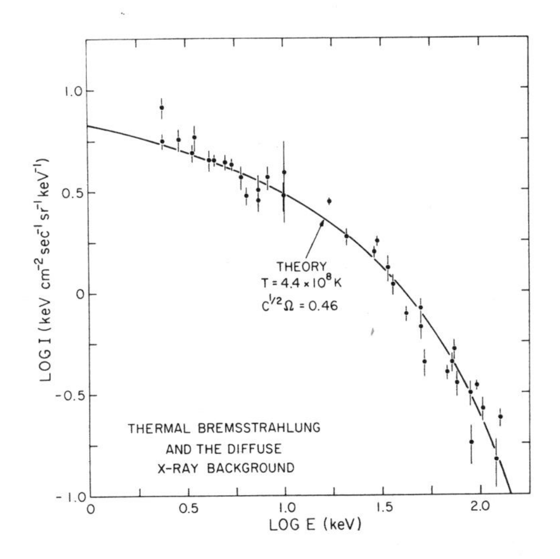

Fig. 20.15 A fit of the Field–Perrenod theory to the observed X-ray background spectrum. The shape determines the present temperature to be 4.4×10^8K; the intensity determines the density parameter Ω for smoothly distributed gas to be about 0.5. (If the gas is clumped by a factor C, Ω can be reduced by a factor $C^{-1/2}$). *Source:* G. B. Field and S. C. Perrenod, Astrophys. J. **215,** 717 (1977).

ferences exist in this respect will be discussed in Sec. 5, I am doubtful that they are profound enough to meet this objection.

3. The amount of intergalactic gas required greatly exceeds that allowed by the constraint of cosmological production of deuterium ($\Omega < 0.1$) as well as that allowed by cosmological production of helium ($\Omega < 0.04$), according to Yang et al.[39] Such considerations of course mitigate against a closed universe generally.

In view of these difficulties I have examined an alternative model, in which a much smaller total amount of gas is gravitationally clumped and thus shock-heated by its own gravitation;[40] the resulting hot clouds might account for the background. Because clumping results in greater emission from the same amount of gas, Ω can be reduced, meeting objection (3). Shock heating avoids objections (1) and (2). Of course, one question arises immediately: To what astronomical objects would the hot clouds correspond? The only candidate I know of is superclusters, groups of galaxy clusters identified by Abell as containing several clusters of galaxies and having radii of about 25 Mpc.[41] Indeed, it has been claimed by Murray et al. that several superclusters are identifiable X-ray sources, although it is not known whether their spectra are consistent with that of the X-ray background.

Such considerations are probably academic anyway, because one can show that such a model cannot at the same time account for the spectrum, intensity, and isotropy of the background.[43] Briefly, the argument is this: The spectrum implies a temperature and, hence, a mass-radius ratio for the hot clouds. The mean square fluctuations around isotropy are inversely proportional to the number density of sources and to the distance of the nearest one. Because the hot clouds are extended, nearby ones do not contribute to the fluctuation, so the distance to the nearest hot cloud effective in causing fluctuations with a given collimator is proportional to its radius and, thus, to its mass. Hence the fluctuations are inversely proportional to the product of the number density and mass of the hot clouds and, thus, to their contribution to Ω. As part of the fluctuations are due to other sources, such as clusters of galaxies and active galaxies, the observed fluctuations imply an upper limit to those due to hot clouds. In this way one derives a lower limit on Ω, which upon putting in the numbers is 12. However, from the intensity of the background we know that Ω would be about 0.5 even for smoothly distributed gas and even less for clumped gas, so there is a contradiction.

In summary, both gravitational heating and quasar heating of a putative intergalactic gas outside of clusters, put forward to explain the diffuse X-ray background, fail. I conclude that one cannot reliably estimate the contribution of such gas to Ω at this time. Progress in this field depends on attempts, currently underway with the NASA's High Energy Astronomy Observatory, or *Einstein,* to image discrete extragalactic sources to very faint levels, to see if the X-ray background can be thus accounted for.

5. Intergalactic Matter and Active Galaxies

I have already had occasion to refer to active galaxies. Discovery of such galaxies is one of the key achievements of astronomy in the postwar period. Radio astronomers were the first to detect such activity by means of the synchrotron radiation (see the chapter by W. K. H. Panofsky in this volume) emitted by relativistic electrons trapped in galactic magnetic fields. The Milky Way emits such radiation at the level of 10^{37} erg/sec as a consequence of fast electrons injected into the interstellar medium by supernova explosions, but a powerful active galaxy emits 10^7 times as much, 10^{44} erg/sec or $3 \times 10^{10} L_\odot$, far more than can be explained by supernovae.

As resolution of radio telescopes improved, it became clear that the electrons are invariably accelerated within the nucleus of the galaxy, often within a region less than 1 parsec across. Extraordinary processes must be at work to generate more energy in relativistic electrons than in the light of all the stars in a galaxy within less than 10^{-10} of its volume. It has been demonstrated that nuclear energy is insufficient; only the energy released when matter collapses gravitationally into compact objects appears to be sufficient. A promising candidate at present for explaining the energy produced in galactic nuclei is accretion onto a black hole. As Dr. Sargent explains in the following chapter, there is now convincing indirect evidence that a massive black hole does in fact exist in the nucleus of the radio source Virgo A, which is the massive elliptical galaxy known as Messier 87.

5.1 *Types of Active Galaxies*

Hubble's morphological classification of galaxies applies, with modification, to active galaxies as well. Elliptical (E) galaxies have smooth brightness profiles, without spiral structure, and appear to be spheroidal in shape. There is very little interstellar matter detected in most ellipticals. Spiral (S) galaxies, on the other hand, are disklike, with spiral arms embedded in the disk; interstellar matter abounds. Current theory suggests that spiral arms are really waves driven by the presence of interstellar matter. Lenticular (*S*0) galaxies are disklike but have neither spiral arms nor gas.

To two of the morphological classes, ellipticals and spirals, there correspond types of active galaxy; I am not aware of a case of an active lenticular galaxy. Active ellipticals are known as radio galaxies. While there is often radio emission from a pointlike source in the nucleus, now confirmed by X-ray studies, radio galaxies usually have very extensive radio lobes, which are giant ($\sim$50 kpc) clouds disposed at great distances (10^2–10^3 kpc) on either side of the galaxy. Powerful radio galaxies are often associated with the brightest and hence most massive ellipticals in rich clusters of galaxies.

Active spirals are known as Seyferts, after the astronomer Carl Seyfert, who discovered their active nuclei by their optical emission. The spiral structure is apparently undisturbed, but a pointlike source in the nucleus emits copiously at radio, optical, and according to recent data, X-ray wavelengths. Seyferts rarely have radio lobes.

Quasars emit so much optical radiation (up to $10^{14} L_\odot$) that they can be observed up to great distances, the record red shift being 3.5. At these distances the optical emission appears pointlike. However, in the local part of the universe ($z < 0.1$, say), they are rather rare, so it is difficult to assess the morphology of quasars with ground-based telescopes, the seeing disk of which corresponds to > 6 kpc at red shifts > 0.1. In the few cases of nearby quasars that have been studied, the evidence is consistent with the presence of a normal galaxy, perhaps a spiral. Furthermore, the optical spectra of Seyferts and quasars have many features in common. The belief is growing, therefore, that quasars are exceptionally bright Seyferts, which, because of their high luminosity and rare occurrence, are normally observed only at great distances where it is difficult to discern the much fainter spiral galaxy in which the active nucleus is embedded, because of glare of that nucleus. If this interpretation is correct, quasars are spirals, in which there are copious amounts of interstellar gas.

Membership in clusters is an important feature. Ellipticals can be found anywhere, inside or outside of clusters, although the brightest ones — sometimes called cD galaxies — have most often been found in rich clusters. Spirals, on the other hand, are rarely found in the centers of rich clusters, but are often observed in small groups outside of clusters.

5.2 *The Role of Intergalactic Matter*

The presence of intergalactic matter may play an important role in our understanding of the foregoing facts. What follows are tentative hypotheses that can serve to organize our thinking about both normal and active galaxies. The basic assumptions:

1. In ellipticals the first generation of stars used virtually all the gas. In the absence of intergalactic matter, the gas produced in the course of stellar evolution is swept out of the galaxy by a supernova-driven galactic wind; in the presence of intergalactic matter the sweeping process may be aided by ram pressure. But if the galaxy is stationary in a cluster the wind may be halted or even reversed by the pressure of the intergalactic matter.
2. In spirals the first generation of stars did not use all the gas. As a consequence of its angular momentum, the gas formed a disk of relatively high density, which is able to absorb and radiate away the energy of supernova explosions and thus prevent the formation of a galactic wind. In the absence of intergalactic matter, the gas remains in the galaxy. In the presence of a high density of intergalactic matter, such as is found in rich clusters, the interstellar gas is swept out of the galaxy by ram pressure, forming a lenticular galaxy.
3. Active galactic nuclei are due to the presence of a black hole, and to the accretion of matter into it. The matter may be either interstellar or intergalactic in origin.

With these hypotheses one can understand certain facts about both normal and active galaxies. Ellipticals owe their appearance to the absence of interstellar matter. They do not depend upon intergalactic gas to sweep out their interstellar matter and are therefore found both in clusters and in the field. Lenticular galaxies do depend on intergalactic matter to sweep them clean, and so are found only in clusters, while their progenitors, spirals, are found mostly outside of clusters.

Now consider an active galaxy that by hypothesis requires matter to power the black hole in its nucleus. If the galaxy is an elliptical, there is little interstellar matter to feed upon, unless the galactic wind can somehow be reversed. This cannot happen outside of clusters where the intergalactic pressure is too low, or even if the galaxy is orbiting in a cluster, for then the intergalactic ram pressure aids the sweeping process. If, however, the elliptical has a very low velocity in the cluster, the intergalactic ram pressure is small and the inward pressure of intergalactic matter may serve to reverse the wind, forcing the gas lost from stars to fall to the center of the galaxy instead. This flow can feed a black hole, if present, creating a radio galaxy. The fact that radio galaxies are often identified with massive ellipticals located in the cores of rich clusters suggests that the required low velocity is a consequence of equipartition of kinetic energy of orbital motion with other galaxies in the cluster.

The high mass also implies a compression of the intergalactic matter near the galaxy. If this compression is high enough, Gunn and Gott[44] suggest that the density will become high enough for radiative cooling to be important, and thus, for the pressure of intergalactic gas pervading the galaxy to fall, initiating an accretion flow into the galaxy; this suggestion has been verified in detail by Mathews.[45]

In fact, M87 (see next chapter) may be an example of this process. Fabricant has obtained an X-ray image of M87 that discloses the presence of vast amounts of gas ($> 10^{12}\ M_\odot$) within the galaxy.[46] Its temperature, about 3×10^7 °K, is measurably lower than the 10^8 °K characteristic of the cluster as a whole.[47] As Mathews demonstrates,[48] these data are consistent with a model in which the gas cools catastrophically as it approaches the nucleus, leading to an accretion flow of about $10\ M_\odot$ years^{-1}. If there is a black hole at the center of M87, as described by W. Sargent in the next chapter, the available energy, calculated as 1 percent of mc^2, is 6×10^{45} erg/sec, easily large enough to supply the nonthermal phenomena associated with the nucleus of M87.

If the galaxy originates as a spiral, everything depends upon whether it is located in a cluster or not. If it is, ram pressure will sweep it clean and convert it to a lenticular, which has no gas to feed a black hole, if present; consequently, there is no activity. If it is not in a cluster, the interstellar medium remains, and is a possible source for feeding the black hole. Lynden–Bell showed that magnetic fields threading the interstellar medium in disk galaxies can transfer the angular momentum of the medium from the inner regions to the outer regions of the galaxy.[49] As a result, the inner gas spirals into the nucleus, where it can feed a black hole. According to this hypothesis, neither Seyferts nor quasars should be found in rich clusters. According to N. Bahcall, this prediction is consistent with observation (private communication, 1978).

5.3 *Effects of Cosmological Evolution*

It is interesting to speculate on how the phenomena described here would depend on cosmological epoch, and hence, on how things should look at high red shift.

Perrenod has shown that because of the dynamical evolution of galaxy clusters, the density of intergalactic matter in them should increase with time (Fig. 20.16). Hence one expects that rich

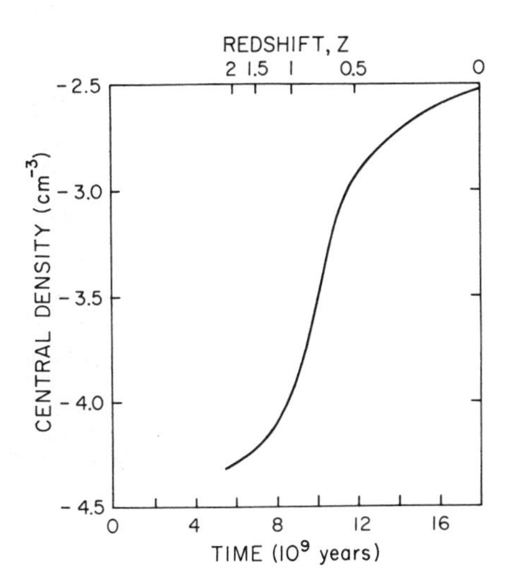

Fig. 20.16 Variation of the logarithm of the density of intergalactic matter at the center of a rich cluster, as a function either of red shift z, or of time on a scale on which the present is 18×10^9 years. The density increases largely because of the progressive deepening of the gravitational well. *Source:* S. C. Perrenod, Astrophys. J. **226**, 566 (1978).

clusters at large red shift should contain more spiral galaxies, and indeed that is the case. Butcher and Oemler have studied two clusters with red shifts ~ 0.4; even though they are rich clusters like Coma, the colors of their galaxies imply a much larger fraction of spirals (Fig. 20.17). Although this is in qualitative agreement with prediction, it remains to be seen whether it can be understood quantitatively. According to Gisler, it can be if one includes the additional effect that spiral galaxies had a larger amount of interstellar matter earlier in their history;[50] the inertia of this gas tends to resist ram-pressure sweeping. The same effect, incidentally, would, according to our hypotheses, be consistent with the well-known fact that quasars are much more frequent at high red shifts than in the local region; there is more gas at earlier epochs available to feed the black hole.

By the same token, however, one predicts that radio galaxies should be less numerous at large red shifts. This does not accord with the facts.[51] It is encouraging that the *Einstein* X-ray observatory will have the capability of studying both the intergalactic matter and the presence of active nuclei in rich clusters at large red shifts.

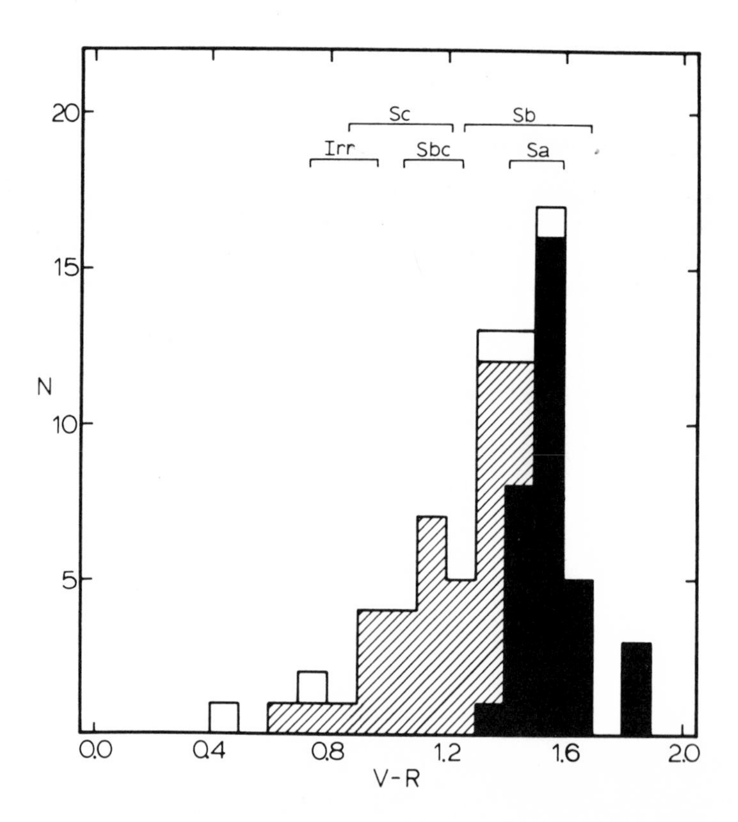

Fig. 20.17 The number of galaxies in the distant cluster 0024 + 1654 (red shift = 0.39) as a function of the V-R color. Based on the colors, Butcher and Oemler propose that the galaxies in the shaded area are ellipticals and lenticulars, while those in the hatched area are spirals. The resulting fraction of spirals (54 percent) is far greater than found in nearby clusters of comparable morphology. *Source:* H. Butcher and A. Oemler, Jr., Astrophys. J. **219,** 18 (1978).

6. Summary

Intergalactic matter comparable in mass to the luminous mass of galaxies is found in clusters of galaxies; the ram pressure of this gas is capable of sweeping the interstellar medium out of spirals, converting them to lenticular galaxies. It is therefore understandable that Seyferts and quasars are rarely found in clusters at the present epoch. The intergalactic matter now present in clusters may account for radio galaxies like M87 by accreting onto black holes in the nuclei of ellipticals. Earlier, when the interstellar density in spirals was higher and the intergalactic density was lower, the ram-pressure sweeping process was less effective, and spirals survived. These effects suggest that quasars (which are believed to be bright Seyferts) may exist in clusters of large red shift even though they are rarely found within clusters in the local region. The larger number of quasars at high red shift may be due to the larger amounts of interstellar matter available to feed a central black hole, whether or not the parent spiral galaxy is in a cluster.

The question of significant amounts of intergalactic gas outside of clusters is controversial at present. The detection of X-ray halos indicates that there are significant amounts of hot gas very far from the centers of clusters. The existence of the X-ray background is consistent with, but does not demand, a smoothly distributed intergalactic medium whose density is half the closure density and whose present temperature is 5×10^8 °K. However, explaining the energy content of such a medium is a major theoretical problem. An alternative attempt to explain the X-ray background as due to clumps of gravitationally heated gas fails. Whether the more reasonable suggestion, that the X-ray background is in reality not diffuse, but due to large numbers of discrete sources, will depend on discovery of such sources with the *Einstein* X-ray observatory and the demonstration that their spectra are consistent with that of the background.

NOTES

1. L. Spitzer, Jr., *Physical Processes in the Interstellar Medium* (John Wiley, New York, 1978).
2. G. B. Field, Astrophys. J. **129,** 525 (1959); Astrophys. J. **135,** 684 (1962); and Ann. Rev. Astr. Astrophys. **10,** 227 (1972).
3. J. E. Gunn and B. A. Peterson, Astrophys. J. **142,** 1633 (1965).
4. S. Weinberg, *Gravitation and Cosmology* (John Wiley, New York, 1972).
5. B. J. T. Jones, Rev. Mod. Phys. **48,** 107 (1976).
6. J. R. Gott et al., Astrophys. J. **194,** 543 (1974).
7. R. Weymann, Astrophys. J. **147,** 887 (1967).
8. R. Giacconi et al., Phys. Rev. Lett. **9,** 439 (1962).
9. G. Abell, Astrophys. J. Suppl. **3,** 211 (1958).
10. H. J. Rood et al., Astrophys. J. **175,** 627 (1972).
11. H. Gursky et al., Astrophys. J. **167,** L81 (1971).
12. R. F. Mushotzky et al., Astrophys. J. **225,** 21 (1978).
13. D. G. Fabricant, Ph.D. Thesis, Harvard University, 1978, and P. Gorenstein et al., Astrophys. J. **230,** 26 (1979).
14. A. Cavaliere and R. Fusco-Femiano, Astron. Astrophys. **49,** 137 (1976).
15. D. S. DeYoung, Astrophys. J. **223,** 47 (1978).
16. J. E. Gunn and J. R. Gott, III, Astrophys. J. **176,** 1 (1972).
17. DeYoung, op. cit. in n. 15.
18. Gunn and Gott, op. cit. in n. 16.

19. A. Yahil and J. P. Ostriker, Astrophys. J. **185,** 787 (1973).

20. W. G. Matthews and J. C. Baker, Astrophys. J. **170,** 241 (1971).

21. Gunn and Gott, op. cit. in n. 16.

22. DeYoung, op. cit. in n. 15.

23. G. B. Field, Mitt. Astron. Ges., in press.

24. C. Jones and W. Forman, Astrophys. J. **224,** 1 (1978).

25. Op. cit. in n. 12.

26. Ibid.

27. As predicted by R. A. Sunyaev and Y. B. Zel'dovich, Astrophys. Space Sci. **7,** 3 (1970), and Comments Astrophys. Space Sci. **4,** 173 (1972).

28. Ibid.

29. W. Forman et al., Astrophys. J. **225,** L1 (1978).

30. G. B. Field, Bull. Am. Astron. Soc. **10,** 675 (1978).

31. D. A. Schwartz, *Proceedings of COSPAR — International Astronomical Symposium on X-ray Astronomy, Innsbruck, 1978,* edited by L. Peterson and W. Baity (Pergamon Press, New York, 1979).

32. Ibid.

33. S. C. Perrenod, Astrophys. J. **226,** 566 (1978).

34. E. A. Boldt, NASA Technical Memo 78106, March 1978; G. B. Field and S. C. Perrenod, Astrophys. J. **215,** 717 (1977).

35. R. Cowsik and E. J. Kobetich in *Cosmic Ray Conference Papers,* 12th International Conference on Cosmic Rays, University of Tasmania, Vol. 1 (1971), p. 38, and Astrophys. J. **177,** 585 (1972).

36. M. Schmidt, Astrophys. J. **176,** 273, 289, 303 (1972).

37. Field and Perrenod, op. cit. in n. 34.

38. Op. cit. in n. 34.

39. J. Yang et al., Astrophys. J. **227,** 697 (1979).

40. Field, op. cit. in n. 30.

41. G. Abell, Astron. J. **66,** 601 (1961).

42. S. S. Murray, Astrophys. J. **219,** L89 (1978).

43. See Field, op. cit. in n. 30.

44. Op. cit. in n. 16.

45. W. G. Mathews, Astrophys. J. **219,** 413 (1978).

46. Op. cit. in n. 13.

47. P. J. N. Davison, Monthly Notices, Roy. Astron. Soc. **183,** 39 p. (1978).

48. Op. cit in n. 45.

49. D. Lynden-Bell, Nature **223,** 690 (1969).

50. G. R. Gisler, Astrophys. J. **228,** 385 (1979).

51. Schmidt, op. cit. in n. 36.

21. COMMENT ON "GALAXIES AND INTERGALACTIC MATTER"

Wallace L. W. Sargent

The attempt to understand the way in which galaxies formed and then evolve is now one of the central problems of astronomy. The study of galactic evolution has now supplanted the study of the evolution of individual stars as the subject in astronomy in which progress is being made over a broad front.

A theory of galactic evolution would seek to explain how the main types of galaxies — ellipticals, S0's, spirals, and irregulars — form, what controls the rate at which they use up their interstellar gas to form stars, and how the stars enrich the remaining interstellar gas with heavy elements.

Following the work of Morgan[1] it was found that there is a relatively simple correlation between the form of a galaxy and its stellar population. The spectra of elliptical galaxies show them to be composed of old, cool K-giant stars, while the integrated light from spirals reveals that they are composed of progressively younger and hotter stars as one goes along the sequence from Sa to Sc. Essentially from this simple observation arose a unified view of the evolution of the stellar populations of galaxies. In the ellipticals and the spheroidal central bulges of spirals the original gas was consumed into stars early — probably in the first 10^9 years or so during the initial collapse of the galaxy. In spirals the initial angular momentum was such that some of the interstellar gas settled into a rotating disk in which star formation has proceeded much more slowly than in the spheroidal component.

In seeking to firm out the details of this picture of galactic evolution, it has been natural to concentrate on such questions as what controls the rate at which the disk gas is used up and what determines the mass distribution of newly formed stars and hence the rate of enrichment of the remaining interstellar gas in heavy elements through supernova explosions and other forms of stellar mass ejection.

In making such considerations it would be simplest to assume that galaxies are isolated systems — that they neither accrete fresh supplies of gas from intergalactic space after they are formed nor eject processed gas into intergalactic space. The main thrust of Dr. George Field's paper (the preceding chapter) is that this assumption of isolation is now being seriously questioned. The development is not completely new because Oort proposed that the so-called "high velocity clouds" of neutral hydrogen, which are detected by their 21-cm emission, might be

Harry Woolf (ed.), Some Strangeness in the Proportion: A Centennial Symposium to Celebrate the Achievements of Albert Einstein
ISBN 0-201-09924-1

primordial clouds falling into our galaxy from the surrounding intergalactic space.[2] Oort estimated that if this hypothesis were correct, it would lead to the mass of gas in the disk of our galaxy being doubled on a time scale of 5×10^9 years, thereby having a profound impact on the evolution of the disk. However, it now seems most likely that the high-velocity clouds represent the result of supernova explosions in the vicinity of the sun. Accordingly, the only direct evidence that we have for intergalactic gas comes from the observations of X-ray emission from the great clusters of galaxies. In this case there now seems to be overwhelming evidence for interactions between galaxies and the external medium that lead to profound effects on the evolution of the disk galaxies within the cluster. The pressure of the hot gas in the X-ray clusters is very high: with typical values of the temperature and density $T_e = 10^8$ °K, $n_e = 10^{-3}$, the product $n_e T_e = 10^5$ far exceeds the product $nT = 1 \times 100$ typical of the neutral gas in the disk of a typical spiral galaxy. Thus, the hot gas would have an important effect on even a static disk galaxy near the center of a large cluster. However, the fact that the virial motions of the galaxies in clusters are at speeds of over 1,000 km/sec (equal to the thermal velocities of the ions in the hot gas) means that, instead of just being squeezed, the gas in a spiral galaxy traversing the central regions of a cluster would be swept out. As Dr. Field explained, this almost certainly accounts for the fact that the only disk galaxies found near the centers of the X-ray clusters are S0's — systems with no interstellar gas, no recent star formation, and no spiral arms. These galaxies are presumably the remnants of spirals that have happened to traverse the central regions of the cluster during their virial motions. The swept-out disks of S0's provided the iron that is seen as an emission line in the X-ray spectra of clusters and that is not expected in primordial gas.

Dr. Field mentioned that it is now widely thought that the ultimate energy source in active galaxies (radio galaxies and Seyfert galaxies) and in quasars is due to the accretion of gas by a supermassive collapsed object, possibly a black hole. He emphasized that the environment of the supermassive object in that case would control both the extent and the precise manifestation of its activity. Dr. Field also mentioned the active galaxy NGC 1275, which lies at the center of the Perseus cluster of galaxies. I should like to dwell on NGC 1275 in order to emphasize how it illustrates the complexity of our present view of the interaction between galaxies and their environment.

NGC 1275 is a Seyfert galaxy — it has a bright nucleus whose spectrum is characterized by broad emission lines of high excitation. This object is unique among Seyfert galaxies in two respects: it is an elliptical galaxy and it lies at the center of a large cluster. Spectroscopic studies by Minkowski, since elaborated by Burbidge and Burbidge and by Kent and Sargent, have shown that NGC 1275 is surrounded by a network of gaseous filaments, which were photographed by Lynds.[3] The filaments have the same red shift (5,400 km/sec) as the main body of NGC 1275, which in turn is the same as the mean velocity of the galaxies in the Perseus cluster. Fabian and Nulsen have suggested that NGC 1275 is situated, almost stationary, in the gravitational potential well at the center of the Perseus cluster and that the observed filaments result from the condensation of the hot gas responsible for the X-ray emission.[4] (As Dr. Field pointed out, the cooling rate of a hot gas is proportional to n_e^2, and the densest gas is at the center of the cluster.) Fabian and Nulsen calculated that the cooling time of the hot gas at the center of the Perseus cluster is comparable to the Hubble time (the characteristic expansion time scale for the universe). Kent and Sargent have shown that the spectrum of NGC 1275 is consistent with this idea and that the filaments are probably reheated by shock waves. NGC 1275 is also peculiar in that unlike the typical elliptical galaxy, its spectrum indicates that it contains A-type stars whose characteristic age is 10^8–10^9 years. It is conjectured that these stars have been born from the material accreting from the hot gas. Moreover, the unique properties of NGC 1275 support the idea that the active nucleus may be fueled by the accretion flow, as Dr. Field suggested in other contexts.

Further evidence that external factors play an important role in producing the nonthermal activity in the nuclei of galaxies has been provided by recent observations of the radio galaxies NGC 4278 and NBC 5128. In general, elliptical galaxies are devoid of detectable interstellar gas. It was, therefore, curious that the first elliptical in which 21-cm neutral hydrogen emission was detected was the well-known radio galaxy NGC 4278. Even more remarkable, Knapp, Kerr, and Williams have shown that the $10^8 M_\odot$ of interstellar hydrogen in this galaxy is distributed in the form of a rotating disk that is inclined at an angle of 50 degrees to the major axis of the whole galaxy.[5] It may be shown that this situation is unstable and that, on a time scale of $\sim 10^8$ years, the rotating disk of gas must align itself along one of the principal axes of NGC 4278. It is, therefore, hard to escape the conclusion that NGC 4278 has captured a large intergalactic cloud within the past 10^8–10^9 years and that this gas is responsible for the nonthermal activity in the galaxy.

NGC 5128, which is the closest example of a radio galaxy (Centaurus A), has also been a puzzle for several years. NGC 5128 is an elliptical object with a pronounced dust lane that, unexpectedly, is roughly along the minor axis of the galaxy. Deep exposures made by Graham show that in fact the outer part of the dust lane is inclined to the inner part.[6] It has been known for some time that the dust lane contains gas and young stars, which are not commonly found in elliptical galaxies, although NGC 5128 does not resemble any kind of spiral galaxy. Graham has recently studied the kinematics of the stars and gas in NGC 5128. He finds that the gas is distributed in a rotating disk inclined at an angle of 23° to the minor axis of the galaxy. Again, this is an unstable situation and again the recent capture of either an intergalactic cloud or a gas-rich, dwarf irregular galaxy by an elliptical galaxy has to be invoked to explain the phenomena.

In the case of NGC 4278 and probably in the case of NGC 5128, we are witnessing the consequences of the capture of intergalactic gas clouds with masses of order $2 \times 10^8 M_\odot$. (In the spectrum of NGC 4278 there is no sign of young, blue stars, which would exist in a captured dwarf irregular galaxy.) However, recent searches for 21-cm neutral hydrogen clouds in nearby groups of galaxies by Lo and Sargent and by Haynes and Roberts have been completely unsuccessful.[7] Thus, although Lo and Sargent detected gas clouds with masses as low as $10^7 M_\odot$, in all cases it was found that they were associated with visible dwarf irregular galaxies: there is thus no direct evidence for primordial clouds of neutral hydrogen at the present epoch.

In summary, although there is ample evidence that the evolution of galaxies in large clusters may be profoundly influenced by an external medium, the situation regarding galaxies in the general field is less clear. In this connection, one should emphasize that:

1. There is still not direct evidence for a general intergalactic medium.
2. There is no evidence for young galaxies. Consequently, whatever general medium does exist must be such that galaxies are no longer condensing out of it.

Abnormally blue galaxies, objects with a high proportion of young, hot, massive stars, have been found in relatively large numbers during the past few years, largely through the work of Markarian and his associates in the Armenian Soviet Socialist Republic. These galaxies, of which Markarian 116 and II Zw 40 are the archetypes, are almost always dwarf systems; detailed spectroscopic studies have shown that they are rich in gas and, moreover, that their interstellar gas is deficient in heavy elements. Those objects could be young systems that have only consumed a small fraction of their initial supply of gas; the remaining gas has only been slightly enriched in heavy elements by stellar nucleosynthesis. However, it now seems almost certain that the abnormally blue dwarf galaxies are in fact old systems that experience intense bursts of star formation, during which time they briefly become both bluer and brighter than they are in the quiescent state.[8] That this is the correct explanation is shown by the fact that no primordial gas clouds of 10^8–10^9 $M_\odot$ have been detected in the recent sensitive 21-cm surveys referred to earlier. A reservoir of 30 per Mpc^3 such clouds would be required in order for galaxies like Mk 116 and II Zw 40 to be forming continuously.

Dr. Field emphasized that radio and optical searches for various manifestations of a general intergalactic medium have been unsuccessful. The Gunn–Peterson test[9] enables a very stringent limit to be placed on the density of any smoothly distributed intergalactic cool neutral hydrogen. The X-ray background could be produced by a hot gas, which Field and Perrenod estimated to have a temperature of 3×10^8 °K and a density comparable to the closure density ($\Omega \sim 0.5$). However, as Dr. Field emphasized, the X-ray background could also be produced by discrete sources — quasars, Seyfert galaxies, and clusters of galaxies; in estimating their contribution it is hard to assess how the number density and the X-ray luminosity of these objects have evolved in cosmic time from the large red shifts of interest.

It has always been hoped that observations of absorption lines in the spectra of quasars would eventually lead to information about the intergalactic medium. The quasars are ideal for this purpose since they are bright and have been found out to red shifts up to $z_{em} = 3.53$. Absorption lines were found in quasar spectra more than a decade ago, and, after several years of effort, it has been possible to establish that there are two distinct kinds of lines that appear to have distinct physical origins. On the one hand, there are absorption red shifts containing lines of heavy ions — Si-II, Si-III, Si-IV, Mg-II, Ca-II, C-IV, as well as the Lα line and other lines of the Lyman series. Red shift systems composed of these lines account for essentially all of the absorption lines longward of the Lα emission line, which is such a prominent feature in the spectra of quasi-stellar objects. Shortward of Lα emission we find many otherwise unidentified absorption lines that have been shown to be single Lα lines. Recent work indicates that the heavy element absorption systems arise in intervening galaxies and that the single Lα lines arise in intergalactic clouds that are not associated with galaxies.[10] My colleagues and I have estimated that the Lα clouds have total hydrogen densities $n_H \sim 10^{-4}$ to 4×10^{-3} cm^{-3}, tempertures near 3×10^4 °K and diameters in the range 10^{20}–10^{23} cm. They are photoionized by the integrated ultraviolet radiation of the quasars (which at the red shifts in question $z \sim 2.5$ have a much larger number density than at the present epoch). The Lα clouds contribute a trifling amount to the overall mass of the universe having $\Omega = 10^{-3}$. If these parameters are correct, then the existence of the clouds places constraints on the general intergalactic medium at these large red shifts. In particular, their existence rules out a hot, dense intergalactic medium ($T_M = 3 \times 10^8$ °K, $\Omega = 0.5$), which, as Dr. Field pointed out, is required if the observed X-ray background radiation is produced by thermal bremsstrahlung. In fact, if the current theory of the evaporation of cool clouds by a surrounding hot gas is correct, then the survival of the Lα clouds demands a relatively cool, tenuous general intergalactic medium with $n_M \sim 10^{-5}$, $\Omega = 0.1$, and $T_M = 3 \times 10^5$ at a red shift of $z = 2.45$. However, the main point is that we are now in a position to make direct observations, via the absorption lines in quasars, of intergalactic clouds at large red shifts. With the advent of the Space Telescope in 1984 it will be possible to extend this kind of investigation to Lα clouds at low red shifts and hence obtain information on the intergalactic medium nearer the present epoch.

It will be evident, both from these particular remarks and from the broad sweep of Dr. Field's lecture, that both the evolution of the stellar populations of galaxies and the existence of non-thermal activity in active galaxies are influenced by their interaction with the intergalactic medium, both in clusters and in the general field. Meanwhile, observations of absorption lines in quasars are at last giving direct information about the intergalactic gas at large red shifts, perhaps close to the epoch of galaxy formation. What we are still lacking are direct observations of the intergalactic medium outside of clusters in our more immediate vicinity and at the present epoch.

NOTES

1. W. W. Morgan, "A Preliminary Classification of the Forms of Galaxies According to Their Stellar Population," Pub. Astron. Soc. Pacific **70**, 364 (1958).
2. J. H. Oort, "The Formation of Galaxies and the Origin of High-Velocity Hydrogen," Astron. Astrophys. **7**, 381 (1970).
3. R. L. Minkowski, "Optical Investigations of Radio Sources," in *Radio Astronomy, International Astronomical Union Symposium 4,* edited by H. C. van der Hulst (Cambridge University Press, Cambridge, England, 1957), p. 107; E. M. Burbidge and G. R. Burbidge, "Optical Evidence Suggesting the Occurrence of a Violent Outburst in WGC 1275," Astrophys. J. **142**, 1351 (1965); S. M. Kent and W. L. W. Sargent, "Ionization and Excitation Mechanisms in the Filaments Around WGC 1275," Astrophys. J. **230**, 667 (1979); R. Lynds, "Improved Photographs of the WGC 1275 Phenomenon," Astrophys. J. Lett. **159**, L151 (1979).
4. A. C. Fabian and P. E. T. Nulsen, "Subsonic Accretion of Cooling Gas in Clusters of Galaxies," Monthly Notices, Roy. Astron. Soc. **180**, 479 (1977).
5. G. Knapp, F. J. Kerr, and B. A. Williams, "H-I Observations of Elliptical Galaxies," Astrophys. J. **222**, 800 (1978).
6. J. A. Graham, "The Structure and Evolution of NGC 5128," Astrophys. J. **232**, 60 (1979).
7. K. Y. Lo and W. L. W. Sargent, "A Search for Intergalactic Neutral Hydrogen in Three Nearby Groups of Galaxies," Astrophys. J. **227**, 756 (1979); M. Haynes and M. S. Roberts, "Are There Really Intergalactic Neutral Hydrogen Clouds in the Sculptor Group?" Astrophys. J. **227**, 767 (1979).
8. L. Searle, W. L. W. Sargent, and W. G. Bagnuolo, "The History of Star Formation and the Colors of Late-Type Galaxies," Astrophys. J. **179**, 427 (1973); J. P. Huchra, "Star Formation in Blue Galaxies," Astrophys. J. **217**, 928 (1977).
9. J. E. Gunn and B. A. Peterson, "On the Density of Neutral Hydrogen in Intergalactic Space," Astrophys. J. **142**, 1633 (1965).
10. W. L. W. Sargent et al. "The Distribution of Lyman Alpha Absorption Lines in the Spectra of Six QSOs — Evidence for an Intergalactic Origin," Astrophys. J. Suppl. **42**, 41 (1980).

Open Discussion

Following Papers by G. B. FIELD and W. L. W. SARGENT

Chairman, D. SCIAMA

E. Segrè (U. Calif., Berkeley): In all this discussion (I'm a physicist, I'm not an astrophysicist) I never heard the mention of antimatter. By common sense you would say at a certain moment there should be at least about as much antimatter as matter. What has happened to all of it?

Sciama: Well, actually, I'll be discussing that in my paper. Thanks for the chance for advertisement.

Y. Ne'eman (Tel-Aviv U.): Considering the failure of all these various models to account for a larger omega, as we turn again to neutrinos or other particles, is there an upper bound on what neutrinos could be doing?

Sciama: Who would know?

Field: I think there has been some speculation that would be interesting in this connection. Of course a zero-rest-mass neutrino would not be expected to contribute if it ever thermalized because the number density of such neutrinos would be comparable to that in photons. However, a finite-rest-mass neutrino — and you'll have to tell me whether that is possible or not — would contribute and in fact would close the universe if its rest-mass exceeded about 10.

Ne'eman: But wouldn't they fall into the clusters of galaxies and then be counted anyhow at once?

Field: I think that is true. There has been some speculation that the missing mass in clusters may be in part due to such neutrinos.

S. Hawking (via Lapedes): Some time ago Stephen Hawking said that he had a bet about the critical density. In view of what George Field has said, are you prepared to concede the bet?

Sciama: I always lose my bets with Stephen Hawking. But I'm not ready to concede yet — no. There could be a distribution of black holes, for instance.

M. J. Rees (Cambridge U.): As a comment on that I would like to recall a symposium some years ago when John Wheeler, I think, gave the summary talk. He conducted an opinion poll in the audience about who wanted the universe closed and who wanted it open. And the one gratifying thing about the poll was that most people were prepared to say they did not yet know.

S. Treiman (Institute for Advanced Study): With regard to the problem of the neutrinos, you can actually rule out the possibility that neutrinos with a rest mass of a few electron volts would provide the missing mass in clusters and groups because the required phase space density would violate the Pauli principle.

G. de Vaucouleurs (U. of Texas): First I would like to remind you that the visible intergalactic light in the Coma cluster, if it has a normal spectrum due to stars, will not contribute more than about 20 or 30 percent of the mass. But it does make a significant contribution. The second point is that I should like to remind you that the possible effect of capture of intergalactic matter by galactic nuclei was discussed long before 1970. I discussed it, for example in 1963 in Canberra as a possible source of energy for nuclear radio sources in galaxies, and I doubt that even then it was being discussed for the first time.

VIII. The Universe (continued)

22. BEYOND THE BLACK HOLE

John Archibald Wheeler

The Sibyline Strangenesses of the Landscape

Arthur Wellesley, Duke of Wellington, in the long years of activity in England that followed Waterloo, from time to time for relaxation would take a companion along for a carriage ride of hours through a distant countryside unfamiliar to them both. The Iron Duke was accustomed to draw his companion into his favorite game. From the look of the terrain up to this moment, predict what new panorama will be seen as the carriage tops the next long hill. Wellington generally produced the winning forecast of the lay of the land. Einstein traveled through a different countryside. His ability to sense ahead of time the upcoming landscape of physics is well known.

Today we find ourselves traversing a new realm. It contains such strange features as the black hole, the gauge or phase field, and complementarity. What lies beyond, over the hill?

If all strangenesses of a landscape made for Wellington the best indicators of the new terrain, the same was true, we know, for Einstein and the same surely holds for physics now. There is no hope of progress, we often say, until we are in possession of a central paradox, a difficulty, a contradiction. However, in our hearts we know it takes more. We need two paradoxes. Only then can we play off one against the other to locate the new point.

Two strangenesses stand out with special prominence in the landscape of the physics of our day: one is the Bounds of Time; the other, the Quantum.

Of the bounds of time the black hole[1] is the one most immediately accessible; then, beyond, the big bang[2] and — if the universe, as Einstein argued,[3] is closed and therefore collapses[4] in time to come — the big crunch.[5] The bounds of time tell us that physics comes to an end. Yet physics has always meant that which goes on its eternal way despite all surface changes in the appearance of things. Physics goes on, but physics stops; physics stops, but physics goes on. That is paradox number one, strange feature number one in the landscape we survey.

Paradox number two, the quantum principle[6] thrusts upon us. In every elementary quantum process the act of observation, or the act of registration,[7] or the act of observer-participancy,[8] or whatever we choose to call it, plays an essential part in giving "tangible reality" to that which we say is happening. Paradox number two is this: The universe exists "out there" independent of acts of registration, but the universe does not exist out there independent of acts of registration.

Harry Woolf (ed.), Some Strangeness in the Proportion: A Centennial Symposium to Celebrate the Achievements of Albert Einstein
ISBN 0-201-09924-1

Preparation for publication assisted by University of Texas Center for Theoretical Physics and by NSF grant PHY 78-26592.

If these are no small paradoxes, they suggest no small questions about what lies over the hill. How can one possibly believe that the laws of physics were chiseled on a rock for all eternity if the universe itself does not endure from everlasting to everlasting? If law, field, and substance come into being at the big bang and fade out of existence in the final stage of collapse,[9] how can a change so all-encompassing take place except through a process, the elementary mechanism of which has already made itself known? In what other way does an elementary quantum phenomenon become a phenomenon except through an elementary act of observer-participancy? To what other foundation then can the universe itself owe its existence except billions upon billions of such acts of registration?[10] What other explanation is there than this for the central place of the quantum principle in the scheme of things, that it supplies the machinery by which the world comes into being?[11]

Laws Derived from Symmetry Considerations but They Hide the Machinery Underlying Law[12]

Before we inspect more closely the two sibyline strangenesses of the landscape, let us look at the laws of physics themselves to recognize how little guidance they give us in forecasting what lies over the hill. Nothing in all the great achievements of science is more beautiful than Maxwell's electromagnetism, Einstein's geometric theory of gravitation, and the Yang–Mills theory of the quark-binding field.[13] Each expressible in a single line, these three theories are the yield of decades of research by hundreds of investigators performing thousands of experiments. However, the more we learn, the more we learn how little we have learned.[14]

Are not these three great theories of our time about to suffer the reappraisal already undergone by the great theory of elasticity of an earlier century?[15] Look at one of the good old textbooks on that subject. In the first chapter or two the laws of elasticity are deduced from elementary symmetry considerations. To say that the energy of deformation of a homogeneous isotropic material goes as the square of the deformation is to be confronted with two alternatives. One must either take the trace of the tensor of deformation and square it, or square the tensor of deformation and take its trace. More generally one makes a linear combination of these two expressions with two disposable constants of proportionality to construct the general expression for the energy of elastic strain. From this reasoning — that there are two constants of elasticity and no more — one goes on to build up all the rest of the great treatise on elasticity of one hundred years ago, complete with theorems, methods of analysis, applications and all kinds of beautiful problems for the student to solve at the end of each chapter. Likewise in our own time we have textbooks on electromagnetism,[16] gravitation,[17] and the Yang–Mills quark-binding field[18] or, more generally, on "gauge" fields, or "phase" fields, as Professor Yang suggests we call them, again complete with symmetry-argument foundations, theorems, applications, and problems for the student.

When we look back to elasticity, however, we recall that the most important fact about the subject — where the forces come from, molecular interactions between dozens of different atoms and molecules, multiplied by appropriate direction cosines — was not revealed one bit by these laws of elasticity. One hundred years of the study of elasticity would not have revealed atomic and molecular forces.[19] Neither would one hundred years of the study of atomic and molecular forces have revealed that these forces went back for their foundations to Schroedinger's equation and the motions of individual electrons and nothing more. We had to learn that we should explain, not electronic motions in terms of elasticity, but elasticity in terms of electronic motions. The very considerations of symmetry that had allowed one to master elasticity so early, taken by themselves, would have hidden from view forever the mechanism of elasticity.

The considerations of symmetry that reveal law hide the mechanism that underlies law. This lesson out of elasticity we today see afresh in electromagnetism, gravitation, and the dynamics of the Yang–Mills field, thanks to considerations of Hojman, Kuchař, and Teitelboim.[20] They consider a spacelike hypersurface σ_1 slicing through space-time (Fig. 22.1). They think of the field in question as given on all points of that hypersurface, along with the initial time rate of change of that field or, equivalently, the "field momentum." They ask: How does one go about predicting what the field will have for values at the points on a later spacelike hypersurface σ_2? The general marching his troops forward from one river to another may move the front ahead faster first on the right and later on the left, or alternatively order the line to advance more rapidly first on the left and then on the right, ending up, however, in the same stance on the same river. So the analyst of the field with his computer calculations can calculate ahead from instant to instant the successive configurations of the field either on the dashed or upon the dotted sequence of intermediate spacelike hypersurfaces in the diagram. He, unlike the general, ordinarily will arrive at different results for the field on σ_2 by the two maneuvers, the two alternative slicings of space-time, the two foliations: two incompatible predictions for one future. The fault, when there is one, is wrong choice of the particular Hamiltonian law assumed to govern the evolution of the field from instant to instant.

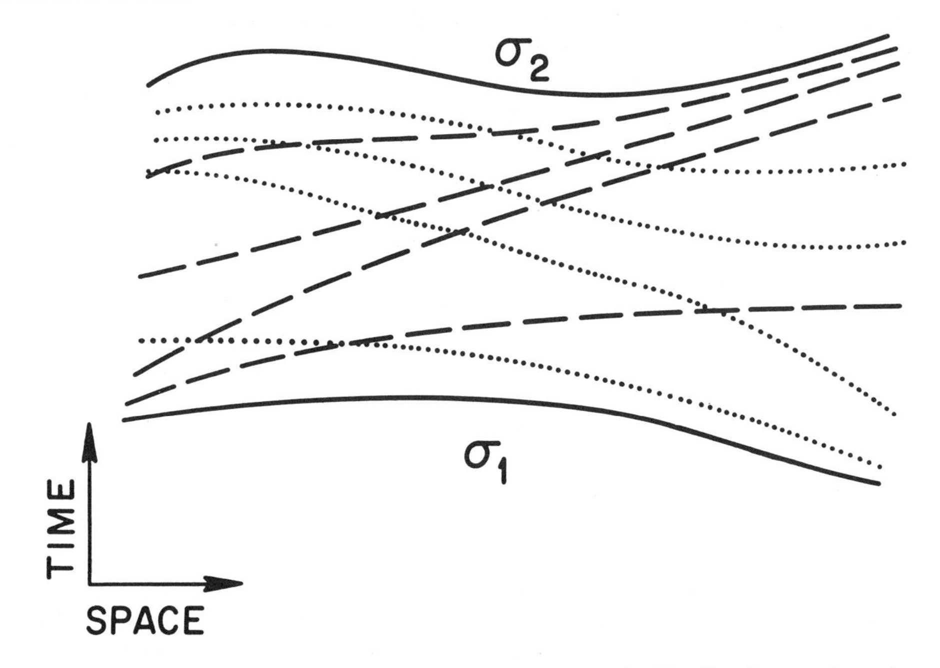

Fig. 22.1 The two alternative ways to calculate physics forward step by Hamiltonian step from the spacelike hypersurface σ_1 to the spacelike hypersurface σ_2 have to give the same result, the central point of the Hojman–Kuchař–Teitelboim "embeddability requirement." This simple demand leads straight to Maxwell electrodynamics, Einstein geometrodynamics, and the Yang–Mills theory of the quark-binding field.

When the field in question is a vector field and we restrict attention to Hamiltonians of the second order, there is only one option that is compatible with consistency. It is Maxwellian electrodynamics. When the field is a tensor field — the metric measuring the distance from point to point on the spacelike hypersurface — the requirement of consistency leads uniquely to Einstein's general relativity theory of gravitation. Any other Hamiltonian conflicts with the requirement that different ways of figuring ahead should fit into, be embeddable in, one and the same space-time manifold. Finally, when we impose this Hojman–Kuchař–Teitelboim demand of embeddability on a vector field that has an internal spin degree of freedom, we get the Yang–Mills theory of the quark-binding field.[21]

All three great theories of physics fall straight out of the utterly elementary demand for embeddability, as epitomized in Fig. 22.1. One does not have to recall Einstein's now abandoned dream of a geometrical unification of the forces of nature.[22] One does not have to have followed the exciting rebirth of this dream within the framework of that new and wider concept of geometry that is forced on us by the discovery in nature[23] — and in mathematics[24] — of "gauge" or "phase" fields, fields possessing at each point of space an "internal spin" degree of freedom. It is enough for the theoretical physicist to demand embeddability to deduce in a few hours what it took great men years of work to establish. Again, the more we learn the more we learn how little we have learned.

Fields in the end remind us more than ever of elasticity: modes of "vibration" of something; modes of a structure quite different from anything that shows on the surface; modes of a substrate, call it pregeometry or call it what one will, that is not and will not be revealed by reasoning from the top down, only from the bottom up[25]; not from the obvious but from the strange. Where better can we turn now for a hint of the new than to the two greatest strangenesses of the present landscape, the bounds of time and the quantum?

The Bounds of Time

In Fig. 22.2 the sequence of circles starting at the lower left symbolizes a closed universe beginning small and in the course of time becoming larger.

Philosophical considerations had guided Einstein to his 1915 geometric and still-standard theory of gravitation in the first place.[26] Considerations of the same kind, spelled out in his book *The Meaning of Relativity,*[27] led him to conclude that the universe must be closed. No one in subsequent years has found any boundary condition comparable in its reasonableness to the requirement of closure.[28] Is it necessary to worry about alternatives? Mass-energy so far located falls short of what is required to curve space up into closure, and the factor of shortfall looks big, but not big in order of magnitude as compared to the uncertainties of, and discrepancies between, estimates of mass-energy based on 1) the mass-luminosity relation,[29] 2) primordial deuterium,[30] and 3) clustering of galaxies.[31]

A closed-model universe with the topology of a three-dimensional sphere comes to a maximum volume, begins to contract, and finally collapses, as indicated in Fig. 22.2. Therefore it might seem that all the particles gather together in a common place at the start of time. However, the phrase "common place" we know to be a bad phrase and we know how to see that it is bad. We "unroll the space"; or in our symbolic model for the space, we unroll the circle from north pole to south pole. Then the separation between two particles measures itself not in miles but in degrees. The space-time history of the particles then allows itself to be displayed, though it is not

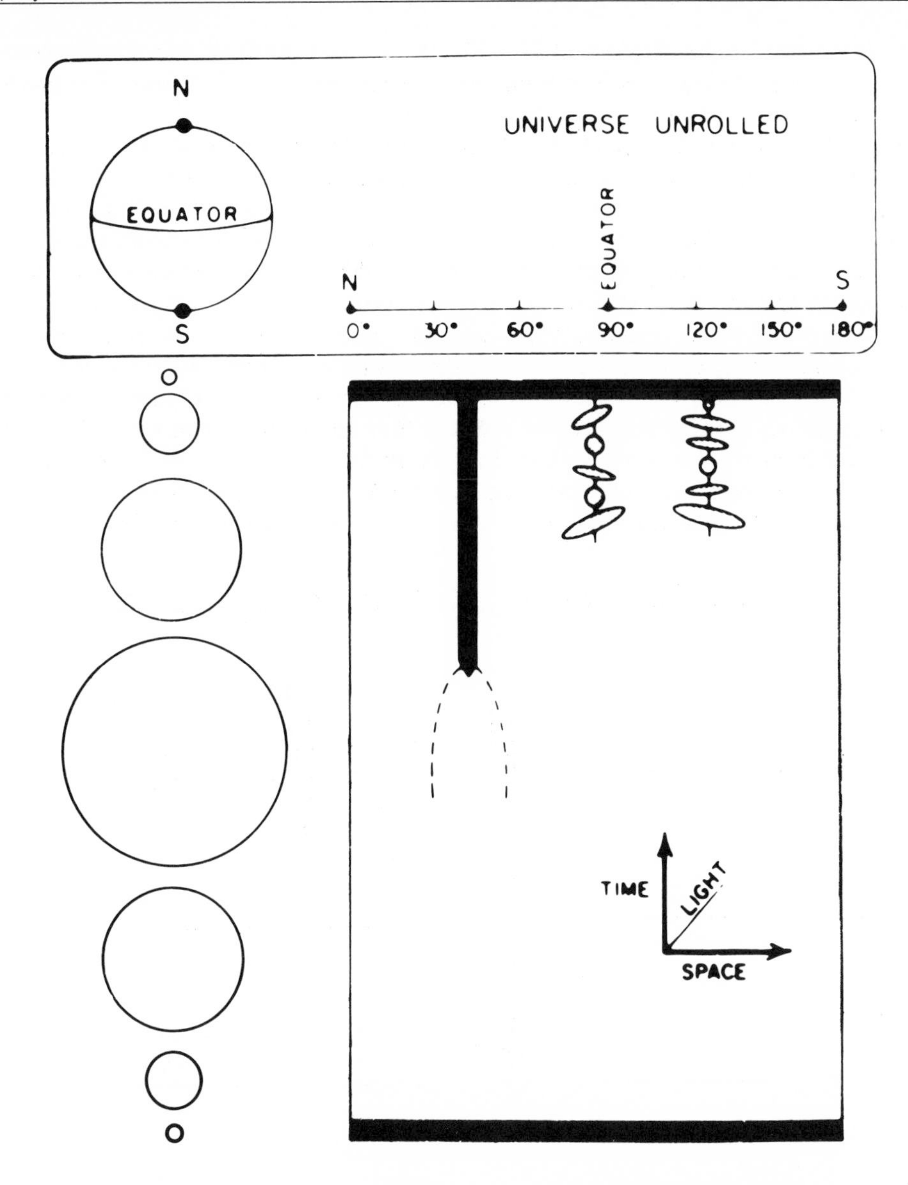

Fig. 22.2 Two symbolic representations of the history of a closed-model universe, idealized as a 3-sphere but depicted as if a 1-sphere (circle). *Lower left:* The universe starts small, expands, reaches a maximum volume, contracts and collapses. *Lower right:* The same history depicted in a rectangular diagram in which particle positions are given in angular measure by the scheme of translation sketched at the top of the page. Time is plotted on such a scale that light rays run at $\pm 45°$, being bounded however by the big bang at the bottom of the rectangle and the big crunch at the top. Two particles separated by 30° at the time of the big bang have to wait $\sim 10^8$ to $\sim 10^9$ years before a signal from the one gets to the other. The simple 45° algorithm is modified when there are inhomogeneities, such as the black hole "spike hanging from the roof" or the symbolically represented "mixmaster oscillations" of the geometry in the final stages of gravitational collapse.

displayed, in the rectangular diagram of Fig. 22.2 as two vertical lines. It can take hundreds of millions of years after the big bang before one particle communicates its presence to another particle that began its life in the same microscopic fireball.

The dashed lines in Fig. 22.2 symbolize the outer boundaries of a cloud of dust that gradually shrinks and eventually collapses to a black hole. The singularity at the center of the black hole is seen to be, not a new and distinct bound of time, but part and parcel of the big crunch.[32]

In a very wide class of models of closed universes compatible with Einstein's field equations — Marsden and Tipler have recently proved[33] — the four-dimensional geometry admits a foliation in one way and in one way only into slices of constant mean extrinsic curvature. One value of the mean curvature gives us one spacelike slice. Another value, zero mean curvature, gives us that instant, that unique spacelike slice, that phase of the history of the universe, for which the volume is the largest. Later slices depict the geometry of the universe, whatever its lumps, bumps, and ripples, with successively smaller and smaller volume. How does this circumstance bear on black holes that, once formed, ultimately coalesce into the final big crunch? The successive hypersurfaces of the foliation englove the singularity of each black hole more and more closely, according to recent calculations by A. Qadir and me,[34] as illustrated schematically in Fig. 22.3. The "last

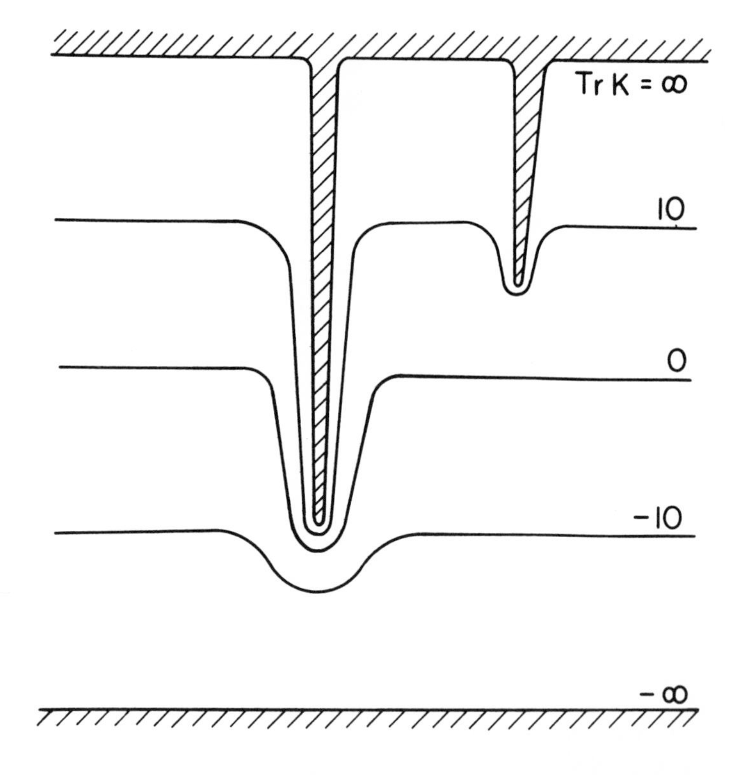

Fig. 22.3 Black hole and big crunch seen as part and parcel of the final singularity. The closed model universe is uniquely foliated by a sequence of spacelike hypersurfaces distinguished one from another by the value — constant over any one hypersurface — of the mean extrinsic curvature; that is, the trace of the tensor of extrinsic curvature or the fractional rate of decrease of volume per second.

hypersurface," the one of infinite mean extrinsic curvature, "establishes contact" simultaneously all along its front with the black hole singularity and the big crunch singularity. No better way could one desire to see that those are not two singularities but one.

The generic way of approach to the final singularity, if Belinsky, Khalatnikov, and Lifshitz are right,[35] proceeds through so-called mixmaster oscillations in the geometry, with the amplitude, phase, and direction of the principle axes of the space deformation varying from point to point of the spacelike hypersurface. Therefore also for the approach to the singularity of the physical black hole, as distinguished from the ideal Schwarzschild "dead" — or Reissner–Nordstrøm charged[36] or Kerr rotating[37] or Kerr–Newman charged *and* rotating[38] — black hole, it is not unreasonable to expect a mixmaster character.

More than one hundred papers[39] of recent years, many of them beautiful in method and in results, deal with the physics outside the "horizon" of a black hole, but almost none with conditions inside. Thanks to this work we have learned in what sense "a black hole has no hair."[40] A "hair," a departure from ideality, a perturbation in the geometry outside the horizon associated with irregularities and turbulence when the black hole formed, washes out by a factor of $1/e = 1/2.718$ in each "characteristic time," a time of the order of magnitude of 10^{-4} seconds for a black hole of ten solar masses. Thus such a black hole, 1 sec after matter has stopped falling in, has attained a fantastic perfection outside.[41] Inside the horizon, however, it is natural to expect the direct opposite: small initial departures from ideal symmetry as matter falls in across the horizon leading to enormous mixmaster curvature fluctuations from point to point as one approaches the singularity.[42]

How far away is that singularity? My watch, the baryons of which came into being at the big bang, has 10 more years of life. When it stops, can we spare its baryons the ignominy of further use? Instead of burying the watch or melting it down can we obliterate it? Can we make those 10 years stretch to the end of time, to the singularity after which there is no after? Yes. Can we even choose whether the place of obliteration shall be a black hole or the big crunch? Yes, if there *is* a big crunch and if it lies at the estimated time[43] in the future. For either purpose we must put the watch aboard a powerful rocket, one that will make the factor of time dilatation of the order of 10^{11} years per 10 years if in 10 years of life we would have it reach the big crunch; or of the order of 10^4 years per 10 years, a black hole located in this galaxy.

Black Hole as Bound of Time

How near the end of time is we are reminded by each new piece of evidence for a black hole, the object of about 10 solar masses in the constellation Cygnus remaining the best studied of presumptive black holes.[44] Bursts of X rays suggest a black hole of 100 to 1,000 solar masses at the center of each of five clusters of stars in our own galaxy.[45] Charles Townes and his colleagues, as well as Jan Oort, and others give considerations arguing for a black hole of about 4×10^6 solar masses at the center of the Milky Way.[46] The Lick Observatory group and others find evidence[47] pointing to the possible existence of a black hole of about 5×10^9 solar masses at the center of the violently active galaxy M87.

Is it clear that the center of a black hole offers obliteration only, not the chance to emerge somewhere else in space? In favor of such a possibility for space travel there exists not the slightest evidence. On the contrary, if at any time there ever were a wormhole or tunnel of appreciable diameter (as distinguished from the dimensions of quantum fluctuations), it would collapse with the speed of light.[48] The matter that falls in does not reappear somewhere else. All its details fade away, but its gravitational attraction remains. Any planet once in circumambient orbit stays in orbit. Mass-energy, an exterior property, remains; matter, an interior property, is obliterated.

Figure 22.4 reminds us that all details of whatever is dropped in are washed away. Provided that nature has no other long-range charge-conserving field than electromagnetism, we have to conclude that the resulting black hole is fully characterized by its mass, charge, and angular momentum, and nothing more. Of course mass implies energy, and therefore also the possibility of another property for a black hole, momentum.[49] However, we think of this momentum, not as an independent feature of the black hole, but as a consequence of our choice of reference frame. There is another feature of the black hole, Claudio Teitelboim tells us,[50] its spinor spin, that — like momentum — can be given one value or another depending on our choice of reference frame, except that now the frame of reference that comes into consideration is not the Lorentz frame but the spinor reference frame. It does not matter for this reasoning whether we use the theory of supergravity as originally developed by Freedman, van Nieuwenhuizen, and Ferrara, and by Deser and Zumino, or whether we follow Teitelboim's beautiful procedure of taking "the Dirac square root" of Einstein's general relativity, for by these two very different routes we come to the same theory with the same "internal spin-$\frac{3}{2}$" or "phase" degree of freedom.[51]

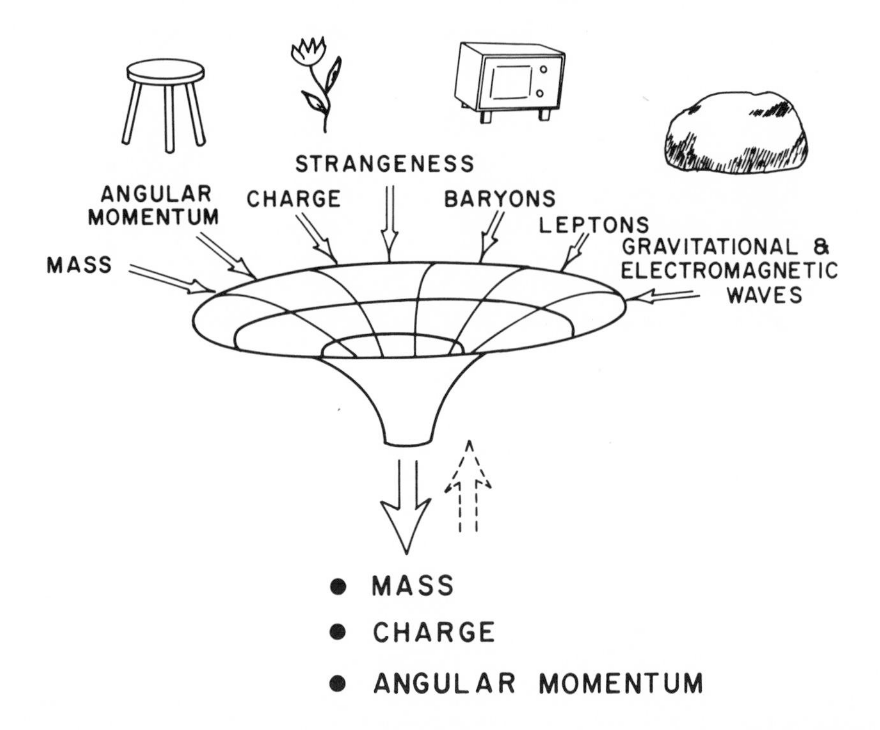

Fig. 22.4 Details of all objects dropped into a black hole are obliterated. The resulting system, according to available theory, is characterized by its linear momentum, angular momentum, mass, and charge and by no other parameter. No detail inside its "horizon," or surface of no return, can be probed from outside. At its center sits the singularity of final crunch.

Of baryon number, lepton number, and strangeness not a trace is left, if present physics is safe as guide.[52] Not the slightest possibility is evident, even in principle, to distinguish between three black holes of the same mass, charge, and angular momentum, the first made from baryons and leptons, the second made from antibaryons and antileptons, and the third made primarily from pure radiation.[53] This circumstance deprives us of all possibility to count, or even define, baryon and lepton number at the end and compare them with the starting counts.[54] In this sense the laws of conservation of baryon and lepton number are not violated; they are transcended.

Up the Staircase of Law and Law Transcended to Mutability

Figure 22.5 pictures the development of physics as a staircase.[55] Each step symbolizes a new law or discovery. Each riser marks the attainment of conditions so extreme as to overcome the usefulness of that law, or transcend it.

Archimedes, discovering how to measure density,[56] could regard it as a constant of nature. However later ages achieved pressures great enough to bring about measurable alterations in density.[57] The concept of valence[58] brought into order the major facts of chemistry, but today we

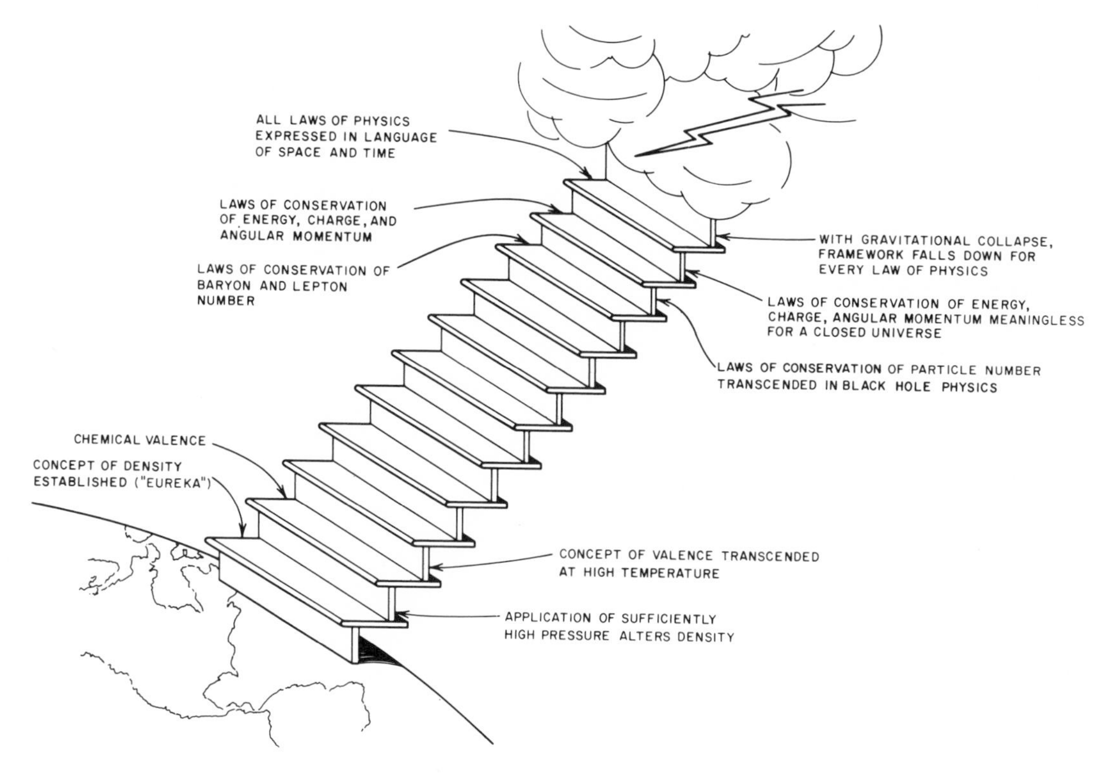

Fig. 22.5 The staircase of law and law transcended. Each step symbolizes the discovery of a new regularity or constancy of nature; each riser, the discovery of a technical means or a natural condition so extreme as to overcome or transcend that regularity.

know we have only to go to very high temperatures to outrun traditional valence considerations.[59] Later came the discovery that every atomic nucleus admits rigid classification by its charge number and its mass number;[60] but the advent of nuclear transmutations[61] destroyed that rigidity. The laws of conservation of baryon number and lepton number are indispensible in accounting for the wealth of experience in elementary particle physics,[62] but they have no application in black hole physics.[63] There they are not violated, but transcended.

In the end can we not at least say that the black hole has mass and therefore mass-energy? And does not the law of conservation of energy stand up against arbitrarily extreme conditions? In an asymptotically flat space, yes; in a closed universe, no. There total energy is not even defined.[64] Thus the local law of conservation allows one to express the total energy in a bounded region as an integral over the two-dimensional frontier of that region. The larger the region subsumed in counting up the energy, the larger at first is the boundary. However, as more and more volume is swept for energy, the boundary pushes on over the great bulge of the universe and begins to shrink. As we complete the sweep through "the other half of space," we push this surface down to extinction. The law of conservation of energy degenerates to the identity $0 = 0$. This lesson of the mathematics physics can be put into other words. To measure the mass-energy of a moon, a planet, a star, or larger system, it says, put a satellite in orbit about it. Measure the period of revolution, apply Kepler's "1–2–3 law" of motion[65] and obtain the mass. In the case of the closed universe, however, there is no "outside," no circumferential highway, in which to orbit a satellite. The idea of "total mass-energy" — and with it the law of conservation of energy — lose all meaning and application.

At the head of the stairs there is a last footstep of law and a final riser of law transcended. There is no law of physics that does not require "space" and "time" for its statement. Obliterated in gravitational collapse, however, is not only matter, but the space and time that envelop that matter.[66] With that collapse the very framework falls down for anything one ever called a law of physics.

Einstein's general relativity gives not the slightest evidence whatsoever for a before before the big bang or an after after collapse.[67] For law no other possibility is evident but that it must fade out of existence at the one bound of time and come into being at the other.[68] Law cannot stand engraved on a tablet of stone for all eternity. Of this strangeness of science we have for symbol the staircase; and for central lesson, "All is mutable."[69]

If the lesson comes in two parts, "Space-time ends" and "Laws are not eternal," each can be pursued further.

Space-time as Idealization

A crystal reveals nowhere more clearly than at a crack[70] that the concept of "ideal elastic medium" is a fiction. Cloth shows nowhere more conspicuously than at a selvedge that it is not a continuous medium, but woven out of thread (Fig. 22.6). Space-time — with or without "phase" or "internal spin" degrees of freedom — often considered to be the ultimate continuum of physics, evidences nowhere more strikingly than at big bang and collapse that it cannot be a continuum.[71]

There is an additional indication that space cannot be a continuum. Quantum fluctuations of geometry and quantum jumps of topology are estimated[72] and calculated[73] to pervade all space at the Planck scale of distances and to give it a foamlike structure.

"Space is a continuum." So bygone decades supposed from the start when they asked, "Why does space have three dimension?"[74] We, today, ask instead, "How does the world manage to give the impression it has three dimensions?" How can there be any such thing as a space-time continuum except in books? How else can we look at "space" and "dimensionality" except as approximate words for an underpinning, a substrate, a "pregeometry,"[75] that has no such property as dimension?

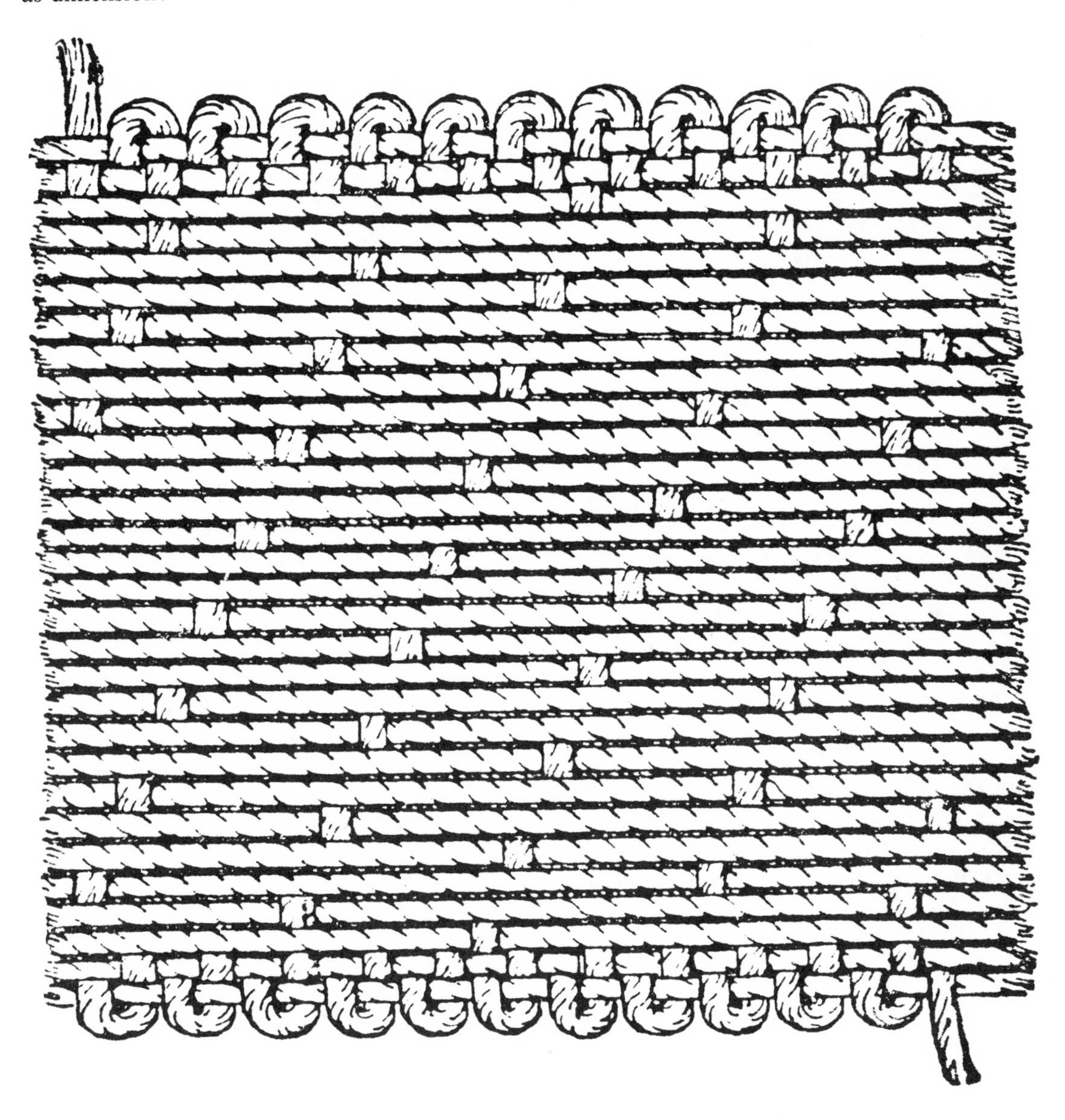

Fig. 22.6 It is disclosed more clearly at the selvedges than anywhere that what is woven is not a continuum.

Law without Law

"Physical space-time is not mathematical space-time" is the one lesson of mutability; the other, "Physical law is not ideal mathematical law." Law that comes into being at the beginning of time and fades away at the end of time cannot be forever 100 percent accurate. Moreover, it must have come into being without anything to guide it into being.

It is not new for a regularity to develop unguided. Thermodynamics, we know, rests upon the random motions of billions upon billions of molecules.[76] Ask any molecule what it thinks about the second law of thermodynamics and it will laugh at the question. All the same the molecules, collectively, uphold the second law. The genera and species of the kingdom of life go back for their foundation to billions upon billions of accidents of mutation.[77] The fantastically elaborate organization of plants and animals is of nothing but higgledy-piggledy origin. The laws of physics themselves, coming into being and fading out of existence: in what else can they have their root but billions upon billions of acts of chance? What way is there to build law without law, field without field, substance without substance except "Individual events. Events beyond law. Events so numerous and so uncoordinated that, flaunting their freedom from formula, they yet fabricate firm form?"[78]

Strangeness Number Two: Quantum and Chance

We have been led to consider chance events, astronomical in number, as the statistical foundation of all the regularities of physics, and this in default of any other way to come to terms with mutability and the bounds of time, strangeness number one of the landscape. What kind of chance event? For a clue it is not clear where else to look except at strangeness number two, the quantum, "God plays dice."

"I cannot believe that God plays dice." Who that has known or read Einstein (Fig. 22.7) does not remember him arguing against chance in nature?[79] Yet this is the same Einstein who in 1905, before anyone, explained that the energy of light is carried from place to place as quanta of energy,[80] accidental in time and space in their arrival; and in 1916, again before anyone, gave us in his A's and B's, his emission and absorption coefficients, the still standard mathematical description of quantum jumps as chance events.[81] How could the later Einstein speak against this early Einstein, against the evidence and against the views of his greatest colleagues? How can our own day be anything but troubled to have to say "nay" to one teaching, "yea" to others of the great Einstein, the man who gave us in his geometric account of gravitation[82] a model, still unsurpassed, for how a physical theory should be founded and what it should do?

Was Einstein's Thinking Constrained by His Philosophical Antecedents?

We are less troubled, more understanding, when we recall the philosophical antecedents of Einstein's thinking. They derived from a cast of characters who seemed to live within his cranium and counsel with him as he spoke: Leibniz and Newton, Hume and Kant, Faraday and Helmholtz, Hertz and Maxwell, Kirchhoff and Mach, Boltzmann and Planck; but above them all Benedictus

Fig. 22.7 Albert Einstein photographed at Princeton during broadcasting by Popperfoto (reproduced with the kind permission of the photographer — copyright reserved by the photographer).

de Spinoza, hero and role-creator to Einstein in youth as well as later life.[83] In earlier centuries no one expressed more strongly than Spinoza a belief in the harmony, the beauty, and — most of all — the ultimate comprehensibility of nature; in our own century, no one more than his admirer, Einstein. Guide Einstein to high goals Spinoza certainly did; but did he not — Hans Küng suggests[84] — on two occasions misguide?

Why was twenty-four-year-old Spinoza excommunicated in 1656 from the Amsterdam synagogue? Because he denied the bible story of an original creation.[85] What was the difficulty with the teaching? In all the nothingness before creation where could any clock sit that should tell the universe when to come into being! Therefore, Spinoza reasoned, the universe must endure from everlasting to everlasting. In contrast, and to Einstein's surprise, general relativity already in its first two years predicted that a static 3-sphere universe is an impossibility. Of necessity it is dynamic. Consequently Einstein reluctantly changed the theory and introduced a so-called cosmological term with the sole point and purpose to hold the universe static, to rule out what Alexander Friedmann later showed was a big-bang-to-big-crunch cosmology.[86] A decade later, when Edwin Hubble established the expansion of the universe, Einstein's chagrin about the cosmological term is well known: "the biggest blunder of his life."[87] Today, looking back, we can forgive him his Spinoza-inspired blunder and give him credit for the theory of gravitation that predicted the expansion. Of all the great predictions that science has ever made over the centuries, was there ever one greater than this, to predict, and predict correctly, and predict against all expectation a phenomenon so fantastic as the expansion of the universe? When did nature ever grant man greater encouragement to believe he will someday understand the mystery of existence?

Spinoza's influence on his thinking about cosmology Einstein could shake off, but not Spinoza's deterministic outlook. Proposition XXIX in *The Ethics* of Spinoza states, "Nothing in the universe is contingent, but all things are conditioned to exist and operate in a particular manner by the necessity of divine nature."[88] Einstein accepted determinism in his mind, his heart, his very bones. What other explanation is there for his later-life position against quantum indeterminacy than this "set" he had received from Spinoza?

No Elementary Phenomenon Is a Phenomenon until It Is a Registered Phenomenon

From Einstein's discomfort we turn to today's assessment of the central lesson of the quantum. In Fig. 22.8 the left-hand view symbolizes the concept of the universe of the old physics. Galaxies, stars, planets, and everything that takes place can be looked at, as it were, from behind the safety of a one-foot-thick slab of plate glass without ourselves getting involved. The right-hand view reminds us that the truth is quite different. Even when we want to observe, not a galaxy, not a star, but something so miniscule as an electron, we have in effect to smash the glass so as to reach in and install measuring equipment. We can install a device to measure the position x of the electron, or one to measure its momentum p, but we cannot fit both registering devices into the same place at the same time. Moreover the act of measurement has an inescapable effect on the future of the electron. The observer finds himself willy-nilly a participator. In some strange sense this is a participatory universe.[89]

A story may symbolize what it means for the observer to find himself a participator.[90] We had been playing the familiar game of twenty questions. Then my turn came, fourth to be sent from the room, so that Lothar Nordheim's other fifteen after dinner guests could consult in secret and agree on a difficult word. I was locked out unbelievably long. On finally being readmitted, I found a smile on everyone's face, sign of a joke or a plot. I nevertheless started my attempt to find the word. "Is it animal?" "No." "Is it mineral?" "Yes." "Is it green?" "No." "Is it white?" "Yes." These answers came quickly. Then the questions began to take longer in the answering. It was strange. All I wanted from my friends was a simple yes or no. Yet the one queried would think and think, yes or no, no or yes, before responding. Finally I felt I was getting hot on the trail, that the word might be "cloud." I knew I was allowed only one chance at the final word. I ventured it: "Is it cloud?" "Yes," came the reply, and everyone burst out laughing. They explained to me there had been no word in the room. They had agreed not to agree on a word. Each one questioned could answer as he pleased — with the one requirement that *he* should have a word in mind compatible with his own response and all that had gone before. Otherwise if I challenged he lost. The surprise version of the game of twenty questions was therefore as difficult for my colleagues as it was for me.

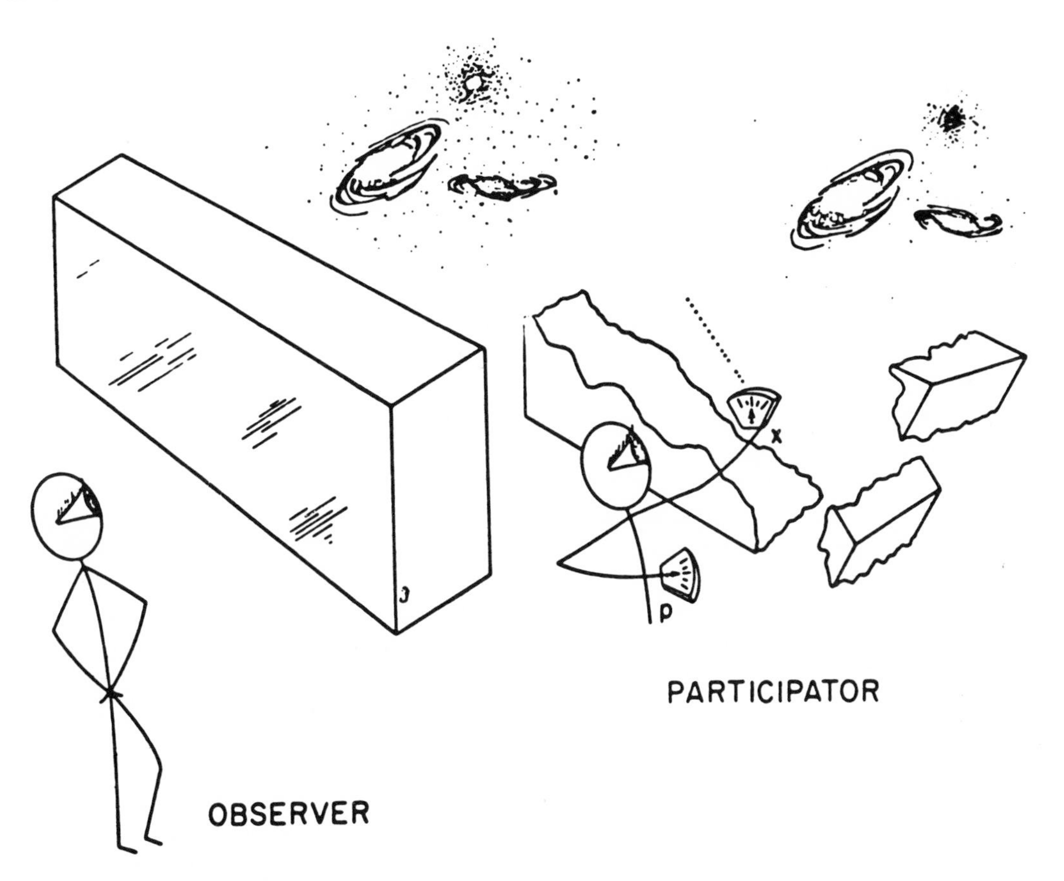

Fig. 22.8 Quantum mechanics evidences that there is no such thing as a mere "observer (or register) of reality." The observing equipment, the registering device, "participates in the defining of reality." In this sense the universe does not sit "out there."

What is the symbolism of the story? The world, we once believed, exists "out there" independent of any act of observation. The electron in the atom we once considered to have at each moment a definite position and a definite momentum. I, entering, thought the room contained a definite word. In actuality the word was developed step by step through the questions I raised, as the information about the electron is brought into being by the experiment that the observer chooses to make; that is, by the kind of registering equipment that he puts into place. Had I asked different questions or the same questions in a different order I would have ended up with a different word as the experimenter would have ended up with a different story for the doings of the electron. However, the power I had in bringing the particular word "cloud" into being was partial only. A major part of the selection lay in the "yes" and "no" replies of the colleagues around the room. Similarly the experimenter has some substantial influence on what will happen to the electron by the choice of experiments he will do on it, "questions he will put to nature"; but he knows there is a certain unpredictability about what any given one of his measurements will disclose, about what "answers nature will give," about what will happen when "God plays dice." This comparison between the world of quantum observations and the surprise version of the game of twenty questions misses much but it makes the central point. In the game no word is a word until that word is promoted to reality by the choice of questions asked and answers given. In the real world of quantum physics, *no elementary phenomenon is a phenomenon until it is a recorded phenomenon.*[91]

Delayed-Choice Experiments[92]

Figure 22.9 recalls the double-slit experiment that did so much to clarify the issues in the great thirty-year dialogue between Bohr and Einstein.[93] All features to the right of the photographic plate — and the slicing of that plate itself to convert it into the slats of a venetian blind[94] — are new and to be postponed a moment. A photon enters from the left and is recorded on the photographic plate by the blackening of a grain of silver bromide emulsion. No matter how great the spacing in time between one photon and the next, the record of arrivals shows[95] the standard two-slit interference pattern, basis for deducing that each photon has "gone through both slits." One can also tell "through which slit" each quantum goes, Einstein argued,[96] by measuring the vertical component of the kick that the photon imparts to the photographic plate. If it comes from the upper hole it kicks the plate down; from the lower hole, up. But for quantum theory to say in one breath "through which slit" and in another "through both" is logically inconsistent, Einstein objected, and shows that the theory itself is inconsistent. Bohr's response is well known. We have to do with two experiments, not one. We can fasten the photographic plate to the apparatus so it will not move up and down. Then we can register the interference fringes. Or we can free it to slide up and down in a slot, not shown. Then we can measure the vertical kick. But we cannot do both at the same time. The experiments are not contradictory. They are complementary.[97]

Now we come to the new feature: delayed choice.[98] We do not have to decide in advance which feature of the photon to record, "through both slits" that pierce the metal screen, or "through which slit." Let us wait until the quantum has *already* gone through the screen before we — at our free choice — decide whether it *shall have* gone "through both slits" or "through one."

We use a carefully timed source. We know when the photon has definitely passed through the metal screen and is on the last lap of its journey toward the photographic plate. At this moment we

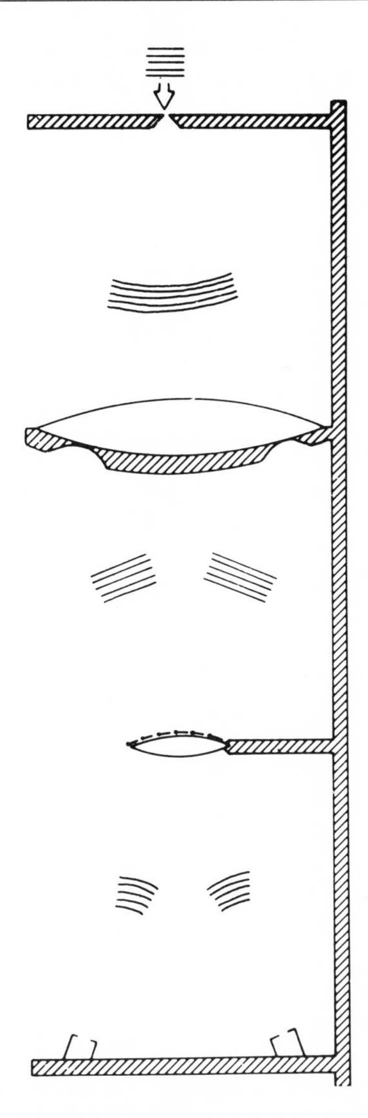

Fig. 22.9 The double-slit experiment both in the familiar version and in the "delayed-choice" version. The familiar layout includes the source of photons at the top, the entering slit, the first lens, the doubly slit metal screen that covers it, and the photographic plate that registers interference fringes. We secure delayed choice by supplements to this classic arrangement. We replace the continuous source of illumination at the top by a source that gives off one photon per timed flash. We slice the photographic plate to make it into a venetian blind. We make a last-minute choice, after the photon has *already* traversed the doubly slit screen, whether to open this blind or close it. Closed, as shown, it registers on a blackened grain of silver halide emulsion the arrival of that photon "through *both* slits." Opened, it allows the light to be focused by the second, or L. F. Bartell, lens on the two photon counters. There being only one photon, only one counter goes off. It tells "through *which* slit" the photon came. In this sense we decide, after the photon has *passed* through the screen, whether it *shall* have passed through only one slit or both.

make our choice: open the venetian blind and record through which slit the photon came; or close the blind, use it as a photographic plate, and add to the interference-pattern record that testifies to photons all going through both slits.

In the delayed-choice experiment we, by a decision in the here and now, have an irretrievable influence on what we will want to say about the past — a strange inversion of the normal order of time. This strangeness reminds us more explicitly than ever that "The past has no existence except as it is recorded in the present"; or more generally, in the words of Torny Segerstedt, "Reality is theory."[99] What we call "reality," that vision of the universe that is so vivid in our minds, we plaster in (Fig. 22.10) between a few iron posts of observation by an elaborate labor of imagination and theory. We have no more right to say "what the photon is doing" — until it is registered — than we do to say "what word is in the room" — until the game of question and response is terminated.

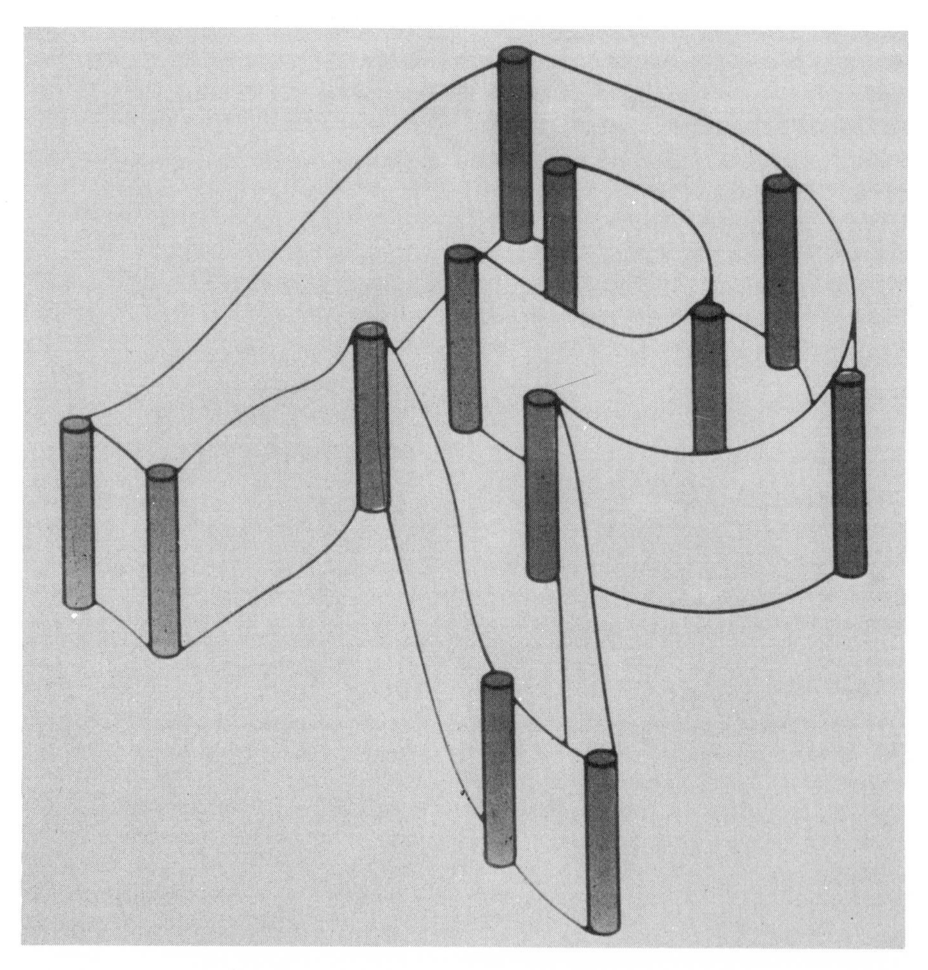

Fig. 22.10 What we call "reality," symbolized by the letter R in the diagram, consists of an elaborate papier-mâché construction of imagination and theory filled in between a few iron posts of observation.

The Central Lesson of the Quantum

"No elementary phenomenon is a phenomenon until it is a registered phenomenon."[100] This summary of the central lesson of the quantum takes its two key words from Bohr. "Registered" as Bohr uses it means "brought to a close by an irreversible act of amplification" and "communicable in plain language."[101] This adjective, equivalent in most respects to "observed," has a special feature as compared to that more frequently seen word. It explicitly denies the view that quantum theory rests in any way whatsoever on "consciousness."[102] The critical word, "phenomenon," Bohr found himself forced to introduce in his discussions with Einstein to stress how different "reality" is from Einstein's "any reasonable conception of reality."[103]

The Building of "Reality": This Participatory Universe

What lies over the hill? What are we to project ahead out of the present landscape's two greatest strangenesses? Of these, one, the "bounds of time," argues for mutability, law without law, law built on the statistics of multitudinous chance events, events that — undergirding space and time — must themselves transcend the categories of space and time. What these primordial chance events are, however, it does not answer; it asks. Unasked and unwelcomed, the other strangeness, the quantum, gives us chance. In "elementary quantum phenomenon" nature makes an unpredictable reply to the sharp question put by apparatus. Is the "chance" seen in this reply primordial? As close to being primordial as anything we know. Does this chance reach across space and time? Nowhere more clearly than in the delayed-choice experiment. Does it have building power? Each query of equipment plus reply of chance inescapably do build a new bit of what we call "reality." Then for the building of all of law, "reality" and substance — if we are not to indulge in free invention, if we are to accept what lies before us — what choice do we have but to say that in some way, yet to be discovered, they all must be built upon the statistics of billions upon billions of such acts of observer-participancy? In brief, beyond the black hole, past the two great strangenesses of the landscape and over the hill, what other kind of universe can we expect to see than one built as "phenomenon" is built, upon query of observation and reply of chance, a *participatory* universe?

If the concept of a participatory universe seems to make the world a never-never land, we can recall Samuel Johnson's remark on kicking the stone. Whatever the theory of reality, the pain in the toe made the stone real enough to him. In recent decades we have judged solid matter no less solid for being made up of electrons, nuclei, and mostly emptiness. It will make the stone no less real to regard it as entirely emptiness.

The Example of Mathematics

Mathematics also is emptiness without emptiness. A familiar theorem tells us that the sum of the three interior angles of a plane triangle is 180°. However, when we review all the definitions, postulates, and axioms that go into proving that theorem, we find that the statement reduces in the end to an identity, equivalent to "0 = 0." No identity? Then no theorem! It may take 300 pages of computer paper to spell out all of the foundation pieces of a theorem, the customary journal proof of which requires only two pages.[104] But packaged as all the parts of the theorem are, and useful as the theorem is, it is still in the end packaged identity. Like the structure of rope in Fig. 22.11, it has only to be pulled on to fall apart into nothingness.

Fig. 22.11 The construction of rope that falls apart on being pulled symbolizes the elaborate theorem that, conformant to the inexorable demand of mathematics, dissolves into an identity on being analyzed. The author thanks William Wootters for engineering this photograph.

Participatory Universe as Self-Excited Circuit

Looking at an empty courtyard, we know that the game will not begin until a line has been drawn across the court to separate the two sides. Where, is not very important; but whether, is essential. "Elementary phenomena" are impossible without the distinction between observing equipment and observed system[105]; but the line of distinction can run like a maze (Fig. 22.12), so convoluted that what appears from one standpoint to be on one side and to be identified as observing apparatus, from another point of view has to be looked at as observed system.

Fig. 22.12 Maze symbolic of the tortuous course through nature of the interface between observing equipment and system observed.

From "nothingness ruled out as meaningless,"[106] to the line of distinction that rules it out; from this dividing line to "phenomenon"; from one phenomenon to many; from the statistics of many to regularity and structure: these considerations lead us at the end to ask if the universe is not best conceived as a self-excited circuit[107] (Fig. 22.13): Beginning with the big bang, the universe expands and cools. After eons of dynamic development it gives rise to observership. Acts of observer-participancy — via the mechanism of the delayed-choice experiment — in turn give tangible "reality" to the universe not only now but back to the beginning. To speak of the universe as a self-excited circuit is to imply once more a participatory universe.

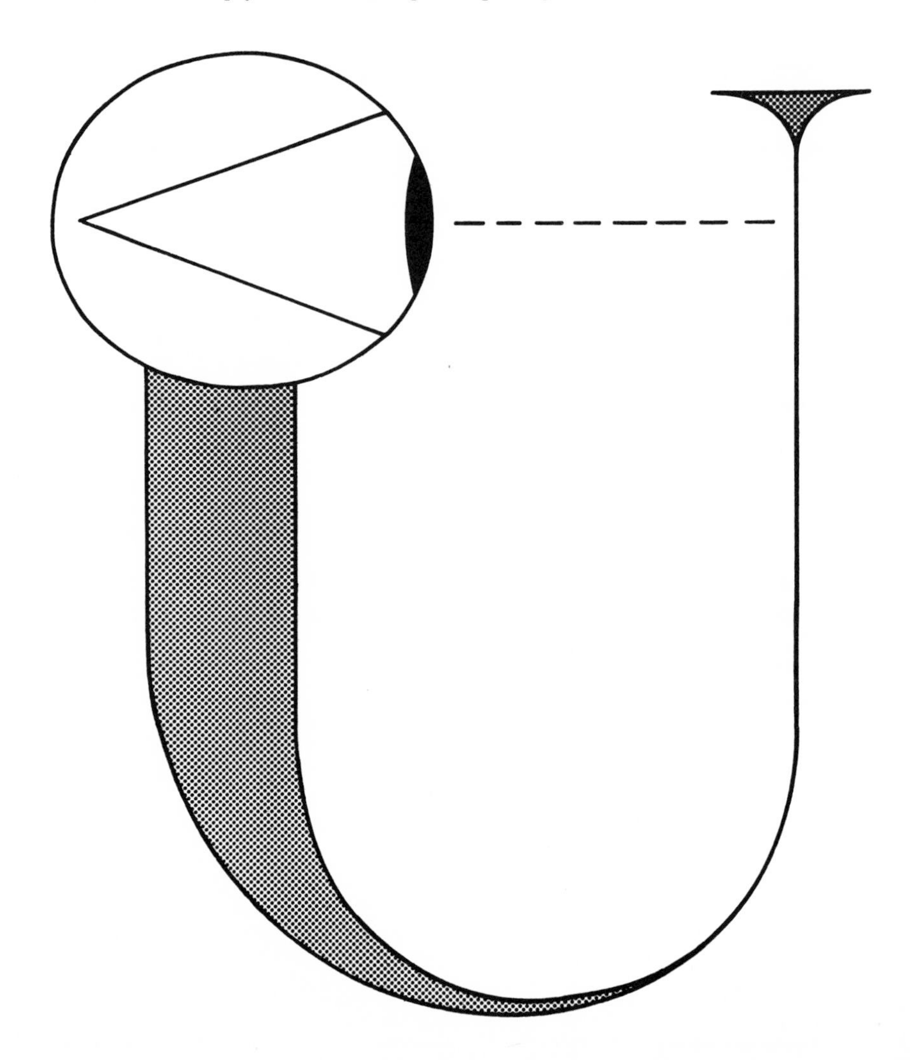

Fig. 22.13 The universe viewed as a self-excited circuit. Starting small (thin U at *upper right*), it grows (loop of U) and in time gives rise (*upper left*) to observer-participancy — which in turn imparts "tangible reality" (cf. the delayed-choice experiment of Fig. 22.9) to even the earliest days of the universe.

If the views that we are exploring here are correct, one principle, observer-participancy, suffices to build everything. The picture of the participatory universe will flounder, and have to be rejected, if it cannot account for the building of law; and space-time as part of law; and out of law substance. It has no other than a higgledy-piggledy way to build law: out of the statistics of billions upon billions of acts of observer-participancy each of which by itself partakes of utter randomness.

Two Tests

No test of these views looks more like being someday doable, nor more interesting and more instructive, than a *derivation* of the structure of quantum theory from the requirement that everything have a way to come into being[108] — as the word "cloud" was brought into being in the surprise version of the game of twenty questions. No prediction lends itself to a more critical test than this, that every law of physics, pushed to the extreme, will be found to be statistical and approximate, not mathematically perfect and precise.

The Challenge of "Law without Law"

We can ask ourselves if it is not absolutely preposterous to put into a formula anything at first sight so vague as law without law and substance without substance. How can we hope to move forward with no solid ground at all under our feet? Then we remember that Einstein had to perform the same miracle. He had to reexpress all of physics in a new language. His curved space seemed to take all definite structure away from anything we can call solidity. In the end physics, after being moved bodily over onto the new underpinnings, shows itself as clear and useful as ever. We have to demand no less here. We have to move the imposing structure of science over onto the foundation of elementary acts of observer-participancy.[109] No one who has lived through the revolutions made in our time by relativity and quantum mechanics — not least through the work of Einstein himself — can doubt the power of theoretical physics to grapple with this still greater challenge.

Acknowledgement

For much help on the references I thank Adrienne Harding.

NOTES

The references in these notes provide, not completeness, but some points of access to the literature on topics in the text and, in some instances close to the central theme, some documentation of the evolution of outlooks.

1. Prehistory of black hole: J. Michell, "On the means of discovering the distance, magnitude, & c. of the fixed stars, in consequence of the diminution in the velocity of their light, in case such a diminution should be found in any of them, and such data should be procured from observations, as would be further necessary for that purpose," Phil. Trans. [Roy. Soc. London] **74,** 35–37 (1784) (read 27 Nov. 1783), cited and discussed in S. Schaffer, "John Michell and Black Holes," J. Hist. Astron. **10**, 42–43 (1979). "Un astre lumineux de même densité que la terre, et dont le diamètre serait deux cents cinquante fois plus grand que celui du soleil, ne laisserait en vertu de son attraction,

parvenir aucun de ses rayons jusqu'a nous; il est donc possible que les plus grands corps lumineux de l'univers, soient par cela meme, invisibles.'' [P.-S. Laplace, *Exposition du systeme du monde,* vol. 2 (Cercle-Social, Paris, 1795), p. 305.] Laplace gives the calculations underlying this statement in *Allgemeine geographische Ephemeriden,* edited by F. X. von Zach, IV, Band I St. (Einer Gesellschaft Gelehrten, Weimar, 1799), May 1798, p. 603; translated, ''Proof of the theorem, that the attractive force of a heavenly body could be so large, that light could not flow out of it,'' in S. W. Hawking and G. F. R. Ellis, *The Large Scale Structure of Space-Time* (Cambridge University Press, Cambridge, England, 1973), pp. 365–68.

First treatment of collapse within the framework of general relativity: ''The total time of collapse for an observer comoving with the stellar matter is finite . . .; an external observer sees the star asymptotically shrinking to its gravitational radius.'' [J. R. Oppenheimer and H. Snyder, ''On Continued Gravitational Attraction,'' Phys. Rev. **56,** 455–59 (1939).]

Coming to terms with gravitational collapse: B. K. Harrison, M. Wakano, and J. A. Wheeler, 'Matter-energy at High Density; End Point of Thermonuclear Evolution,'' in Onzième Conseil de Physique Solvay, *La structure et l'évolution de l'univers,* (Stoops, Brussels, 1958), pp. 124–46. In particular: white dwarfs and neutron stars shown for the first time to be two sectors of one continuous family giving ''the absolutely lowest state possible for an A-nucleon system under the dual action of nuclear and gravitational forces,'' pp. 137–38; the equilibrium state of ''cold matter ideally catalyzed to the end point of thermonuclear evolution,'' p. 138; ''What is the final state of an A-nucleon system under gravitational forces when A is large? Perhaps there is no equilibrium state when A is large: this is the proposal of Oppenheimer and Snyder,'' pp. 139–40. ''If we are to reject as physically unreasonable the concept of an indefinitely large number of nucleons in equilibrium in a finite volume of space, it seems necessary to conclude that the nucleons above a critical number convert themselves to a form of energy that can escape from the system: radiation . . . [C]onditions of superdensity would seem to be particularly favorable for altering the number of nucleons in the universe'' (a proposed 1958 rejection of complete gravitational collapse in favor of an as then undiscovered mechanism of radiation; Y. B. Zel'dovich [Zh. eksp. teor. Fiz. **42,** 641 (1962)], English translation [''The Collapse of a Small Mass in the General Theory of Relativity,'' Soviet Physics JETP **15,** 446 (1962)] notes, ''By prescribing a sufficiently large density we can obtain for any given number N of particles a configuration with mass as close to zero as we please, and clearly less than the mass of the static solution. Such a solution obviously cannot go over into the state of equilibrium (into the static solution), and consequently can only contract without limit.'' J. A. Wheeler, ''Geometrodynamics and the Issue of the Final State,'' *Relativity, Groups and Topology,* edited by C. DeWitt and B. DeWitt, pp. 315–520 (Gordon and Breach, New York, 1965); in particular, ''Thus there exists a second crushing point, the 'Landau-Oppenheimer-Volkoff crushing point,' with central density $\sim 10^{16}$ g/cm^3, and mass $\sim 0.7\ M_{\odot}$. One cannot add matter to the system without initiating collapse. . . . Cannot one save the day by assuming that matter becomes incompressible at a sufficiently high density? No!'' The relativistic equation of hydrostatic equilibrium ''has the remarkable feature that it provides a mechanism for multiplying pressure . . . ('divergent chain reaction'),'' p. 321; ''No matter how small the number of nucleons that one starts with, in principle they can be pressed from outside with enough pressure to initiate collapse,'' p. 325; gravitational collapse of a toroidal bundle of magnetic lines of force, pp. 445–49; Schwarzschild and geon geometry as unstable with respect to gravitational collapse, p. 500–501; and a collapsing ''cloud of matter may be of dust and so dilute that its density is 10^{-3} g/cm^3 or less at the moment when its radius decreases to the order of the Schwarzschild value. Therefore no details of any equation of state can save it from gravitational collapse,'' pp. 502–503. Also, ''It is difficult to escape the conclusion that the creation or destruction of matter goes on in regime IV [where quantum effects dominate]. At issue here is not the familiar process of a positive electron annihilating a negative electron, or an antiproton disappearing by union with a proton. Instead, one is concerned about a process in which ordinary matter — composed of protons, neutrons and electrons — is crushed out of existence, or brought into being, by a mechanism intimately connected with gravitation and with the curvature of space. . . . [P]rocesses of baryon creation or destruction would seem unavoidable,'' pp. 513–16, discussion of relation between the quasi-stellar objects discovered in 1963 and gravitational collapse. ''[No] escape is now known . . . from a new physical process. In this process baryons disappear . . . [G]ravitational collapse must occur for a subcritical mass as well as for a supercritical mass [via] a quantum mechanical tunneling process. . . . [A]ll matter must manifest, however weakly, a new form of radioactivity, in which baryon number changes.'' [B. K. Harrison et al. *Gravitation Theory and Gravitational Collapse* (University of Chicago Press, Chicago, Illinois, 1965), p. vii, viii.] Name ''black hole'': J. A. Wheeler, ''Our Universe: The Known and the Unknown,'' address before the American Association for the Advancement of Science, New York, 29 Dec. 1967 in Am. Scholar **37,** 248–74 (1968) and Am. Scientist **56,** 1–20 (Spring 1968). See also R. Ruffini and J. A. Wheeler, ''Introducing the Black Hole,'' Phys. Today **24,** 30–36 (1971), and ''The Black Hole,'' in *Astrophysics and Gravitation: Proceedings of the Sixteenth Conference on Physics at the University of Brussels, September 1973* (Editions de l'Université de Bruxelles, 1040 Bruxelles, Belgium, 1974), pp. 279–316. In black hole physics the laws of conservation of particle number are transcended; see J. A. Wheeler, ''Transcending the Law of Conservation of Leptons,'' in *Atti del Convegno Internazionale sul Tema: The Astrophysical Aspects of the Weak Interactions; Quaderno N.157* (Accademia Nazionale dei Lincei, Roma, 1971), pp. 133–64. Gravitational collapse implies that "there is no law except the law that there is no law" [J. A. Wheeler, "From Relativity to Mutability," in *The Physicist's Conception of Nature,* edited by J. Mehra (Reidel, Dordrecht, Holland, 1973), pp. 202–47].

Proof that gravitational collapse is inescapable under assumptions more and more elementary: A. Avez, "Propriétés globales des espace-temps périodiques clos," Acad. des Sci., Paris, Comptes Rend. **250,** 3583–87 (1960); R. Penrose, "Gravitational Collapse and Spacetime Singularities," Phys. Rev. Lett. **14,** 57–59 (1965); S. W. Hawking, "The occurrence of singularities in cosmology," Proc. Roy. Soc. London **A294,** 511–21 (1966); R. P. Geroch, "What is a Singularity in General Relativity?" Ann. Phys. (U.S.A.) **48,** 526–40 (1960); S. W. Hawking and G. F. R. Ellis, *The Large Scale Structure of Space-time* (Cambridge University Press, Cambridge, England, 1973); J. E. Marsden and F. J. Tipler, "Maximal Hypersurfaces and Foliations of Constant Mean Curvature in General Relativity," preprint, Mathematics Department, University of California at Berkeley, 1979.

Theorems on the uniqueness of the geometry around a black hole: B. Carter, "An Axisymmetric Black Hole Has Only Two Degrees of Freedom," Phys. Rev. Lett. **26,** 331–36 (1970); papers by B. Carter and others in *Black Holes,* Proceedings of 1972 session of École d'été de physique théorique, edited by C. DeWitt and B. S. DeWitt (Gordon and Breach, New York, 1973).

Quantum aspects of the black hole: "wormholes" continually being produced and annihilated at the Planck scale of distances, giving rise to a "foam-like structure" of space [J. A. Wheeler, "On the Nature of Quantum Geometrodynamics," Ann. Phys. **2,** 604–14 (1957)]; calculation of same by the method of sum over histories [G. W. Gibbons and S. W. Hawking, "Action Integrals and Partition Functions in Quantum Gravity," Phys. Rev. **D15,** 2752–57 (1977)]; surface area and surface gravity of black hole not merely analogous to, but identical with, entropy and temperature [J. Bekenstein, "Black Holes and Entropy," Phys. Rev. **D7,** 2333–46 (1973)]; thermal radiation associated with this effect calculated by S. W. Hawking, ["Particle Creation by Black Holes," Comm. Math. Phys. **43,** 199–220 (1975)].

2. A. Friedmann, "Über die Krummung des Raumes," Z. f. Phys. **10,** 337–86 (1922); E. P. Hubble, "A Relation between Distance and Radial Velocity among Extragalactic Nebulae," Proc. Nat. Acad. Sci. U.S. **15,** 169–73 (1929); R. A. Alpher, H. A. Bethe, and G. Gamow, "The Origin of Chemical Elements," Phys. Rev. **L 73,** 803–804 (1948); R. H. Dicke et al., "Cosmic-black-body Radiation," Astrophys. J. **142,** 414–19 (1965); A. A. Penzias and R. W. Wilson, "A Measurement of Excess Antenna Temperature at 4080 *Mc/s,*" Astrophys. J. **142,** 419–21 (1965).

3. "Thus we may present the following arguments against the conception of a space-infinite and for the conception of a space-bounded universe [(1) simplicity (2) Machian] [A. Einstein, *The Meaning of Relativity,* 3rd ed. (Princeton University Press, Princeton, N. J., 1950), pp. 107–108]; "In my opinion the general theory of relativity can only solve this problem [of inertia] satisfactorily if it regards the world as spatially self-enclosed," [A. Einstein, *Essays in Science* (Philosophical Library, New York, 1934), translated from *Mein Weltbilde,* (Querido, Amsterdam, 1933)]; J. A. Wheeler, "Conference Summary: More Results Than Ever in Gravitation Physics and Relativity," in *General Relativity and Gravitation,* edited by G. Shaviv and J. Rosen (Wiley, New York, 1975), pp. 299–344, especially status of the "mystery of the missing mass," lens effect and its difficulties as a way to check on closure, difficulties with alternatives to closure [pp. 320–24].

4. Mathematical investigation from which one concludes there are not any periodic closed model universes — an indirect argument that a closed universe necessarily collapses [A. Avez, op. cit. in n. 1]; S. W. Hawking and R. Penrose, "The Singularities of Gravitational Collapse and Cosmology," Proc. Roy. Soc. London **A314,** 529–48 (1969); all "W model universes" are closed and have an upper limit to the time from big bang to big crunch [J. E. Marsden and F. J. Tipler, op. cit. in n. 1].

5. J. R. Gott, III, et al. "Will the Universe Expand Forever?" Sci. Am. **234,** 62–79 (March 1976), esp. p. 69 for term "big crunch"; see also J. R. Gott, III, et al. "An Unbound Universe?" Astrophys. J. **194,** 543–53 (1974). Principle that the universe must have a way to come into being and to fade out of existence proposed, not as a deduction from cosmology, but as a requirement for cosmology [C. M. Patton and J. A. Wheeler, "Is Physics Legislated by Cosmology?" in *Quantum Gravity: An Oxford Symposium,* edited by C. J. Isham, R. Penrose, and D. W. Sciama (Clarendon, Oxford, 1975)].

6. N. Bohr, *Atomic Theory and the Description of Nature* (Cambridge University Press, Cambridge, England), 1934; *Atomic Physics and Human Knowledge* (Wiley, New York, 1958); and *Essays 1958–1962 on Atomic Physics and Human Knowledge* (Wiley, New York, 1963).

7. N. Bohr, "Discussion with Einstein on Epistemological Problems in Atomic Physics," in *Albert Einstein: Philosopher-Scientist,* edited by P. A. Schilpp (Library of Living Philosophers, Evanston, Ill., 1949), pp. 201–41, "registration," p. 238; "irreversible amplification," p. 88, and "closed by irreversible amplification," p. 73, in Bohr, *Atomic Physics and Human Knowledge,* op. cit. in n. 6.

8. J. A. Wheeler, "Genesis and Observership," *Foundational Problems in the Special Sciences,* edited by R. E. Butts and K. J. Hintikka (Reidel, Dordrecht, 1977), pp. 3–33, esp. "direct involvement of observership in genesis," p. 26;

J. A. Wheeler, *Frontiers of Time* [North-Holland, Amsterdam (for the Società Italiana di Fisica, Bologna), 1979]; also appears from the same two houses in *Problems in the Foundations of Physics,* edited by G. Toraldo di Francia (1979), "billions upon billions of acts of observer-participancy," p. 5ff; add 394 for pagination in latter.

9. Come into being, fade out of existence: Patton and Wheeler, op. cit. in n. 5; Wheeler, op. cit. in n. 8; in gravitational collapse the framework falls down for everything one ever called a law [J. A. Wheeler, "From Relativity to Mutability," in *The Physicist's Conception of Nature,* edited by J. Mehra, Reidel, Dordrecht, 1973), pp. 202–47]; to be contrasted with "Beyond the End of Time," in *Gravitation,* by C. W. Misner, K. S. Thorne, and J. A. Wheeler (Freeman, San Francisco, 1973), ch. 44, pp. 1196–1217, where gravitational collapse was envisaged as precipitating a reprocessing of the universe, except for a penultimate paragraph foreshadowing the concept of genesis through observer-participancy.

10. "Billions," op. cit. in n. 8, n. 9.

11. Challenge to derive the quantum principle from the requirements that the universe should have a way to come into being: Patton and Wheeler, op. cit. in n. 5, p. 564; Wheeler, "Genesis and Observership," op. cit. in n. 8, p. 29; Wheeler, *Frontiers of Time,* op. cit. in n. 8, p. 8.

12. Derivation from symmetry principle hides the machinery underlying physical law: Wheeler, "Genesis and Observership," op. cit. in n. 8, pp. 15–16; Wheeler, *Frontiers of Time,* op. cit. in n. 8, sec. 4

13. J. C. Maxwell, "A Dynamical Theory of the Electromagnetic Field," Trans. Roy. Soc. London **155,** 459 ff. (1865), and *A Treatise on Electricity and Magnetism,* 1873; 3rd ed. (Clarendon, Oxford, 1892); A. Einstein, "Die Feldgleichungen der Gravitation," Preuss. Akad. Wiss. Berlin, Sitzber., 844–47 (1915); C. N. Yang and R. L. Mills, "Conservation of Isotopic Spin and Isotopic Gauge Invariance," Phys. Rev. **96,** 191–95 (1954).

14. To be distinguished from the "I believe that it would be worth trying to learn something about the world even if in trying to do so we should merely learn that we do not know much," of K. Popper, *Conjectures and Refutations, The Growth of Scientific Knowledge,* 2nd ed. (Routledge and Kegan Paul, London, 1972), p. 29.

15. A. E. H. Love, *The Mathematical Theory of Elasticity,* (Cambridge University Press, London, 1892); 4th ed. (Dover, New York, 1944).

16. J. D. Jackson, *Classical Electrodynamics,* 2nd ed. (Wiley, New York, 1975); L. D. Landau and E. M. Lifshitz, *Electrodynamics of Continuous Media* (Pergamon, Oxford, 1960).

17. S. Weinberg, *Gravitation and Cosmology: Principles and Applications of the General Theory of Relativity* (Wiley, New York, 1972); Misner, Thorne, and Wheeler, op. cit. in n. 9.

18. L. D. Faddeev and A. A. Slavnov, *Gauge Fields: Introduction to Quantum Theory,* translated by D. B. Pontecorvo (Benjamin/Cummings, Advanced Book Program, Reading, Mass., 1980).

19. The discussion here comes from Wheeler, "Genesis and Observership," op. cit. in n. 8, p. 16, and Misner, Thorne, and Wheeler, op. cit. in n. 9, pp. 1206–1208.

20. S. Hojman, K. Kuchař, and C. Teitelboim, "New Approach to General Relativity," Nature Phys. Sci. **245,** 97–98 (1973); C. Teitelboim, "How Commutators of Constraints Reflect Spacetime Structure," Ann. Phys. **79,** 542–57 (1973), and "The Hamiltonian Structure of Spacetime," doctoral dissertation, unpublished, Princeton University, 1973, available from University Microfilms, Ann Arbor, Mich.; K. Kuchař, "Canonical Quantization of Gravity," in *Relativity, Astrophysics and Cosmology,* edited by W. Israel (Reidel, Dordrecht, Holland, 1973), pp. 238–88, and "Geometrodynamics regained: a Lagrangian approach," J. Math. Phys. **15,** 708–15 (1974); C. Teitelboim, "Surface Deformations, Spacetime Structure and Gauge Invariance" in *Relativity, Fields, Strings and Gravity: Proceedings of the Second Latin American Symposium on Relativity and Gravitation SILARG II held in Caracas, December 1975,* edited by C. Aragone (Universidad Simon Bolivar, Caracas, 1976); S. A. Hojman, K. Kuchař, and C. Teitelboim, "Geometrodynamics Regained," Ann. Phys. **76,** 88–135 (1976).

21. Teitelboim, op. cit. in n. 20.

22. A. Einstein, "Autobiographical Notes," pp. 2–95 in *Albert Einstein: Philosopher-Scientist,* op. cit. in n. 7, where Einstein discusses his attempts to find a unified field theory [pp. 89–95]; M. A. Tonnelat, *La Théorie du Champ Unifié d'Einstein et Quelque-uns de ses Developpments* (Gauthier-Villars, Paris, 1955), and *Les Théories Unitaires de l'Électromagnetisme et de la Gravitation* (Gauthier-Villars, Paris, 1965).

23. Gravitation recognized for the first time as a guage theory [R. Utiyama, "Invariant Theoretical Interpretation of Interaction," Phys. Rev. **101,** 1597–1607 (1956)].

Gauge theory of gravitation step by step recognized as equivalent to the metric-plus-torsion theory of gravitation originally urged by Élie Cartan on geometrical grounds: "Sur une généralisation de la notion de courbure de Riemann et les espaces à torsion," Acad. Sci., Paris, Compt. Rend. **174,** 593–95 (1922), for more on the development of which see *Élie Cartan–Albert Einstein Letters on Absolute Parallelism 1929–1932,* edited by R. Debever (Princeton University Press, Princeton, N. J., 1979). The history and bibliography of this growth in understanding to what is today known as the Einstein–Cartan–Sciama–Kibble (or U_4) theory of gravitation, derivable in its spin-torsion parts from a Lorentz gauge, is recounted by A. Trautman, "On the Einstein–Cartan Equations," Bull. Acad. Polon. Sci., Ser. Sci. Mat. Ast. et Phys. **20**, 185–90, and in "Theory of Gravitation," in *The Physicist's Conception of Nature,* edited by J. Mehra (Reidel, Dordrecht, Holland, 1973), pp. 179–98; also see F. W. Hehl et al., "General Relativity with Spin and Torsion: Foundations and Prospects," Rev. Mod. Phys. **48**, 393–416 (1976), who clarify the gauge associated with translations; for a still more recent perspective, see Y. Ne'eman, "Gravity Is the Gauge Theory of the Parallel-Transport Modification of the Poincaré Group," in *Differential Geometrical Methods in Mathematical Physics II,* edited by K. Bleuler, H. R. Petry, and A. Reetz (Springer, New York, 1978), pp. 189–215.

Electromagnetic field as gauge field: H. Weyl, "Gravitation und Elektrizität," Preuss. Akad. Wiss., Berlin, Sitz'ber., 465–80 (1918); brief summary of subsequent developments, including 1) recognition that the right concept is not gauge, but phase, 2) the Bohm–Aharonov experiment, and 3) "$f_{\mu\nu}$ underdescribes electromagnetism, . . . $\oint A_\mu dx^\mu$ overdescribes electromagnetism . . . [and the] phase factor $\exp(ie/\hbar c) \oint A_\mu dx^\mu$ is just right to describe electromagnetism" [T. T. Wu, "Introduction to Gauge Theory," in *Differential Geometrical Methods in Mathematical Physics II,* op. cit., pp. 161–69].

Introduction of non-Abelian phase field by Yang and Mills, op. cit. in n. 13.

Physical evidence for gauge theory: reviewed in L. O'Raiffeartaigh, "Hidden Gauge Symmetry," Rep. Prog. Phys. **42**, 159–223 (1979); also, for example, in R. E. Taylor, "Introduction, σ_L/σ_T," pp. 285–286, and H. Fritzsch, "Flavordynamics," pp. 593–603 *Proceedings of the 19th International Conference on High Energy Physics, Tokyo, August 1978,* edited by S. Homma, M. Kawaguchi, and H. Miyazawa (Phys. Soc. of Japan, 1979); also reviewed in Y. Ne'eman, *Symétries jauges et variétés de group* (Les Presses de l'Université de Montréal, 1979), in particular phenomenological findings of the quark model and scaling [p. 33], Yang–Mills field meets four out of five requirements of the observational evidence [pp. 34–35, 44–46] and perhaps the fifth, confinement of quarks [more on pp. 46ff.]. Proof that matrix elements go down at high energy as required by observation rather than continuing to go up as predicted by Fermi theory of the weak interaction, D. Gross and F. Wilczek, "Ultraviolet Behavior of Non-Abelian Gauge Theories," Phys. Rev. Lett. **30**, 1343–46 (1973), and H. D. Politzer, "Reliable Perturbative Results for Strong Interactions?" Phys. Rev. Lett. **30**, 1346–49 (1973). Discovery of neutral currents as predicted by the Weinberg-Salam model, six experiments in 1973 reviewed in *Proceedings of the International Conference on High Energy Physics 1974,* Rutherford Laboratory, Didcot, U.K., 1975. D. J. Sherlen and nineteen others, "Observation of Parity Violation in Polarized Electron Scattering," pp. 267–90 in *Proceedings of the Summer Institute on Particle Physics July 10–21, 1978: Weak Interactions — Present and Future,* Report 215, Stanford Linear Accelerator Center, Stanford, Calif., 1978. Observation of parity violation in conformity with the Weinberg-Salam model. SLAC experiment on electron-deuteron scattering and polarization fits the predictions, including the predicted angle of symmetry breaking of the Weinberg-Salam theory. It is not clear whether the theory will give correctly the observed confinement of quarks. Renormalizability of the theory has been proved (literature in O'Raiffeartaigh, op. cit. p. 194).

24. For a survey of the mathematics of non-Abelian gauge fields today, see for example M. E. Mayer, "Gauge Fields as Quantized Connection Forms," in *Differential Geometrical Methods in Mathematical Physics I,* edited by K. Bleuler and A. Reetz (Springer, New York, 1977), and "Characteristic Classes and Solutions of Gauge Theories," pp. 81–104 in *Differential Geometrical Methods in Mathematical Physics II,* op. cit. in n. 23, as well as other papers in the latter volume. In the second paper Mayer notes the intensive interaction now taking place in this subject between mathematics and physics, between "the classification of algebraic vector bundles of rank 2 over P_3" and Yang–Mills theory [p. 98]. Central to this subject is the Atiyah-Singer index theorem: M. F. Atiyah and I. M. Singer, "The Index of Elliptic Operators, I, III," Ann. Math. **87**, 484–530, 546–604 (1968); R. S. Palais, *Seminar on the Atiyah–Singer Index Theorem* (Princeton University Press, Princeton, N. J., 1965); M. F. Atiyah, R. Bott, and V. K. Patodi, "On the Heat Equation and the Index Theorem," Inventiones Math. **19**, 279–330 (1973), and errata, **28**, 277–80 (1975).

25. Misner, Thorne, and Wheeler, op. cit. in n. 9, p. 1207, Fig. 44.2.

26. "Last Lecture of Albert Einstein," as recorded by J. A. Wheeler, in *Albert Einstein: His Influence on Physics, Philosophy and Politics,* edited by P. C. Aichelburg and R. U. Sexl (Vieweg, Braunschweig/Wiesbaden, 1979), pp. 207–11; Einstein, "Autobiographical Notes," pp. 2–95 in *Albert Einstein, Philosopher-Scientist,* op. cit. in n. 7; A. Einstein, "Über das Relativitätsprinzip und die aus demselben gezogenen Folgerungen," Jahrb. Radioakt. **4**, 411–62 (1908); and A. Einstein, letter to Ernst Mach, 26 June 1913, reproduced as Fig. 21.5 in Misner, Thorne, and Wheeler, op. cit. in n. 9, pp. 544–545; for this and other correspondence between Einstein and Mach in the original German, see "Die Beziehungen zwischen Einstein und Mach, dokumentarisch dargestellt," in F. Herneck, *Einstein und sein Weltbild,* (*Der Morgen,* Berlin DDR, 1979), pp. 109–15.

27. Op. cit. in n. 3.

28. "Mach's principle" interpreted as the requirement that the universe be closed, thus to provide a boundary condition to separate allowable solutions of Einstein's equation from physically inadmissible solutions: J. A. Wheeler in Onzième Conseil de Physique Solvay, op. cit. in n. 1, pp. 49-51; H. Hönl, in *Physikertagung Wien,* edited by E. Brüche (Physik Verlag, Mosbach/Baden, 1962); J. A. Wheeler, "Geometrodynamics and the Issue of the Final State," in DeWitt and DeWitt, eds., op. cit. in n. 1, pp. 363-399, 411-425, esp. p. 425, "every so-called 'asymptotically flat' geometry taken up hereafter will be considered to be a part of a closed universe"; J. A. Wheeler, "Mach's Principle as Boundary Condition for Einstein's Equations," pp. 303-349 in *Gravitation and Relativity,* edited by H.-Y. Chiu and W. F. Hoffmann (W. A. Benjamin, New York, 1964). "Mach's Principle and the Origin of Inertia," sec. 21.12, pp. 543-49 in Misner, Thorne, and Wheeler, op. cit. in n. 9. J. Isenberg and J. A. Wheeler, "Inertia Here Is Fixed by Mass-Energy There in Every W Model Universe," in *Relativity, Quanta, and Cosmology in the Development of the Scientific Thought of Albert Einstein* (in English: Johnson Reprint, New York, 1979), pp. 267-93, and in *Astrofisica e Cosmologia Gravitazione, Quanti Relatività negli Sviluppi del Pensiero Scientifico di Albert Einstein* (in Italian: Giunti Barbera, Firenze, 1979), pp. 1099-1136, both edited by M. Pantaleo and F. De Finis.

29. J. H. Oort, "Distribution of Galaxies and the Density of the Universe," in Onzième Conseil de Physique Solvay, op. cit. in n. 1, pp. 163-81. Density of matter in the universe as estimated from abundance of primordial deuterium and other methods: J. R. Gott, III, et al., "An Unbound Universe?" op. cit. in n. 5; see also J. R. Gott, III, et al. "Will the Universe Expand Forever," also op. cit. in n. 5.

30. J. R. Gott, III, et al., ibid.

31. "With $\Omega = 1$ [density equal to that required for closure] we end up with a power law $\psi(\theta)$ [for correlations of galactic density as a function of apparent angular separation] terminated by a sharp break, in agreement with the observations . . . [and in] conflict with a low density cosmological model" [M. Davis, E. J. Groth, and P. J. E. Peebles, "Study of Galaxy Correlations: Evidence for the Gravitational Instability Picture in a Dense Universe," Astrophys. J. **212,** L 107-11 (1977)]. E. J. Groth and P. J. E. Peebles, "Statistical Analysis of Catalogs of Extragalactic Objects VII. Two- and Three-point Correlation Functions for the High Resolution Shane-Wirtanen Catalog of Galaxies," Astrophys. J. **217,** 385-405 (1977).

32. "From the point of view of the topological or causal structure, all the 'stalactites' which represent black hole singularities could be straightened out" (Fig. 7, p. 270, R. Penrose, "Singularities in Cosmology," in *Confrontation of Cosmological Theories with Observational Data,* edited by M. S. Longair (Reidel, Dordrecht, Holland, 1974), pp. 263-71. A. Qadir and J. A. Wheeler, "Black Hole Singularity as Part of Big Crunch Singularity," preprint, Center for Theoretical Physics, University of Texas, 1980.

33. Marsden and Tipler, op. cit. in n. 1.

34. Op. cit. in n. 32.

35. V. A. Belinsky and I. M. Khalatnikov, "On the Nature of the Singularities in the General Solution of the Gravitational Equations," Zh. Eksp. Teor. Fiz. **56,** 1700-12 (1969) [English translation in Sov. Phys. JETP **29,** 911-17 (1969)], and "General Solution of the Gravitational Equations with a Physical Singularity," Zh. Eksp. Teor. Fiz. **57,** 2163-75 (1969); [English translation in Sov. Phys. JETP **30,** 1174-80 (1970)]. I. M. Khalatnikov and E. M. Lifshitz, "General Cosmological Solutions of the Gravitational Equations with a Singularity in Time," Phys. Rev. Lett. **24,** 76-79 (1970); E. M. Lifshitz and I. M. Khalatnikov, "Oscillatory Approach to Singular Point in the Open Cosmological Model," Zh. Eksp. Teor. Fiz. Pis'ma **11,** 200-203 (1970) [English translation in Sov. Phys. JETP Lett. **11,** 123-25 (1971)]. V. A. Belinsky, I. M. Khalatnikov, and E. M. Lifshitz, "Oscillatory Approach to a Singular Point in the Relativistic Cosmology," Usp. Fiz. Nauk **102,** 463-500 (1970) [English translation in *Advances Phys.* **19,** 525-73 (1970)]; V. A. Belinsky, E. M. Lifshitz, and I. M. Khalatnikov, "Oscillatory Mode of Approach to a Singularity in Homogeneous Cosmological Models with Rotating Axes," Zh. Eksp. Teor. Fiz. **60,** 1969-79 (1971) [English translation in Sov. Phys. JETP **33,** 1061-66 (1971)]. J. D. Barrow and F. J. Tipler, "Analysis of the Singularity Studies by Belinski, Khalatnikov and Lifshitz," *Phys. Reports* **56,** 371-402 (1979).

36. K. Schwarzschild, "Über das Gravitationsfeld eines Massenpunktes nach der Einsteinschen Theorie," Preuss. Akad. Wiss. Berlin, Sitzungsb., Kl. Math.-Phys.-Tech., 189-96 (1916); H. Reissner, "Über die Eigengravitation des elektrischen Feldes nach der Einsteinschen Theorie," Ann. d. Physik **50,** 106-20 (1916); G. Nordstrøm, "On the Energy of the Gravitational Field in Einstein's Theorem," Proc. Kon. Ned. Akad. Wet. **20,** 1238-45 (1918).

37. R. P. Kerr, "Gravitation Field of a Spinning Mass as an Example of Algebraically Special Metrics," Phys. Rev. Lett. **11,** 237-38 (1963); B. Carter, "Global Structure of the Kerr Family of Gravitational Fields," Phys. Rev. **174,** 1559-71 (1968).

38. E. T. Newman et al., "Metric of a Rotating, Charged Mass," J. Math. Phys. **6,** 918-19 (1965).

39. For the literature on black holes see for example the bibliographies in H. L. Shipman, *Black Holes, Quasars and the Universe* (Houghton Mifflin, Boston, 1976); C. DeWitt and B. DeWitt, eds., *Black Holes, Les Houches 1972, Lectures Delivered at the Summer School of Theoretical Physics of the University of Grenoble* (Gordon and Breach, New York, 1973); H. Gursky and R. Ruffini, *Neutron Stars, Black Holes and Binary X-ray Sources* (D. Reidel, Dordrecht, Holland, 1975); R. Giacconi and R. Ruffini, *Physics and Astrophysics of Neutron Stars and Black Holes* (North-Holland, Amsterdam, 1978).

40. "A black hole has no hair — or particularities" [J. A. Wheeler, "From Mendeleev's Atom to the Collapsing Star," in *Atti del Convegno Mendeleeviano,* edited by M. Verde (Accad. delle Sci. di Torino, 1971), pp. 189–233, esp. pp. 191–92].

    "Hair" of most extreme character — the "spike" of density formed by a particle falling into a black hole — fades away exponentially: R. Ruffini and J. A. Wheeler, "Relativistic Cosmology and Space Platforms," in *Proceedings of the Conference on Space Physics* (European Space Research Organization, Paris, 1971), pp. 45–174; for adaptation of curve of fall see Misner, Thorne, and Wheeler, op. cit. in n. 9, Fig. 25.5, p. 667. B. Carter, "An Axisymmetric Black Hole Has Only Two Degrees of Freedom," op. cit. in n. 1, and "Black Hole Equilibrium States," in *Black Holes,* op. cit. in n. 1, esp. Sec. 12, "The Pure Vacuum No Hair Theorem," pp. 205–209.

    No hair in the sense of a weak-interaction force caused by leptons that have fallen into the black hole: J. B. Hartle, "Long Range Neutrino Forces Exerted by Kerr Black Holes," Phys. Rev. **D3**, 2938–40 (1971), and "Can a Schwarzschild Black Hole Exert Long-Range Neutrino Forces?" in *Magic without Magic: John Archibald Wheeler,* edited by J. Klauder (Freeman, San Francisco, 1972), pp. 259–75, esp. pp. 271–74; C. Teitelboim, "Nonmeasurability of the Lepton Number of a Black Hole," Nuovo Cimento II **3,** 397–400 (1972), and "Nonmeasurability of the Quantum Numbers of a Black Hole," Phys. Rev. **D5,** 2941–54 (1972).

    No hair left from baryons that have gone down the black hole: J. Bekenstein, "Nonexistence of Baryon Number for Static Black Holes," I, Phys. Rev. **D5,** 1239–46 (1972); II, Phys. Rev. **D5,** 2403–12 (1972); C. Teitelboim, "Nonmeasurability of the Baryon Number of a Black Hole," Nuovo Cimento Lett., II, **3,** 326–28 (1972); see however M. J. Perry, "Black Holes are Coloured," Phys. Lett. **71B,** 234–36 (1977), giving reasons to believe that a black hole may act as the source of a Yang–Mills field.

41. Rate of attainment of perfection outside black hole: W. H. Press, "Long Wave Trains of Gravitational Waves from a Vibrating Black Hole," Astrophys. J. Lett. **170,** 105–108 (1971); M. Davis et al. "Gravitational Radiation from a Particle Falling Radially into a Schwarzschild Black Hole," Phys. Rev. Lett. **27,** 1466–69 (1971); S. W. Hawking and J. B. Hartle, "Energy and Angular Mementum Flow into a Black Hole," Comm. Math. Phys. **27,** 283–90 (1970).

42. N. Zamorano, *Interior Reissner-Nordstrøm Black Holes,* doctoral dissertation, The University of Texas at Austin, August 1979, available from University Microfilms, Ann Arbor, Mich.; J. Pfautsch, unpublished results extending those of N. Zamorano, op. cit.

43. Misner, Thorne, and Wheeler, op. cit. in n. 9, p. 738, Box 27.4.

44. V. M. Lyutyi, R. A. Sunyaev, and A. M. Cherepashchuk, "Nature of the Optical Variability of Hz Herculis (Her X-1) and BD + 34° 3.815 (Cyg X-1)," Sov. Ast. AJ **17,** 1–6 (1973); U. Avni and J. N. Bahcall, "Ellipsoidal Light Variations and Masses of X-ray Binaries," Astrophy. J. **197,** 675–88 (1975), where the possibility of Cyg X-1 secondary being either a black hole or a normal early-type star is discussed; C. R. Canizares and M. Oda, "Observations of Rapid X-ray Flaring from Cygnus X-1," Astrophy. J. **214,** L119–22 (1977); J. N. Bahcall, "Masses of Neutron Stars and Black Holes in X-ray Binaries," Ann. Rev. Astron. Astrophys. **16,** 241–64 (1978), esp. pp. 261–63, for discussion of the data on Cyg X-1 and references; D. M. Eardley et al., "A Status Report on Cygnus X-1," Comments Atrophys., Comments Mod. Phys. part C **7,** 151–60 (1978) (a review article with sixty-one references).

45. J. E. Grindlay, "Two More Globular Cluster X-ray Sources?" Astrophy. J. Lett. **224,** L107–11 (1977).

46. Works by Towne and colleagues include E. R. Wollman, "Ne II 12.8 μ Emission from the Galactic Center and Compact H-II Regions," doctoral thesis, University of California, Berkeley, 1976, available from University Microfilms, Ann Arbor, Mich.; E. R. Wollman, et al., "Spectral and Spatial Resolution of the 12.8 Micron Ne II Emission from the Galactic Center," Astrophy. J. Lett. **205,** L5–9 (1976); E. R. Wollman and C. H. Townes, "Ne II Micron Emission from the Galactic Center. II," Astrophy. J. **218,** L103–07 (1977); J. H. Lacy et al., "Observations of the Motion and the Distribution of the Ionized Gas in the Central Parsec of the Galaxy," Astrophy. J. **227,** L17–20 (1979). J. H. Oort, "The Galactic Center," Astron Astrophys. **15,** 295–362 (1977), see esp. pp. 341, 343, 347, 349, 352, and on p. 353, "It is therefore possible that practically the whole of the 4–6 $\times$ $10^6 M_\odot$ deduced from the Ne II observations would be concentrated in the ultracompact radio source; it might be the mass of a black hole in the center." K. I. Kellermann et al., "The small radio source at the galactic center," Astrophy. J. **214,** L61–L62 (1977); L. F. Rodriguez and E. J. Chaisson, "The Temperature and Dynamics of the Ionized Gas in the Nucleus of Our Galaxy," Astrophy. J. **228,** 734–39 (1979).

47. P. J. Young et al., "Evidence for a Supermassive Object in the Nucleus of the Galaxy M87 from SIT and CCD Area Photometry," Astrophy. J. **221,** 721–30 (1978); W. L. W. Sargent and P. J. Young, "Dynamical Evidence for a Central Mass Concentration in the Galaxy M87," Astrophy. J. **221,** 737–44 (1978); J. Stauffer and H. Spinrad, "Spectroscopic Observations of the Core of M87," Astrophy. J. Lett. **231,** L151–56 (1979).

48. For dimensions of quantum fluctuations, see references listed under "Quantum aspects of the black hole" in n. 1, and J. A. Wheeler, "Superspace and the Nature of Quantum Geometrodynamics," in *Battelle Rencontres: 1967 Lectures in Mathematics and Physics,* edited by C. DeWitt and J. A. Wheeler (Benjamin, New York, 1968), pp. 242–307, in particular, pp. 263–68 and 286–90, including Fig. 5 on p. 265 and Fig. 8 on p. 289, or, in German, J. A. Wheeler, *Einsteins Vision* (Springer, Berlin, 1968), esp. pp. 39–46 (including Fig. 8) and pp. 68–78 (esp. Fig. 10).
For collapse within the black hole, see R. W. Fuller and J. A. Wheeler, "Causality and Multiply Connected Spacetime," Phys. Rev. **128,** 919–29 (1962).

49. D. Christodoulou, "Reversible and Irreversible Transformations in Black-Hole Physics," Phys. Rev. Lett. **25,** 1596–97 (1970); D. Christodoulou and R. Ruffini, "Reversible Transformations of a Charged Black Hole," Phys. Rev. **D4,** 3552–55 (1971). The resulting formula for the energy E of the black hole in terms of its mementum p, charge Q, intrinsic angular momentum S, and irreducible mass M_{ir} [= (surface area of horizon/16π)$^{1/2}$], all being measured in geometric units, is

$$E^2 = p^2 + (M_{ir} + Q^2/4M_{ir})^2 + S^2/4M_{ir}^2 \; .$$

50. P. Cordero and C. Teitelboim, "Remarks on Supersymmetric Black Holes," Phys. Lett. **78B,** 80–83 (1978): the most general black hole with supercharge is equivalent, under supersymmetry transformation, to a black hole without supercharge — as a black hole with momentum is equivalent under Lorentz transformation to a black hole without momentum.

51. D. L. Freedman, P. van Nieuwenhuizen, and S. Ferrara, "Progress Toward a Theory of Supergravity," Phys. Rev. **D13,** 3214–18 (1976); S. Deser and B. Zumino, "Consistent Supergravity," Phys. Lett. **62B,** 335–37 (1976); C. Teitelboim, "Supergravity and Square Roots of Constraints," Phys. Rev. Lett. **38,** 1106–10 (1977); R. Tabensky and C. Teitelboim, "The Square Root of General Relativity," Phys. Lett. **69B,** 453–56 (1977).

52. Carter, "Black Hole Equilibrium States," in *Black Holes,* op. cit. in n. 1, esp. Sec. 12, pp. 205–209; also, see the references cited in the "No hair" paragraphs of n. 40.

53. Lepton number of black hole not measurable: see references in third "No hair" paragraph of n. 40 and, also, the references cited under "Theorems on the uniqueness of the geometry around a black hole" in n.1. See the latter as well as the fourth "No hair" paragraph of n. 40 on the subject of immeasurability of baryon number of black holes.

54. J. A. Wheeler, "Transcending the Law of Conservation of Leptons," op. cit. in n. 1., and "From Relativity to Mutability," op. cit. in n. 9, esp. p. 202: "Baryon number and lepton number are well defined quantities for a normal star; but when this star collapses to a black hole, the well established laws of conservation of particle number lose all applicability."

55. The staircase is described in Wheeler, "From Relativity to Mutability," op. cit. in n. 9, p. 241.

56. The method of determining density of an object by weighing it, first in air, then under water is reputed to have been found in answer ("*Eureka*" — I have found it) to the question of Hieron, king of Syracuse, whether his "gold" crown did not contain an admixture of silver. [Archimedes of Syracuse (287–212 B.C.), *Peri ochoumenon* (On Floating Bodies)].

57. Fixity of density transcended: J. A. Morgan, "The Equation of State of Platinum to 680 GPa," High Temperature-High Pressure **6,** 195–201 (1974), which reports density approximately doubled via use of gun; H. K. Mao et al., "Specific Measurements of Cu, Mo, Pd, and Ag and Calibration of Ruby R_1 Fluorescence Pressure Gauge from 0.06 to 1 *M* Bar," J. Appl. Phys. **49,** 3276–83 (1978), for use of nuclear explosion to go to extreme density; C. E. Ragan, III, M. G. Silbert, and B. C. Diven, "Shock Compression of Molybdenum to 2.0 TPa by Means of a Nuclear Explosion," J. Appl. Phys. **48,** 2860–70 (1977), for achievement of pressures in the 30 megabar range and determination of increase of density by a *measured* factor of about 3; L. V. Al'tshuler et al., "Shock Adiabats for Ultrahigh Pressures," Soviet Phys. JETP **45,** 167–71 (1977), which reports the extreme of published pressures, 50 megabars, where *any* measurements have been made. Thanks are expressed here to William Deal for guidance to this literature.

58. The first table of atomic weights, and idea that chemical combination takes place between atoms of different weights, law of combination in multiple proportions, and atomic theory, are included in outline form by J. Dalton's consent (6 Sept. 1803) in T. Thomson, *System of Chemistry,* 3rd ed. (1807) and in the first volume of Dalton's own *New System of Chemical Philosophy* (S. Russell, Manchester, England, 1808). Under identical physical conditions

equal volumes of gas contain the same number of molecules: A. Avogadro, "Essai d'une manière de déterminer les masses relatives des molécules élémentaires des corps, et des proportions selon lesquelles elles entrent dans les combinaisions," J. de Phys. (1811). Simplest molecular formulas for organic compounds and systematic use of "rational" chemical formulas: C. F. Gerhardt, *Introduction a l'étude de la chimie par le systéme unitaire* (1848). Distinction between molecular and atomic weights and deduction of atomic weights from vapor density or specific heat: S. Cannizzaro, *Sunto di un corso di filosofia chemica* (1858). Atom of each elementary substance can combine only with a certain limited number of the atoms of other elements, foundation of theory of valency: E. Frankland, "On the Dependence of the Chemical Properties of Compounds upon the Electrical Characters of Their Constituents," Roy. Inst. Proc. **1,** 451–54 (1852). Carbon tetravalence with some of the affinities of the generic carbon atom bound by atoms of other kinds, some bound by other carbon atoms: F. A. Kekulé, "Über die Constitution und die Metamorphosen der chemischen Verbindungen und über die Natur der Kohlenstoffs," Liebig, Annal. **106,** 129–59 (1858); result also obtained independently and reported by A. S. Cooper, Acad. des Sci., Paris, Comptes Rend. (1858).

59. Carbon was found to have valence 3 in the first free radical discovered, triphenyl methyl [M. Gomberg, "An Instance of Trivalent Carbon," J. Am. Chem. Soc. **22,** 757–71 (1900)], which opened way to the realization that valence can be transcended in a few cases at room temperature and for every atom at a sufficiently high temperature.

60. Mass number and isotopes: F. Soddy, "Radio Elements and the Periodic Law," Chem. News **107,** 97–99 (1913), elucidates the chemical identity discovered by B. B. Boltwood in 1906 and by H. N. McCoy and W. H. Ross in 1907 between radioactively distinct ionium, radiothorium, and thorium, and discovered in other cases by Soddy himself and other workers in the subject.

    Charge number: H. G. J. Moseley, "High-Frequency Spectra of the Elements," Phil. Mag. **26,** 1024–34 (1913), and same title, II, Phil. Mag. **27,** 703–13 (1914), reported determination of number of elementary charges on the nucleus, and the place of the corresponding element in the periodic table, by measuring the wavelengths of the strongest lines in the X-ray spectrum of the atom and applying Bohr's theory of the atom.

61. Discovery of natural radioactivity: H. A. Becquerel, "Sur les radiations émises par phosphorescence," Acad. Sci., Paris, Compt. Rend. **122,** 420–21 (1896). First artificial transmutation: E. Rutherford, "Collision of α Particles with Light Atoms. IV. An Anomalous Effect in Nitrogen," Phil. Mag. **37,** 581–87 (1919).

    J. D. Cockroft and E. T. S. Walton, "Experiments with High-Velocity Positive Ions. Part II. Disintegration of Elements by High-Velocity Protons," Proc. Roy. Soc. London **137,** 229–42 (1932), and M. A. Tuve and L. R. Hafstad, "The Emission of Disintegration-Particles from Targets Bombarded by Protons and by Deuterium Ions at 1200 Kilovolts," [Letter] Phys. Rev. **45,** 651–53 (1934).

62. Baryon conservation: F. Reines, C. L. Cowan, Jr., and M. Goldhaber, "Conservation of the Number of Nucleons," Phys. Rev. **96,** 1157–58 (1954); F. Reines, C. L. Cowan, Jr., and H. W. Kruse, "Conservation of the Number of Nucleons," Phys. Rev. **109,** 609–10 (1957); G. N. Flerov et al., "Spontaneous Fission of Th^{232} and the Stability of Nucleons," Sov. Physics Doklady **3,** 79–80 (1958); G. Feinberg and M. Goldhaber, "Microscopic Tests of Symmetry Principles," U.S. Nat. Acad. Sci., Proc. **45,** 1301–12 (1959); G. Feinberg and M. Goldhaber, "Experimental Tests of Symmetry Principles," Science **129,** 1285 (1959); H. S. Gurr et al., "Experimental Test of Baryon Conservation," Phys. Rev. **158,** 1321–30 (1967). J. Learned, F. Reines, and A. Soni, "Limits on Nonconservation of Baryon Number," Phys. Rev. Lett. **43,** 907–909 (1979), report nucleon lifetime greater than about 10^{30} years at 90 percent confidence level.

    Baryon conservation seriously questioned: S. Weinberg, "Cosmological Production of Baryons," Phys. Rev. Lett. **42,** 850–53 (1979).

    Lepton conservation: M. Goldhaber, "Weak Interactions: Leptonic Modes — Experiment," in *Proceedings of the 1958 Annual International Conference on High Energy Physics at CERN,* pp. 233–50; V. R. Lazarenko, "Double Beta Decay and the Properties of the Neutrino," Usp. Fiz. Nauk **90,** 601–22 (1961) [English translation in Sov. Phys. Uspekhi **9,** 860–73 (1967)]; B. Pontecorvo, "Neutrino Experiments and the Problem of Conservation of Leptons," Sov. Phys. JETP **26,** 984–88 (1968); K. Boher et al., "Untersuchung über die Erhaltung der μ-Leptonenzahl," Helv. Phys. Acta **43,** 111–32 (1970).

    R. I. Steinberg et al., "Experimental Test of Charge Conservation and Stability of the Electron," Phys. Rev. **D12,** 2582–86 (1975), reported mean life of electron against decay into nonionizing particles greater than 5.3×10^{21} years.

63. Transcended, see n. 54.

64. Energy not defined in a closed universe: "There is no such quantity as total energy, for example, in a closed universe; there the integrated conservation laws reduce to the trivial identity, zero equals zero" [J. Weber and J. A. Wheeler, "Reality of the Cylindrical Gravitational Waves of Einstein and Rosen," Rev. Mod. Phys. **29,** 509–15 (1957), p. 512].

    The key point for defining mass is the existence of a region where the geometry goes over asymptotically to the Schwarzschild character. When there is no such region, then it is not known how to give an unambiguous

> meaning to the term "mass." This is particularly the case for a closed universe. There is no asymptotically flat region in which to measure the pull of the system by the bending of light or by the periods of planetary orbits and their precession. If there is no experimental way to *measure* mass for a closed universe, and no theoretical way to *define* mass, this is happily compatible with the circumstance that no one knows any *use* for the concept of the mass of a closed universe. Therefore it would appear appropriate to reject this phrase as being physically meaningless as well as being subject to misunderstanding.

[J. A. Wheeler, "Geometrodynamics and the Issue of the Final State," op. cit. in n. 1, pp. 434–35.]

Misner, Thorne, and Wheeler, op. cit. in n. 9, pp. 457–58.

65. Misner, Thorne, and Wheeler, op. cit. in n. 9, p. 450: $(\text{mass of center of attraction})^1 = (2\pi/\text{orbital period})^2 (\text{semi-major axis of ellipse})^3$.

66. "The dimensions of the collapsing system in a finite proper time are driven down to indefinitely small values. The phenomenon is not limited to the space occupied by matter. It occurs also in the space surrounding the matter." [Wheeler, "Superspace and the Nature of Quantum Geometrodynamics," op. cit. in n. 48, p. 254].

67. No before, no after: "[A]t small distances and in the final phase of collapse" 'spacetime' is nonexistent, 'events' and the 'time ordering of events' are without meaning, and the question 'what happens after the final phase of gravitational collapse' is a mistaken way of speaking" (Ibid., p. 254]. "*[T]here is no such thing as spacetime in the real world of quantum physics* . . . complementarity forbids. [S]uperspace leaves us space but not spacetime and therefore not time. With time gone the very ideas of 'before' and 'after' also lose their meaning." [Wheeler, "From Relativity to Mutability," op. cit. in n. 1, p. 227.] "Nowhere more clearly than in the ending of spacetime are we warned that time is not an ultimate category in the description of nature" [Wheeler, *Frontiers of Time,* op. cit. in n. 8, p. 6]. " 'Before' and 'after' don't rule everywhere, as witness quantum fluctuations in the geometry of space at the scale of the Planck distance. Therefore 'before' and 'after' cannot legalistically rule anywhere. Even at the classical level, Einstein's standard closed-space cosmology denies all meaning to 'before the big bang' and 'after the big crunch.' Time cannot be an ultimate category in the description of nature. We cannot expect to understand genesis until we rise to an outlook that transcends time" [Ibid., p. 20]. "Not the slightest warrant does Einstein's equation give for thinking there can be any such thing as a 'before' before the big bang or an 'after' after the big crunch or after the collapse of a star to a black hole. These three processes mark three 'gates of time' " [Ibid., p. 75]. "Little escape is evident from these words: there is no 'before' before the big bang and no 'after' after the big crunch. Time ends with spacetime. The universe does not endure from everlasting to everlasting. Everything came from 'nothing' " [Ibid., p. 85].

68. "There never was a law of physics that did not require space and time for its statement. With collapse the framework falls down for everything one ever called a law. The laws of physics were not installed in advance by a Swiss watchmaker, nor can they endure from everlasting to everlasting. They must have come into being. They could not always have been accurate. They are derivative and superficial, not primary and revelatory." [Wheeler, *Frontiers of Time,* op. cit. in n. 8, p. 20]. This position, based on the conclusion that the category of "time" is itself not primordial, but secondary, derivative and approximate, differs in that respect from the position of Peirce, who tacitly accepted the primordiality of time: "May they [these forces of nature] not have naturally grown up?" *The philosophy of [Charles S.] Peirce: Selected Writings,* edited by J. Buchler (Routledge and Kegan Paul, London, 1940); available also as a paperback reprint under the title *Philosophical Writings of Peirce* (Dover, New York, 1955). The quotation is from p. 358 in the paperback; see further on that page; also see pp. 335–37 and p. 353.

69. Mutability: see the references in n. 9.

70. Crack in crystal: A. Joffe, "On the Cause of the Low Value of Strength," pp. 72–76, and "On the Mechanism of Brittle Rupture," pp. 77–80 in *International Conference on Physics, London 1934. A Joint Conference Organized by the International Union of Pure and Applied Physics and the Physical Society. Papers and Discussions in Two Volumes. Vol. 2. The Solid State of Matter* (Cambridge University Press and The Physical Society, Cambridge, England, 1935).

71. Collapse inevitable: see paragraph entitled "Proof that gravitational collapse is inescapable . . ." in n. 1. Theorem on inescapability of singularity: Hawking and Penrose, op. cit. in n. 4. Consideration of details of approach to singularity in the generic case: see the references listed in n. 35.

72. Quantum fluctuations in topology and geometry predicted and estimated at the Planck scale of distances: Wheeler, "On the Nature of Quantum Geometrodynamics," op. cit. in n. 1; Wheeler, "Superspace and the Nature of Quantum Geometrodynamics," op. cit. in n. 48; and Misner, Thorne, and Wheeler, op. cit. in n. 9, pp. 1190–94.

73. These fluctuations calculated: Gibbons and Hawking, op. cit. in n. 1. Further calculations: M. J. Perry, S. W. Hawking, and G. W. Gibbons, "Path Integrals and the Indefiniteness of the Gravitational Action," Nucl. Phys. **B 138,** 141–50 (1978).

74. Three dimensions of space (as other laws of nature) as a "precondition for the possibility of phenomena" [I. Kant, *Critique of Pure Reason,* translated by F. M. Muller (Anchor, Garden City, N. Y., 1966), p. 24]; see, however, A. Grünbaum, *Philosophical Problems of Space and Time* (Knopf, New York, 1963), ch. 11.

    Space as analyzed, not metrically, but via analysis situs (topology in the large) shows itself to be three-dimensional: H. Poincaré, *Mathematics and Science: Last Essays,* translated by J. W. Bolduc (Dover, New York, 1963), ch. 3, pp. 27–28.

    Only in three-dimensional space is sense made by the laws of gravitation and planetary motion, the duality of 1) translation and rotation, of 2) force and pair of forces, and of 3) electric field and magnetic field: P. A. Ehrenfest, "In What Way Does It Become Manifest in the Fundamental Laws of Physics That Space Has Three Dimensions?", Proc. Amsterdam Acad. **20,** 200–209 (1917).

    Only in a space with an odd number of dimensions "will darkness follow the extinction of a candle"; gauge invariance holds only for three dimensions; other considerations and reference to others who have asked, "Why three dimensions?": H. Weyl, *Philosophy of Mathematics and Natural Science* (Princeton University Press, Princeton, N. J., 1949; paperback reprint, Atheneum, New York, 1963), p. 36.

    In three-dimensions Einstein's field equation requires space to be flat, thereby making geodesics be straight lines; but gravitation nevertheless shows itself in the global equivalent of curvature produced by conelike singularities. A. Staruszkiewicz, "Gravitation Theory in Three-Dimensional Space," Acta Phys. Polonica **24,** 735–40 (1963).

75. Pregeometry:

    > [T]he number of dimensions should not be assumed in advance; it should be *derived* to be four. . . . [A]ny derivation of the 4-dimensionality of spacetime can hardly start with the idea of dimensionality. . . . [O]ne can imagine probability amplitudes for the points in a Borel set to be assembled into manifolds with this, that and the other dimensionality. . . . [D]efine an action principle over a collection of points of undefined dimensionality. One might also wish to accept to begin with the idea of a distance, or edge length, associated with a pair of these points, even though this idea is a very great leap, and one that one can conceive of later supplying with a foundation of its own. . . . [T]here must be a connection in the appropriate action principle between *every* point and every other point. . . . Try therefore a propagator of the form
    >
    > $$\sum_{\text{diagram}} \exp i\, S_{\text{diagram}}$$
    >
    > Here the sum goes over all conceivable ways of connecting the given number of vertices up into nearest neighbors, whatever the dimensionality or lack of dimensionality of these "wiring diagrams." How this phase depends upon the topology of the diagram is to be deduced — in whole or in part — from natural combinatorial principles.

    [J. A. Wheeler, "Geometrodynamics and the Issue of the Final State," pp. 315–520 in *Relativity, Groups and Topology,* edited by C. DeWitt and B. DeWitt (Gordon and Breach, New York, 1964), pp. 495–99.]

    Pregeometry, including discussion of "pregeometry as the calculus of propositions": Misner, Thorne, and Wheeler, op. cit. in n. 9, pp. 1203–12. "[The] concept of 'ideal mathematical geometry' is too finalistic to be final and must give way to a deeper concept of structure [- . . .] 'pregeometry' " [C. M. Patton and J. A. Wheeler, "Is Physics Legislated by Cosmogony?" pp. 538–605 in *Quantum Gravity: An Oxford Symposium,* edited by C. J. Isham, R. Penrose, and D. W. Sciama (Clarendon, Oxford, 1975), p. 573]. "[W]e have to give up the idea that pregeometry is the calculus of propositions, or the statistics of propositions, or the mathematical machinery of any formal axiomatic system" [Report on the Search for Pregeometry, February-March-April 1974, op. cit., Appendix B, pp. 589–91].

    Pregeometry viewed as the statistics of billions upon billions of acts of observer-participancy: J. A. Wheeler, "Pregeometry: Motivations and Prospects," in *Quantum Theory and Gravitation,* edited by A. R. Marlow (Academic Press, New York, 1980).

76. Thermodynamics rests upon the random motions of billions upon billions of molecules: for key selections from the original literature see S. G. Brush, *Kinetic Theory,* Vol. 1, *The Nature of Gases and of Heat,* and Vol. 2, *Irreversible Processes* (Pergamon, Oxford, 1965, 1966).

77. C. Darwin, *Origin of Species by Means of Natural Selection* (London, 1859); T. Dobzhansky, *Genetics and the Origin of Species* (New York, 1937); M. Eigen, "The Origin of Biological Information," *The Physicists' Conception of Nature,* edited by J. Mehra (Reidel, Dordrecht, 1973), pp. 594–632; M. Eigen and R. Winkler, *Das Spiel: Naturgesetze steuern den Zufall* (Piper, Munich, 1975).

78 Wheeler, *Frontiers of Time,* op. cit. in n. 8, p. 6.

79. "The statistical character of the present theory would then have to be a necessary consequence of the incompleteness of the description of the systems in quantum mechanics, and there would no longer exist any ground for the supposition that a future basis of physics must be based upon physics" [Einstein, "Autobiographical Notes," in *Albert Einstein, Philosopher-Scientist,* op. cit. in n. 7, p. 87]. "[I]t [quantum mechanics] seems to make the world quite nebulous unless somebody, like a mouse, is looking at it" ["Last Lecture of Albert Einstein," op. cit. in n. 26].

80. A. Einstein, "Über einen die Erzeugung und Verwandlung des Lichtes betreffenden heuristischen Gesichtspunkt," Ann. d. Phys. **17,** 132–48 (1905).

81. A. Einstein, "Strahlungs-emission und -absorption nach der Quantentheorie," Deutsche physikalische Gesellschaft, Verhandlungen **18,** 318–23 (1916); "Quantentheorie der Strahlung," Physikalische Gesellschaft, Zürich, Mitteilungen **16,** 47–62 (1916).

82. A. Einstein, "Die Feldgleichungen der Gravitation," op. cit. in n. 13.

83. "Ich habe keinen besseren Ausdruck als den Ausdruck [religiös] fur dieses Vertrauen [von Spinoza] in die vernünftige und der menschlichen Vernunft wenigstens einigermassen zugängliche Beschaffenheit der Realität" [A. Einstein, letter to Maurice Solovine, 1 January 1951]. "Wo stünden wir wenn Leute wie Giordano Bruno, Spinoza, Voltaire und Humboldt so gedacht und so gehandelt hätte?" [A. Einstein, letter to Max von Laue, May 1933, regarding the capitulation of intellectuals to the advent of gangsterism]. The latter two letters are reproduced in F. Herneck, *Einstein und sein Weltbild* (Der Morgen, Berlin, DDR, 1979), p. 35 and p. 87. "[I]gnoramuses who use their public positions of power to tyrannize over professional intellectuals must not be accepted by intellectuals without a struggle. Spinoza followed this rule when he turned down a professorship at Heidelberg and (unlike Hegel) decided to earn his living in a way that would not force him to mortgage his freedom" [Einstein, letter to *The Reporter,* published 5 May 1955, a few days after his death]. "Just in this appears the moral side of our nature — that internal striving towards the attainment of truth, which under the name *amor intellectualis* was so often emphasized by Spinoza" [Einstein statement, Forum **83,** 373–374 (1930)].

84. Appreciation is expressed here to Professor Hans Küng for emphasizing in June 1978 at Tübingen the influence of Spinoza on Einstein's outlook.

85. "But Spinoza rejected the idea of an external Creator suddenly, and apparently capriciously, creating the world at one particular time rather than another, and creating it out of nothing" ["Spinoza," in *Encyclopaedia Brittanica,* Vol. 21 (Chicago, Ill., 1959), p. 235].

86. A. Einstein, "Kosmologische Betrachtungen zur allgemeinen Relativitätstheorie," Preuss. Akad. Wiss., Berlin, Sitzber. 142–52 (1917); Friedmann, "Über die Krummung des Raumes," op. cit. in n. 2.

87. Hubble, "A Relation between Distance and Radial Velocity among Extragalactic Nebulae," op. cit. in n. 2; Einstein as quoted by G. Gamow, *My World Line* (Viking, New York, 1970).

88. B. de Spinoza, *Ethics,* finished at The Hague 1675 and circulated privately; English translation by H. White and A. H. Stirling (1899).

89. Participatory: see n. 8.

90. Story of twenty questions: in Wheeler, *Frontiers of Time,* op. cit. in n. 8.

91. Phenomenon: Introduced by N. Bohr to meet and overcome the objections of Einstein, op. cit. in n. 7, p. 230. Preliminary account of stages in Bohr's evolution of this term: A. Petersen, *Quantum Mechanics and the Philosophical Tradition* (M.I.T. Press, Cambridge, Mass., 1968). "No phenomenon is . . . until it is . . . ," used by J. A. Wheeler in Varenna lectures of 1977; revised in printed version, *Frontiers of Time,* op. cit. in n. 8, to read "No elementary phenomenon . . ." to exclude macroscopic phenomena. The ending used there, "until it is an observed phenomenon" is revised here to "until it is a registered phenomenon" to exclude any suggestion that quantum mechanics has anything whatsoever directly to do with "consciousness" and to recall Bohr's point that an irreversible act of amplification is required to bring an elementary phenomenon to a close.

92. J. A. Wheeler, "The 'Past' and the 'Delayed-Choice' Double-Slit Experiment," in *Mathematical Foundations of Quantum Theory,* edited by A. R. Marlow (Academic, New York, 1978).

93. N. Bohr, op. cit. in n. 7; chapters on the Bohr–Einstein dialogue in M. Jammer, *The Philosophy of Quantum Mechanics* (Wiley, New York, 1974).

94. This "venetian blind" and other experimental arrangements, alternative to that depicted in n. 92, have been devised and generously communicated to the author by Professor L. F. Bartell of the University of Michigan at Ann Arbor.

95. "No significant change in the correlation was observed over separations of up to 2.5 m" [A. R. Wilson, J. Lowe, and D. K. Butt, "Measurement of the Relative Planes of Polarization of Annihilation Quanta as a Function of Separation Distance," J. Phys. G: Nuc. Phys. **2,** 613–24 (1976)] is a distinct but related finding.

96. N. Bohr, op. cit. in n. 7.

97. Ibid. Complementarity defined: N. Bohr, *Atomic Theory and the Description of Nature* (Cambridge University Press, Cambridge, England, 1934), p.5.

98. Wheeler, op. cit. in 92.

99. "Past" exists only in the present: Wheeler, *Frontiers of Time,* op. cit. in n. 8, p. 21; T. Segerstedt as quoted in ibid.

100. Cf. n. 91.

101. Closed by irreversible amplification: Bohr, *Atomic Physics and Human Knowledge,* op. cit. in n. 6, pp. 73, 88. Unambiguously communicable in plain language: *Essays 1958–1962 . . .,* op. cit. in n. 6, pp. 3, 6, 7.

102. E. P. Wigner, "Are We Machines?" Proc. Am. Philos. Soc. **113,** 95–101 (1969), p. 97, and "The Philosophical Problem," *Foundations of Quantum Mechanics,* edited by B. d'Espagnat (Academic, New York, 1971), p. 3.

103. N. Bohr, n. 91; A. Einstein, B. Podolsky, and N. Rosen, "Can Quantum-Mechanical Description of Physical Reality Be Considered Complete?" Phys. Rev. **47,** 777–80 (1935).

104. See for example D. Gorenstein, "The Classification of Finite Simple Groups. I. Simple Groups and Local Analysis," Bull. Am. Math. Soc. **1,** 43–199 (1979), esp. ch. 1, sec. 3, "Why the Extreme Length?": "There exists an often expressed feeling in the general mathematical community that the present approach to the classification of simple groups must be the wrong one — no single theorem can possibly require a 5,000 page proof!"; also the discussion on pp. 50–52 of problems and progress in completing the classification.

105. For a study of this distinction between and interaction of observed system and observing equipment see especially M. M. Yanase, "Optimal Measuring Apparatus," Phys. Rev. **123,** 666–68 (1961); E. P. Wigner, "The Problem of Measurement," Am. J. Phys. **31,** 6–15 (1963), "Interpretation of Quantum Mechanics," 93 pages of mimeographed notes of lectures delivered at Princeton University in 1976, on deposit in Fine Library, Princeton University, Princeton, N. J.; A. Peres, "Can We Undo Quantum Measurements?" 1979 preprint, Center for Theoretical Physics, The University of Texas at Austin, to appear in Phys. Rev. D (1980).

106. "Nothingness" ruled out as meaningless: "There are three ways of research, and three ways only. Of these, one asserts '*It is not,* and there must be not-being.' This is utterly forbidden: what is not cannot even be thought of. A second way [is] that of mortals without wisdom, who say of what is that 'it is and is not,' 'is the same and not the same.' In contrast to them the way of truth starts from the proposition '*It is,* and not-being is impossible.' " [Parmenides of Elia, poem (~502 B.C.) *Nature,* part 2 "Truth," as summarized in article "Parmenides, p. 327–28 in *Encyclopaedia Brittanica,* Vol. 17 (Chicago 1959), p. 327.]

107. Universe as a self-excited circuit: in Wheeler, "Is Physics Legislated by Cosmogony," in op. cit. in n. 75, p. 565 and in *Frontiers of Time,* op. cit. in n. 8, p. 11.

108. "Towards the finding of this 'pregeometry' no guiding principle would seem more powerful than the requirement that it should provide the universe with a way to come into being. It is difficult to believe that we can uncover this pregeometry except as we come to understand at the same time the necessity of the quantum principle, with its 'observer-participator,' in the construction of the world" [Patton and Wheeler, op. cit. in n. 75, p. 575]. "No test of these views looks more like being someday doable, nor more interesting and more instructive, than a derivation of the structure of quantum theory from the requirement that everything have a way to come into being out of nothing" [Wheeler, *Frontiers in Time,* op. cit. in n. 8, p. 28]. See also J. A. Wheeler, "Pregeometry: Motivations and Prospects," and W. K. Wooters, "Information Is Maximized in Photon Polarization Measurements," in A. R. Marlow (ed.), op. cit. in n. 75.

109. Move over onto the new "foundation of elementary acts of observer-participancy." For three steps towards this development see R. M. F. Houtappel, H. Van Dam, and E. P. Wigner, "The Conceptual Basis and Use of the Geometric Invariance Principles," Rev. Mod. Phys. **37,** 595–632 (1965), esp. secs. 4.1–4.5 on pp. 610–16; W. Wooters, "Information Is Maximized in Photon Polarization Measurements," in A. R. Marlow (ed.), op. cit. in n. 75; and A. Peres, op. cit. in n. 105.

23. COMMENT ON THE TOPIC "BEYOND THE BLACK HOLE"

Freeman Dyson

It is a pity we failed to get Dick Feynman to give this paper. He would be the right person to respond to John Wheeler. I am a poor substitute.

Thirty-one years ago, Dick Feynman told me about his "sum over histories" version of quantum mechanics. "The electron does anything it likes," he said. "It just goes in any direction at any speed, forward or backward in time, however it likes, and then you add up the amplitudes and it gives you the wave-function." I said to him, "You're crazy." But he wasn't.

Now John Wheeler is saying similar things at an even more basic level. He says, "The whole universe does what it likes, and then you observe it and it gives you the laws of physics. Freedom from law produces law." I say again, "You're crazy." But he isn't.

Einstein once wrote of Faraday, "Dieser Mann liebte die rätselhafte Natur wie ein Liebhaber die ferne Geliebte. Es gab noch nicht das öde Spezialistentum, das mit Hornbrille und Dünkel die Poesie zerstört." I am quoting from the wonderful collection of Einstein writings that Helen Dukas has picked out of the files and published this month with Banesh Hoffmann. Here is Hoffmann's translation: "This man loved mysterious Nature as a lover loves his distant beloved. In his day there did not exist the dull specialization that stares with self-conceit through horn-rimmed glasses and destroys poetry." Fortunately for us, dull specialization has never destroyed the poetry of John Wheeler's imagination. Even more than Faraday, Wheeler loves the mysteries of nature and uses the whole universe as his playground. For more than forty years, Wheeler has been astounding the narrow specialists of science with his speculations. The really astounding thing about Wheeler's speculations is that so many of them have turned out in the end to be right. I quote from Wheeler himself his own explanation of how it happens that his imaginative visions often lead to truth: "We will first understand how simple the universe is when we recognize how strange it is. The simplicity of that strangeness, Everest summit, so well directs the eye that the feet can afford to toil up and down many a wrong mountain valley, certain stage by stage to reach someday the goal." That is Wheeler's style, a style inseparable from the substance of his thinking, as poetry is inseparable from science in his mind.

Wheeler's philosophy of science is much more radically relativistic than Einstein's. Wheeler would make all physical law relative to observers. He has us creating physical laws by our existence. In principle, if the role of observers in the universe is as essential as he imagines, life may

---

Harry Woolf (ed.), Some Strangeness in the Proportion: A Centennial Symposium to Celebrate the Achievements of Albert Einstein
ISBN 0-201-09924-1

even create physical laws by conscious decision. This is a radical departure from the objective reality in which Einstein believed so firmly. Einstein thought of nature and nature's laws as transcendent, standing altogether above and beyond us, infinitely higher than human machinery and human will.

One of the questions that has always puzzled me is this: Why was Einstein so little interested in black holes? To physicists of my age and younger, black holes are the most exciting consequence of general relativity. With this judgment the man-in-the-street and the television commentators and journalists agree. How could Einstein have been so indifferent to the promise of his brightest brain-child? I suspect that the reason may have been that Einstein had some inkling of the road along which John Wheeler was traveling, a road profoundly alien to Einstein's philosophical preconceptions. Black holes make the laws of nature contingent on the mechanical accident of stellar collapse. John Wheeler embraces black holes because they show most sharply the contingent and provisory character of physical law. Perhaps Einstein rejected them for the same reason.

Let me now come down to the details of what Wheeler has been saying. He says:

> Law without law. It is difficult to see what else than that can be the plan of physics. It is preposterous to think of the laws of physics as installed by a Swiss watchmaker to endure from everlasting to everlasting when we know that the universe began with a big bang. The laws must have come into being. Therefore they could not have been always 100 percent accurate. That means that they are derivative, not primary . . . Events beyond law. Events so numerous and so uncoordinated that, flaunting their freedom from formula, they yet fabricate firm form. . . . The universe is a self-excited circuit. As it expands, cools, and develops, it gives rise to observer-participancy. Observer-participancy in turn gives what we call tangible reality to the universe. . . . Of all strange features of the universe, none are stranger than these: time is transcended, laws are mutable, and observer-participancy matters.[1]

The idea of observer-participancy is for Wheeler central to the understanding of nature. Observer-participancy means that the universe must have built into it from the beginning the potentiality for containing observers. Without observers, there is no existence. The activity of observers in the remote future is foreshadowed in the remote past and guides the development of the universe throughout its history. The laws of physics evolve from initial chaos into the rigid structure of quantum mechanics, because observers require a rigid structure for their operations. All this sounds to a contemporary physicist vague and mystical. But we should have learned by now that ideas that appear at first sight to be vague and mystical sometimes turn out to be true.

Wheeler is building on the work of Bob Dicke and Brandon Carter, who were the first to point out that the laws of physics and cosmology are constrained by the requirement that the universe should provide a home for theoretical physicists. Brandon Carter has shown that the existence of a long-lived star such as the sun, giving steady warmth to allow the slow evolution of life and intelligence, is only possible if the numerical constants of physics have values lying in a restricted range. Carter calls the requirement that the universe be capable of breeding physicists the "anthropic principle." Dicke and Carter used the anthropic principle to set quantitative limits to the structure of the universe. Wheeler carries their idea much further, conjecturing that the laws of nature are not only quantitatively constrained but qualitatively molded by the existence of observers.

Wheeler unified two streams of thought that had before been separate. On the one hand, in the domain of astronomy and cosmology, the anthropic principle of Dicke and Carter constrains the structure of the universe. On the other hand, in the domain of atomic physics, the laws of quantum mechanics take explicitly into account the fact that atomic systems cannot be described independently of the experimental apparatus by which they are observed. In atomic physics, the reaction of the process of observation upon the object observed is an essential part of the description of the object. The famous Einstein–Rosen–Podolsky paradox showed once and for all that it is not possible in quantum mechanics to give an objective meaning to the state of a particle, independently of the manner in which the state is measured. Wheeler has made an interpolation over the enormous gap between the domains of cosmology and atomic physics. He conjectures that the role of the observer is crucial to the laws of physics, not only at the two extremes where it has hitherto been noticeable, but over the whole range. He conjectures that the requirement of observability will ultimately be sufficient to determine the laws completely. I think he is probably right, though it may take a little while for particle physicists and astronomers and mathematicians to fill in the details in his grand picture of the cosmos.

That is all I have to say in direct response to Wheeler. The rest of my remarks will be concerned with some speculations of my own that are partly inspired by Wheeler's ideas. I have been concerned as he has with the ultimate role of life and intelligence in the universe. I have not tried as he did to deduce the laws of physics from the existence of life and intelligence. I have only been playing with a much easier problem, trying to deduce from the known laws of physics what the ultimate scope of intelligence and observation may be. As ground rules for my game I assume that the laws of physics do not change with time and that the relevant laws of physics are already known to us. It is of course highly improbable that the presently known laws of physics are the final and unchanging truth. But we have learned from Wheeler that it is better to be too bold than too timid in extrapolating our meager knowledge from the known into the unknown.

Looking at the past history of life, we see that it takes about 10^6 years to evolve a new species, 10^7 years to evolve a genus, 10^8 years to evolve a class, 10^9 years to evolve a phylum, and less than 10^{10} years to evolve all the way from the primeval slime to *Homo sapiens*. If life continues in this fashion in the future, it is impossible to set any limit to the variety of physical forms that life may assume. What changes could occur in the next 10^{10} years to rival the changes of the past? It is conceivable that in another 10^{10} years life could evolve away from flesh and blood and become embodied in an interstellar black cloud, as described by Fred Hoyle, or in a sentient computer, as described by Karel Čapek.

Is the basis of consciousness matter or structure? This is a deep question that we do not know how to answer. Let me spell out its meaning more explicitly. My consciousness is somehow associated with a collection of organic molecules inside my head. The question is whether the existence of my consciousness depends on the actual substance of a particular set of molecules, or whether it only depends on the structure of the molecules. In other words, if I could make a copy of my brain with the same structure but using different materials, would the copy think it was me? I assume as a working hypothesis that the answer to this question is yes, that the basis of consciousness is structure. Then it is possible to talk about life and intelligence in abstract terms, independent of the details of organic chemistry and the physiological properties of flesh and blood.

How can we characterize quantitatively the structural essence of a living creature independently of its substance? One attribute of a creature that may be useful for this purpose is a number Q that I call the "complexity" of the creature. Q is the quantity of entropy that the creature produces through its metabolism during each instant of consciousness. If entropy is measured in information units, or bits, Q is a pure number without physical dimensions, expressing the amount of information that must be processed in order to keep the creature alive long enough to say "Cogito. Ergo sum." For example, a human being dissipates about 200 watts of power at a temperature of 300 degrees Kelvin, with each moment of consciousness lasting about 1 second. A human being therefore has complexity Q of the order of 10^{23}. It is probably no accident that this number is of the same order of magnitude as the number of molecules in a human brain. The hypothesis that the basis of consciousness is structure rather than substance can then be stated quantitatively as follows. If A and B are any two material embodiments of life, and a creature exists in A with a certain complexity Q, then a subjectively equivalent creature can exist in B with the same Q.

This statement of life's adaptability may sound pale and abstract, lacking in color and detail. But it has far-reaching consequences. It implies that the universe becomes more and more hospitable to life as space becomes colder and quieter. It implies that the prime requirement for life is not an abundance of energy but a good signal-to-noise ratio. It means that, at least if we are lucky enough to be living in an open universe that continues to expand forever, the material basis exists for life also to persist and grow indefinitely. A society of any given complexity can survive forever and constantly increase its store of information and experience, using a finite total quantity of energy. The actual amount of energy required for life to exist forever turns out to be less than 0.01 joule per unit of complexity.

In conclusion, I would like to put Wheeler's speculations about the role of observers in determining the laws of physics together with my own speculations about the inexhaustible potentialities of life in the universe. According to Wheeler, the laws of physics evolve progressively in such a way as to make the universe observable. According to me, the scope of the observer expands forever as time proceeds. These two speculations together give support to the conjecture that the laws of physics themselves are inexhaustible.

Forty-eight years ago, Kurt Gödel, who afterward became one of Einstein's closest friends, proved that the world of pure mathematics is inexhaustible. No finite set of axioms and rules of inference can ever encompass the whole of mathematics. Given any finite set of axioms, we can find meaningful mathematical questions that the axioms leave unanswered. This discovery of Gödel came at first as an unwelcome shock to many mathematicians. It destroyed once and for all the hope that they could solve Hilbert's "Entscheidungsproblem," the problem of deciding by a systematic procedure the truth or falsehood of any mathematical statement. After the initial shock was over, the mathematicians realized that Gödel's theorem, in denying them the possibility of a universal algorithm to settle all questions, gave them instead a guarantee that mathematics can never die. No matter how far mathematics progresses and no matter how many problems are solved, there will always be, thanks to Gödel, fresh questions to ask and fresh ideas to discover.

It is my hope that we may be able to prove the world of physics as inexhaustible as the world of mathematics. Some of our colleagues in particle physics think that they are coming close to a complete understanding of the basic laws of nature. They have indeed made wonderful and astonishing progress in the last five years. But I hope that the notion of a final and complete statement of the laws of physics will prove as illusory as Hilbert's notion of a formal decision process for all of mathematics. If it should turn out that the whole of physical reality can be described by a finite set of equations, I would be very disappointed. I would feel that the Creator had been uncharacteristically lacking in imagination. I would have to say, as Einstein once said in a similar context, "Da könnt' mir halt der liebe Gott leid tun" ("Then I would have been sorry for the dear Lord").

Fortunately, the ideas of Wheeler give us a basis for believing that the world of physics may be truly inexhaustible. Especially if the universe is open and infinite in time, the world of life and consciousness is inexhaustible too. Then there will always be new worlds to explore, new effects of observer-participation reacting back upon the laws of physics, new connections for the Einsteins and John Wheelers of the future to speculate upon.

NOTE

1. J. A. Wheeler, *Frontiers of Time* [North-Holland, Amsterdam (for the Società Italiana di Fisica, Bologna), 1979], p. 11.

Open Discussion

Following Papers by J. A. WHEELER and F. DYSON

Chairman, P. A. M. DIRAC

Dirac: I'd like to make one remark. Wheeler was speaking many times about the big bang and the big crunch. I want to emphasize that there is really no symmetry between the big bang and the big crunch. The big bang is very well established — I think pretty well everybody believes it. The big crunch is completely unestablished and very doubtful. I do not believe in it myself. I do not think Einstein believed in it either, because he gave a model, jointly with De Sitter in 1932, and this model is satisfactory so far as agreement with observation goes. This model definitely requires a big bang — it does not require a big crunch. It is an open universe although only just opened.

E. Segrè (U. of California, Berkeley): I would like to remind Freeman Dyson of a more specialized statement concerning the inexhaustibility of physics in a book by Ivanenko and Sokolow on classical field theory — namely, a quote, not that physics, but that the electron, was inexhaustible. And this was a quote from Lenin.

E. Wigner (Princeton U.): I will discuss elsewhere [Part X of this volume] as many of you know, some of the more philosophical questions, but I have a question within physics that always bothers me in connection with the black hole, and it is this: The laws of relativity are time inversion invariant. As a result, it is clear that the same amount of information is contained after the black hole collapses as before. This is not evident because just as, even though the laws of physics are also time inversion invariant, thermodynamics shows that things come to an equilibrium. But in that case we can see what it means. It means that the information that time reversal will push the system back into its original state is not relevant because we cannot reverse the velocity of every particle. Now the question is: What is a similar problem in the black hole situation? What does become uninteresting? Tell us, John.

Wheeler: One has to recognize right from the start that in all the theory of black hole physics that we do today, we feed in the assumption about the direction of time — we do not deduce it. We put it in as an initial condition. We could perfectly well talk of holes that work the other way around, just as we could perfectly well talk of heat going from a cold teacup to a hot teacup. But it would violate all our understanding that it is initial conditions that count and not final conditions. And so, this one-sidedness we feed in as a generalization of our experience — not understanding, at least I don't understand — why the world is built that way, with a one-sided initial condition.

Wigner: In thermodynamics we say, when deriving the law of entropy, that the initial conditions are irregular except for those which are macroscopically constrained. Otherwise initial conditions are fully irregular. This is the basis of the increase of entropy. What is a similar postulate for the black hole? Clearly, since the laws are reversible, the final information is just as great as the initial one.

Wheeler: It is that the final information has become useless; it's . . .

Wigner: But in what way? I mean, what part of it is useless?

Wheeler: It has gone behind the horizon where we cannot grab hold of it.

Wigner: I did not see a horizon.

Wheeler: That is to say, when everything falls in, we cannot get any signal back out again. So we have no possibility of looking at the state of the system. We cannot look at the details of what has gone inside of the horizon of the black hole: all the matter that has fallen in, all the particularities it has carried with it, all the complications that are there. Or, if we want to put it, not in terms of the frame of reference of the material falling in, which goes beyond the horizon, but if we want to put it in terms of the standpoint of the person who is looking at it all from outside and getting signals from outside, then we recognize that everything approaches asymptotically to the horizon of the black hole, to the surface of the black hole, and all these details get washed out.

Wigner: What do you mean by details? That is the question.

Wheeler: The churning about of the molecules, the atoms, the radiation, everything that has fallen in, all the complications of the star, or, if it is a galaxy . . .

Wigner: Complications — that is not contained in the equation. The equation has g_{ik} in it. What happens to the g_{ik} that makes it so uninteresting?

Wheeler: Yes. Well, all the occupation numbers of the various quantum states.

Wigner: There are no quantum states in general relativity.

Wheeler: Yes. Well, we are talking about the states of motion of these particles that are going in.

Wigner: I know the theory of — perhaps I should stop — but . . .

Wheeler: I know this famous statement — Never listen to anything Eugene Wigner says, but when he asks a question, listen very carefully.

Wigner: This makes it easy for me not to speak tomorrow.

Dirac: This discussion had better be completed in private. [Everyone in the audience says, "No."] Well then, let it go on.

Wigner: You see, I understand the increase of the entropy and the approach to equilibrium in a gas that has originally different temperatures at different locations, and one can specify what is becoming uninteresting. Namely, what becomes uninteresting is the structure of an increasing number of domains of phase space. Originally, if I take a small part of phase space, it will be either occupied or unoccupied., but the occupied regions will spread out over the whole space and if I want to see an unoccupied region, I have to go very, very carefully to tiny fractions of phase space. What is the analog of this in the collapse?

Wheeler: Everything that you have dropped in of all the matter that has fallen in is there around the horizon if you will, and . . .

Wigner: What is the horizon?

Wheeler: That is the point $r = 2M$.

Wigner: Oh! Oh, good!

Wheeler: The Schwarzschild radius.

Wheeler: Everything is coming up to the Schwarzschild radius and decreasing its distance from the Schwarzschild radius by a factor e in every 10^{-5} seconds, so it is an unbelievably compact amount of material there as viewed from a faraway observer, and the possibility of his seeing or ever hoping to know all those details has got lost, so that is where the information has disappeared.

Wigner: Is it — I'll ask one more question, then I'll shut up — is it possible to imagine the explosion of a black hole? I often think that the big bang was an explosion of a black hole. Is that nonsense?

Wheeler: Zel'dovich and Novikov proposed it some years ago — the idea that there are in addition to black holes, white holes. And that they should vomit forth matter the way a black hole sucks in matter. But then, in the meantime, Eardley has shown that such things are so unstable that in the very earliest 10^{-5} seconds of the universe, such a thing would have blown up and nothing of that kind would remain until today.

Dirac: Ne'eman — if I understand the loss of information that is generally quoted is about the baryon number that also came up in the discussion following Professor Hawking's paper. In the example that you quoted, you mentioned the fact that you could have a black hole made of baryons, one made of antibaryons, and one made of radiation, and they would look the same from the outside. On the other hand, I can imagine the scattering of such black holes as a *Gedanken* experiment, and I assume there would be surface effects that would depend on whether it is matter, antimatter, black holes colliding, or radiation black holes colliding, or ordinary matter — I don't know. I think one could find ways of distinguishing between them because of the fact that finally there is a surface connecting them with the outside world or with each other.

Wheeler: In principle, if one were to collide two black holes sufficiently soon after the act of formation — within 10^{-5} seconds — then there would be enough detail left. However, all the analysis that is conducted today, for example, by Smarr on the collison of two black holes, presupposes a time lapse great enough to have washed away all these details because, after all, one day is something like 10^{10} powers of e, and if you have washed away a detail to 10^{10} powers of e, there is not much left of it to show up in the collision.

Ne'eman (Tel-Aviv U.): To the outside observer it is also washed away?

Wheeler: Yes. It is to the outside observer. It has disappeared.

J. Ostriker (Princeton U.): A question to John Wheeler on causality. It seems to me that the universe is open, on the best astronomical evidence. Thus, things are always appearing over the horizon that simply by causality we never could have affected. There has not been time for a signal to get from us to them and for the results to get back to us. So that it seems very difficult, if the universe is open, for participatory democracy not to violate causality in the usual way. Is that true or false?

Wheeler: Far more in violation of any ordinary notion of causality and yet again in the end not in violation of it, is this delayed-choice double-slit experiment, where a choice made in the here and the now produces an irretrievable effect on the past as to whether the electron or photon went through both holes in the metal plate or through only one. The electron or photon has already gone through the metal plate at the moment when you decide, and yet despite that fact, you can have your choice as to which way you will have it. It is in this sense that quantum mechanics does strange things to what we call causality if we examine it with sufficient care.

Ostriker: I'm afraid I do not follow the relevance of that example to the question with regard to galaxies appearing at the horizon — how it would be possible for an observer to influence them.

Wheeler: I have mentioned the delayed-choice double-slit experiment but perhaps I have not spoken sufficiently vividly about it in the sense that the experiment in which we were talking about it was the one in which the interval occurred between the metal plate with the two holes in it and the place with the venetian blinds where we make our choice — we'll say 1 foot or time of 1 nanosecond — but there is nothing in principle to prevent it being a billion light years in the sense that we in the here and the now, by observing a photon of the primordial cosmic fireball radiation, have an irretrievable consequence for what we have the right to say about that photon. In that sense, we are operating in a reverse direction of the sense of time and it is in that sense that anything having to do with things swimming up over the horizon also comes into this category of events.

G. Field (Harvard-Smithsonian): This question is a very mundane one compared to the discussions that we have been having. In a paper by Leonard Parker some years ago, he posited a peculiar quantum state of mesons that would have a negative pressure about 10^{-23} seconds before the final collapse of the universe. He was able to show that that negative pressure, according to Einstein's equations, with a classical metric, would in fact result in a bounce. Since that time, I have wondered whether the final collapse is truly inevitable, or is it an artifact of the particular physics that we are assuming?

Dirac: You are talking about the big crunch — are you?

Field: Sorry, yes. The big crunch.

Wheeler: It is certainly true that the conclusions that people have drawn about the inevitability of the big bang and the big crunch in a closed universe are most completely covered in a paper that is shortly to be submitted by Marsden and Tipler. It is certainly true that those conclusions go back for their foundation to an assumption about the nonnegative, definite character of the stress-energy tensor. The minute one accepts the stress-energy tensor that can be negative, then everything is changed.

D. Laughton (Princeton U.): I have a vaguely theological question to ask John Wheeler. You and your friends have experienced many things including a cloudy day, at some time before you decide to play your game. Out of the nothingness that is in the room there is actually a potentiality for anything that you have experienced or can imagine. It just so happens that out of all those potentialities you manage to retrieve, through your own active participation, that one concept of a cloud. I am wondering about another way of phrasing your question about what is on the horizon. Is it possible to rephrase it as: What is it that we all have in common, before we have our first moments of consciousness, that allows us to actively participate in the creation of this universe around us?

Wheeler: I am afraid I am no better off at answering questions like that than I am at answering the question that somebody raised in a discussion of what is the status of mathematics before the universe comes into being.

Ne'eman: Considering the fact that we are undoubtedly influencing the past in the examples you are quoting, I wonder about the reluctance to adopt models of the universe in which there is, let us say, a circular time variable and so forth, because one sticks to causality. If one is abandoning causality anyhow in the sense that you just explained, why does one stick to it so much, let us say, against a Gödel universe, and so forth?

Wheeler: The question — why should one stick to causality in the semiclassical large scale when at the small scale these quantum fluctuations come about that take away any real use of the concept of time in our ordinary sense, which deny the concepts of before or after us . . .

Ne'eman: At 10 billion years.

Wheeler: And even at 10 billion years and this participatory character of things, we just have to learn to struggle our way through these problems. I do not have any magic answer.

Dirac: We could have another short question if anyone wants to propose one.

L. Cooper (Brown U.): I had not meant to bring this up, but I think that if you take the Everett–Wheeler interpretation of the quantum theory seriously, you do not really have to violate causality in your shutter experiment. I admit I have not had a chance to analyze this particular experiment completely, but in other experiments of the same type, with your own interpretation, I do not believe you have to violate causality. Do you have any comment on that?

Wheeler: The interpretation of quantum mechanics that Hugh Everett gave us some years ago and that Leon Cooper himself elucidated considerably more is one in which branches, if I can put it that way, in the act of observation are simultaneously present. I confess that I have reluctantly had to give up my support of that point of view in the end — much as I advocated it in the beginning — because I am afraid it creates too great a load of metaphysical baggage to carry along. But they say nobody knows sin like a sinner.

24. ISSUES IN COSMOLOGY

Dennis W. Sciama

The discovery of the 3 °K microwave background has completely transformed our understanding of cosmology. With its help we can now discern a number of fundamental issues that, although they remain unresolved, are probably closely related:

1. Why is the universe isotropic?
2. Why is the universe homogeneous?
3. What is the origin of the 3 °K background?
4. Why is the universe baryon-asymmetric?
5. Was the big bang a singularity?

I shall discuss these issues in turn. In keeping with the occasion, my discussion will emphasize the fundamentals and is not intended to provide a scholarly account of the fine points, reservations, and so on, which would be appropriate in a more detailed survey. I hope, of course, not to be misleading, and I have provided references to the more technical discussions.

1. **Why Is the Universe Isotropic?**

It has been evident since the early days of observational cosmology that both the distribution of galaxies and the Hubble law of expansion are roughly isotropic. This provided empirical justification for the procedure initiated by Einstein and Friedmann of constructing exactly isotropic models of the universe conforming to general relativity.[1] This procedure was codified in the 1930s by Milne, Robertson, and Walker, and led to the well-known Robertson–Walker models (also often named after Friedmann).[2] Although these models have dominated the subject ever since, it must be recognized that the early observational evidence for isotropy was not very precise. Moreover anisotropic models were extensively investigated since the 1960s.[3] Nevertheless, as we shall see, the modern evidence shows that the isotropy *is* very precise, to such an extent that it has become urgent to understand why it should be so.

An important attempt to quantify the isotropy of the universe was made by Kristian and Sachs before this modern evidence was clearly established.[4] They made a kinematic analysis of the

Harry Woolf (ed.), Some Strangeness in the Proportion: A Centennial Symposium to Celebrate the Achievements of Albert Einstein
ISBN 0-201-09924-1

different types of anisotropy that can exist, and this analysis is the one that is used today. They distinguished the following possibilities:

1.1 *Peculiar Velocity of a Galaxy*

From the distribution of galaxies and their motion one can, by a suitable averaging process, define a preferred rest-frame at each point. If a galaxy is moving with respect to its local rest-frame, then observations made from such a galaxy would reveal an anisotropic universe.

1.2 *Shear*

If one expands the velocity field in the neighborhood of a point in a Taylor series, the first-order terms can be classified, as in fluid mechanics, into expansion, shear, and vorticity. In the cosmological case the expansion would correspond to the Hubble effect, while the shear would represent a change of shape without any accompanying rotation. The expansion rate for a volume V is measured by

$$\theta = \frac{1}{V} \frac{dV}{dt} ,$$

and its present value is observed to be

$$\begin{aligned} \theta_0 &= 3\, H_0 \\ &\sim (5 \times 10^{-18}\ \mathrm{sec}^{-1}) \quad (H_0/50\ \mathrm{km/sec/Mpc}) , \end{aligned}$$

where H_0 is probably within a factor 2 of 50 km/sec/Mpc.

The presence of shear σ in the galaxy flow would manifest itself both in the form of transverse velocities and in a direction dependence of the radial expansion rate. If in any one direction φ we have the Hubble law

$$\mathrm{v} = H_0 \gamma ,$$

then qualitatively

$$\sigma \sim \frac{d}{d\varphi} H_0 .$$

As the universe expands the shear will change, and in a large variety of cases one has in the absence of appreciable dissipation the simple relation

$$\sigma \propto V^{-1} . \tag{24.1}$$

This is not always true, and in particular it is possible for the shear to increase with time. A good compact discussion of the various possibilities is given by Olson.[5] For brevity we shall adopt (24.1), with the warning that there are other cases to be considered.

The main assumption underlying (24.1) is that at any one cosmic epoch the universe is homogeneous. The assumption will be further discussed in Sec. 2. Here I note simply that we are thereby dealing with *large-scale* shear. One consequence of this is that the shear could be important not only kinematically but also dynamically in that it would influence the rate of expansion of the universe via Einstein's field equations. We shall be mainly concerned with this effect at an early cosmic epoch ($t \sim 100$ sec) during the nucleosynthesis of some of the light elements. At that epoch it is a good approximation to write

$$\theta^2 = 3\,(8\pi G\rho + \sigma^2)\,,$$

where ρ is the total energy density. Thus the dynamical effect of the shear is to contribute an "anisotropy energy density" that always acts to increase θ, that is, to speed up the expansion.[6] The relative importance of σ^2 and ρ will itself be time-dependent since we have

$$\rho \propto \frac{1}{V} \quad \text{(pressure-free matter, } p = 0\text{)}$$

$$\rho \propto \frac{1}{V^{4/3}} \quad \text{(radiation-dominated matter, } p = \frac{1}{3}\rho\text{)}$$

$$\rho \propto \frac{1}{V^2} \quad \text{(ultra-stiff matter, } p = \rho\text{)}$$

while for the shear energy-density

$$\sigma^2 \propto \frac{1}{V^2}\,.$$

Thus if any large-scale shear at all is present it would dominate dynamically at sufficiently early times except in the ultra-stiff case.

1.3 *Vorticity*

The possibility that the universe might possess large-scale vorticity is of particular interest in relation to the question of absolute rotation, which so concerned Einstein when he was developing general relativity. The existence of such vorticity would correspond to a rotation of the universe with respect to the "compass of inertia," a possibility that would, of course, have to be excluded if one adopted Mach's principle. The existence of an exact cosmological solution of Einstein's field equations representing a rotating universe was pointed out in a famous paper by Gödel in 1949.[7] This possibility disturbed Einstein, not so much for its anti-Machian character, since by then Einstein had lost his attachment to Mach's principle, but because the Gödel solution contained closed timelike lines, implying that one could travel into one's past. In the present paper it will be regarded in the first place as an empirical question what is the value of the large-scale vorticity.

The Gödel model was actually nonexpanding, but one can also have expanding models with large-scale vorticity in general relativity. The change of vorticity ω with time would then be determined by the law of conservation of angular momentum, with the result (for small ω and no dissipation):

$$\omega \propto \frac{1}{V^{2/3}} (p = 0)$$

$$\omega \propto \frac{1}{V^{1/3}} (p = \frac{1}{3}\rho)$$

$$\omega \propto V^{1/3} (p = \rho)$$

(for the latter case, see Barrow[8]). The vorticity also has a dynamical influence that can be represented by introducing a *negative* vorticity-energy density $-\omega^2$. This would *slow down* the expansion of the universe.

Let us now consider the present observational evidence concerning these various forms of anisotropy.

1.4 *Observational Evidence*

1.4.1 *Peculiar velocity of a galaxy.* The observed scatter in the red shift–distance relation (Fig. 24.1) enabled Hubble and later Sandage to place a limit of ~100 km/sec on the peculiar velocity of an individual galaxy. By contrast, galaxies in rich clusters can have peculiar velocities 10 times greater. Attempts have been made to detect the peculiar motion of our own galaxy from the resulting anisotropy in the 3 °K and X-ray backgrounds. The early attempts were not very convincing and are best interpreted as providing an upper limit on our motion of a few hundred kilometers per second. Two recent measurements of the 3 °K background[9] agree in giving for our motion ~300 km/sec towards the constellation Leo. When allowance is made for the motion of the sun around the center of our galaxy, and for the motion of the galaxy relative to the other galaxies in the local group, one obtains a net motion for the local group of ~600 km/sec in the direction of galactic longitude 261 ° and galactic latitude 33 °. Some workers consider this velocity to be on the high side and difficult to explain. It needs confirming by further observations (Smoot et al. observed only two-thirds of one hemisphere).* To avoid contamination of the data from radio emission by the galaxy one has to observe at short wavelengths $\lesssim$ one cm, which would be best done from a satellite. Plans exist for such measurements to be performed in the next few years.

1.4.2 *Shear.* The red shift–distance relation of Fig. 24.1 shows that there is no gross anisotropy in the value of the Hubble constant. As Kristian and Sachs pointed out, one obtains thereby only a weak limit on the shear, namely

$$\frac{1}{\sigma} \gtrsim 3\tau ,$$

where $\tau = 1/H_0 \sim 10^{10}$ years. To obtain a stronger limit we can appeal to the observed isotropy of the 3 °K background. This radiation density decreases as the universe expands, and if the expansion rate were anisotropic the radiation would continue to have a blackbody spectrum in each direction, but the temperature itself would vary with direction.

*Note added in proof. Further confirming of observations have now been carried out (E. S. Cheng, et al., Astrophys. J. Lett. **232**, L139 (1979), G. F. Smoot and P. M. Lubin, Astrophys. J. Lett. **234**, L83 (1979), R. Fabbri, et al., Phys. Rev. Lett. **44**, 1563 (1980). Their interpretation has been discussed by D. W. Sciama, Mem. Acad. Lincei (1981).

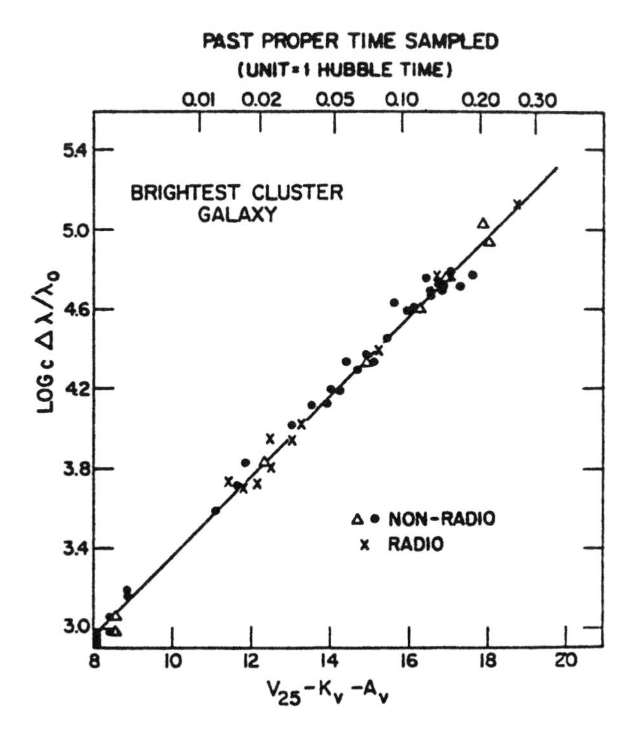

Fig. 24.1 Hubble diagram for first-ranked cluster galaxies in forty-one clusters. Abscissa is the photoelectric V magnitude corrected for aperture effect, K-dimming, and galactic absorption. Dots represent galaxies that are radio-quiet to 9 flux units at 178 MHz; crosses, radio sources from the 3CR catalog that are first-ranked cluster members; triangles, Baum's data transformed to the V-magnitude system of the dots and crosses. Ordinate is red shift, corrected for galactic rotation. *Source:* A. Sandage, Astrophys. J. (Lett.), **152,** L152 (1968).

A number of observational studies have been made of the angular distribution of the microwave background. If one subtracts out the anisotropy attributed to the peculiar motion of the earth, one finds no residual anisotropy with a precision of about 1 part in 3,000 on a variety of angular scales.[10] Earlier work had placed a limit three times weaker, and this limit was used by Collins and Hawking to derive a corresponding limit on the large-scale shear (and vorticity) of the universe.[11] Their discussion was complicated by the necessity of distinguishing between different anisotropic models of the universe. These models lead to different predictions for the angular distribution of the temperature, and also for the time-dependence of the shear. To permit a surveyable discussion they made the common assumption that the universe is exactly homogeneous (more precisely, admits a family of homogeneous spacelike 3-surfaces), but is anisotropic. The different possible cases were classified by Bianchi in 1890 into nine types, of which five admit Robertson–Walker models as limiting cases. (For a modern discussion see Ellis, MacCallum, and Ryan and Shepley.[12]) In addition, the limit on the present shear of the universe depends on the length of time during which the microwave radiation has been propagating without being scattered. The surface of "last scattering" has a red shift that varies from 7 if the universe contains the densest permitted ionized intergalactic gas to 1,000 in the low-density case. This latter value is determined by the consideration that at that epoch the temperature of the blackbody background was great enough to ionize all the (pregalactic) material in the universe.

The various cases that arise are discussed in the paper by Collins and Hawking. Here we simply note that the weakest limit for the present value of the large-scale shear is given by

$$\frac{1}{\sigma} > 700\tau ,$$

while the strongest limit is

$$\frac{1}{\sigma} > 1.4 \times 10^{7} \tau .$$

These limits represent a considerable improvement on the one obtained from the observed motions of the galaxies.

A still stronger limit for many of the cases can be derived if we accept that most of the helium we observe in the universe was formed by thermonuclear processes occurring 100 seconds after the hot big bang.[13] The reason is that the amount of helium formed depends sensitively on the time scale of the expansion at that epoch, and, as we have seen, this time scale would be influenced by the anisotropy energy density. This argument was developed by Barrow whose calculations were simplified by Olson.[14] If one restricts the discussion to Bianchi types which admit Robertson–Walker limiting cases, all the possibilities can be discussed together. Assuming that the helium/hydrogen ratio by mass lies between 0.25 and 0.33[15] and that the lepton number is small, one obtains a strong upper limit for the present value of the shear, namely

$$\frac{1}{\sigma} > 2 \times 10^{11} \tau .$$

This limit is four orders of magnitude more stringent than the best limit derivable from the isotropy of the 3 °K background.

So far in this discussion we have neglected the possibility that the shear might be reduced by the action of dissipative processes. The importance of this possibility was pointed out by Misner, in relation both to accounting for the low shear observed and to linking the heat thereby produced to the heat present in the 3 °K background.[16] If the dissipation rate is large, then one can again limit the shear by requiring that the heat produced should not exceed that observed in the 3 °K background. This argument is particularly powerful if the dissipation rate is large at very early times, since the shear energy density increases into the past faster than radiant energy density by the two-thirds power of the volume factor $V(t)$. Thus the earlier a given amount of shear is dissipated the more heat it produces.[17]

We are therefore impaled on what might be called the Barrow–Matzner fork. If the dissipation rate is low the early shear must be small so that the observed abundance of helium is obtained. If the dissipation rate is large, the early shear must be small to ensure that too much heat is not produced. If, for example, efficient dissipation occurred at or before a time $\sim$1 second, then at present

$$\frac{1}{\sigma} \gtrsim 10^{13}\ \tau\ ,$$

which is a 50-times-stronger limit than the one derived directly from the helium abundance.

This kind of argument also limits the initial shear of the universe. If, say, there was strong dissipation due to quantum processes at 10^{-43} seconds, the initial shear would have had to be so small that even without dissipation we would call it essentially zero (the undissipated shear energy density today would be less than 10^{-39} of the present matter energy density[18]).

From this discussion it can be concluded that, in all probability, the shear of the universe has always been very small.

1.4.3 *Vorticity.* If the universe as a whole were rotating relative to our local inertial frame we would expect to see a *transverse* Doppler shift in the spectra of distant galaxies, except for those lying in the direction of the rotation axis. Since the transverse effect is of order v^2/c^2, this method of detecting rotation is not very sensitive. The absence of any clear effect in the data of Fig. 24.1 tells us only that $l \lesssim c$ at a Hubble radius.[19] The rotation period P associated with such motion then satisfies only the weak inequality

$$P = \frac{2\pi}{\omega} \gtrsim \tau\ .$$

Nevertheless this is a more stringent limit on ω than one can deduce from the consideration that the flattening of our galaxy due to its rotation is compatible with observations of the proper motions of slowly rotating outlying stars and of extragalactic objects. This tells us only that $P \gtrsim 10^9$ years, which is a 10 times weaker limit than the Kristian–Sachs one. However, this limit is still stronger than one could derive using the best available gyroscopes to determine the local nonrotating frame.

We again enter a new regime if we use the isotropy of the 3 °K background. If the universe has large-scale vorticity the last-scattering surface of the 3 °K background would be rotating around us, giving rise to a transverse Doppler effect whose magnitude would depend on the angle between the direction of observation and the rotation axis. This question has been analyzed by Hawking and by Collins and Hawking.[20] The limits obtained again depend on the Bianchi type and the red shift of the last-scattering surface. The extreme cases give for the present value of ω

$$\frac{2\pi}{\omega} > 1.5 \times 10^3\ \tau\ ,$$

$$\frac{2\pi}{\omega} > 6 \times 10^{11}\ \tau\ .$$

These limits are very strong, but one must remember that during the matter-dominated phase of the expansion the ratio of rotation period to expansion time scale increases with the age of the universe. To some extent, then, the large value of this ratio at the present epoch simply reflects the advanced age the universe has today. If we want to say that the universe is rotating slowly, or not at all, we must consider what limits can be placed on the total number of rotation periods that may have occurred since the big bang. This question was also considered by Collins and Hawking. Their strongest limit arises for a closed universe (Bianchi type IX), for which they found that the universe could have rotated through only 2×10^{-4} seconds of arc since the big bang. We do not get a stronger limit by considering helium formation because in the radiation-dominated phase the vorticity energy density increases into the past slower than the radiation energy density.

We now return to our original question, why is the universe isotropic? Five possible answers have so far been proposed:

(α) Initial conditions
(β) Dissipative processes
(γ) The anthropic principle
(δ) Ultra-stiff early universe
(ε) Mach's principle

(α) The usual objection to simply attributing the isotropy of the universe to the initial conditions at the big bang is the very special nature of the conditions assumed. It is not clear to me what the rules of the game are here. I shall therefore confine my remarks to a new point recently made by Penrose.[21] He takes his cue from the discovery that a gravitational field can easily possess an enormous amount of entropy. For example, a solar mass black hole has an entropy that is 10^{18} times greater than the entropy of the actual sun.[22] Penrose also attributes an entropy to a general gravitational field. Since the gravitational field has a negative specific heat,[23] its entropy is greater when it is irregular than when it is smooth. Penrose proceeds to relate these ideas to the second law of thermodynamics. He pictures the universe as beginning in a state of low (or zero) entropy and therefore in a state in which the gravitational field is highly regular. This initially regular state can be characterized by the vanishing of the Weyl tensor, which would correspond to the universe being initially exactly Robertson–Walker. Considerable entropy then develops in the gravitational field as a result of its attractive character, and some of this entropy is converted into more familiar forms, such as heat radiation. Since the final state of the universe would not be gravitationally regular, one would be dealing here with a time-asymmetrical boundary condition on the initial singularity. Penrose relates this condition to the postulated existence of a local microscopic law that is noninvariant under both T and CPT. This is an interesting suggestion, but the relation between the noninvariant law and the vanishing of the Weyl tensor needs futher elucidation. As we shall see, these ideas are also related to the problem of baryon-asymmetry of the universe.

(β) Misner[24] suggested that the universe was initially chaotic, and that dissipative processes later smoothed it out, making the 3 °K background in the process. He was particularly impressed with the difficulty, which is still unresolved, produced by the existence of particle horizons in the standard Robertson–Walker models.[25] The existence of these horizons means that at any time a particle can only have been in causal contact since the big bang with the finite number of other particles lying within its particle horizon. In particular, points on the last-scattering surface of the 3 °K background would be out of causal contact with one another if they subtend at us quite a modest angle ($\sim 30°$ in some models). Yet their temperatures are observed to be the same to at least 1 part in 3,000.

The motivation for Misner's chaotic cosmology program lay partly in the hope that in models that were initially highly anisotropic, the particle horizon would be eliminated. Thus were born the famous mixmaster universes in which complicated oscillations occurred so that, first in one direction, and then in another, causal communication was possible, so that the isotropy of the universe, and in particular of the 3 °K background, could, it was hoped, be established by dissipative processes. Unfortunately, it later emerged that such efficient mixmaster models represented a small set in the space of all initial conditions,[26] and hence were just as special as models with initial isotropy. Moreover the Barrow–Matzner fork suggests that a dissipative model leading precisely to the heat observed in the 3 °K background would have to be regarded as very contrived.

(γ) The anthropic principle[27] was invoked in this context by Collins and Hawking.[28] They argued that galaxies could form only in an isotropic universe and, therefore, that any human being would have to find that his universe is isotropic. To quote Collins and Hawking, "The answer to the question, why is the universe isotropic appears to be, because we are here."

This prompts the further question: Why are we here? If our existence requires a carefully contrived universe, why did the universe have this careful contrivance? An extreme form of the anthropic principle answers this question by invoking the existence of all conceivable logically self-consistent universes. Intelligent life would then develop in that subset of these universes that would permit such a development. Intelligent life in its turn would find present in its universe all those features that are needed for its development. The existence of these features would then not require further explanation, although further explanations might exist in some cases.

I have considerable sympathy for this extreme form of the anthropic principle. However, its application to the isotropy of the universe must be regarded as uncertain until it can be shown in detail that the formation of galaxies does depend critically on this isotropy. We understand so little about galaxy formation[29] that this must be regarded as an open question.

(δ) It has recently been pointed out by Barrow[30] that if the equation of state for baryons in the early universe is of the ultra-stiff kind, characterized by

$$p = \rho$$

one might be able to understand the low anisotropy of the early universe. The reason springs from the fact that for such an extreme equation of state the time-dependence of any asymmetry is quite different from that holding for more conventional equations of state. For example, the early shear energy density would change with time in the same way as the radiation energy density rather than dominating it. According to the present state of elementary particle physics it seems unlikely that the early universe was ultra-stiff, so that we shall not discuss this proposal any further here.

(ε) It was Einstein himself who named Mach's principle the proposition that the inertial forces acting on accelerated bodies have their origin in distant masses. As is well known, he proposed that this influence of the distant masses is gravitational in origin, and this idea played a key part in the development of general relativity. However, when that theory was established he came to feel that the g_{ij} field existed in its own right, and that the question of its sources was incidental. Thus he writes in his autobiographical notes, "So, if one regards as possible, gravitational fields of arbitrary extension which are not initially restricted by spatial limitations, the concept, 'acceleration relative to space,' then loses every meaning and with it the principle of inertia together with the entire problem of Mach."[31]

This proposal to give reality to the g_{ij} field independently of its sources is usually regarded as having been strengthened by the subsequent blurring introduced by quantum mechanics between the concepts of field and of matter. I recognize the force of this objection to Mach's principle, but I remain unhappy with the point of view that confers physical reality on, for example, Minkowski space-time, despite the fact that, according to the field equations, there are no sources present to determine which reference frames are the preferred inertial ones.

I had the privilege of discussing this question with Einstein only a week before he died, in April 1955. It was at the end of my year's visit to the Institute for Advanced Study. To help ease my tension I started the discussion with a prepared sentence, "Professor Einstein, I would like to talk about Mach's principle, and I have come to defend your former self against your later self." Fortunately he laughed uproariously at this rather feeble beginning, perhaps to put me at my ease, and said, "That is gut, ja." Our subsequent discussion was rather inconclusive and soon wandered onto other topics such as steady-state cosmology, unified field theory, and the interpretation of quantum mechanics.

According to the point of view that I and others have advocated for many years, one should regard Mach's principle as a rule for choosing boundary conditions for those solutions of the field equations that one would accept as properly Machian. Apart from the conceptual problems involved in such a program, there are severe technical problems, arising in particular from the nonlinearity of the field equations. I believe that these problems have now been largely overcome. This is not the place to go into these questions, and I would refer the interested reader to a paper by Raine.[32]

The relevance of this discussion to the problem of the isotropy of the universe is that one would expect any Machian model to possess zero large-scale vorticity. Indeed from a Machian point of view one could regard the stringent observational limits that we can now place on the vorticity of the universe as a triumphant empirical verification of Mach's principle. In fact Raine goes further. With his "natural" choice of boundary conditions he finds that all spatially homogeneous models with a perfect fluid equation of state have to be isotropic, that is, would have vanishing shear as well as vanishing vorticity. Thus the stringent observed limits on large-scale shear also find a natural explanation if one accepts Mach's principle as a fundamental restriction on the universe arising from the absence of absolute space.

2. Why Is the Universe Homogeneous?

So far in this paper I have assumed that on a large-scale the universe is homogeneous. Let us now consider whether this is in fact the case, and if so what is the reason for it. There clearly exist substantial inhomogeneities (galaxies, clusters, and probably superclusters) on a scale up to, say, 50 megaparsecs, but since the "radius" of the universe is about 3,000 megaparsecs, we may regard these as localized disturbances. There is no obvious gradient in the distribution of galaxies on a larger scale, but it is not possible to quantify this in detail. However, we can argue from the observed isotropy, that if we are in a typical position in the universe, that is, if the universe is isotropic at every point, then it must also be homogeneous (Schur's theorem). Again for real precision we must turn to observations of the 3 °K background. If our line of sight passes through a density fluctuation we would expect to find a change in the temperature of the microwave background.[33] This change arises from a number of causes that in general cannot be clearly separated from one another. However, if the effects are small, we can split them approximately into the following:

1. A Doppler shift, associated with our own pecular velocity, resulting from the gravitational action of the fluctuation
2. A gravitational shift arising from the difference between the time-dependence of the gravitational potential of the fluctuation and that of the mean universe
3. A gravitational time-delay associated with propagation through the fluctuation, implying that we would be observing the last-scattering surface of the radiation at a different time and so at a different temperature
4. A Doppler shift associated with the peculiar velocity of the electrons in the last-scattering surface resulting from the gravitational action of the fluctuation.

The detailed calculation of all these effects is quite complicated, and in particular allowance must be made for the smearing produced by the finite depth of the last-scattering "surface." Since no positive effect has been observed in the background I shall not describe these calculations here but simply give an indication of the implied upper limits on possible large-scale density inhomogeneities. The interested reader should consult the papers of Sachs and Wolfe, Rees and Sciama, Dyer, and Anile and Motta, where references are given to other work.[34]

Let us consider a possible concentration with density contrast 3 to 1 at a red shift of 1.5 having a mass of $3 \times 10^{19} M_{\odot}$. This concentration can be modeled in a zero-pressure Robertson–Walker universe by removing a co-moving sphere of dust and replacing it with a smaller sphere of equal mass. This replacement would not affect the dynamics of the universe outside the sphere. In an Einstein–de Sitter universe (that is, one that would just expand forever) the resulting hole would subtend an angle of about 20° at the observer and its present diameter would be about 750 megaparsecs. According to Dyer, the center of such a hole would be about 0.2 percent colder than the surrounding universe away from the hole, while the radiation from the limb would be about 0.1 percent hotter. In a low-density universe (with deceleration parameter $\sim$0.05) the central temperature would be about 0.1 percent hotter than in the surrounding universe.

Temperature fluctuations of this order of magnitude are not observed. We can therefore conclude that density fluctuations with the proposed parameters do not exist. On the other hand, the present observations would not rule out 10 percent density fluctuations on the same length scale. This gives us a general order of magnitude of the currently permitted large-scale irregularities in the distribution of matter in the universe.

The question now arises: Why is the universe so homogeneous? It must be admitted that at present this is a very obscure problem. In a general way one can see that Mach's principle would require there to be enough matter everywhere to ensure that the universe is not asymptotically flat, but the principle does not seem to enforce large-scale homogeneity. One might appeal to dissipation, but most investigations of this possibility suggest that dissipation is more efficient on small length scales than on large ones. In addition, special problems arise for disturbances whose length-scale exceeds the particle horizon. For such disturbances the density contrast, in the absence of dissipation, is given by Sachs and Wolfe:[35]

$$\frac{\delta\rho}{\rho} = A(x)t^{\frac{\gamma-2}{2}} + B(x)t^{\frac{2(3\gamma-2)}{2}}$$

where γ is determined by the equation of state:

$$p = (\gamma - 1)\rho .$$

For $\gamma < 2$, the A perturbation decreases with time and the B perturbation increases with time. The increasing perturbation has been much discussed in connection with the formation of galaxies,[36] while the decreasing perturbation, which diverges at the origin, is normally excluded ad hoc. Again the ultra-stiff case $\gamma = 2$ would be of interest here, since for that case the A perturbation would not diverge.[37] In addition no shocks could occur since the velocity of sound would equal the velocity of light. Accordingly no entropy would be produced by nonlinear hydrodynamical processes,[38] and so the constraint imposed by the observed entropy in the 3 °K background could perhaps be satisfied.

3. What Is the Origin of the 3 °K Background?

A convenient way of characterizing the intensity of the background is in terms of its entropy per baryon in the universe, or equivalently by the photon-baryon ratio. This ratio is independent of time during any epoch in which dissipation is unimportant and baryon number is conserved. Curiously, the present photon density is known more accurately than the mean baryon density, which is uncertain to a factor $\sim$30. If galaxies make the main contribution to the baryon density, the photon-baryon ratio is about 10^9, although given the uncertainty in galaxy masses, this number could be as low as 10^8.

The earliest time this heat could have been produced is in the big bang itself, unless as will be discussed later, quantum effects or new physical laws enable one to speak of an earlier phase of the universe. The latest time is determined by the requirement that thermalization of the radiation be guaranteed. The thermalization time today would be much longer than the Hubble time, and if we appeal to free-free and free-bound interactions of photons with ionized gas, then the last epoch at which thermalization would be more rapid than the expansion time scale is about 300 years after the big bang. However, even this estimate is not secure, since a later generation of pregalactic stars could have produced both the necessary heat and enough dust particles and molecules to thermalize it, at a time of about 3 million years after the big bang.[39]

However, such a scheme would throw into jeopardy our understanding of the formation of helium and deuterium 100 seconds after the big bang.[40] In particular, the agreement with observation of the abundance of helium predicted by the standard model, in which all the heat observed today was present at 100 seconds, is rather remarkable. It would be interesting to investigate a dual model, in which enough heat was present then to account for the helium, and yet in which a substantial amount of heat was made later by pregalactic stars. Such a model might seem artificial, but it must be remembered that there are other reasons for invoking such a generation of stars (having mainly to do with the distribution of heavy elements), and if they did form, then the heat they produced must in any case be allowed for.

The origin of the early heat could be ascribed either to a pre-big bang phase of the universe or to initial conditions at the big bang itself. More fruitful at the present time is to appeal to dissipative processes occurring after the big bang. Doubt has already been shed on the proposal that large-scale shear or vorticity has been dissipated. The idea that is attracting considerable current interest is that quantum effects are involved, either through the Hawking radiation produced by primordial mini-black holes or by pair production arising from the rapidly changing gravitational field in the very early stages of the universe.[41]

The radiation produced by black holes is described in this volume by Hawking himself. The time scale for a black hole to lose an appreciable fraction of its mass by the radiation of one particle species of zero rest mass $\sim(M/10^{15}$ grams$)^3$ 10^{10} years. Thus mini-black holes of mass $\ll 10^{15}$ grams must be invoked. How many of these were formed in very early stages of the universe is unknown, but preliminary discussions of this problem have been given by Carr and by Barrow and Carr.[42] This problem is under investigation at the present time.[43]

Cosmological particle production is also under investigation. One of the main problems here is that naive estimates can lead to an infinite rate of particle production. To avoid this, one must include in the calculation the back reaction of the created particles on the metric of the space-time. This is a difficult problem that has been studied in a preliminary way by Hartle and his colleagues.[44]

4. Why Is the Universe Baryon-Asymmetric?

We know from direct observation that not more than one particle in $\sim 10^7$ in our galaxy can be made of antimatter. We are less sure about the constitution of other galaxies. It is sometimes argued that half the universe could be made of antimatter if clusters of galaxies are always of one sign.[45] Others argue that probably nearly all the universe is made of matter.[46] On the first view one could suppose that the hot big bang was symmetric between matter and antimatter, and that subsequently an efficient separation process occurred on a scale of clusters of galaxies. The observed 10^8 photons per baryon would then be a measure of the amount of baryon–antibaryon annihilation that occurred before the separation process became effective. Attractive as this view is, it has not gained many adherents, chiefly because an efficient separation mechanism has not yet been discovered.[47]

If we adopt the alternative point of view, we need to explain why there are so few antibaryons in the universe. A variety of suggestions has been put forward, although there is as yet no detailed theory. Most of them aim to explain also the 10^8 photons per baryon by invoking a process that makes $1 + 10^{-8}$ baryons for every antibaryon under circumstances in which the antibaryons would later be completely annihilated. In this way our present problem becomes related to the previous one of the origin of the 3 °K background.

Any process that makes more baryons than antibaryons must involve the noninvariance of some or all of baryon number *B, C, P, T,* and their combinations.[48] This consideration is usually combined with some nonequilibrium process associated with the rapid expansion of the early universe and involving collisions or decays of particles, pair-production, or Hawking radiation by primordial mini-black holes. In view of the speculative nature of the existing proposals I shall not give a detailed discussion here, and will confine myself to the three options described by Ellis, Gaillard, and Nanopoulos.[49] These options involve *B* violating effects that arise in grand unified theories of elementary particles or in gravitational interactions (for example via virtual black holes and Hawking radiation). There appear to be three possible regimes in which these effects are important. In the very early universe the *B* violating interaction rates may have been less than the expansion rate, so that thermal equilibrium would not have prevailed and the net baryon number could have been nonzero. This may have been followed by a phase in which thermal equilibrium was achieved and the baryon asymmetry partially or completely wiped out. Finally the universe would have come out of equilibrium again and a new baryon asymmetry be established. The three options involve differing possibilities for these three regimes.

(α) The equlibrium regime may not have completely destroyed the baryon asymmetry of the first regime, which would partially survive to the present day.[50]

(β) Essentially complete thermal equilibrium may have been established in the second regime, while the present baryon asymmetry was established in the third regime.[51]

(γ) The baryon asymmetry may have been negligible in the first regime, and the observed asymmetry produced in the third regime. This option differs from the previous one in that it does not require the B violating effects ever to have come into thermal equilibrium. Ellis et al. prefer this option and develop a detailed model of it. They obtain more than 10^{11} photons per baryon for the present universe, but in view of the uncertainties in their model they are not too worried by this discrepancy with observation.

5. Was the Big Bang a Singularity?

If the universe were exactly Robertson–Walker, then the big bang would represent a singularity of infinite density and curvature involving the whole universe. In the old days this conclusion was often avoided by the introduction of a cosmical λ term in Einstein's field equations. Einstein himself later called this the biggest blunder of his life, but some workers still prefer to retain it, regarding it as an empirical question what its value is. However, if we accept the argument from the thermality of the 3 °K background, that the universe was once many powers of 10 denser than it is today, then no empirically allowed λ term could influence significantly the dynamics of the expansion at that and earlier times, and the existence of a singularity would be guaranteed in Robertson–Walker models.

The next suggestion made to avoid a singularity was that the presence of irregularities in the distribution of matter and radiation in the universe would suffice to "defocus" the motions of expansion (or collapse if we use the reversed sense of time). As is now well known, the celebrated singularity theorem of Penrose for a collapsing star, adapted by Hawking and by Geroch to the expanding universe, shows that self-gravitation is so strong in Einstein's theory, that even in the presence of irregularities the big bang was singular.[52]

Of course such a powerful result does depend on the validity of certain assumptions, and there exist a variety of singularity theorems based on different assumptions. In some versions the assumptions include some that, though plausible, are difficult or impossible to verify in practice (for instance that the universe should admit a well-defined global Cauchy surface, that is, a spacelike surface on which initial-value data can be defined). The most economical theorem in this respect uses the 3 °K background in two different ways.[53]

Theorem: Space-time is not singularity-free if the following conditions hold:

a. Einstein's field equations.
b. A positive energy condition ($T_{ab}\, W^a W^b \geq \frac{1}{2} T$, where T_{ab} is the energy-momentum tensor and W^a any timelike unit vector).
c. Strong causality (every neighborhood of a point contains a neighborhood of that point that no nonspacelike curve intersects more than once).
d. There exists a point P such that all past-directed timelike geodesics through P start converging again within a compact region in the past of P.

The crucial conditions are (b) and (d). The energy condition (b) guarantees that gravity has always been attractive; (d) is a precise statement of the requirement that there was enough gravitation to produce a singularity. It turns out that we can use observations of the 3 °K background to show that in the actual universe (d) is satisfied, with the point P corresponding to the earth. The procedure is to use these observations to show that the actual universe is sufficiently like an exactly isotropic one, where the reconvergence certainly does occur. As we have seen, the universe is isotropic to better than 1 part in 1,000 back to the last-scattering surface. There are two possibilities:

1. The red shift z_s of this surface is small ($\sim$7). This would require a relatively large amount of intergalactic scattering material, and direct calculation then shows that the gravitational effect of this material would produce the required reconvergence before the red shift z_s is reached.
2. z_s is appreciably larger than 7. The influence of intergalactic matter may now itself be unable to produce reconvergence. On the other hand, the isotropy of the universe would remain close to ideal out to z_s, that is, to a much greater red shift than 7. In this case one can show that the energy density of the microwave background alone is enough to cause reconvergence.

To avoid a singularity we must therefore challenge conditions (a), and (b), or (c). A violation of (c) would perhaps be worse than a singularity, being a global rather than a local breakdown of our ordinary physical concepts. Altering Einstein's field equations should perhaps be a last resort, so at the moment people are concentrating on challenging the energy condition.[54] It is generally agreed that nonquantum matter probably always does satisfy this condition. However, in the early universe one would expect the quantum properties of matter (and of gravitation) to be important, and we now know that in some circumstances the quantum energy-momentum tensor in a gravitational field can violate the energy condition.

In fact this violation can be regarded as the driving force for the Hawking radiation emitted by black holes.[55] This can be seen from the following perturbation relation, for the effect on the area A of a black hole's horizon, of a flux of matter into the black hole:

$$\frac{\mathrm{d}A}{\mathrm{d}t} \sim \frac{8\pi}{\varkappa} \int T_{ij}\, l^i l^j \, dA \, ,$$

where $\varkappa$ is the surface gravity of the black hole (proportional to its temperature)[56] and l^i is the null vector that generates the horizon. For classical matter we would have

$$T_{ij}\, l^i l^j > 0$$

by the energy condition, so that the area would increase with time. This is an example of Hawking's classical theorem of increasing area for black holes.[57]

Now suppose that we treat the matter field quantum mechanically, writing the semiclassical equation:

$$\frac{dA}{dt} \sim \frac{8\pi}{\varkappa} \int < T_{ij}\, l^i l^j > dA \, ,$$

where $< >$ means quantum expectation value in the appropriate state. Just as in Minkowski space $< T_{ij}l^il^j >$ is divergent, but it can be argued that the Minkowski contribution does not produce gravitational effects.[56] We therefore subtract out the Minkowski divergence, and what is left turns out to be finite. We therefore have

$$\frac{dA}{dt} \sim \frac{8\pi}{\varkappa} \int < T_{ij}l^il^j >_{\mathrm{reg}} dA.$$

The crucial point is that the regularized expression is negative. In other words, the vacuum fluctuations near a black hole have *less* energy than in Minkowski space. The energy condition is thereby violated, and the area of the black hole decreases. This precisely corresponds to the Hawking effect, and it can be shown[58] that the rate of loss of area matches up with the radiation rate at infinity, as calculated by other, asymptotic, techniques.[59]

We might expect that in the very early stages of the universe quantum effects would be important and might also lead to a violation of the energy condition. Two significant epochs are 10^{-23} seconds, when the particle horizon was of the same order as the radius of an elementary particle, and 10^{-43} seconds, which is the so-called Planck time $(Gh/c)^{1/2}$. At the Planck time we would expect effects arising from quantum gravity to be important. Unfortunately there exists as yet no reliable theory of quantum gravity, and we do not know whether any violation of the energy condition, at either 10^{-23} seconds or 10^{-43} seconds, would be great enough to counteract the attractive effects arising when the energy condition is satisfied, thus eliminating the singularity and permitting a rational study of the universe in its pre-big bang phase.[60] This problem is under investigation at the moment. Its solution would usher in a new regime that would transform our understanding of the main issues in cosmology.

Acknowledgement

I am grateful to Dr. J. D. Barrow for his comments on a draft of this paper.

NOTES

1. A. Einstein, "Cosmological Deductions from the General Theory of Relativity," S. B. Preuss. Akad. **142**, (1917); A. Friedmann, "On the Curvature of Space," Z. Phys. **10**, 377 (1922).
2. E. A. Milne, *Relativity, Gravitation and World-Structure* (Oxford University Press, London, 1935); H. P. Robertson, "Kinematics and World Structure," Astrophys. J. **83**, 257 (1936); A. G. Walker, "On Milne's Theory of World-Structure," Proc. London Math. Soc. **42**, 90 (1936).
3. G. F. R. Ellis, "Relativistic Cosmology," in *General Relativity and Cosmology*," edited by R. K. Sachs (Academic Press, New York, 1971); M. A. H. MacCallum, "Cosmological Models from a Geometric Point of View," in *Cargese Lectures in Physics*, Vol. 6, edited by E. Schatzman (Gordon and Breach, New York, 1973); M. P. Ryan and L. C. Shepley, *Homogeneous Relativistic Cosmologies* (Princeton University Press, Princeton, N.J., 1975).
4. J. Kristian and R. K. Sachs, "Observations in Cosmology," Astrophys. J. **143**, 379 (1966).
5. D. W. Olson, "Helium Production and Limits on the Anisotropy of the Universe," Astrophys. J. **219**, 777 (1978).
6. C. W. Misner, "The Isotropy of the Universe," Astrophys. J. **151**, 431 (1968).
7. K. Gödel, "An Example of a New Type of Cosmological Solution of Einstein's Field Equations of Gravitation," Rev. Mod. Phys. **21**, 447 (1949).
8. J. D. Barrow, "Light Elements and the Isotropy of the Universe," Monthly Notices, Roy. Astron. Soc. **175**, 359 (1976).

9. B. E. Corey and D. T. Wilkinson, "A Measurement of the Cosmic Microwave Background Anisotropy at 19 GHz," Bull. Astron. Astrophys. Soc. **8**, 351 (1976); G. F. Smoot, M. V. Gorenstein, and R. A. Muller, "Detection of Anisotropy in the Cosmic Black Body Radiation," Phys. Rev. Lett. **39**, 898 (1977).
10. Ibid.
11. C. B. Collins and S. W. Hawking, "Why Is the Universe Isotropic?" Astrophys. J. **180**, 317 (1973), and "The Rotation and Distortion of the Universe," Monthly Notices, Roy. Astron. Soc. **162**, 307 (1973).
12. Op. cit. in n. 3.
13. H. Reeves, "On the Origin of the Light Elements," Ann. Rev. Astron. Astrophys. **12**, 437 (1974).
14. Barrow, op. cit. in n. 8; Olson, op. cit. in n. 5. See also J. D. Barrow and F. J. Tipler, "Eternity Is Unstable," Nature **276**, 453 (1978).
15. J. Yang et al., "Constraints on Cosmology and Neutrino Physics from Big Bang Nucleosynthesis," Astrophys J. **227**, 697 (1979).
16. Misner, op. cit. in n. 6.
17. J. D. Barrow and R. A. Matzner, "The Homogeneity and Isotropy of the Universe," Monthly Notices, Roy. Astron. Soc. **181**, 719 (1977).
18. Ibid.
19. Kristian and Sachs, op. cit. in n. 4.
20. S. W. Hawking, "On the Rotation of the Universe," Monthly Notices, Roy. Astron. Soc. **142**, 129 (1969); Collins and Hawking, op. cit. in n. 11.
21. R. Penrose, "Singularities of Space-Time," in *Theoretical Principles in Astrophysics and Relativity*, edited by N. R. Leibovitz, W. H. Reid, and P. O. Vandervoot (University of Chicago Press, Chicago, 1978), and "Singularities and Time Asymmetry," in *General Relativity*, edited by S. W. Hawking and W. Israel (Cambridge University Press, Cambridge, England, 1979).
22. J. D. Bekenstein, "Generalised Second Law of Thermodynamics in Black-Hole Physics," Phys. Rev. **D 9**, 3292 (1974).
23. D. Lynden-Bell and R. M. Lynden-Bell, "On the Negative Specific Heat Paradox," Roy. Astron. Soc. **181**, 405 (1977).
24. Op. cit. in n. 6.
25. W. Rindler, "Visual Horizons in World-Models," Monthly Notices, Roy. Astron. Soc. **116**, 662 (1956).
26. A. B. Doroshkevich, V. Lukash, and I. D. Novikov, "Impossibility of Mixing in a Cosmological Model of the Bianchi IX Type," JEPT **60**, 1201 (1971); D. M. Chitre, Ph.D. thesis, University of Maryland, 1972.
27. R. H. Dicke, "Dirac's Cosmology and Mach's Principle," Nature **192**, 440 (1961); B. Carter, "Large Number Coincidences and the Anthropic Principle in Cosmology," in *International Astronomical Union Symposium No. 63* (Reidel, Dordrecht, Boston, 1974); J. D. Barrow and F. J. Tipler, op. cit. in n. 14 and *The Anthropic Cosmological Principle* (Oxford University Press, New York, 1980); B. J. Carr and M. J. Rees, "The Anthropic Principle and the Structure of the Physical World," Nature **278**, 605 (1979).
28. In Astrophys. J. **180**, 317 (1973).
29. B. J. T. Jones, "The Origin of Galaxies: A Review of Recent Theoretical Developments and Their Confrontation with Observation," Rev. Mod. Phys. **48**, 107 (1976); W. H. McCrea and M. J. Rees, "The Origin and Early Evolution of the Galaxies," Phil. Trans., Roy. Soc. **296**, 269–435 (1980).
30. J. D. Barrow, "On the Origin of Cosmic Turbulence," Monthly Notices, Roy. Astron. Soc. **179**, 47P (1977), and "The Origin and Early Evolution of the Galaxies," Nature **272**, 211 (1978).
31. A. Einstein, "Autobiographical Notes," in *Albert Einstein, Philosopher-Scientist*, edited by P. A. Schilpp (Library of Living Philosophers, Evanston, Ill., 1949), p. 67.
32. D. J. Raine, "Mach's Principle in General Relativity," Monthly Notices, Roy. Astron. Soc. **171**, 507 (1975).
33. R. K. Sachs and A. M. Wolfe, "Perturbations of a Cosmological Model and Angular Variations of the Microwave Background," Astrophys. J. **147**, 73 (1967).
34. Ibid.; M. J. Rees and D. W. Sciama, "Large Scale Density Inhomogeneities in the Universe," Nature **217**, 511 (1968); C. C. Dyer, "The Gravitational Perturbation of the Cosmic Background Radiation by Density Concentrations,"

Monthly Notices, Roy. Astron. Soc. **175**, 429 (1976); A. M. Anile and S. Motta, "Perturbations of the General Robertson–Walker Universes and Angular Variations of the Cosmic Black Body Radiation," Astrophys. J. **207**, 685 (1976).

35. Op. cit. in n. 33.

36. Jones, op. cit. in n. 29; J. D. Barrow, "Galaxy Formation: the First Million Years," Phil. Trans., Roy. Soc. **296**, 273 (1980).

37. Barrow 1978 op. cit. in n. 30.

38. E. P. T. Liang, "A Note on the Zeldovich $p = \rho$ Cold Big Bang," Monthly Notices, Roy. Astron. Soc. **171**, 551 (1975), and "Dynamics of Primordial Inhomogeneities in Model Universes," Astrophys. J. **204**, 3 235 (1976).

39. M. J. Rees, "Origin of Pregalactic Microwave Background," Nature **275**, 35 (1978), where references are given to earlier such discussions.

40. Reeves, op. cit. in n. 13.

41. Hawking radiation: S. W. Hawking, "Black Hole Explosions?" Nature **248**, 30 (1974), and Comm. Math. Phys. **43**, 107 (1975). Pair production: V. N. Lukash et al. "Particle Creation by Black Holes," Nuovo Cimento **35B**, 293 (1976); L. Parker, "Quantum Effects and Evolution of Cosmological Models," in *Asymptotic Properties of Space-Time*, edited by F. P. Esposito and L. Witten (Plenum, New York, 1977).

42. B. J. Carr, "The Primordial Black Hole's Mass Spectrum," Astrophys. J. **201**, 1 (1975); J. D. Barrow and B. J. Carr, "Primordial Black Hole Formation in an Anisotropic Universe," Monthly Notices, Roy. Astron. Soc. **182**, 537 (1978).

43. G. V. Bicknell and R. N. Henriksen, "Self-Similar Growth of Primordial Black Holes," Astrophys. J. **219**, 1043 (1978), and **225**, 237 (1978).

44. J. B. Hartle and B. L. Hu, "Quantum Effects in the Early Universe III," Phys. Rev. **D21**, 2756 (1980). J. B. Hartle in *Quantum Gravity II*, edited by C. J. Isham, R. Penrose and D. W. Sciama, Oxford (1981).

45. R. Omnes, "The Possible Role of Elementary Particle Physics in Cosmology," Phys. Reps. **C 3**, 1 (1972).

46. G. Steigman, "Observational Tests of Antimatter Cosmologies," Ann. Rev. Astron. Astrophys. **14**, 339 (1976).

47. A. Ramani and J. L. Puget, "Coalescence and 2.7K Black Body Distortion in Baryon Symmetric Big Bang Cosmology," Astron. Astrophys. **51**, 411 (1976); J. J. Aly, "Matter-Antimatter Hydrodynamics: Computation of the Annihilation Rate," Astron. Astrophys. **67**, 199 (1978).

48. M. Yoshimura, "Unified Gauge Theories and the Baryon Number of the Universe," Phys. Rev. Lett. **41**, 281 (1978), and **42**, 746E (1979).

49. J. Ellis, M. K. Gaillard, and D. V. Nanopoulos, "Baryon Number Generation in Grand Unified Theories," Phys. Lett. **80 B**, 360 (1979).

50. S. Dimopoulos and L. Susskind, "Baryon Number of the Universe," Phys. Rev. **D 18**, 4500 (1978).

51. D. Toussaint et al., "Matter-Antimatter Accounting, Thermodynamics and Black Hole Radiation," Phys. Rev. **D 19**, 1036 (1979); S. Weinberg, "Cosmological Production of Baryons," Phys. Rev. Lett. **42**, 850 (1979).

52. R. Penrose, "Gravitational Collapse and Space-Time Singularities," Phys. Rev. Lett. **14**, 57 (1965); S. W. Hawking, "The Occurrence of Singularities in Cosmology," Proc. Roy Soc. **A 294**, 511 (1966); R. P. Geroch, "Singularities in Closed Universes," Phys. Rev. Lett. **17**, 445 (1966); S. W. Hawking and G. F. R. Ellis, *The Large Scale Structure of Space-Time* (Cambridge University Press, Cambridge, England, 1973).

53. S. W. Hawking and G. F. R. Ellis, "The Cosmic Black Body Radiation and the Existence of Singularities in Our Universe," Astrophys. J. **152**, 25 (1968).

54. F. J. Tipler, "Energy Conditions and Space-Time Singularities," Phys. Rev. **D 17**, 2521 (1978).

55. P. Candelas and D. W. Sciama, "Irreversible Thermodynamics of Black Holes," Phys. Rev. Lett. **38**, 1372 (1977).

56. Bekenstein, op. cit. in n. 22; D. W. Sciama, "Black Holes and Their Thermodynamics," Vistas in Astronomy **19**, 385 (1976); P. C. W. Davies, "Thermodynamics of Black Holes," Rep. Prog. Phys. **41**, 1313 (1978).

57. S. W. Hawking, "Black Holes in General Relativity," Comm. Math. Phys. **25**, 152 (1972).

58. P. Candelas, "Vacuum Polarization in Schwarzschild Space-Time," Phys. Rev. **D21**, 2185 (1980).

59. Hawking, op. cit. in n. 41.

60. Tipler, op. cit. in n. 54.

25. SYMMETRY PARADOXES AND OTHER COSMOLOGICAL COMMENTS

Charles W. Misner

I shall try to comment briefly on three different topics that have in common a tie to Einstein's cosmological insights. They are

Symmetry paradoxes
Initial conditions for the universe
A proposed "Einstein orrery"

The preceding paper by Dennis Sciama has given a masterfully broad and balanced survey of the major cosmological puzzles that are visible at our intellectual horizons. We now view the very early beginnings of the universe from the vantage point of a generally accepted "standard big bang" theory that is held in the tentative way prior revolutions have taught us to regard any current theory. This standard big bang will, however, probably join Newtonian mechanics, special relativity, the Bohr atom, and similar insightful simplifications as part of the permanent archives of science even after cosmology is refounded upon new ideas. Of the current puzzles that Sciama has surveyed, some will no doubt be solved within the general framework of the standard big bang, although perhaps with the help of revolutionary advances in particle physics, whereas others may demand new insights on the specifically cosmological level, or even in our characterization of the science of physics. In my first comment below I try to enlist Einstein's example as a guide for recognizing the problems that are most promising as pointers toward a new world view, and argue that "paradoxical isotropy" is such a problem. My second comment suggests an answer (No!) to Sciama's issue "Was the big bang a singularity?". The connection to Einstein here is that, as with special relativity, it is sometimes possible to leave prior calculations intact and yet dispose of a problem merely by a change in the physical or philosophical interpretation of the standard equations. The weakness of my suggestions below about the singularity is that, at least as yet, they have not proved to be fruitful for clarifying other problems and are therefore merely one of many open speculations concerning the ultimate significance of the initial singularity. My third comment is a suggestion for a memorial to Einstein that would help bring the main ideas of cosmology, in the geometrical language he has taught us, to a much wider public, with reasonable accuracy, and with as much beauty as we shall succeed in requiring.

Harry Woolf (ed.), Some Strangeness in the Proportion: A Centennial Symposium to Celebrate the Achievements of Albert Einstein ISBN 0-201-09924-1

1. Symmetry Paradoxes

Einstein introduced special relativity with the remark that the then current understanding of Maxwell's electrodynamics "leads to asymmetries which do not appear to be inherent in the phenomena." Our present understanding of cosmology suffers from this same defect, as Sciama has elaborated; perhaps, under Einstein's tutelage, we may at least recognize important problems even if we do not dispatch them as quickly and decisively as Einstein might have.

First recall the theoretical asymmetry to which Einstein objected. When there was relative motion between a bar magnet and a loop of wire, only an accidental conspiracy between the coefficients in the Lorentz force law and the Faraday induction law allowed the same current to flow regardless of which object was moving. For if the magnet moved, an electrical force caused the current, whereas if the wire moved, a magnetic force caused it. Einstein objected to two agencies that were at that time quite distinct — electricity and magnetism — getting credit for what he saw as indistinguishable results. He then, of course, proceeded to introduce the current viewpoint. This makes electricity and magnetism so little distinct that only with great difficulty can a physicist now see the problem that Einstein solved.

It is possible to view Einstein's enduring contribution from his first cosmology paper as being the resolution of a similar problem of asymmetries in the theory where none are evident in the phenomena. (Peebles indicates (in this volume) more historically the light in which Einstein approached his work on cosmology.) In addition, just because this work is infamous for Einstein's having missed an opportunity to "predict" the expansion of the universe, it is useful to restate from a current viewpoint the problem that it did solve definitively. The symmetrical phenomenon that enters here is the roughly homogeneous universe that has no special or central position for the earth. Given a spatially uniform matter density ρ, the Newtonian gravitational field equation $\nabla^2\phi = -4\pi\rho$ admits no uniform or homogeneous solution. The most symmetrical solution

$$\phi = \phi_0 + (2\pi\rho/3)\,(r - r_0)^2$$

insists that there be some preferred central position where the gravitational field vanishes, and this violates scientific instincts that have become firmly rooted since Copernicus removed man from the geographical center of the world. Scientific cosmology without Einstein therefore faces a baffling predicament. Gravitational forces are the controlling interaction between galaxies, but Newton's gravitational theory cannot allow galaxies to be uniformly distributed everywhere without selecting (gravitationally) an absolute center point. Thus Newtonian gravitation theory demands a center of symmetry that is not inherent in the distribution of the galaxies.

Einstein introduced the concept that gravitation is the curvature of space-time. This new view solved the Newtonian cosmological predicament by banishing, as fiction, the gravitational acceleration field g in the cosmos. It was banished as decisively as special relativity theory had banished the electromagnetic aether. Unlike the banished gravitational field

$$g = -\nabla\phi = -\,(4\pi\rho/3)\,(r - r_0)$$

the space-time curvature (or field gradient)

$$dg = -(4\pi\rho/3)\,dr$$

can be uniform in the presence of a uniform matter distribution. Only the relative motion between galaxies is then meaningful, and it is no longer necessary to designate some one galaxy as unaccelerated. Thus, in the view we have absorbed from Einstein, the appearance of the constant r_0 in the foregoing equation for the cosmological gravitational field g is merely an artifact of the distortion to which the cosmos must be subjected when we choose to represent it in familiar language, just as the geometry of the globe gets distorted in various ways when one tries to display it on a flat page. Continuing this cartographical analogy, consider the circumstance that, in some map projections, the meridian through Greenwich will be drawn as a straight line, whereas all other meridians will be curved. The Newtonian cosmological view (g with an r_0) corresponds to an interpretation of these meridians as indications of a peculiarity of geology near Greenwich, while the Einsteinian cosmological view corresponds to an interpretation that regards them as evidence for a limitation in the language of flat maps that says nothing about Greenwich, except possibly that a cartographer worked there.

The cosmologically significant part of Einstein's lesson is not so much the mathematics of space-time curvature that he learned and used, but his insight that the geometry of space-time is a better conceptual model of the gravitational field than anything known previously. This view then shows that the curvature or field gradient dg is a less distorted representation of the geometry than is the field g itself, and lets us see that the geometry (or gravitational structure) of the universe shares the same large-scale regularity or homogeneity that we assume or see in the distribution of matter in the universe.

Einstein's rejection of the reality of gravitational acceleration fields is more convincing in the cosmological context than for, say, solar system gravitation. In an isolated system, a gravitational acceleration field g can reasonably be reconstituted from the space-time curvature dg by successively adding up all the field gradients, working in from an asymptotic region where a zero acceleration field is assigned:

$$g = \int_{\infty}^{r} dg \quad .$$

In the homogeneous cosmological context there is no asymptotic region to serve as a natural zero reference, and Einstein's curvature language is compelling. Therefore, even at the level of Einstein's first (static) cosmological model, cosmology provides important support for general relativity because only curvature concepts open cosmology to plausible mathematical modeling — that is, to models without spurious centers.

A problem of the same character as those discussed thus far — theoretical distinctions that appear to correspond to no observable distinction — faces us now in the isotropy of the microwave background radiation. The radiation measured from different sectors of the sky is indistinguishable (after correction for the earth's motion), yet drastically disjoint agents are, in our present unsatisfactory understanding, assigned credit for producing it. This is the problem of "particle horizons" that I will illustrate geometrically with a "map of the universe" a little further on. For now let me assemble three remarks. The first and central remark is that the *agents producing the microwave radiation can be too far apart to have synchronized their activities since the beginnings of the unvierse.* This difficulty is made more serious by the second point (which follows from the singularity theorems Hawking and Sciama both reviewed in this volume) that standard theory forbids a discussion of epochs prior to the big bang, and thus forbids synchronization's being explained then. The third remark is an opinion that this paradox has not yet been successfully treated by new principles that require synchronization at the initial singularity (such as Sciama's views on Mach's principle, or Penrose's suggestions concerning the entropy of the initial state) because these speculations are not fully developed and have not shown convincing power and generality by application to a variety of other problems. Current theory can easily describe and admit the equality of the microwave temperatures measured as independent phenomena seen in different places, but it has no inherent known symmetry that requires or explains this equality. I am hopeful that this "paradoxical isotropy" of the microwave background will provoke new insights in our generation with the efficacy that one thinks Olber's paradox should have had toward the end of the nineteenth century. (At that time thermodynamics was so firmly established that the darkness of the night sky should have been an argument against a static universe — but in fact this powerful line of argument appears to have been useful only much later as a pedagogical device.)[1] Further effort on this question would be very appropriate, especially any yielding relevant observational evidence.

2. Initial Conditions for the Universe

The vocabulary Einstein has provided for cosmology will no doubt suffice for the development, in the normal course of science, of an accepted picture of the evolution of the universe — a picture that will probably elaborate the current standard model of the "post-big bang" universe. But when an explanation of the initial conditions is demanded, one must echo Sciama's remark: "It is not clear to me what the rules of the game are here." Within either Newtonian or Schroedinger mechanics, the experimenter chose (by art, design or caprice) the initial conditions; then physical theory computed the knowable future evolution. But not everyone is content to accept his own universe as one such experiment, especially when the Experimenter grants no interviews. The plausible solubility of the evolution problem for the universe presents us with the deeper mystery of the initial conditions. Both scientific impertinence and a proper religious sense then demand that the mystery of the initial conditions be put into humanly comprehensible form so that, like Einstein, we can respond with an enobling awe. Sciama mentions (in the preceding chapter) several approaches aiming toward this desired comprehension. Some seek to avoid choosing initial conditions. These include the apparently unsuccessful initial chaos idea that in a sense wants to accept all initial conditions simultaneously and the anthropic principle that accepts all initial conditions in series or in parallel but locates the observer selectively in the resultant plethora of universes. Initial conditions can also be avoided (or at least deferred) by anticipating that new physical laws needed under extreme conditions will evade the singularity theorems and allow the big bang expansion to be merely the dynamical evolution of a prior phase of contraction. Alternatively one may meet the initial conditions head-on with principles capable of organizing them, such as new generations of Mach's principle, or a thermodynamic imperative. Sciama has surveyed these alternatives, and the initial condition question remains open but fundamental. The isotropy paradox forces it upon us as a current observational mystery.

I wish to suggest yet another speculative alternative that seems not to attract the attention I think it deserves in view of its roots in Einstein's equations. This I call the steady-state big bang. It is a suggestion that the initial conditions of the universe might, by a shift in language and philosophy, be removed to the infinite past where time would be available either to resolve chaos, or to prepare a thermodynamic state, or where the initial conditions could at least be viewed with the same detachment accorded asymptotic past conditions in models that speculate upon an almost periodic universe or a pre-bang collapse. Again one faces a theoretical (philosophical) asymmetry not inherent in the (plausibly anticipated) phenomenon — a singularity described as looming only a finite time into the past where plausible models, if run backward on a computer, would run on forever. Why do we insist on frightening ourselves with a theoretical disaster in the finite past when a numerical computation to explore it would, if sufficiently accurate, run on happily forever? I propose that we may not have chosen our words carefully enough when we say that current theory demands a singularity a finite time in the past. I challenge the mathematics of no computation or theorem, just the implicit interpretations, reactions, and discomfits of a common language ''restatement'' of these results.

''The universe is 20 billion years old,'' but what does that mean? Not, certainly, that a physical year in the sense of the earth orbiting the sun has been repeated 20 billion times. That, we believe, has occurred only 5 billion times, marking the first portion of a converse Xeno paradox. Xeno questioned the philosophical possibility of Achilles completing a single stride because that action was conceptually decomposable into its first half, plus half the remainder, and again half that remainder infinitely many times, and so never fully accomplished. We reject this infinity as a figment of a mathematical imagination and accept the completed single stride as a finitely realized physical phenomenon. Approach the claimed finite age of the universe with a similar physical criterion. Is the finiteness inherent in the (postulated) pheomena, or is it introduced by extraneous mathematical manipulations? Five billion earth orbits accomplish the first physical steps back toward the singularity. Before that use Bohr orbits — the electron's motion in a hydrogen atom — to tick off the pulse of time; some 10^{31} of these can be executed serially before we have delved so far into the past that hydrogen is ionized and yet another clock is needed. This progression has no known end in a big bang universe where temperatures and densities are unlimited in the past, and we begin to realize that the famous finite age is a complicated convergent series. It is not Xeno's series

$$1 = 1/2 + 1/4 + 1/8 + \ldots + 1/2^n + \ldots$$

but another one:

$$\begin{aligned} 2 \times 10^{10} \text{ years} = {} & 5 \times 10^{9} \text{ earth orbits} && \times 1 \text{ year/orbit} \\ & + 10^{31} \text{ Bohr orbits} && \times 10^{-14} \text{seconds/orbit} \\ & + 10^{38} \text{ quark orbits} && \times 10^{-25} \text{ seconds/orbit} \\ & + (\quad \ldots \quad) && \times \quad ???\text{seconds/orbit} \\ & + \ldots \end{aligned}$$

Here the more immediately physical phenomena are the counted motions of prevalent matter in the separate terms on the right, and convergence is achieved only by discounting these accomplished physical acts into a common currency of earth orbit years. But if every imaginable series of clocks capable of serially recording counts all the way back to the beginning should necessarily yield an infinite number of counts, why do we stand fearing finiteness just because an infinite number of mathematical manipulations succeeds in discounting an infinite sequence of actual phenomena into a finite sum of imaginary earth orbit years? Proper time is an unambiguous technical word with a clear relevance to phenomena (for example, reaction rates) in locally flat regions of space-time. But it is not *proper* time, but some older, broader connotation of the word time that excites us with a sense of mystery, adventure, and possible intellectual conquest when the finite age of the universe is cited as a singular paradox for current science. For these heuristic and motivational purposes, relativistic cosmology suggests an infinite past age for the universe. The most plausibly generic solutions of the Einstein equations near the initial singularity[2] incorporate in the oscillating curvature of space-times a built-in physical clock that does continue ticking all the way back to the singularity, but it ticks infinitely many times. Figure 25.1 is a schematic, simplified sketch of the means by which the curvature succeeds in oscillating infinitely many times since the singularity. I could imagine this "initial state" with its infinitely structured historical depth being relevant even to a low-entropy initial quantum state. There the quantum fluctuations about a highly symmetric average would tick off an infinitely structured path back toward an "initial singularity" and thus tame it, in a calmer view, into an asymptotically stationary and therefore nonsingular beginning.

3. The Einstein Orrery

Although his first example of a cosmological model was faulty in detail, Einstein's cosmological vocabulary (space-time curvature) is now standard even in dissident theories and appears to have a significant cultural impact even when scantily understood. I would like to propose

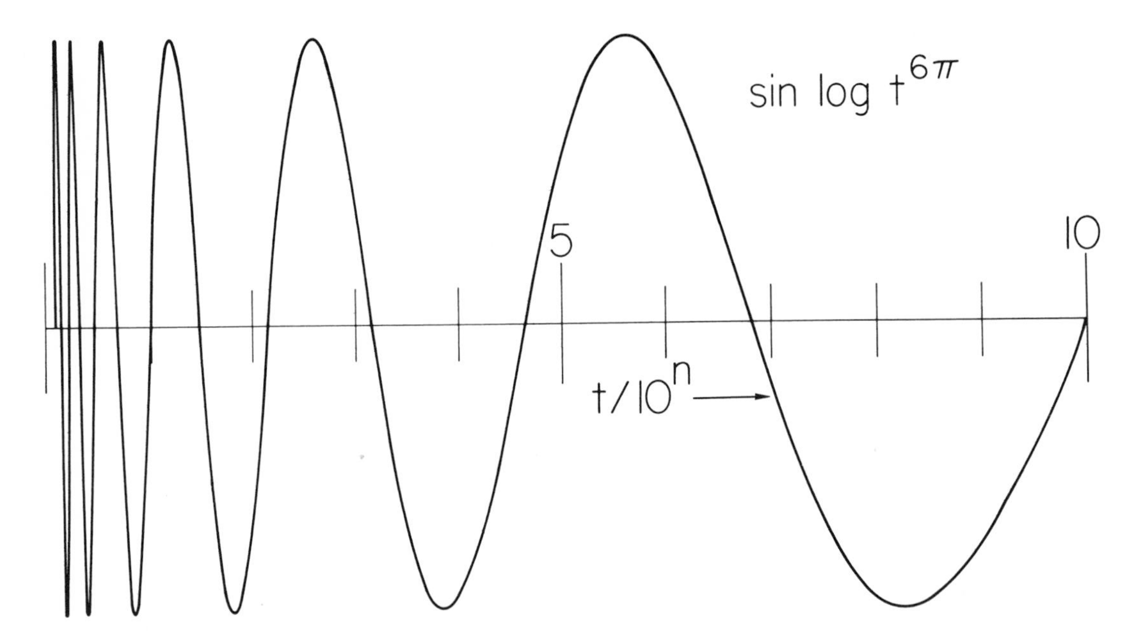

Fig. 25.1 The curvature clock. The most plausibly generic solutions of Einstein's equations currently available for study show oscillations of space-time curvature near the initial singularity that are qualitatively similar to those of the function $f = \sin \log t^{6\pi}$ that is plotted here. This function approaches no limit for $t \rightarrow 0$, but shows infinitely many cycles of oscillation in every neighborhood of its singularity at $t = 0$. The oscillations can, however, be considered as a stationary state since the function has exactly the same behavior in every decade of t. The graph for $0.00001 \leqslant t \leqslant 0.0001$ is identical to that for $0.1 \leqslant t \leqslant 1$, and the plot shows any two decades selected by one's choice of $n = \ldots -2, -1, 0, 1 \ldots$ in the horizontal scale factor.

that in homage to Einstein on a future occasion some reasonably accurate and more readily accessible presentation of this geometrical conception of the universe be opened to a wide audience. What I have in mind is a four-dimensional "map of the universe" measuring perhaps (3 meters)^3 $\times$ (5 minutes) in space-time extent. A two cent two-dimensional version will be given below.

After the earth became the globe in the centuries of exploration, the 2- or 3-foot library globe became a common symbol of scientific culture. Similarly, the scientific revolution from Copernicus to Newton allowed for its ornamental and pedagogical symbolization in mechanical orreries that were not quite common but were widely known. The Rittenhouse orrery near the Princeton University Astronomy Library is a beautiful and accessible example. Einstein's exploration of the universe is as significant and heroic as Columbus's exploration of the globe, or Newton's conquest of the solar system. But a visible, pedagogical symbol, an Einstein orrery, is still lacking. Einstein's cosmological viewpoint is a treasure of the twentieth century; a museum visitor should be able to see it and learn.

What I hope may become practical in the next decade is a holographic motion picture, produced by animation and computer, that would exhibit on the scale of reasonable human extravagance, a suitably scaled-down diagram, in four dimensions, of the evolving universe. Although a two-dimensional version lacks the showmanship and popular appeal that I would expect for a four-dimensional map, and my sketch lacks the artistry that good craftmanship could supply, let me present a minimal map of the universe to suggest the scientific content that a map can convey (Fig. 25.2). The first thing to note in the figure is that the mariner's compass rose has been replaced on this chart by Minkowski's light cone. It reminds us that the vertical directions in the diagram represent time, while the horizontal direction represents the space directions. The scale factors are chosen so that light rays, represented by slightly wavy lines, are drawn at 45° corresponding to a light velocity of 1 light year per year. The vertical lines on the diagram represent galaxies, with one near the center of the figure marked "us." The future (top) endpoints of these lines are merely the edge of the diagram, but the past (lower) endpoints are meaningful and signify the formation of the galaxies. These world lines of galaxies are straight since we neglect any peculiar velocity that a galaxy might have toward one or another direction in the universe, and each galaxy is taken to be merely evolving in place, subject to no external forces. The symbols q mean quasar; they appear most prevalent at an epoch near the time when galaxies were forming. Continuing progress can be expected in our understanding of the nature of quasars and their possible relation to stages in the life of young galaxies, but the fact that they were more prevalent at this earlier epoch than now seems to be established. Lower in the diagram, before galaxies and quasars, there is a transparent region when the universe was filled with neutral hydrogen and helium gas. Still earlier this gas was hotter and opaque; it was ionized as indicated by the caption "plasma" in the diagram. At a still earlier and hotter stage, temperatures and densities allowed nucleosynthesis to occur, primarily the formation of helium and deuterium nuclei from the protons and neutrons present in the hotter and denser plasma of prior epochs. The bottom of the diagram is the cosmological singularity, the edge of the universe, a limit prior to time itself where theory currently sheds no plausible light.

One of the features of standard models of the universe that this map illustrates rather clearly is spatial homogeneity. None of the galaxies behaves differently than any other; none is in a preferred position. In particular, no galaxy is near the outside of an exploding or expanding ball of matter, and one important job that this map should perform is to counteract the misimpression that "big bang" means a lump of matter expanding in an otherwise empty universe. At no time is

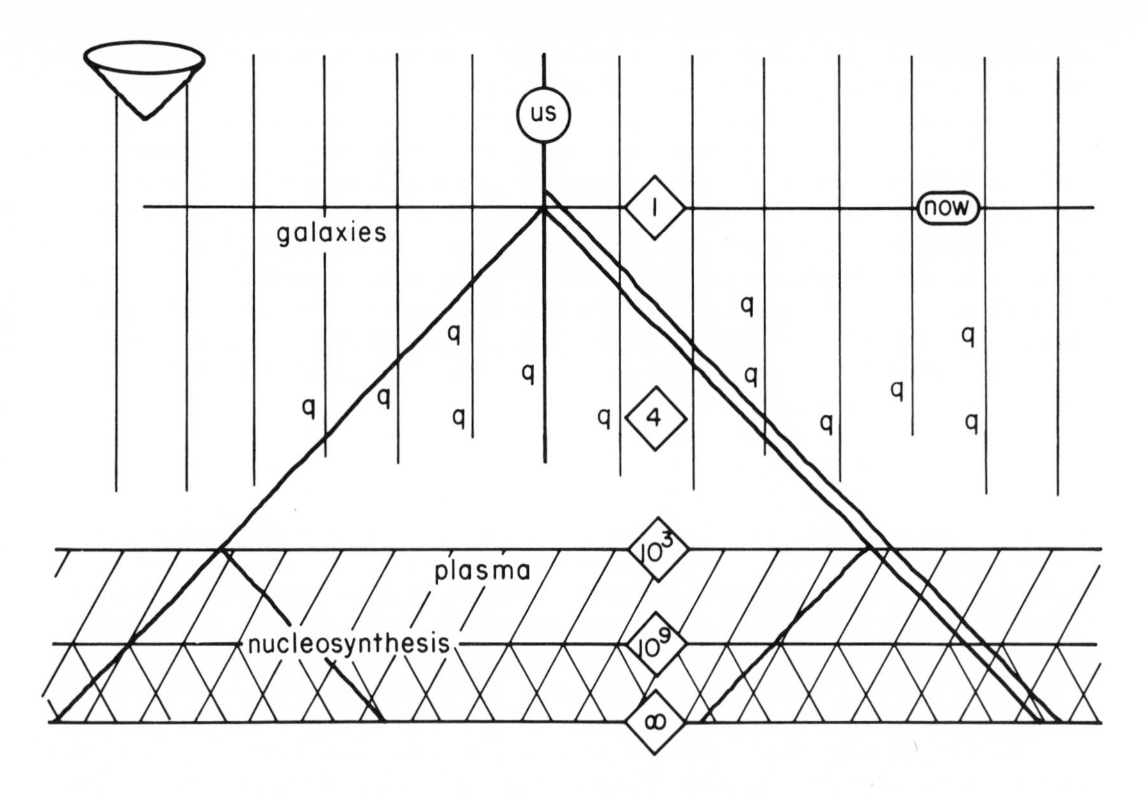

Fig. 25.2 Map of the universe. This is a space-time diagram on which the horizontal coordinate x marks spacial differences while the vertical coordinate distinguishes different times. The map scale varies with time (cf. Mercator projections of the globe where the map scale varies with latitude), and proper time or proper distance intervals $d\tau$ or ds are given by

$$-d\tau^2 = a^2(y)\,[-dy^2 + dx^2] = ds^2$$

where dx and dy are horizontal and vertical intervals on the map. The scale factor $[a(y)]^{-1}$ is given by the Einstein equations and is shown within diamond boxes at several points on the map. For further discussion see the main text.

there any empty space in this universe apart from the interstices between particles or galaxies that everywhere have the same average density. The map itself is not infinitely wide, but this is a technical limitation in the printing process. The map in the reader's mind should not stop at the page margins, but continue forever to the left and right of the sector shown in Fig. 25.2. Each galaxy should have neighbors both to its left and right (and in the two omitted spatial dimensions). This model universe was never small; it is infinite in spatial extent at the earliest time $t > 0$ that one cares to consider, just as it is at the time slice marked "now." (A map representing a closed universe would appear very similar, except that it could stop at each side along a line marked "cosmological longitude 180°" and a galaxy located there would be drawn twice on the map, once at each edge. The special treatment of this galaxy would be removed if one bent the map into a cylinder and glued the two sides together at this 180° line. Again there is no empty space, and each galaxy has a set of neighbors comparable to any other's.)

How, then, does this map exhibit the expansion of the universe, if this universe is always infinite in spatial extent? Of course it is not the boundaries of the universe that expand — there are no boundaries — but the distances between galaxies. At early epochs the expansion would be seen in the expansion of distances between pairs of particles rather than pairs of galaxies. On this particular map the expansion has been deemphasized to allow for a more prominent and simple representation of the homogeneity of the universe and of the propagation of light rays. The expansion information is contained in the varying scale factors on the map that are shown in diamond-shaped boxes at several points. Thus, along the horizontal line marked "now" there is a scale of "1" indicated, which may be taken as approximately 1 millimeter per million light years. Galaxies plotted as 1 centimeter apart along this line on the map would in reality be 10^7 light year apart. At an earlier epoch, when these same galaxies are shown shortly after formation, still 1 centimeter apart on the map, the relevant scale factor of 4 millimeters per million light years shows that they were only 2.5×10^6 light years apart then. In the intervening time their separation has expanded by a factor of 4, and this change in separation implied by the varying scale factors is the Hubble expansion. The red shift by which the Hubble expansion is observed is also most directly deduced directly from the scale factors on the map. Consider the two parallel 45° light rays on the right-hand side of the map to represent the beginning and the end of a single wavelength of light. The spacial separation between these two rays is the same at every epoch as measured on the map, but the implied actual separation (one wavelength) is continually changing because the map scale is different at different times. Suppose, for instance, that this light was emitted from the hot ionized primordial plasma in its final stages as it recombined into a transparent gas of neutral atoms. This epoch is a horizontal line on the map that shows a scale factor of 10^3. If the wavelength then were 1 micron, and if the rays on the sketch could be spaced to correspond to this at the 10^3 scale, then this same spacing on the map when interpreted at the present epoch with a scale of 1 would correspond to a wavelength of 1 millimeter. In this way the optical radiation from the primordial "fireball" becomes the microwave background radiation observed today. (The same argument applies, of course, to light emitted more recently from distant galaxies. The more distant the galaxy is, the further in the past it will have emitted the radiation we collect from it now, and thus the greater the difference in map scales between emission and detection times. This gives the Hubble variation of red shift with distance.)

Einstein's most captivating insight — that space-time is curved — can also be recognized in this map of the universe. This map is drawn with a variable scale factor because there is no way to draw it using a constant scale factor. Some space or time relations must be distorted, and not just rigidly scaled down, to be represented on a flat piece of paper. The universe has a different shape, not just a different size, than this page (even after we have simplified matters by thinking of a universe of two dimensions instead of four). In short, the universe is curved. The varying scale factor shows that we had to stretch the universe more some places (or times) than others in order to iron it out onto a flat page. The difficulty, and the method of dealing with it, is exactly the same as that met in attempting to represent the curved surface of the earth on a page in a book instead of upon the surface of a globe. To make a flat map of more than a small section of the globe it is necessary to stretch it and iron it flat, resulting in a map with scale factors that differ from one region to another. The two-dimensional map given here does not allow the possibility that space itself (now one-dimensional) can be curved. Thus the distinctions between positive and negative space curvature that arise in four-dimensional cosmology cannot be exhibited without a higher-dimensional map, but it is space-time curvature that is most directly related to gravity and to Einstein's original insights, and space-time curvature is evident even in this simple two-dimensional map.

I will conclude this discussion of a map of the universe (intended as a sales talk for the eventual construction of a wonderful four-dimensional map, an Einstein orrery) by using the map in Fig. 25.2 to describe the paradoxical isotropy mentioned in Sec. 1. We place ourselves at the "us"-"now" point on the map and look left and right out into the sky. The light rays reaching us are drawn on the map; and we trace them back to the points where this (microwave) light we see was emitted from the cooling plasma. The observation we seek to explain is the very precise equality in temperature of the two distinct portions of plasma seen in these two different directions. Without the detailed view that the map provides, this equality may not seem implausible. Everything in the universe began, did it not, at a single point ("the singularity") at the same time and was therefore in the same initial state. Why should it not evolve to equal temperatures later? But on the map we cannot find this single initial point, and we recognize that the scale factor of ∞ shown on the infinite line marking the initial singularity gives an indeterminate ∞/∞ size to the initial singularity. In fact, "singularity" means either nonexistent or not describable by the mathematics at hand, so we are forced to limit the discussion to times after the singularity.

The question is: Was there ever any contact between the portions of matter in the two bits of plasma we see now in different distant parts of the sky? Could they have arranged to compromise their differences and settle upon a mutually agreed temperature? If such "negotiations" (physical interactions) were possible, we should not be surprised to observe equal temperatures. But the map shows that no such interactions could ever have occurred. All known physical interactions reach from one bit of matter to another by propagating through space-time at speeds not greater than the local speed of light. Therefore light rays drawn on the map will mark the limits on causal interactions. On the map we have drawn backward from each bit of plasma that we see directly, a pair of left- and right-moving light rays to extrapolate back toward all those particles of matter that could possibly have influenced the state of this plasma. The paradoxical fact is that, in this standard model, there is not one particle of matter than can have influenced both observed bits of plasma, and the matter that became one observed plasma glob cannot have sent any signals that would help the other observed plasma reach a similar state. Any difference in the initial state of one region compared to the other would not only go uncorrected, it would even go unnoticed until the universe was sufficiently old (now, for instance) that some single observer could simultaneously see both regions. The situation is quite similar in the four-dimensional standard model of the universe, except that the observer's sky has more than the two directions (left and right) of the map in Fig. 25.2. Any direction on a two-dimensional celestial sphere can be observed. But the paradox arises as soon as one observes (via the microwave background radiation) two bits of primordial plasma that are even a small distance apart in the sky (for example, 1°). Their past light cones, calculated all the way back toward the initial singularity, again do not overlap, so that — in spite of their remarkably equal observed temperatures — these two bits of plasma have evolved from completely isolated and totally independent portions of the universe. We therefore expect that our map is inaccurate near the singularity and needs to be revised, but we currently lack the evidence needed to produce a plausable revision.

Acknowledgment

This research was supported in part by the National Science Foundation.

NOTES

1. S. L. Jaki, *The Paradox of Olbers' Pardox* (Herder and Herder, New York, 1969). While this book is an essential source for the history of the ideas called "Olbers' paradox," and abundantly demonstrates that "the proverbial respect of scientists for the facts of the laboratory does not necessarily include respect for the facts of scientific history . . ." (p. 243), it is flawed by a technical misunderstanding in claiming that "The only solution for Olbers' paradox is the willing acceptance of the concept of a universe finite in space . . ." (p. 240). In fact, all Friedmann universes (as well as the now defunct steady state universe) obtain a dark sky by cooling radiation through the action of the Hubble expansion. In contrast, Einstein's static (spatially finite) universe would have a blindingly bright night sky if the stars are assumed to have been eternally luminous.
2. V. A . Belinsky, E. M. Lifshitz, and I. M. Khalantnikov, Zh. E. T. F. **60** (1969) or Sov. Phys. JETP **33,** 1061 (1971).

26. ON THE EXTRAGALACTIC DISTANCE SCALE AND THE HUBBLE CONSTANT

Gerard de Vaucouleurs

The pass or fail test of physical theories is their degree of quantitative agreement with experiment or observation. A minute but stubborn discrepancy between the observed and calculated secular rate of precession of the orbit of Mercury was the first irremediable failure of Newton's law of gravitation and the first successful test of Einstein's general theory.

In general, such tests depend on the prior knowledge of some auxiliary constants not provided by the theory. Unfortunately it is not yet possible to derive the numerical values of the fundamental constants of nature from first principles — numerous attempts in this direction have all ended in failure or numerology. A critical confrontation of theory and observation requires, therefore, that we first provide the theory with precise empirical determinations of the fundamental constants, such as *h, c, g, e/m,* and so on. In the realm of cosmology the corresponding independent constants are the kinematic and dynamic scale factors represented by H_0, the Hubble "constant" measuring the current, local expansion rate, and q_0, the deceleration parameter; other physical parameters such as the mean mass and radiation densities, isotopic abundances, and so forth are more difficult to define, measure, and interpret.

The scale factor H_0 is particularly important because of its dual role: 1) as a velocity-distance ratio $H_0 = V_0/\Delta$ may be used as length scale factor since velocity can be measured directly via the Doppler shift, then $\Delta = V_0/H_0$ gives the distance, possibly with better precision than direct estimates; 2) as a relative expansion ratio $H_0 = (\dot{R}/R)_0$, its inverse H_0^{-1} is also a time scale factor (the Hubble time), which is the "age of the universe" if $q_0 = 0$, and an upper limit to it if $q_0 > 0$ (for $\Lambda = 0$).

In practice H_0 is difficult to determine with precision for two reasons:

1. The cosmological expansion rate is reduced by local gravitation in regions of space where the mean density $<\rho>$ is significantly greater than in the universe at large. Because of the clumpy, hierarchical distribution of galaxies on all observable scales (up to perhaps 300 megaparsecs $\simeq$ 1 billion light years), this leads to the concept that the Hubble *ratio* $H = V_0/\Delta$ is really a stochastic variable function of the local density of matter $<\rho>$, and to a redefinition of the Hubble *constant* H_0 as the asymptotic limit of H when $<\rho> \to 0$. For all we know the mean expansion rate in a

---

Harry Woolf (ed.), Some Strangeness in the Proportion: A Centennial Symposium to Celebrate the Achievements of Albert Einstein
ISBN 0-201-09924-1

volume some hundreds of megaparsecs across, where $<\rho> << 10^{-28}$ g/cm^3, may be a close enough approximation to H_0, but there is growing evidence that significant departures exist on scales of some tens of megaparsecs, such as in and near the Local Supercluster, where $<\rho> \gtrsim 10^{-28}$ g/cm^3.[1]

2. Velocities can be measured with adequate precision — say, better than 1 percent — at all distances, but distances can be measured indirectly (and with lower precision, say $\sim$10 percent) only out to the nearest galaxies, well within the Local Supercluster (say, $\Delta < 10$ Mpc) and even the Local Group ($\Delta < 1$ Mpc). It is from such a short base line and small sample that we must extrapolate by steps the extragalactic distance scale out to distances of at least several tens of megaparsecs where peculiar (random or systematic) motions are presumably negligible compared to the cosmological red shift and the mean density is not significantly in excess of the cosmic average. Herein lies the difficulty of the problem.

Estimates of the numerical value of H_0 in the practical units of km/sec/Mpc have ranged all the way from an early estimate of 560 by Hubble and Humason to values as low as 50 after Sandage and Tammann.[2] It is doubtful, however, that the latter can be accepted as final. The difficulty is mainly in the construction of a reliable scale of extragalactic distances. I have recently rediscussed the problem in great detail and come to the conclusion that the low-density or asymptotic value of the Hubble constant is $H_0 = 100 \pm 10$ (*me*).[3] The corresponding Hubble time is $H_0^{-1} = (9.5 \pm 1.0) \cdot 10^9$ years.

As an aid to understanding the sources of the disagreement with the previously favored value of 50 and to help the reader assess the validity of the methodology and the reliability of the data base underlying each determination a schematic comparison of two approaches to the Hubble constant is presented in Figs. 26.1 and 26.2.

Both involve three major steps: 1) a choice of galactic extinction model; 2) a selection of primary, secondary, and tertiary distance indicators and of calibration methods; 3) the derivation of the mean velocity-distance ratio with allowance for selection effects and for the influence of the Local Supercluster.

1. Galactic Extinction Model

Because the sun is near the equatorial plane of the gas-and-dust-filled galactic disk, comparison of stars in external galaxies with similar stars within our galaxy used as distance indicators must make proper allowance for interstellar extinction. This requires a choice of galactic extinction model and extinction law. The Sandage–Tammann (ST) model, inspired by the low color excesses of stars near the galactic poles, assumes that by a fortunate accident in the distribution of nearby individual dust clouds the sun is at the common apex of two opposite dust-free cones of $\sim 90°$ apertures centered at the North and South galactic poles, so that there is no extinction at all latitudes $|b| > 50°$ and a rather low value at all $|b| < 40°$. I prefer a variant of the standard model (plane parallel approximation) with 0.2 mag extinction (in blue light) at the poles; the evidence for this model rests on galaxy counts, galaxy colors, and hydrogen-luminosity ratios; the extinction formula takes into account a dependence on both latitude *and* longitude. The different assumptions of the two models introduce a mean zero point difference of 0.2 mag in the distance moduli (or 10 percent in distances) in the calibration of all extragalactic distances.[4]

The reddening law (or differential extinction ratio R) used by ST is classical, but is only correct to first-order; neglect of the second-order terms dependent on intrinsic color $(B - V)_0$ and color excess $E(B - V)$ introduces another zero point error of 0.2 mag in the calibration of the cepheids in galactic clusters, the *only* primary distance indicator used by ST.

COMPARISON OF TWO APPROACHES TO HUBBLE CONSTANT

	SANDAGE-TAMMANN 1974-75	DE VAUCOULEURS 1978-79
GALACTIC EXTINCTION MODEL	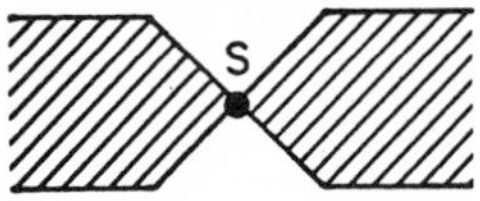 [hatched diagram, pinched at S]	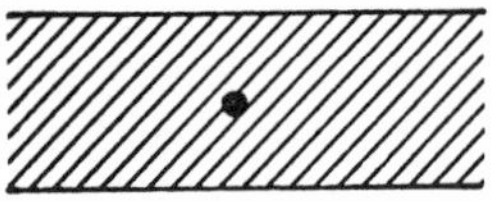 [hatched diagram, uniform slab]
EXTINCTION LAW	$A_B(\lvert b\rvert > 50^\circ) = 0$	$A_B(90^\circ) = 0.20$ MAG
	$A_B = 0.13\lvert \csc b - 1\rvert$ $(\lvert b\rvert < 40^\circ)$	$A_B = F(\ell, b)$
REDDENING LAW	$A = 3\ E(B-V)$	$A_V = R \cdot E$
		$R = 3.1 + 0.2(B-V)_o + 0.05(B-V)$
PRIMARY INDICATORS	1	5
CALIBRATION METHODS	1	10
SECONDARY INDICATORS	3	6
TERTIARY INDICATORS	1	3
LINEARITY CHECKS	0	2
GALAXIES	∿ 100	∿ 300
VELOCITY FIELD	ASSUMED LINEAR -ISOTROPIC	ALLOWANCE FOR LOCAL SUPERCLUSTER
H_o	50 ± 5 (?)	100 ± 10 (M.E.)
SUPPORTING EVIDENCE	1	5
	SN theory ?	HI line widths - H mag velocity dispersion in HII regions M-C and M-D relations in E,L superluminal velocities in quasars ?

Fig. 26.1 Comparison of two approaches to Hubble constant.

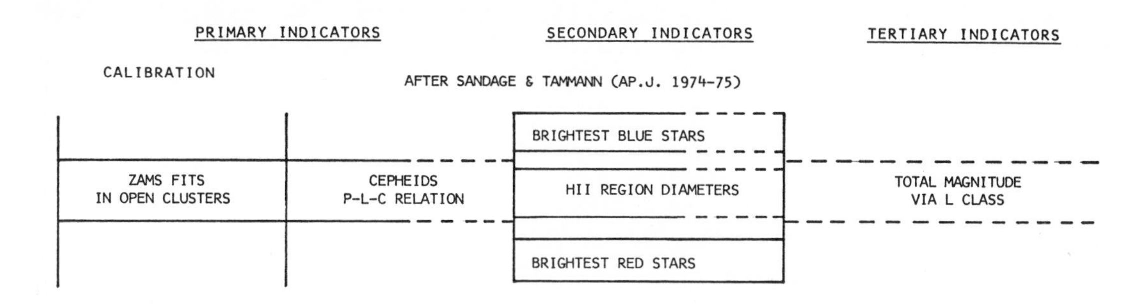

Fig. 26.2 Primary, secondary and tertiary indicators (Dashes indicate extrapolations).

2. Distance Indicators and Calibration

Primary indicators are stars that can be observed individually in the nearest galaxies, that is, in the Local Group ($\Delta < 1$ Mpc), and whose absolute magnitudes can be calibrated by fundamental methods in our galaxy. ST used only *one* primary indicator — the cepheid variables — and relied on a single calibration method (a comparison of the nearby Hyades cluster to eight more distant galactic clusters including cepheids). I used *five* different indicators (Novae, cepheids, RR Lyrae variables, AB supergiants, eclipsing variables) and no less than ten different methods of calibration (of which nine are independent). Because each indicator and each calibration method is subject to uncertainty, it is risky to depend entirely on one only; it is prudent, indeed necessary to use many to "spread the risks" and minimize possible systematic errors. A posteriori tests show that all five indicators are in systematic agreement within $\pm$ 0.15 mag and that the absolute calibration of distances in the Local Group has a zero point mean error of 0.1 mag. The mean systematic difference between the two approaches is 0.3 mag in the distance moduli of Local Group galaxies (that is, ST distances are systematically larger by 15 percent).

Secondary indicators are stars or other objects (star clusters, H-II regions) that can be calibrated in the Local Group and remain observable well beyond (say, $\Delta < 10$ Mpc). ST used *three* such indicators (brightest blue and red supergiants, apparent diameters of H-II regions). Unfortunately the last one that has a greater range than the other two is both ill-defined and subject to large systematic errors (the diameter measured visually on the photographs is not inversely proportional to distance — that is, it is not a metric length), and its application to more distant supergiant galaxies (ScI) requires an uncertain extrapolation.

I used *six* indicators (brightest blue and red stars, brightest variables, globular clusters, metric diameters of largest H-II rings, and velocity dispersion in H-II regions) that are not subject to such systematic errors and require little or no extrapolation. Again, the variety of methods and indicators provides consistency checks and tends to minimize cumulative errors. A posteriori tests show that all six indicators are in systematic agreement within $\pm$ 0.3 mag and that the total zero-point mean error is 0.15 mag. The mean systematic difference between the two approaches is 1.0 mag (that is, ST distances are systematically larger by $\sim$60 percent) for galaxies within 10 Mpc.

Tertiary indicators are metric or photometric properties (magnitudes, diameters) of the galaxies themselves that can be calibrated among the nearer galaxies ($\Delta < 10$ Mpc) and that are easily observable at much greater distances. These indicators usually require an independent estimate of the absolute luminosity of a galaxy, such as its luminosity class *LC*[5] or luminosity index Λ.[6]

ST used only *one* such indicator, the magnitude via its correlation with luminosity class calibrated by means of the absolute magnitudes derived from their H-II regions distances. Unfortunately, this directly carries forward the nonlinearity and systematic errors of the H-II regions calibration and extrapolation. The final result — and, therefore, the derived value of H_0 — depends critically on this extrapolation.

I used *three* indicators: the effective (or metric) and isophotal (photometric) diameters and the total magnitude (all corrected for extinction and inclination effects) via their correlations with the luminosity index (also corrected for inclination effects). All correlations were established by several different methods and were used only for reduction to a central standard value ($\Lambda_c = 1$ — that is, an "average" spiral), no extrapolation is involved. Tests demonstrate that the final results and the corresponding value of H_0 are insensitive to the analytical form of the interpolating function. A posteriori comparisons of the different tertiary indicators show them to be in systematic agreement within $\pm$ 0.5 mag and that they jointly define the zero point with a mean error of 0.2 mag (that is, 10 percent in distance). The mean systematic difference between the two approaches is 1.0 to 1.5 mag out to distances of several tens of megaparsecs.

Because of the precarious, piecemeal procedures used to construct the extragalactic distance scale it is prudent, even mandatory to obtain some independent check of the linearity of the resultant scale. Unfortunately *no* such test was presented by ST. I checked the distance scale derived from the primary, secondary, and tertiary indicators against *two* additional indicators not previously used in its construction: the brightest superassociation in galaxies (a new indicator) and the maxima of type I supernovae. Both confirm the linearity of the distance scale within a few percent over the whole 15 mag interval of distance moduli from 18 to 33, corresponding to distances ranging from 0.05 to 40 Mpc.*

***Note** (April, 1980): The linearity of the scale was verified again by comparison with distances of several hundred spirals derived independently from the width of the 21-cm H-I line width measuring their rotation velocity (Tully–Fisher relation).

3. Derivation of the Hubble Constant

Derivation of the Hubble constant requires red shifts and distances for a substantial sample of galaxies, well distributed over the whole sky, and proper allowance for selection effects and other sources of bias. In particular, a sample dominated by northern hemisphere objects (because of the preponderance of observatories in this hemisphere and of our outlying location in the Local Supercluster) will give too low an estimate of H_0.

ST used a sample of about 100 spiral galaxies and assumed the velocity field to be linear and isotropic about the Local Group (that is, they neglected the gravitational slowing down of the expansion within the supercluster). This approach leads to $H_0 = 50$ with an estimated 10 percent *internal* error, but has no way to estimate the true (external) error because of the lack of redundancy and external checks.

I used a sample of about 300 spirals and made an approximate allowance for the conspicuous anisotropy of the red shift law related to the supercluster; from an analysis of the velocity-distance relations in different directions, and particularly in the low-density regions (defined by the anti-center and polar cap regions of the supercluster), the asymptotic value of the Hubble ratio is found to be $H_0 = 100$, with a calculated *external* mean error of 10 percent based on a careful evaluation of all known causes of error.[7]

Other methods of estimating extragalactic distances may be called upon to support either the low or the high value of H_0. Two are theoretical and clearly speculative; one, the blackbody model of type I supernovae, tends to support the low value of H_0, the other, Lynden-Bell's theory of superluminal velocities in quasars, agrees with the high value. Both could, in principle, provide an independent absolute calibration, but neither carries much weight at present and obviously they cannot both be correct.

Four empirical methods are presently under investigation, none will provide an absolute calibration of the distance scale, but all can serve to test the linearity of the scale and strengthen the distance determinations. Two are applicable to spirals and rest on correlations between the absolute luminosities of galaxies and either the maximum rotation velocity (measured by the 21-cm H-I line widths) or the velocity dispersion in their brightest H-II regions. Two are applicable to elliptical and lenticular galaxies and depend on correlations between absolute luminosities and either color or diameter of the galaxies. Preliminary comparisons suggest that all tend to confirm the linearity of the distance scale leading to $H_0 = 100$ km/sec/Mpc.

Rapid progress may be expected in the next few years toward the resolution of the current disagreement in the absolute calibration of the extragalactic distance scale and a more reliable determination of the Hubble constant.

NOTES

1. G. de Vaucouleurs, "Further Evidence for a Local Supercluster of Galaxies: Rotation and Expansion," Astron. J. **63,** 253 (1958), and G. de Vaucouleurs, "Structure, Dynamics and Statistical Properties of Galaxies" in *The Formation and Dynamics of Galaxies,* International Astronomical Union Symposium Number 58 (Reidel, Dordrecht, Holland, 1974), p. 46.

2. E. Hubble and M. L. Humason, "The Velocity Distance Relation among Extragalactic Nebulae," Astrophys. J. **74,** 77 (1931).

   A. Sandage and G. A. Tammann, "Steps toward the Hubble Constant. I to VI," Astrophys. J. **190,** 525 (1974); **191,** 603 (1974); **194,** 223, and **559** (1975); **196,** 313 (1975); **197,** 265 (1975).

3. G. de Vaucouleurs, "The Extragalactic Distance Scale. I to VI," Astrophys. J. **223,** 351 and 730 (1978); **224,** 14 and 710 (1978); **227,** 380 and 729 (1979).

4. This distance modulus is, by definition, the difference between apparent and absolute magnitudes m,M and, numerically, $m - M = 5 \log \Delta + 25$, if Δ is the distance in megaparsecs.

5. S. van den Bergh, "A Preliminary Luminosity Classification of Late-type Galaxies," Astrophys. J. **131,** 215 and 558 (1960).

6. G. de Vaucouleurs, "Qualitative and Quantitative Classifications of Galaxies" in *The Evolution of Galaxies and Stellar Populations,* Yale University Conference, Yale University Observatory, New Haven, Conn. (1977), p. 43, and G. de Vaucouleurs, "The Local Supercluster" in *The Large Scale Structure of the Universe,* International Astronomical Union Symposium No. 79, (Reidel, Dordrecht, Holland, 1978), p. 205.

7. G. de Vaucouleurs and G. Bollinger, "The Extragalactic Distance Scale. VII," Astrophys. J. **233**, 433.

Open Discussion

Following Papers by D. W. SCIAMA and C. W. MISNER

Chairman, P. A. M. DIRAC

Unruh (U. of British Columbia): I direct this question to Dennis Sciama. In particular, in your arguments about homogeneity, the argument you mentioned is really just exactly the same as the argument for isotropy clothed as though it were an argument for homogeneity. And if you do not buy the cosmological principle and believe that there could be inhomogeneities that are spherically symmetrically distributed around us, it does not seem that one could tell at all that this were the case. In particular, as George Ellis has pointed out, our universe models already do place us in a preferred instant in time in our universe and you could equally well explain the observations by placing us in a preferred position in space instead and leaving the universe to be homogeneous in time. I was wondering what your comments to that would be?

Sciama: Yes, I think I agree with that. I was taking the conventional astrophysicist's view that there might be these very-large-scale lumps around the place. In fact, we thought about this some years ago when there was slight evidence then that the quasars on a large scale were clumped together. And if that were so and the quasar clumps represented also a localized density increase and they were just dotted about the universe and we were placed somewhere in that system, we would expect anisotropies in the three-degree background that were not observed and that led us to calculate effects of this kind. Well, the evidence for quasar clumping went away anyway, but we can still say that that kind of irregularity would be absent, and that already is something hard for us to understand, but other than that, I agree with what you say.

Dirac: Any more questions or comments?

J. Stachel (Boston U.): I preface this question by noting that Emerson remarked that consistency was a virtue of small minds. It will be clear why I said that. You consider vacuum fluctuations as a possible way out of the singularities and the use of the infinity of the Minkowski space-time as sort of a reference background relative to which one can calculate deviations and get negative values for the expectation value of the stress-energy tensor dotted into the null vector field generating the horizon. Now, this implies, it seems to me, using Minkowski space-time as a reference background for any gravitational field and regarding all other gravitational fields as deviations from that. Isn't that intolerable from a Machian point of view?

Sciama: There is another quotation along the lines of the one you said. Walt Whitman said, I believe, "Do I contradict myself? Very well then, I contradict myself." However, I do not think I have in fact contradicted myself. I think it is clear that if we do the black hole calculation, it is just a straightforward technical problem at that level, and I think the experts agree that you get this negative value in that case. Now, I am not quite sure what your question is —

Stachel: I'll clarify it. You pointed out that, just accepting the Machian point of view, one can eliminate anisotropies or an absolute rotation of the universe; and that on the same Machian basis one would have to deny physical reality to Minkowski space-time. Shouldn't you then just a priori reject a regularization argument that is based on ascribing an independent reality to Minkowski space-time, if you accept the Machian argument as a good valid ground for rejecting something?

Sciama: I think the point is that the zero-point fluctuations in Minkowski space do give you a response of the preferred inertial frames. In fact, when Bill Unruh discussed the fact that when you accelerate through the zero-point fluctuations you see a heat bath, of course, that is another example like when you rotate, you have centrifugal and coriolis forces acting on you. You can tell you are accelerating relative to the inertial frame by feeling the effect of the zero-point fluctuations on you. I agree with that; but I think from my point of view, what I would say is that the preferred inertial frames that the zero-point fluctuations are responding to in a kind of a rough Minkowski picture of the universe — of course, Minkowski space itself is not allowed, I agree, by Mach's principle — but the idea would be, if you have matter as sources and that matter is coupled to the radiation field, then the zero-point fluctuations of the matter would drive zero-point fluctuations in the radiation field. So that you could regard the zero-point fluctuations of the radiation field not as something sui generis without sources but actually produced by the distant matter. And I think, if you use that idea, which itself is involved in the most physical explanation I have for the Unruh effect, then you can get around this difficulty.

Dirac: Any further remarks?

Sciama: I would like to elaborate my point about the number of photons per baryon as a constant of nature. I am thinking of a particularly simple situation where you have a photon flux produced somehow previously and you have a baryon assembly, and you just let the universe expand and there are no further processes, say making heat, changing the number of photons, and there are none of these baryon-violating processes that might change the baryon structure. Then, in that simplest of all situations, it is an elementary result that the number of photons per baryon is constant. But, of course, if black holes form or baryon-violating processes occur or heat is produced by one process or another, then that ratio is changed. But it is just convenient to have a ratio that is time-independent under those simple assumptions I made as a way of specifying what I am trying to do, what my task is in making the heat in the photons.

Dirac: Any further comments?

H. R. Pagels (Rockefeller U.): I have a question for Dennis Sciama. Is there any evidence, cosmological evidence, at this time that the universe is not a fluctuation of the vacuum of some kind? The total charge of the universe is consistent with zero. If you can solve this problem with baryon asymmetry, it is not inconceivable that the baryon number of the universe was zero at one time, and it might be that the other quantum numbers of the universe are zero. Is there any evidence against this idea? That the universe is in fact degenerate with the vacuums?

Sciama: I think there is no evidence. It has been much discussed. I am really only repeating what I said before. I do not think I have anything new to add, but since you asked me, let me repeat it. It has been much discussed whether other galaxies could be antimatter. Probably other galaxies in the local group could not be, because there is also gas between the galaxies and there are enough interactions to prevent that being so, but I believe it is correct to say that if you take whole clusters of galaxies at a time and do not perhaps have too much intergalactic gas between them, then it is quite possible that there are as many anticlusters as there are clusters. The difficulty is only, as I say, we have not yet found an efficient separation mechanism. So that, with the mechanisms we are aware of at the moment in the early universe, these baryons and antibaryons would simply have annihilated. According to some people, including Roland Omnes in Paris, the precision with which baryons and antibaryons annihiliate is measured by this number 10^{-8} that keeps coming up. But to achieve that 10^{-8}, you need the separation mechanism, whose efficiency is in fact specified by this 10^{-8}, but the separation mechanisms he has worked with colleagues of his, in Paris particularly, have never been successful and so, of course, if you could think now of a more efficient separation mechanism that would actually be perhaps the most elegant theory, but no one has been able to come up with a sufficiently efficient separation mechanism.

IX. Quantum Gravity

27. QUANTUM GRAVITY AND SUPERGRAVITY

Yuval Ne'eman

For sixty years the theory of gravitation steered a separate course, majestically careless about the evolution in our understanding of the other fundamental interactions. Conscious of its beauty, well aware of its guaranteed superiority — no danger of there appearing more than one Einstein in a century — it let its enraptured adherents spend their happiest years in the uncovering of the innumerable further facets of all that splendor. While all other interactions were undergoing painful transmigrations, first into their quantum realization, and then materializing anew as relativistic quantum fields, gravitation serenely enjoyed that peculiar state of bliss it had reached right at birth, the *geometric* phase. Considering the method of physics, this had to be a "final" stage, and it seemed therefore inconceivable that one should try to undo all that perfection in dealing with the quantum aspects.

Yet there were intrusions. Spin was the first such message received from the quantum world.[1]

1. Spin as a Quantum Manifestation

Weyl, Fock, and Ivanenko found first a quiet niche in the cotangent manifold, supported by local frames

$$u^a = u^a{}_\mu dx^\mu \,, \tag{27.1}$$

and the dual tangent vector field base,

$$n_a = n_a{}^\mu D_\mu \tag{27.2}$$

$$u^a(n_b) = \eta^a{}_b \,, \qquad u^a{}_\mu n^\mu{}_b = \eta^a{}_b \,, \qquad u^a{}_\mu n_{a\nu} = g_{\mu\nu} \,. \tag{27.3}$$

That local flat frame had always existed in the background, since it was the subject of the "principle of equivalence." However, the principle itself, together with the "principle of relativity," were taken to imply that there was nothing special about it. Here however we had the first quantum effect — and it did emphasize just that "flat" local frame. This was one reason that spin was con-

Harry Woolf (ed.), Some Strangeness in the Proportion: A Centennial Symposium to Celebrate the Achievements of Albert Einstein
ISBN 0-201-09924-1

sidered as alien to general relativity, and the majority of relativists still prefer to disregard it, together with its originator, the Hilbert space Lorentz group.[2] Let us just note at this point that the principle of equivalence itself had to be "retouched"[3] so as to include the gravitational effects of spin.[4] It amounts to the replacements in the matter Langrangian,

$$\begin{aligned} &\delta^a{}_\mu \rightarrow u^a{}_\mu \\ &\partial_a = \delta_a{}^\mu \partial_\mu \rightarrow D_a = n_a{}^\mu(\partial_\mu + w_\mu{}^{cd} r_{cd}) \\ &\mathcal{L}(\psi, \partial_\mu \psi, \eta_{\mu\nu}) \rightarrow e\mathcal{L}(\psi, D_a \psi, \eta_{ab}) \\ &e = \det u^a{}_\mu \quad , \end{aligned} \tag{27.4}$$

where $w_\mu{}^{cd}$ is the Lorentz-connection field, r_{cd} a representation of the Lorentz generators, so that

$$w = w_\mu{}^{cd} r_{cd} dx^\mu \tag{27.5}$$

is the Lorentz connection. The appropriate theory is then known as the Einstein–Cartan version of general relativity, with nonvanishing but nonpropagating torsion in the presence of spinning matter.[5] Note that one could not help but require invariance of the frames under local Lorentz transformations, that is, gauging the group generated by

$$[J_{ab}, J_{cd}] = \eta_{ad} J_{bc} - \eta_{ac} J_{bd} + \eta_{bc} J_{ad} - \eta_{bd} J_{ac} \,. \tag{27.6}$$

Should one really be looking for quantum gravity? The late Leon Rosenfeld, for one, was not convinced of that necessity, and held that gravitation could stay as a "background" classical theory for all other interactions, perhaps in the spirit of one version of the Copenhagen interpretation of quantum mechanics (that is, classical apparatus). It seems that since 1970 the question has been answered in the affirmative as a result of a number of developments:

1. The emergence of possible macroscopic effects of quantum gravity, especially tunneling through Schwarzschild barriers,[6] for example, the decay of black holes.
2. Renewed hopes of a unification with other interactions, in the wake of the success of the weak-electromagnetic merger.[7]
3. Success in the renormalization of "internal" gauge theories,[8] a program initiated in fact as a "pilot project" for the treatment of gravity itself, which appeared much more complicated.[9] Note that the original "analogy" features (invariance under local diffeomorphisms → invariance under local Lie group transformations) have since been reinforced by the geometrization of internal gauge theories in fiber-bundle manifolds. I have just used the latter feature in deriving an aesthetic and irreducible mathematical hypothesis yielding the Salam–Weinberg phenomenological unification and constraining it severely.[10]
4. The discovery of supergravity[11] and the resulting progress in the quantum gravity program itself, together with what might become a unifying structure. Supergravity provided a completion of the mechanism through which spin is embedded in gravity, beyond the otherwise aesthetic merger provided by twistors. For example, spin 3/2 fields were

notorious for their indefinite metric difficulties and their acausal propagation of signals when coupled with electromagnetism. Supergravity showed how the gravitational field itself has to be present and how it resolves these paradoxes! This is a case in which the coupling of gravity and spin becomes all-important. It is interesting that supergravity also resolves the issue of positive-definiteness of the gravitational field's energy. Considering in addition the advances in renormalizability induced in this theory, it seems fair to state that supergravity offers for the first time a version of the "first-order formalism" of gravity that is as aesthetic and physically potent as Einstein's original theory.

2. Renormalization of Internal Gauge Theories

A Yang–Mills gauge[12] can be identified with the geometry of a principal bundle $P(M,G,\pi,\cdot)$ with M the base manifold corresponding to Minkowski space or some subspace, G the internal symmetry group, π the (vertical) projection, and the dot defining the right action of G on P.

$$\left.\begin{array}{ll} P \times G \rightarrow P\,, & \text{or } (p,g) \rightarrow p\cdot g \\ & \pi(p\cdot g) = \pi(p) \end{array}\right\} \forall\, p \in P\,, \quad g \in G \tag{27.7}$$

I have recently demonstrated[13] that a Yang–Mills gauge with spontaneous symmetry breakdown corresponds to such a principal bundle with G a supergroup (internal). The connection 1-forms ω^A on P can be rewritten for the generic case in a factorized form for $u = xy$

$$\begin{aligned} \omega^A(u) &= ad\,(y^{-1})^A_B\phi^B(dx) + (y^{-1}, dy)^A \\ &= \phi^{A-}(dx) + \chi^A(y, dy) \end{aligned} \tag{27.8}$$

for G a Lie group (y is the coordinate in G). The χ^A are the left-invariant 1-forms. A gauge transformation in P will make χ^A an x^μ-dependent field (the renormalization ghost[15] as we shall see) and taking x^μ as the lift from the base-space coordinates onto a section, $\phi^A = \phi_\mu{}^A dx^\mu$ with $\phi_\mu{}^A$ the Yang–Mills potential (for the odd subalgebra in a supergroup, $\phi_\mu{}^J$ is a ghost vector meson). Note that $\phi_\mu{}^A(x,y)$ depends on y since any $\phi_\mu{}^A(x)$ can be transformed by $e^{-i(y^B\Lambda_B)}\ \phi\ e^{i(y^B\Lambda_B)}$. Matter fields ψ "reside" in associated vector bundles $E(P,M,G,r(G))$ where $r(G)$ is the representation of G the ψ realize. The gauge Lagrangian is given by the following equations, ($C^A{}_{BE}$ are the structure constants of the algebra of G):

$$\mathcal{L}_0 = R^A \wedge {}^*R_A + \mathcal{L}\text{ matter } (d\psi \rightarrow D\psi) \tag{27.9}$$

$$R^A = d\omega^A - \tfrac{1}{2}C^A{}_{BE}\omega^B \wedge \omega^E \tag{27.10}$$

$${}^*R_A = \tfrac{1}{2}g_{AB}\varepsilon_{\mu\nu\rho\sigma}R^B{}_{\mu\nu}dx^\rho \wedge dx^\sigma \tag{27.11}$$

d is the horizontal exterior derivative, D the horizontal covariant derivative (that is, parallel-transporting vertical vectors) and R^A is the curvature, a horizontal 2-form with a vertical anholonomic index$A$. The equations of motion are

$$DR^A = 0 \qquad \text{(Bianchi)} \tag{27.12}$$

$$D{*}R_A = {*}j_A \tag{27.13}$$

where ${*}j_A$ is the current 3-form,

$$ {*}j_A = \tfrac{1}{6}\varepsilon_{\mu\nu\rho\sigma} j^\mu{}_A dx^\nu \wedge dx^\rho \wedge dx^\sigma \tag{27.14}$$

The gauge transformation is given (for an infinitesimal variation) by

$$\delta\phi^A = D\varepsilon^A, \qquad \text{for } \psi' = \exp(i\varepsilon^A \Lambda_A)\psi \tag{27.15}$$

The covariant renormalization procedure[16] adds two new terms to $\mathcal{L}_0$,

$$\mathcal{L} = \mathcal{L}_0 + \mathcal{L}_F + \mathcal{L}_G \tag{27.16}$$

where $\mathcal{L}_F$ is the gauge-fixing term, essential for inversion and the existence of a propagator. $\mathcal{L}_G$ involves "ghost fields" χ^A, $\bar{\chi}^A$ with variations such as to render $\mathcal{L}$ invariant again, this time under a global transformation.[17] The requirements on the ghosts have been reformulated as a specific gauge transformation by Becchi, Rouet, and Stora (BRS), and were recently rederived geometrically. In this derivation, it was shown that the ghost field χ^A are indeed the semivertical components of ω^A, that is, orthogonal to the section defined by $\mathcal{L}_F$. The χ^A-anticommute (since they are 1-forms). In the case of a supergroup, the odd χ^A represent the Goldstone–Higgs fields. The BRS equations result directly from the structural equations of the bundle, that is, from the conditions that R^A in (27.10) have no nonhorizontal component.

$$s\chi^A - \tfrac{1}{2}C^A{}_{BE}\chi^B \wedge \chi^E = 0 \tag{27.17}$$

$$s\phi^A - \not{D}\chi^A = 0 \tag{27.18}$$

$$\not{d}\phi^A - \tfrac{1}{2}C^A{}_{BE}\phi^A \wedge \phi^E = R^A \tag{27.19}$$

The antighost $\bar{\chi}^A$ is nongeometric and its variation depends on $\mathcal{L}_0$ and $\mathcal{L}_F$. Classically, $s\bar{\chi}^A = 0$. The "global" transformation corresponds to the action of Λs, Λ a Grassman element. BRS used $\Lambda s = \delta^A$, a gauge transformation with "parameter" $\Lambda\chi^A$. We see that s is in fact a component (semivertical) of d, with $\not{d}$ and $\not{D}$ denoting the (complementary) component of d and D in the section itself. $\phi_\mu{}^A$ is thus indeed the physical (horizontal) gauge potential.

BRS invariance of $\mathcal{L}$ is automatic since as a 4-form it is closed,

$$d\mathcal{L} = s\mathcal{L} = 0 \tag{27.20}$$

BRS invariance or (27.20) guarantee unitarity, and the noninvolvement of the ghosts χ in physical processes (which is evident in our geometric picture).

The remaining task in renormalizing consists in classifying the necessary counterterms, needed for a finite S-matrix. If such counterterms vanish, or if they can be absorbed in the original

$\mathcal{L}$ through a rescaling, or if they reduce to a finite set of terms, for all orders of loop diagrams — the theory is renormalizable. These loop diagrams have any number of external lines attached to them, that is, they end up with on-mass-shell fields. Computation of n-loop diagrams can be an impossible task, and it is mostly with the help of symmetry arguments that one can perform this census.

3. Noninternal Gauges and the Soft Group Manifold

Using P for gravity, supergravity, and so on causes double counting: space-time appears once in M, as the base space and again in G, as the parameter manifold of the translations. This method has been applied in the past ("affine connections" or "Cartan connections") with "soldering" — that is, putting both M^4 manifolds in tandem. It becomes even more confusing in supergravity or extended supergravity: space-time is doubled although the complementary spinor dimensions are not. Moreover, the Lagrangian is only invariant under a Lorentz local gauge, both in gravity and in supergravity; it is only after applying the equations of motion that it becomes invariant under a Poincaré gauge in gravity, and a graded-Poincaré gauge in supergravity.

To describe a noninternal gauge, Tullio Regge and I have introduced the "soft" group manifold.[19] The rigid manifold of an n-dimensional Lie group G with left-invariant forms ω^A fulfills the Cartan–Maurer equation

$$d\omega^A - \tfrac{1}{2}C^A{}_{BE}\omega^B \wedge \omega^E = 0 \tag{27.21}$$

and is generated by the left-invariant algebra

$$[D_B, D_E] = C^A{}_{BE}D_A \tag{27.22}$$

with orthonormality

$$\omega^A(D_B) = \delta^A{}_B \tag{27.23}$$

In the soft group manifold, the ω^A are replaced by an n-form ρ^A,

$$d\rho^A - \tfrac{1}{2}C^A{}_{BE}\rho^B \wedge \rho^E = R^A \tag{27.24}$$

$$R^A = \tfrac{1}{2}\rho^B \wedge \rho^E \wedge R^A{}_{BE}$$

or therefore,

$$d\rho^A - \tfrac{1}{2}\rho^B \wedge \rho^E(C^A{}_{BE} + R^A{}_{BE}) = 0 \tag{27.25}$$

The D_A are replaced by $\widetilde{D}_A$,

$$\rho^A(\widetilde{D}_B) = \delta^A{}_B \tag{27.26}$$

$$[\widetilde{D}_B, \widetilde{D}_E] = (C^A{}_{BE} + R^A{}_{BE})\widetilde{D}_A \tag{27.27}$$

Looking at the group multiplication law between two n-dimensional elements $x \cdot y$, with a mapping between the two copies $G_x \xrightarrow{\lambda} G_y$, can be regarded as the gauged action of the rigid group transformation y on the (soft) manifold G_x. This yields

$$\Gamma(\lambda)\rho_x{}^A = \lambda^*\omega_y{}^A + \rho_x{}^B \operatorname{adj}(\lambda^{-1}(x))_B{}^A \tag{27.28}$$

$$\Gamma(\lambda)R_x{}^A = R^B{}_x \operatorname{adj}(\lambda^{-1}(x))_B{}^A \tag{27.29}$$

and a covariant derivative

$$D\eta^A = d\eta^A - C^A{}_{BE}\rho^B \wedge \eta^E \tag{27.30}$$

$$(D(D\eta))^A = -C^A{}_{BE}R^B \wedge \eta^E \tag{27.31}$$

$$(DR)^A = 0 \tag{27.32}$$

Infinitesimally,

$$\delta\rho^A = D\varepsilon^A \tag{27.33}$$

However, if we "gauge" the "softened" generators $\widetilde{D}_A$ of (27.26) and (27.27) we find[20]

$$\delta\rho^A = D\varepsilon^A - 2(\varepsilon, R^A) \tag{27.34}$$

This is in fact the Lie-derivative $L_\varepsilon\rho^A$.

The second contribution can be shown[21] to derive from the diffeomorphisms on G, with

$$\varepsilon^M = \varepsilon^A\Delta_A{}^M, \qquad \widetilde{D}_A = \Delta_A{}^M J_M \tag{27.35}$$

(an "anholonomized general coordinate transformation"). They would have appeared in P as well, had we included variations under horizontal transformations besides those generated by the structure group. Here they have to appear in all directions at this stage, since there is as yet no choice of a "vertical" direction.

If the algebra **G** of G in the soft group manifold (SGM) is weakly reducible; that is,

$$\mathbf{G} = \mathbf{H} + \mathbf{F}, \qquad [\mathbf{F},\mathbf{F}] \subset \mathbf{F}, \qquad [\mathbf{F},\mathbf{H}] \subset \mathbf{H} \tag{27.36}$$

a transformation (27.28) can be found,[22] such that all connections ρ^A and curvatures R^A are reduced to forms on H only; that is, they are horizontal. This is a factorization of the F variables. At this point, the contraction (ε, R^A) in (27.34) involves only H variables, and (27.34) can be rewritten as

$$\begin{aligned} \delta\rho^E &= D\varepsilon^E - \rho^I\varepsilon^J R^E{}_{IJ} \\ \delta\rho^K &= D\varepsilon^K - \rho^I\varepsilon^J R^K{}_{IJ} \\ E,F,G &\in \{F\}\,; \qquad I.J.K \in \{H\} \end{aligned} \tag{27.37}$$

Inserting the values for the Poincaré group $\mathscr{P}$, with the Lorentz group (indices *ab, cd,* ...) as F and the translations (indices $a,b,\ldots$) as H, we get the variations of Hehl et al.[23] with R^E for the curvature and R^K the torsion. Note that using (27.27) for $[\tilde{D}_a, \tilde{D}_b]$ corresponds to the correct physical notion of a parallel translation in a curved space. For $\mathscr{G}$, the graded Poincare group, F is the Lorentz group again, and H is "superspace"

$$R^{4/4} = \mathscr{G}/SO(1,3) \tag{27.38}$$

Factorization is not equivalent to fibration, which would require in addition that $R^K = 0$. By choosing a Lagrangian $\mathcal{L}$ which is F-gauge invariant but not G-gauge invariant, spontaneous fibration[24] does occur over a suitable range of groups, provided they are weakly reducible and symmetric (WRS)

$$[\mathbf{H},\mathbf{H}] \subset \mathbf{F} \tag{27.39}$$

$\mathscr{P}$ is WRS, so that

$$R^I \overset{\text{o}}{=} 0 \tag{27.40}$$

where $\overset{\text{o}}{=}$ denotes the application of the equations of motion. $\mathcal{L}$ is then G-invariant on-mass-shell. In $\mathscr{G}$, (27.39) is not obeyed since (α,β are spinor; a,b vector and ab cd tensor 'vielbein" indices)

$$\{D_\alpha, D_\beta\} = (C_\gamma{}^a)_{\alpha\beta} D_a \tag{27.41}$$

and indeed $R^\alpha \neq 0$ even on mass shell. For pure supergravity,

$$R^a \overset{\text{o}}{=} 0$$

but the presence of matter superfields in a non-WRS group may invalidate any such specific cancellations. This nonvanishing of curvature components is why workers in supergravity came across situations that were described as "nonclosure" of the local algebra. This was nonclosure as compared to (27.22), but the relevant algebra (which does close) is (27.27), with R^A contributions. Recently, a version with "closed" algebra was suggested[25] in which the components $R_{\alpha\beta}{}^A$, $R_{a\alpha}{}^A$ were calculated as "auxiliary fields." Because of the equations of motion, there exist relations between these components and it was shown[26] that they included only six independent elements (an axial vector, a scalar, and a pseudoscalar). In the first reference cited in note 19 it was shown that for pure supergravity, there were still several subsisting R^A contributions in the variations (27.37)

$$\delta\rho^{ab} \overset{\text{o}}{=} D\varepsilon^{ab} - \rho^c\varepsilon^d R_{cd}{}^{ab} - \rho^c\bar{\varepsilon}^\alpha R_{c\alpha}{}^{ab} \tag{27.42}$$

$$\delta\rho^a \overset{\text{o}}{=} D\varepsilon^a \tag{27.43}$$

$$\delta\rho^\alpha \overset{\text{o}}{=} D\varepsilon^\alpha - \rho^c\varepsilon^d R^\alpha{}_{cd} \tag{27.44}$$

$\bar{\varepsilon}^\alpha$ of local supersymmetry involves only $R_{ca}{}^{ab}$ (the supersymmetric variation of the "connection" in what amounts to a first-order formalism[27]). However, $R^\alpha{}_{cd}$, the spinor-torsion, enters the translation gauge in (27.44), and when matter is present, other components will appear. The detailed formulas for the curvatures and torsions are (the $[\ \ ]_{SG}$ bracket terms that have to be added when going from gravity to supergravity)

$$R^{ab} = d\rho^{ab} + \rho^{at} \wedge \rho^{tb} \tag{27.45}$$

$$R^a = d\rho^a + \rho^{at} \wedge \rho^t [+ \tfrac{1}{2}\bar{\rho}\gamma^a\rho]_{SG} \tag{27.46}$$

$$[R^\alpha = d\rho^\alpha + \tfrac{1}{2}((\rho^{ab}\sigma^{ab})\rho)^\alpha]_{SG} \tag{27.47}$$

where due to factorization, $z \in R^{4/4}$

$$\rho^{ab} = \Gamma_M{}^{ab} dz^M, \qquad \rho^a = u_M{}^a dz^M, \qquad \rho^\alpha = \psi_M{}^\alpha dz^M \tag{27.48}$$

Nevertheless, both (27.45–27.47) and (27.42–27.44) hold when restricted to space-time M^4. This corresponds to the zero term in the expansions in the Grassman variable. Note that restrictions to the two different M^4 submanifolds are related to each other by a parallel-transport gauge transformation (27.42–27.44) with parameter ε^α ("local supersymmetry"). In a recent work applying our SGM method,[28] it was shown that the restrictability to M^4 for the $\delta\rho^A$ can be used as a set of constraints on the $R^A{}_{IJ}$ for larger groups.[29]

4. Status of Quantization Program

Following the checks on unitarity in particular gauges,[30] BRS transformations were also written for gravity. They were derived first for holonomic ("Einsteinian") gravity,[31] where the invariance is just that of Diff M^4:

$$\tilde{g}^{\mu\nu} = g^{\mu\nu}\sqrt{-g} = \eta^{\mu\nu} + \varkappa h^{\mu\nu} \tag{27.49}$$

$$\delta\chi_\nu - \chi^\mu \partial_\mu \chi_\nu \Lambda = 0 \tag{27.50}$$

$$\delta h_{\mu\nu} - D_{\mu\nu\lambda}\chi^\lambda \Lambda = 0 \tag{27.51}$$

and $\delta\bar{\chi}^\nu = -\partial_\mu h^{\mu\nu}\Lambda/\alpha$ for the nongeometric "antighost." Compare with (27.17–27.18) (α is a parameter given by the gauge fixing).

For "Einstein-Cartan" gravity, that is the first-order form of Einstein gravity, including a Lorentz gauge on tetrad indices, the BRS equations were then shown to be[32]

$$\delta\chi^\lambda + \chi^\nu \partial_\nu \chi^\lambda \Lambda = 0 \tag{27.52}$$

$$\delta\chi^{ab} + \chi^\nu \partial_\nu \chi^{ab}\Lambda - \chi^{ac}\chi_c{}^b \Lambda = 0 \tag{27.53}$$

$$\delta u^a{}_\mu - (\chi^\nu \partial_\nu u^a{}_\mu + \partial_\mu \chi^\nu u^a{}_\nu + \chi^a{}_b u^b{}_\mu)\Lambda = 0 \tag{27.54}$$

$$\delta\Gamma_\mu{}^{ab} - (\chi^\nu\partial_\nu\Gamma_\mu{}^{ab} + \partial_\mu\chi^\nu\Gamma_\nu{}^{ab} - D_\mu\chi^{ab})\Lambda = 0 \qquad (27.55)$$

The use of two independent and intertwining local gauges does not help in identifying the geometric structure of (27.52–27.55). They are supplemented by

$$\delta\bar{\chi}^a = -\eta^{\mu\nu}\partial_\nu u^a{}_\mu\Lambda$$

$$\delta\bar{\chi}^{ab} = -\eta^{\mu\nu}\partial_\nu\Gamma_\mu{}^{ab}\Lambda$$

We have now proved[33] that the geometric identification of the ghosts and of the BRS variation can be extended to the algebra of (27.34). The BRS equations (27.17), (27.18) acquire a right-hand side allowing for off-mass-shell curvature structure functions. For the Poincaré gauge $\mathscr{P}$, which we have treated as in (27.37). We find[33]

$$\delta\chi^a + \chi^{at}\chi^t\Lambda + \tfrac{1}{2}\chi^c\chi^d R_{cd}{}^a\Lambda + \tfrac{1}{2}\chi^{cd}\chi^{ef}R_{cd\,ef}{}^a\Lambda + \chi^c\chi^{de}R_{c\,de}{}^a\Lambda = 0 \qquad (27.56)$$

$$\delta\chi^{ab} + \chi^{at}\chi^{tb}\Lambda + \tfrac{1}{2}\chi^c\chi^\delta R_{cd}{}^{ab}\Lambda + \tfrac{1}{2}\chi^{cd}\chi^{ef}R_{cd\,ef}{}^{ab}\Lambda + \chi^c\chi^{de}R_{c\,de}{}^{ab}\Lambda = 0 \qquad (27.57)$$

$$\delta u_\mu{}^a - (\partial_\mu\chi^a + \Gamma_\mu{}^{at}\chi^t - u_\mu{}^t\chi^{at} - u_\mu{}^c\chi^d R_{cd}{}^a - u_\mu{}^c\chi^{de}R_{c\,de}{}^a -$$

$$\Gamma_\mu{}^{cd}\chi^e R_{cd\,e}{}^a - \Gamma_\mu{}^{cd}\chi^{ef}R_{cd\,ef}{}^a)\Lambda + (\partial_\mu\chi^\nu u_\mu{}^a + \partial_\mu\chi^{\rho\sigma}u_{\rho\sigma}{}^a)\Lambda = 0 \qquad (27.58)$$

$$\delta\Gamma_\mu{}^{ab} - (\partial_\mu\chi^{ab} + \Gamma_\mu{}^{at}\chi^{tb} - u_\mu{}^c\chi^d R_{cd}{}^{ab} - u_\mu{}^c\chi^{de}R_{c\,de}{}^{ab} - \Gamma_\mu{}^{cd}\chi^e R_{cd\,e}{}^{ab} -$$

$$\Gamma_\mu{}^{cd}\chi^{ef}R_{cd\,ef}{}^{ab})\Lambda + (\partial_\mu\chi^\nu\Gamma_\nu{}^{ab} + \partial_\mu\chi^{\rho\sigma}\Gamma_{\rho\sigma}{}^{ab})\Lambda = 0 \qquad (27.59)$$

The holonomic-summation last terms in (27.58–27.59) correspond to an additional "alias" transformation restoring the original value of x. These terms vanish in the geometric identification because $\chi = dx^\nu$ and $\chi^{\rho\sigma} = d\Xi^{\rho\sigma}$ ($\Xi^{\delta\sigma}$ is the Lorentz group parameter). For a noninternal gauge theory of a weakly reducible group G, as in (27.26), with factorized subgroup $F \subset G$, the Lagrangian is written as

$$\mathcal{L} = R^A \wedge \zeta_A \qquad (27.60)$$

and we get equations of motion, which determine the curvature terms

$$D\zeta_A + R^B\delta\zeta_B/\delta\rho^A = 0 \qquad (27.61)$$

For gravity, $\zeta_{ab} = \varepsilon_{abcd}\rho^c \wedge \rho^d$, $\rho^a = u_\mu{}^a dx^\mu$. Writing $\delta^{\text{BRS}} = s\Lambda$, we can reproduce equations (27.56–27.59) for any noninternal factorized gauge group,

$$s\chi^A - \tfrac{1}{2}(C^A{}_{BE})\chi^B \wedge \chi^E = 0 \qquad (27.62)$$

$$s\tau^A - \not{D}\chi^A = \tau^B\chi^E R_{BE}{}^A + (\not{d}\chi,\rho^A) \qquad (27.63)$$

Equation (27.63) has been taken for $\rho^A \to \tau^A$, the restriction to the quotient G/F.

In supergravity, in the nongeometric interpretation in which the algebra doesn't close off-mass-shell, new BRS transformations were suggested[34] involving a $(\chi)^4$ term in L. They were adapted to first-order formalism.[35] These nongeometric transformations also involved making a choice with respect to the commutation relations between the spinor gauge field $\psi_\mu{}^\alpha$ and χ^ν or χ^{ab}. The traditional prescription would be to have these anticommute, but with our identification of χ^ν and χ^{ab} as 1-forms on the internal variable $(\Xi^{\rho\sigma})$ of the even part of the superalgebra, their anticommuting properties do not carry over to $\psi_\mu{}^\alpha$, and they should commute with it. We can now rewrite the BRS equations for supergravity, using (27.37), (27.62), and (27.63). Indeed, the closure of the algebra[36] had already been able to remove the $(\chi)^4$ terms,[37] at the price of introducing auxiliary fields for closure. We find, using the curvatures, supersymmetric BRS equations:

$$\delta\chi^a = \{\chi^{at}\chi^t + \tfrac{1}{2}\bar{\chi}_\gamma{}^a\chi + \tfrac{1}{2}\chi^c\chi^d R_{cd}{}^a + \bar{\chi}^\alpha\chi^c R_{\alpha c}{}^a + \tfrac{1}{2}\chi^{cd}\chi^{ef}R_{cd\,ef}{}^a$$

$$+ \chi^c\chi^{de}R_{cd\,e}{}^a + \chi^{cd}\bar{\chi}^\alpha R_{cd\,\alpha}{}^a + \bar{\chi}^{cd}\chi^\alpha R_{cd\,\alpha}{}^a + \tfrac{1}{2}\chi^\alpha\chi^\beta R_{\alpha\beta}{}^a\}\wedge \tag{27.64a}$$

$$\delta\chi^{ab} = -\{\chi^{at}\chi^{tb} + \tfrac{1}{2}\chi^c\chi^d R_{cd}{}^{ab} + \bar{\chi}^\alpha\chi^c R_{\alpha c}{}^{ab} + \tfrac{1}{2}\chi^{cd}\chi^{ef}R_{cd\,ef}{}^{ab} +$$

$$\chi^c\chi^{de}R_{c\,de}{}^{ab} + \chi^\alpha\bar{\chi}^\alpha R_{cd\,\alpha}{}^{ab} + \tfrac{1}{2}\bar{\chi}^\alpha\bar{\chi}^\beta R_{\alpha\beta}{}^{ab}\}\wedge \tag{27.64b}$$

$$\delta\chi^\alpha = -\{\chi^{ab}\sigma^{ab}\chi + \tfrac{1}{2}\chi^c\chi^d R_{cd}{}^\alpha + \bar{\chi}^\beta\chi^c R_{\beta c}{}^\alpha + \tfrac{1}{2}\chi^{cd}\chi^{ef}R_{cd\,ef}{}^\alpha +$$

$$\chi^c\chi^{de}R_{c\,de}{}^\alpha + \chi^{cd}\bar{\chi}^\beta R_{cd\,\beta}{}^\alpha + \tfrac{1}{2}\bar{\chi}^\beta\bar{\chi}^\gamma R_{\beta\gamma}{}^\alpha\}\wedge \tag{27.64c}$$

$$\delta\hat{\tau}_\mu{}^a = \{\partial_\mu\chi^a + \hat{\tau}_\mu{}^{at}\chi^t - \hat{\tau}_\mu{}^t\chi^{at} - \hat{\tau}_\mu{}^c\chi^d R_{cd}{}^a - \hat{\tau}^\alpha\chi^d R_{\alpha d}{}^a -$$

$$\hat{\tau}^\alpha\chi^{cd}R_{\alpha\,cd}{}^a - \hat{\tau}_\mu{}^c\chi^{de}R_{c\,de}{}^a - \hat{\tau}_\mu{}^{cd}\chi^{ef}R_{cd\,ef}{}^a - \hat{\tau}_\mu{}^c\bar{\chi}^\beta R_{c\beta}{}^a -$$

$$\hat{\tau}_\mu{}^\alpha\bar{\chi}^\beta R_{\alpha\beta}{}^a + \tau_\mu{}^{cd}\chi^e R_{cd\,e}{}^a - \tau_\mu{}^{cd}\bar{\chi}^\alpha R_{cd\,\alpha}{}^a\}\wedge \tag{27.65a}$$

$$\delta\hat{\tau}_\mu{}^{ab} = \{\partial_\mu\chi^{ab} + \hat{\tau}_\mu{}^{at}\chi^{tb} - \hat{\tau}_\mu{}^{tb}\chi^{at} - \hat{\tau}_\mu{}^c\chi^d R_{cd}{}^{ab} - \hat{\tau}_\mu{}^\alpha\chi^d R_{\alpha d}{}^{ab} -$$

$$\hat{\tau}_\mu{}^c\bar{\chi}^\alpha R_{c\,\alpha}{}^{ab} - \hat{\tau}_\mu{}^c\chi^{de}R_{c\,de}{}^{ab} - \hat{\tau}_\mu{}^{cd}\chi^{ef}R_{cd\,ef}{}^{ab} - \hat{\tau}_\mu{}^c\chi^\beta R_{c\beta}{}^{ab} -$$

$$\hat{\tau}_\mu{}^\alpha\bar{\chi}^\beta R_{\alpha\beta}{}^{ab} - \hat{\tau}_\mu{}^{cd}\chi^e R_{cd\,e}{}^{ab} - \hat{\tau}_\mu{}^\alpha\chi^{cd}R_{\alpha\,cd}{}^{ab}\}\wedge \tag{27.65b}$$

$$\delta\hat{\tau}_\mu{}^\alpha = \{\partial_\mu\chi^\alpha - \tfrac{1}{2}(\hat{\tau}_\mu{}^{ab}\sigma^{ab}\chi)^\alpha + \tfrac{1}{2}(\sigma^{ab}\hat{\tau}_\mu{}^\alpha\chi^{ab} - \hat{\tau}_\mu{}^{cd}\chi^{ef}R_{cd\,ef}{}^\alpha -$$

$$\hat{\tau}_\mu{}^c\chi^{de}R_{c\,de}{}^\alpha - \hat{\tau}_\mu{}^c\chi^d R_{cd}{}^\alpha - \hat{\tau}_\mu{}^\beta\chi^c R_{\beta c}{}^\alpha - \hat{\tau}_\mu{}^c\bar{\chi}^\beta R_{c\beta}{}^\alpha -$$

$$\hat{\tau}_\mu{}^\beta\bar{\chi}^\gamma R_{\beta\gamma}{}^\alpha - \hat{\tau}_\mu{}^{cd}\chi^e R_{cd\,e}{}^\alpha - \hat{\tau}_\mu{}^{cd}\bar{\chi}^\beta R_{cd\,\beta}{}^\alpha - \hat{\tau}_\mu{}^\beta\chi^{cd}R_{\beta\,cd}{}^\alpha\}\wedge \tag{27.65c}$$

The $\hat{\tau}_\mu{}^A$ are the space-time restrictions of the superspace $\hat{\tau}_\mu{}^A$. We have omitted the alias terms.

The components $R_{AB}{}^{G}$ are related by the equations of motion. For local supersymmetry, they are the ones involving a spinor χ^{α}. These are $(R_{c\beta}{}^{a}, R_{\alpha\beta}{}^{a}, R_{c\beta}{}^{ab}, R_{\alpha\beta}{}^{ab}, R_{c\beta}{}^{\alpha}, R_{\beta\gamma}{}^{\alpha}, R_{cd\ \beta}{}^{a}, R_{cd\ \beta}{}^{ab}, R_{cd\ \beta}{}^{\alpha})$, that is, $3 \times 3 = 9$ terms. The auxiliary fields that have appeared in the literature are in fact functions of this nonindependent set. Our curvatures may be factorized, reducing to superspace expressions; however, the entire set of variations is valid for their space-time restrictions, that is, the zero-term in a θ^{α} expansion. Clearly, our SGM approach, with the curvature-torsion terms in (27.37) has the advantage of replacing the guesswork involved in the nonlinear "tensor calculus" by directly computable quantities.

The second issue is that of the counterterms necessary for a finite S-matrix. From a study of one-loop diagrams in gravity, it had been shown[38] that pure gravity was indeed one-loop renormalizable (using a result of DeWitt's in the early sixties) but that the presence of matter (whether scalar, vector, or spinor[39]) destroyed renormalizability. With the advent of supergravity, it was shown that both one-loop and two-loop diagrams are renormalizable.[40] The proof for the latter case is somewhat incomplete at this stage since the possible effect of anomalies has not yet been fully understood, and only one-loop diagrams have been checked explicitly. The one- and two-loop results hold for extended supergravity, with an internal $0(N)$. Note that all of this also means that *Gravity is one- and (probably) two-loop renormalizable in the presence of matter*, provided the matter fields and their couplings correspond to that extended supergravity multiplet that includes the graviton $h^{\mu\nu}$. This constitutes a considerable advance, brought about by Supergravity. However, in three-loop diagrams,[41] counterexamples have been found; these candidate counterterms obey "extended local supersymmetry" and can thus not be excluded by present symmetry arguments, (these have since been shown to occur in all higher orders), though they may yet cancel in explicit calculation, but such calculations are extremely cumbersome.

Another result due to supergravity[42] is the positive definiteness of the energy in pure gravity. From (27.27) we have

$$\{\widetilde{D}_{\alpha}, \widetilde{D}_{b}\} = (C_{\gamma}{}^{a})_{\alpha\beta}\widetilde{D}_{a} \tag{27.66}$$

since no $R^{A}{}_{\alpha\beta}$ survives after applying the equations of motion. This is isomorphic to (27.41) and implies that the Hamiltonian is the square of the $\widetilde{D}_{\alpha}$. The result has been shown *to hold for pure quantum gravity*, by a proper choice of physical states (without ψ_{μ} spin 3/2 quanta) and in the "tree" limit.[43]

2. Extended Supergravity

Work on extended supergravity aims at unification with the internal symmetries of the other interactions. From a study of the representations of global extended supersymmetry and the relevant (holonomic) graviton multiplet, it was found that the internal symmetry group is $O(N)$, and $N \leqslant 8$. $O(8)$ is too small to contain $U(2)_{w} \subset SU(2/1)$ of the weak-electromagnetic gauge and $SU(3)_{\text{color}}$. It is possible that only one of the two systems should finally be embedded in supergravity. Alternatively, using the graded-conformal group, one can have the analog super-Weyl-gravity theory.[44] The internal symmetry is then a parity nonconserving $U(N)$. Conformal supergravity has the advantage of a dimensionless coupling, an important point for renormalizability. However, it does not yield Einstein's theory. This may be remedied through spontaneous breakdown, as suggested for dilations in the recently revived version[45] of Weyl's theory.[46]

Extended supergravity has been studied in terms of $0Sp(4/N)$. The orthosymplectic supergroups $0SP(2r/N)$ consist in those transformations of a graded manifold with $2r$ Grassmann and N ordinary dimensions that preserve the metric (an antisymmetric metric in $2r$ and symmetric in N). The even subalgebra is $SP(2r) \times 0(N)$, and there are in addition $2rN$ odd generators. $0SP(4/N)$ has $[SP(4) = \bar{0}(5)] \times 0(N)$ as even subgroup, which becomes $\mathscr{P} \times \bar{0}(N)$ under contraction of the rotations into the fifth dimension. Taking $F = \bar{0}(4) \otimes 0(N) \times S_{\text{Left}}$ one has a WRS group, since $H = T \quad S_{\text{Right}}$ (T-translations, J-Lorentz generators, O-internal generators, S-spinorial generators)

$$\{S_L, S_R\} \subset T + J + O\,, \qquad [S_L, T] \subset S_R\,, \qquad [S_R, T] \subset S_L \tag{27.67}$$

so that spontaneous fibration occurs in this parity-violating mode. One then contracts $H \to \lambda H$, yielding for $\lambda \to 0$

$$\{S_L, S_R\} \subset \mathrm{T}\,, \qquad [S_L, T] \subset S_R\,, \qquad [S_R, T] = 0 \tag{27.68}$$

A second contraction $S_L \to \mu S_L$, $T \to \mu T$, $\mu \to 0$ yields

$$\{S_L, S_R\} \subset T\,, \qquad [S_L, T] = 0\,, \qquad [S_R, T] = 0 \tag{27.69}$$

that is, the algebra of supersymmetry. However, if one starts from an F-invariant Lagrangian, these contractions are not symmetric and the resulting Lagrangian breaks parity. But one can now start again with $F' = 0(4) \otimes 0(N) \times S_R$, $H' = T \otimes S_L$ and contract first $H' \to \lambda H'$ and then $S_R \to \mu S_R$, $T \to \mu T$. When the two resulting Lagrangians are added, we get (parity-conserving) supergravity for $N = 1$. Supergravity is then pseudoinvariant (that is, G-gauge invariant on-mass-shell).[47] It also fulfills (27.20) on-mass-shell. For $N>1$, the $\mathcal{L}_L$ or $\mathcal{L}_R$ have "cosmological terms," a form of spontaneous breakdown of the symmetry in which the equations of motion are not fulfilled by the "flat" group; that is, the vacuum is not on the tangent (rigid) group manifold. This is in contradistinction to the cases of gravity and $N = 1$ supergravity, where the equations of motion are solved by the rigid group manifold. Writing the action

$$A = \int \zeta_A \wedge R^A, \qquad A \in \mathrm{F} \subset G \tag{27.70}$$

the equations of motion are

$$D\zeta_A + R^B \frac{\delta \zeta_B}{\delta \rho^A} = 0 \tag{27.71}$$

and the condition for the vacuum to correspond to flatness is

$$\underset{\rho \to \omega}{D\zeta_A = 0} \tag{27.72}$$

(ζ_A are "pseudo-curvatures"[48]).

In $\mathcal{L}_L$ or $\mathcal{L}_R$, the double contraction does away with the $O(N)$ group, except for the indices on the spinor connections. $O(N)$ thus acts externally and is not gauged.[49] Taking $\mathcal{L} = \mathcal{L}_L + \mathcal{L}_R$, the cosmological term cancels, but its existence is still manifested by a violation of (27.72) for both $\zeta_{\alpha i}$ (spinor) and ζ_{ij} ($O(N)$ term, even if it is not there). Moreover, (27.20) does not hold even on-mass-shell, and the Lagrangian for parity-conserving supergravity is thus "pseudo-closed"[50] for $N = 1$ only.

It seems clear that much of this will depend on developments in our understanding of spontaneous symmetry breakdown. In our results relating to internal gauges[51] this was caused by connections in the odd part of a supergroup. Here, we have cosmological terms that look like a different type of symmetry breakdown, occasioned by a contracted gauge group F under which L is invariant.

One interesting application of $N = 1,2$ supergravity consists in the solution of the difficulties relating to spin 3/2 fields.[52] Such fields were known to suffer from an ambiguity in the definition of the spins for various components[53] and from faster-than-light propagation.[54] All these difficulties are removed in the supergravity case, which can be reread as just a coupled interaction of the $J = 3/2$, the graviton and the photon fields. Gravity comes in here in an entirely new role: it is necessary for the consistency of a $J = 3/2$ field, whether neutral or charged. This should be regarded in the same spirit as the finding that the only consistent $J = 2$ theory is gravity.[55]

To conclude, the last eight years, and even more so the last two years — since the discovery of supergravity — have brought about an intensive development of quantum gravity: Would it be too daring to predict that the problem will be solved within the next decade?[56]

NOTES

1. Spin is a quantum phenomenon: the only reason for its appearance in the physical world is that physical observables involve only norms, and thus allow double-valued representations of the rotation and Lorentz groups. See for example V. Bargmann, Ann. Math. **59**, 1 (1954).
2. With the exception of the Newman–Penrose formalism and of twistors, which integrated spin in an elegant way, at the expense of replacing space-time by an abstract manifold. The Newman–Penrose frames in fact lead in a natural way to supergravity, when one adheres to the spin-statistics correlation.
3. P. Von der Heyde, Lett. Nuovo Cimento **14**, 250 (1975).
4. A. Papapetrou, Proc. Roy. Soc. **A209**, 248 (1951).
5. D. W. Sciama, *Recent Developments in General Relativity* (Pergamon, Oxford, 1962), p. 415; T. W. B. Kibble, J. Math. Phys. **2**, 212 (1961); F. W. Hehl, Abh. Braunschweig. Wiss. Ges. **18**, 98 (1966); A. Trautman, Bull. Acad. Polon. Sci., Ser. Sci. Math. Astr. Phys. **20**, 185, 503, 895; **21**, 345 (1972).
6. J. D. Bekenstein, Phys. Rev. **D9**, 3292 (1974); S. W. Hawking, Comm. Math. Phys. **43**, 199 (1975).
7. S. Weinberg, Phys. Rev. Lett. **19**, 1264 (1967); A. Salam, in *Proceedings of the 8th Nobel Symposium*, edited by N. Svartholm (Almquist and Wiksells, Stockholm, 1968). See also Y. Ne'eman, Phys. Lett. **B891**, 190 (1978).
8. R. P. Feynman, Acta Phys. Polon. **26**, 697 (1963); B. S. DeWitt, *Dynamical Theory of Groups and Fields* (Gordon and Breach, New York, 1965); L. D. Fadeev and V. N. Popov, Phys. Letters, **25B**, 29 (1967); G. 't Hooft, Nucl. Phys. **B35**, 167 (1971); B. W. Lee, Phys. Rev. **D5**, 823 (1972); G. 't Hooft and M. Veltman, Nucl. Phys. **B44**, 189 (1972).
9. Feynman, op. cit. in n. 8.
10. Ne'eman, op. cit. in n. 7.
11. D. Z. Freedman, P. van Nieuwenhuizen, and S. Ferrara, Phys. Rev. **D13**, 3214 (1976); S. Deser and B. Zumino, Phys. Letters **62B**, 335 (1976).
12. C. N. Yang and R. L. Mills, Phys. Rev. **96**, 191 (1954); C. N. Yang, Phys. Rev. Lett. **33**, 445 (1974); C. N. Yang and T. T. Wu, Phys. Rev. **D12**, 3845 (1975).
13. Ne'eman, op. cit. in n. 7. Y. Ne'eman and J. Thierry-Mieg, Proc. Natl. Acad. Sci. USA **77**, 720 (1980).

14. See for example S. Kobayashi and K. Nomizu, *Foundations of Differential Geometry* Vol. 1 (Wiley Interscience, New York, 1963), p. 69. Y. Ne'eman and T. Regge, Riv. del Nuovo Cimento in (5) (1958).
15. J. Thierry-Mieg, J. Math. Phys. (forthcoming); J. Thierry-Mieg and Y. Ne'eman, Ann. Phys. **123,** 247 (1979).
16. Feynman, op. cit. in n. 8; DeWitt, ibid.; Fadeev and Popov, ibid.
17. A. A. Slavnov, Teor. Mat. Fiz. **10,** 99 (1973); J. C. Taylor, Nucl. Phys. **B33,** 436 (1971).
18. Reformulation of requirement on ghosts: C. Becchi, A. Rouet, and R. Stora, Comm. Math. Phys. **42,** 127 (1975); J. Dixon, Nucl. Phys. **B99,** 420 (1975); I. V. Tyutin, Report FIAN 39, 1975. Geometric rederivation: Thierry-Mieg, op. cit. in n. 15, and Thierry-Mieg and Ne'eman, ibid. and in n. 13.
19. The name is in the spirit of "Soft Self-Portrait" and some of the watches of S. Dali. Y. Ne'eman and T. Regge, Phys. Lett. **74B,** 54 (1978), and op. cit. in n. 14.
20. Ne'eman and Regge, op. cit. in n. 14.
21. Ibid.
22. Thierry-Mieg and Ne'eman, op. cit. in n. 15.
23. F. W. Hehl et al., Rev. Mod. Phys. **48,** 393 (1976).
24. Y. Ne'eman in Proceedings of the 19th International Conference on High Energy Physics, Tokyo (1978), edited by S. Homma, S. Kawaguchi, and H. Miyazawa (Physical Society of Japan, 1979), 552.Thierry-Mieg and Ne'eman, op. cit. in n. 15.
25. K. S. Stelle and P. C. West, Phys. Lett. **B74,** 330; S. Ferrara and P. van Nieuwenhuizen, Phys. Lett. **B74,** 333 (1978).
26. M. Gell-Mann and Y. Ne'eman, unpublished; op. cit. in n. 25.
27. Freedman, van Nieuwenhuizen, and Ferrara, op. cit. in n. 11; Deser and Zumino, ibid.
28. Ne'eman and Regge, op. cit. in n. 19 and op. cit. in n. 14.
29. S. W. MacDowell, Phys. Lett. **80B,** 212 (1979).
30. 't Hooft, op. cit. in n. 8.
31. R. Delbourgo and M. Ramon Medrano, Nucl. Phys. **B110,** 467 (1976).
32. P. K. Townsend and P. van Nieuwenhuizen, Nucl. Phys. **B120,** 301 (1977).
33. Y. Ne'eman and E. Takasugi, Phys. Rev. (forthcoming).
34. R. E. Kallosh, JETP Letters **26,** 575 (1977).
35. G. Sterman, P. K. Townsend, and P. van Nieuwenhuizen, Phys. Rev. **D. 17,** 1501 (1978).
36. Stelle and West, op. cit. in n. 25; Ferrara and van Nieuwenhuizen, ibid.
37. K. S. Stelle and P. C. West, Nucl. Phys. **B140,** 285 (1978), and op. cit. in n. 25.
38. G. 't Hooft and M. Veltman, Ann. Inst. HP **20,** 69 (1974).
39. Ibid.; S. Deser and P. van Nieuwenhuizen, Phys. Rev. **D10,** 401, 411 (1974).
40. M. T. Grisaru, J. A. M. Vermaseren, and P. van Nieuwenhuizen, Phys. Rev. Lett. **37,** 1662 (1976); S. Deser, J. H. Kay, and K. S. Stelle, Phys. Rev. Lett. **38,** 527 (1977); M. T. Grisaru, Phys. Lett. **66B,** 75 (1977); P. van Nieuwenhuizen and J. A. M. Vermaseren, Phys. Rev. **D16,** 298 (1977); S. Deser and J. H. Kay, Phys. Lett. **76B,** 400 (1978).
41. Ibid. [M. K. Fung, P. van Nieuwenhuizen, and D. R. T. Jones (ITP–SB–80–5) have just announced (May 1980) completion of the proof of two-loop finiteness.]
42. S. Deser and C. Teitelboim, Phys. Rev. Lett. **39,** 249 (1977).
43. M. T. Grisaru, Phys. Lett. **73B,** 207 (1978).
44. M. Kaku, P. K. Townsend, and P. van Nieuwenhuizen, Phys. Lett. **69B,** 304 (1977); with S. Ferrara, Nucl. Phys. **B129,** 125 (1977).
45. F. Englert et al. Phys. Lett. **57B,** 73 (1975).
46. K. Johnson and E. C. G. Sudarshan, Ann. Phys. **13,** 126 (1961).
47. J. Thierry-Mieg, Lett. Nuovo Cimento **23,** 489 (1978).

48. Ne'eman and Regge, op. cit. in n. 14.

49. Note that the $N = 2$ theory is a partial solution to Einstein's unification program. It works perfectly, except that the electromagnetic coupling is nonminimal (since the $0(2)$ *gauge* vanishes). Alternatively, one may impose $0(2)$ gauging, which then modifies the geometry and introduces a Planck-size cosmological term. The latter may yet turn out to fit quantization requirements, in accordance with S. Hawking's views of the Wheeler "foam."

50. Freedman, van Nieuwenhuizen, and Ferrara, op. cit. in n. 11; Deser and Zumino, ibid.

51. Ne'eman, op. cit. in n. 7.

52. Johnson and Sudarshan, op. cit. in n. 46; G. Velo and D. Zwanziger, Phys. Rev. **186,** 1337 (1969).

53. Johnson and Sudarshan, op. cit. in n. 46.

54. Velo and Zwanziger, op. cit. in n. 52.

55. S. Deser in *General Relativity and Gravitation,* Proceedings of the Seventh International Conference on General Relativity and Cosmology (Tel-Aviv), edited by G. Shaviv and J. Rosen (Wiley, New York, and Israel University Press, Jerusalem, 1974), pp. 1-18.

56. Recently a model of $N = 8$ supergravity has been constructed and appears to allow for larger internal symmetry groups (up to E(7)). See E. Cremmer and B. Julia, Phys. Lett. **80B,** 48 (1978), and Nucl. Phys. **B159,** 141 (1979).

28. SUPERGRAVITY

Peter van Nieuwenhuizen

Ignorance is the condition necessary, I do not say for happiness, but for life itself. If we knew all, we could not endure life for an hour. The feelings which make it sweet, or at least tolerable, are born of a lie and are nourished on illusions.

—Anatole France, *Le jardin d'Épicure*

If Grosman had known Grasmann, Einstein might have studied anticommuting numbers and invented supergravity. After all, general relativity is simply $N = 0$ supergravity. Supergravity with $N > 0$ is an extension of general relativity, not a replacement. It contains a new symmetry principle: Fermi–Bose symmetry (also called supersymmetry), in addition to the space-time symmetries that Einstein, Cartan, Weyl, and others studied. It is said that Einstein, when not engaged in actual research, liked to repeat for himself derivations of theorems he knew, for the sheer pleasure of seeing how pretty they were. In that vein I will begin by showing how pretty *and simple* supergravity has become three years after its birth.[1,2] But for those who agree with Pauli that "Eleganz ist nur fur den Schneider" ("Elegance is only for the tailor"), I will then go on and present exciting new work.

The reason that a field theory with a local symmetry between bosons and fermions is at the same time a theory of gravity, can be traced back to the derivative ∂_μ which, for simple dimensional reasons to be explained below, must be present in the transformation laws for the boson and fermion fields

$$\delta B(x) \sim F(x)\, \varepsilon(x)\,, \qquad \delta F(x) \sim \partial_\mu B(x)\, \varepsilon(x)\,. \tag{28.1}$$

Indeed, if one performs two successive "rotations" on a given field, with "angles" $\varepsilon_1(x)$ and $\varepsilon_2(x)$, respectively, one obtains a translation ∂_μ of the original field over a distance $\varepsilon_1(x)\varepsilon_2(x)$ that differs from point to point. This is the notion of a general coordinate transformation, and hence local supersymmetry leads to gravity. The converse is also true: starting with a theory of gravity that has a global supersymmetry with constant parameters ε, the space-time symmetries transform ε into $\varepsilon(x)$, so that one ends up with a local Fermi–Bose symmetry. It is clear that the origin of

Harry Woolf (ed.), Some Strangeness in the Proportion: A Centennial Symposium to Celebrate the Achievements of Albert Einstein ISBN 0-201-09924-1

supergravity lies in the derivative ∂_μ in Eq. (28.1), and thus in the different dimensions of bosons and fermion fields.

One can find some properties of the rotation angle ε by requiring covariance of Eq. (28.1).

1. *Dimensions.* Boson fields have dimensions 1 and fermion fields 3/2, as follows from their actions. Thus the dimension of ε is $-1/2$, and this explains the derivative in Eq. (28.1).
2. *Statistics.* This shows that ε is an anticommuting number. One treats ε as a Grasmann variable, and not as a Clifford number. Thus ε may be viewed as the $h \to 0$ limit of a quantum fermion field, and $\{\varepsilon_1, \varepsilon_2\} = \{\varepsilon, F\} = [\varepsilon, B] = 0$. Note that $\varepsilon_1, \varepsilon_2, \varepsilon_3 \ldots$ are considered independent, and form an infinite dimensional Grasmann algebra.
3. *Spin.* This shows that ε is a spinorial parameter. The simplest choice, spin 1/2, is realized in supergravity. Thus $\varepsilon_\alpha(x)$, $\alpha = 1,4$ is a spinor field. One takes all spinor fields in supergravity real (Majorana); this is equivalent to two-component van der Waerden formalism.
4. *Representation.* There are many field theories with a global invariance of the kind of Eq. (28.1) with constant ε. From the relation $\delta B = [\varepsilon Q, B]$, and idem for F, one finds charges Q_α, $\alpha = 1,4$. These spinor charges, with the generators P_μ and $M_{\mu\nu}$ of the Poincaré group, form a closed algebra, the graded Poincaré algebra.

In addition to the Poincaré algebra the relations are

$$[Q_\alpha, P_\mu] = 0, \qquad [Q_\alpha, M_{\mu\nu}] = (\sigma_{\mu\nu})_{\alpha\beta} Q_\beta \qquad \{Q_\alpha, \bar{Q}_\beta\} = (\gamma^\mu)_{\alpha\beta} P_\mu\,, \tag{28.2}$$

where the bar is Dirac's bar and the γ^μ and $\sigma_{\mu\nu}$ are Dirac matrices. Salam and Strathdee have shown that the irreducible representations of this algebra are given by doublets of fermion-boson states with adjacent spins $(J, J+\frac{1}{2})$ or $(J, J-\frac{1}{2})$. Thus any supergravity model, even if it contains massive fields, is always constructed from these elementary building blocks. The astonishing thing is not that one needs fermions and bosons differing by spin ½ — that follows from Eq. (28.1) with ε a spin 1/2 parameters — but that one only needs one boson and one fermion.

Two Fermi–Bose rotations lead to a translation, as we have seen. Thus Fermi–Bose symmetry can be viewed as the square root of the translation group, and local supersymmetry as the square root of general coordinate transformations. Hence, supergravity is the square root of general relativity, just as the Dirac equation is the square root of the Klein–Gordon equation, and one might say that supergravity describes spinning space.[3] Because this square root leads to a fermionic rather than a bosonic symmetry, one has an explanation why there are only two kinds of particles in nature, bosons and fermions. If there were three kinds of particles, one would need a modification of Eq. (28.1) such that three consecutive rotations would yield a translation.

The road from gravity to supergravity leaves classical field theory and enters quantum field theory, because fermions are present. Deviations from general relativity do not occur at macroscopic distances, so that such predictions as light bending and the perihelion motion are unaffected. However, at small distances and in particular in the ultraviolet problem of quantum gravity there is great improvement. Infinities that are present in Einstein gravity cancel in the S-matrix of supergravity, due to the extra symmetry between fermions and bosons. This was first discovered in an explicit calculation of a given process at the first-order (one-loop) quantum corrections, and a theoretical proof was given.[4] Subsequently, also the two-loop corrections in a certain class of models were shown to be finite[5] (the so-called extended supergravities; see below). However, at the three loop-level an invariant was shown to exist, at least up to order ψ^2.[6] With the advent of a tensor calculus for simple ($N = 1$) supergravity, it was possible to extend this result to all loop orders and, worse, to prove rigorously that at *any* higher loop level new dangerous invariants exist.[7] For the N-extended supergravities (see below) no tensor calculus yet exists and no such exact results are known, but also for the $N = 2,3$ models an invariant has been shown to exist up to order ψ^2.[8] When also here a tensor calculus will be constructed, we will know whether also here the same situation holds, or that for example the $N = 8$ theory is more finite than the others; this is at present an open question. If supergravity is the correct theory of quantum gravity, the coefficients of all these invariants must vanish. This seems unlikely, but it is in principle possible. At any rate, supergravity is an improvement over Einstein gravity, where a world of only gravitons is one-loop (and perhaps two-loop) finite, but where nonrenormalizability sets in as soon as one couples to matter. Those models of supergravity in which all particles can be transformed into each other and into the graviton (the extended supergravities), are one- and two-loop finite. One has again a world with only supergravitons, as one might call these multiplets containing the graviton. But if one forms a superparticle out of matter, which does not contain the graviton, and couples supergravitons to such supermatter, then finiteness is again lost.[9] A moment of thought will reveal that such conclusions can only be reached by explicit computations and not by a theoretical proof. These computations have been done with the stated results in all extended supergravities up to $N = 8$.[10] Before explaining what the extended supergravities are, let us first consider simple supergravity.

Simple, or nonextended, or $N = 1$ supergravity contains a spin 2 field because gravity is present. Its fermionic partner can have either spin 3/2 or 5/2, but the simplest choice, spin 3/2, is the correct one. The action is the sum of the Hilbert and Rarita–Schwinger actions for spin 2 and 3/2, because these are unique if one requires absence of ghosts, plus three nonpropagating fields S, P, A_a.

$$
\begin{aligned}
\mathcal{L} &= -\tfrac{1}{2}(\det e)\kappa^{-2} R - \tfrac{1}{2}\bar{\psi}_\mu R^\mu - \quad , \\
&\tfrac{1}{3}(\det e)\kappa^{-2}(S^2 + P^2 - A_a^{\,2}) \quad , \\
R &= e^{a\nu} e^{b\mu}(\partial_\mu \omega_{\nu ab} + \omega_{\mu ac}\omega_\nu{}^c{}_b - \mu \leftrightarrow \nu) , \\
R^\mu &= \varepsilon^{\mu\nu\rho\sigma}\gamma_5\gamma_\nu D_\rho \psi_\sigma , \qquad D_\rho\psi_\sigma = \\
&(\partial_\rho + \tfrac{1}{2}\omega_{\rho ab}\sigma^{ab})\psi_\sigma \quad .
\end{aligned}
\tag{28.3}
$$

The symbol κ is the gravitational constant, the vierbein field $e^a{}_\mu$ and not the metric $g_{\mu\nu} = e^a{}_\mu e_{a\nu}$ describes the spinor field since fermions are present, the real vectorial spinor (or spinorial vector) $\psi_\mu{}^\alpha$ (with $\alpha = 1,4$ and $\mu = 1,4$) describes the spin 3/2 field, and γ_5 and $\sigma^{ab} = \tfrac{1}{4}[\gamma^a, \gamma^b]$ are constant matrices ($\gamma_5^{\,2} = 1$), but $\gamma_\nu = e^a{}_\nu\gamma_a$ is not constant. No Christoffel symbol is needed in ($D_\varrho\psi_\sigma - D_\sigma\psi_\rho$), nor present. One may view $e^a{}_\mu$ as gauging P_a, $\omega_{\mu ab}$ as gauging M^{ab} and Q_α as gauging $\psi^\alpha{}_\mu$; this gives added support for the choice of $\psi_\mu{}^\alpha$ as the fermionic graviton, henceforth called gravitino. As in Einstein gravity, one eliminates $\omega_{\mu ab}$ by means of its nonpropagating field equation $\delta I/\delta\omega_{\mu ab} = 0$. The result is

$$\omega_{\mu ab} = \omega_{\mu ab}(e) + \tfrac{1}{4}\kappa^2 \, (\bar\psi_\mu \gamma_a \psi_b - \bar\psi_\mu \gamma_b \psi_a + \psi_a \gamma_\mu \psi_b)\,, \tag{28.4}$$

where $\omega_{\mu ab}(e)$ is the Einsteinian value of the spin connection, and this result shows that there is torsion. Substituting (28.4) into (28.3) yields complicated ψ terms, but it is better not to expand $\omega_{\mu ab}$.

The first two terms in the action in Eq. (28.3) were found in 1976, but the last two were found more recently.[11] They are not needed for the invariance of the action, but they are needed if one wants to obtain a tensor calculus (see below) and if one wants to quantize supergravity just as Yang–Mills theories and Einstein gravity. Without these auxiliary fields one needs to add four-ghosts interactions of the Feynman–De Witt–Faddeev–Popov ghosts in order to restore unitarity at the two- and higher-loop level.[12] That S, P and A_a should be present can be inferred from a simple counting argument. Off-shell there should be as many boson as fermion *fields*, just as on-shell there should be equal numbers of *states*. There are:

(16-4-6)	vierbein fields
6	auxiliary fields and
(16-4)	gravitino fields

(subtracting the components that are absent from the action because of local gauge invariance). Superspace leads also to these fields and shows, moreover, that $\omega_{\mu ab}$ is not an independent field, in agreement with Eq. (28.4).

The transformation laws under which the action in Eq. (28.3) is invariant are expected to be of the form of Eq. (28.1). However, since the gravitino is the gauge field of supersymmetry, it should contain a term $\partial_\mu \varepsilon$; hence, covariantizing with respect to gravity, a term $D_\mu \varepsilon = (\partial_\mu + \tfrac{1}{2}\omega_{\mu ab}\sigma^{ab})\,\varepsilon$. The gravitino states transform in fact under global supersymmetry as $\delta e^a{}_\mu = \kappa\bar\varepsilon_\gamma{}^a\psi_\mu$ and $\delta\psi_\mu = \kappa^{-1}(\omega_{\mu ab})^{\rm lin}(\sigma^{ab}\varepsilon)$, exactly as one can deduce from the algebra in Eq. (28.2).[13] Since field equations rotate into field equations, and $S = P = A_a = 0$ are field equations, the variations of S, P and A_a are expected to be proportional to the gravitino field equation R^μ. They are, provided one introduces the important concept of a supercovariant derivative.[14] Supercovariant derivatives are objects whose variations do not contain $\partial_\mu \varepsilon$ terms. The transformation rules are very simple:

$$\begin{aligned} \delta e^a{}_\mu &= \kappa\bar\varepsilon\gamma^a\psi_\mu\, \quad \delta\psi_\mu = (2\kappa^{-1}D_\mu - \gamma_\mu\eta + iA_\mu\gamma_5)\varepsilon\,, \\ \delta(S + iP) &= \tfrac{1}{2}\bar\varepsilon\,(1 + \gamma_5)\,\gamma_a R_{\rm cov}{}^a\,, \delta A_a = \frac{3i}{2}\bar\varepsilon\gamma_5(\delta_{ab} - \tfrac{1}{3}\gamma_a\gamma_b)\,R^b{}_{\rm cov}\,, \end{aligned} \tag{28.5}$$

where the symbol $\eta = -\tfrac{1}{3}(S - i\gamma_5 P - i\,A\!\!\!/\,\gamma_5)$ can be understood from conformal supergravity. Conformal supergravity is a supersymmetric extension of Weyl's conformal gravity theory, but we have no time to discuss it here. We turn now to the most interesting models of supergravity — the extended supergravities.

The first of the extended supergravities was discovered by Ferrara and van Nieuwenhuizen.[15] They constructed the so-called $N = 2$ model, which unifies electromagnetism with gravitation and which comes closest to the unified field theories Einstein had in mind. The unification is achieved by adding two gravitinos to the photon and the graviton, and the resulting theory has many local invariances: electromagnetic and space-time symmetries, two supersymmetries (one for each gravitino), and, finally, a global 0(2) symmetry that rotates the gravitinos into each other. This model does not contain spin 1/2 and spin 0 fields. Other extended supergravities exist, labeled by a parameter $N(2 \leqslant N \leqslant 8; N = 1$ is simple supergravity) with N gravitinos, $\frac{1}{2}N(N - 1)$ vector fields, and more spin 1/2 and spin 0 according to a binomial series. These theories have a global $0(N)$ invariance that rotates particles with the same spin into each other, as well as N local supersymmetries that rotate particles of adjacent spins into each other. The $N = 8$ model will be discussed further below.

Since there are as many vector fields as there are generators of the group $0(N)$, one might wonder whether the $0(N)$ symmetry can be made local and become a Yang–Mills symmetry that is gauged by the vector fields. This was shown to be possible by Freedman.[16] Thus the extended supergravities have two coupling constants, the gravitational constant $\varkappa$ and the Yang–Mills coupling constant g. Strangely enough, one needs at the same time a cosmological constant (det e) $g^2 \varkappa^{-4}$ in the action, and if g is to be the electric coupling constant, the value of the cosmological constant is much larger than its experimental value. This is the moment to look for another uncomfortable property of supergravity, the degeneracy in masses of fermions and bosons in the same multiplets. Nobody has seen a massless gravitino, although it may have electric or even weak or strong interactions (for $N > 1$ it can be complex).

In order that the $0(N)$ symmetry splits up into other symmetries that describe the electro-weak and strong interaction without losing the good quantum properties of massless supergravity, one needs spontaneous symmetry breaking. For $N = 1$ supergravity this result was obtained by writing down the most general action for the coupling of the action in Eq. (28.3) to the supersymmetric Wess–Zumino action

$$\mathcal{L} = -\tfrac{1}{2}(\partial_\mu A)^2 - \tfrac{1}{2}(\partial_\mu B)^2 - \tfrac{1}{2}\bar{\lambda}\partial\lambda + \tfrac{1}{2}F^2 + \tfrac{1}{2}G^2 . \tag{28.6}$$

It was found[17] that one could choose the coupling such that the following are true:

1. The gravitino eats the would-be Goldstone spinor λ and becomes massive with four helicities $\pm 3/2$, $\pm 1/2$ (the latter from the erstwhile λ),
2. A cosmologial term could be avoided, if and only if $\langle S \rangle = \langle P \rangle = 0$. Perhaps one might reverse the situation and find a symmetry in Eq. (28.3) that leads to $\langle S \rangle = \langle P \rangle = 0$. This would be the first theoretical explanation why our universe seems to have no cosmological constant.
3. Also the scalars get masses and a mass formula holds: $4m_\psi^2 = m_A^2 + m_B^2$.

There exist other schemes to break local supersymmetry spontaneously in extended supergravities. One of them uses dimensional reduction. In this approach one starts in $D + E$ dimensions and wraps the E coordinates into little circles. In the limit of vanishing radius, only the lowest excitations in E space remain, and because one spinor in $D + E$ dimensions breaks up into several spinors in D dimensions, one ends up with extended supergravity models. There are even speculations that such a compactification of space may occur spontaneously. The spontaneous symmetry breaking as obtained by Scherk and Schwarz[18] is best explained by considering first the action in Eq. (28.3). Writing for the gravitino ψ_μ in $D + E = 4$ dimensions as

$$\psi_\mu(x,y,z,t) = \exp(im\gamma_5 x)\, \phi_\mu(y,z,t)\,, \tag{28.7}$$

where m is arbitrary, $x\varepsilon E$, while $y,z,t\varepsilon D$, the $\partial/\partial x$ derivative now yields a mass term, and upon compactification no x-dependence remains, since the $D + E = 4$-dimensional model has the global chiral symmetry $\delta\psi_\mu = i\gamma_5\psi_\mu$. Using this approach, the authors have started to reduce the $D + E = 11$-dimensional model with $N = 1$ supergravity down to $D = 4$, avoiding a cosmological constant, and find four arbitrary masses and a well-behaved potential proportional to m.

The last of the extended supergravities (the $N = 8$ model) was recently constructed by some very interesting methods by Cremmer and Julia.[19] This $N = 8$ model *predicts* the usual four quarks with charges (2/3, $-1/3$, 2/3, $-1/3$) in $SU(3)$ color triplets and one sextet with charges ($-\frac{1}{3}$).[20] However, it does not contain the intermediate vector bosons W^+W^-, nor the μ and τ leptons. If it is to be the model of the world, all these particles have to be excitations of other particles. The full $N = 8$ model was obtained from a $D + E = 11$-dimensional model by dimensional reduction. In 11 dimensions the vierbein fields e_{AM} have as symmetry groups the local Lorentz group 0(10,1) for the index A and general coordinate transformations for M. The global group $GL(11,R)$ of linear coordinate transformations then splits up into a global group $GL(4,)R) \otimes GL(7,R)$ and the latter can even be extended to a global exceptional group $E(7)$. (The reduction from $D + E = 11$ to $D = 5$ yields $E(6)$ and to $D = 3$ yields $E(8)$.) On the other hand, the local Lorentz group 0(10,1) in 11 dimensions splits upon dimensional reduction into $0(3,1) \otimes 0(7)$, and the latter can again be extended, now to a local $SU(8)$ whose gauge fields are nonpropagating (just as the spin connection gauges 0(3,1) in Eq. (28.3) and is nonpropagating). Since auxiliary fields become propagating at the quantum level [this is known to be the case[21] with S,P and A_a in Eq. (28.3)], the authors of the reference in note 19 speculate that W^+W^- and possible μ and τ might be some of these auxiliary fields. The group $E(7)$ is noncompact, but it does not lead to ghosts, contrary to the lore. For example, scalar kinetic terms can be viewed as internal vierbein fields g_{MA} with $A\varepsilon SU(8)$ and $M\varepsilon E(7)$, and read

$$+\, Tr(g^{-1}\partial_\mu g - h_\mu)^2 \Big|_{//\, SU(8)} \qquad - Tr(g^{-1}\partial_\nu g)^2 \Big|_{\perp\, SU(8)} \;. \tag{28.8}$$

The h_μ are the auxiliary gauge connections of the local $SU(8)$ and appear in kinetic terms with wrong signs. In this equation, the group $E(7)$ has been split into its maximal subgroup (also an $SU(8)$ group, but global, this time) and its orthogonal complement, just like splitting $SL(8,R)$ into antisymmetric $SO(8)$ matrices and its complement, symmetric matrices. Solving the h_μ field equations and substituting the result back into the action, it is obvious from Eq. (28.8) why the wrong-sign kinetic terms disappear.

Until now, it was generally believed that $N = 8$ is the roof on the extended supergravities, because for $N > 8$ there are spin 5/2 particles for which no consistent field theory was thought to exist. (Another problem is that for $N > 8$ one has several gravitons, but spontaneous symmetry breaking could make all of the massive except one). It came thus as somewhat of a surprise when recently the high-spin barrier was cracked by Fang and Fronsdal.[22] Applied to spin 5/2, the field equations read[23]

$$\partial_\mu \gamma \cdot \psi_\nu + \partial_\nu \gamma \cdot \psi_\mu - \not\partial \psi_{\mu\nu} = J_{\mu\nu} \,. \tag{28.9}$$

This equation also appears in Schwinger's[24] work, but he requires $\partial^\mu J_{\mu\nu} = 0$, whereas the field equation only leads to

$$\partial_\mu J_{\mu\nu} = \not\partial (\tfrac{1}{2}\gamma_\mu J_{\mu\nu} - \tfrac{1}{4}\gamma_\nu J_{\mu\mu}) \,. \tag{28.10}$$

The remarkable thing is that there are no ghosts in the field equation in Eq. (28.9), nor in the propagator derived from (28.9), despite the weakened conservation condition on $J_{\mu\nu}$.[25] The theory has a gauge invariance $\delta\psi_{\mu\nu} = \partial_\mu \varepsilon_\nu + \partial_\nu \varepsilon_\mu$ but with ε_μ restricted by $\gamma^\mu \varepsilon_\mu = 0$. No Weinberg inconsistencies follow, since the matrix element for spin 5/2 emission $\bar{u}^+ \varepsilon^+{}_\mu \varepsilon^+{}_\nu J^{\mu\nu}$ is still Lorentz invariant, because the k_μ produced by $\varepsilon^+{}_\mu$ upon boosting, turns into a $\not k$ according to Eq. (28.10) and vanishes because u^+ satisfies the Dirac equation. There are also only two dynamical modes. Coupling to dynamical sources leads to problems, however. For example, coupled to gravity[26] the invariance of the spin 5/2 action under $\delta\psi_{\mu\nu} = D_\mu \varepsilon_\nu + D_\nu \varepsilon_\mu$ requires that the full Riemann tensor be the spin 2 field equation. Also problems with coupling to other systems have been found. It is a challenge to find a successful interaction for this spin 5/2 field, because only for $N \geqslant 9$ does the group $0(N)$ contain as subgroup the group $SU(3) \times SU(2) \times U(1)$.

I have already mentioned a tensor calculus for supergravity a few times. Just as in Lorentz invariant field theories, such a calculus exists for simple supergravity[27] and allows us to add, multiply, contract, and build actions from multiplets, which are the analogies of 4-vectors. This frees supergravity from the tedious Noether calculations of the past, because one can write down invariants just as easily as one writes for example in Einstein gravity $R_{\mu\nu}R^{\mu\nu}$ = invariant. This tensor calculus was obtained by taking the flat space tensor calculus of global supersymmetry, which is entirely equivalent to superspace (see below), and first extending it to conformal supergravity. No guesswork is involved, since *in conformal supergravity (and only there) are all fields gauge fields and couple minimally.* Then one proceeds to Einstein gravity using the remarkable relation[28]

$$\delta_Q{}^{\text{Einstein}}(\varepsilon) = \delta_Q{}^{\text{conf}}(\varepsilon) + \delta_S{}^{\text{conf}}(\eta\varepsilon) \tag{28.11}$$

where η was given below Eq. (28.5), and Q is the square root of the translation generators P, while S is the square root of the conformal boosts K. As an example of a multiplet, consider the multiplet containing the fields occurring in Eq. (28.5)

$$W = [S, P, \tfrac{1}{2}\gamma_a R_{\text{cov}}{}^a, \tfrac{1}{2}R_{\text{cov}} - \tfrac{1}{3}\kappa^2 \; (S^2 + P^2 - \tfrac{1}{2}A_a{}^2), \; - D_a{}^{\text{cov}} A^a] \tag{28.12}$$

Note how simple this result is, due to the use of supercovariant derivatives[29] (R_{cov} is the supercovariantized scalar curvature). The action in Eq. (28.3) is now obtained immediately from this multiplet, but also invariants starting with powers of R follow at once. The general results for all invariants at any loop level in $N = 1$ supergravity have been obtained in this way.

The tensor calculus is entirely equivalent to superspace methods,[30] as has been stressed repeatedly. In superspace one considers fields depending on x^μ and extra fermionic coordinates θ^α. Expanding $\phi(x,\theta)$ in terms of θ, one gets a finite set of terms, since $\theta^\alpha\theta^\beta = 0$ if $\alpha = \beta$, and the coefficients of these powers of θ are the components of multiplets as in Eq. (28.12). There is one difference, however. In superspace one usually has many more fields, which are redundant and whose elimination is difficult but desirable in order to be able to do such things as spontaneous symmetry breaking, and so on. Tensor calculus uses the minimal set of fields — but it could be that one may sometimes need the nonminimal set of superspace.

Another result of the tensor calculus[31] is that it has led to the supertopological invariants of supergravity. First the multiplet was constructed that contains the chiral, conformal, and supersymmetry local anomalies. Then the actions were constructed that contain in the bosonic part $R^{*}_{\mu\nu ab}\, {}^{*}R^{\mu\nu ab}$ and $R^{*}_{\mu\nu ab}\, R^{\mu\nu ab}$. Surprisingly, the result turned out to be equal to the topological invariants of ordinary gravity; all extra terms due to supergravity cancelled in the end. This result has also been arrived at from a different point of view by Kostant and Singer. Supercohomology is the same as ordinary De Rham cohomology (Kostant). One can define supercharacteristic classes in terms of a vector potential. It turns out that supercharacteristic classes are the same as ordinary classes (Singer). In dimension 4, the supercharacteristic classes are represented by the usual Euler and Pontragin classes, in agreement with our results.

An interesting speculation by Christensen and Duff concerns the connection between duality-chirality and supergravity counterterms.[32] It is known that in supergravity the duality properties of the Maxwell system are replaced by combined duality-chirality transformations. This was first observed in an explicit calculation up to order ψ^2,[33] and later exactly established for the $N = 2,3,4$ models.[34] The connection with counterterms is, as suggested to Christensen and Duff by their study of superindex theorems, that only those counterterms are realized in supergravity on-shell, which vanish if the Weyl tensor is dual or anti self-dual. Unfortunately, there are infinitely many of such invariants.

The success of supergravity lies more in the deeper theoretical structures discovered than in phenomenological applications. That should not worry us; some of the most useful physical theories needed a very long incubation time. Supergravity even attracts attention from outside our narrow scientific community, as the telex I received recently shows: "I may have discovered the unified field theory. I theorize that mass is the hypercomplex fifth dimension." (name deleted). I do not know who the author is, probably somebody from a patent office or so. Due to the short time available, many other areas of research could not be discussed. Let us hope that Nature is aware of our efforts.

NOTES

1. D. Z. Freedman, P. van Nieuwenhuizen, and S. Ferrara, Phys. Rev. **D13**, 3214 (1976), and **D14**, 912 (1976).
2. S. Deser and B. Zumino, Phys. Lett. **62B**, 335 (1976).
3. C. Teitelboim, Phys. Rev. Lett. **38**, 1106 (1977).
4. M. T. Grisaru, P. van Nieuwenhuizen, and J. A. M. Vermaseren, Phys. Rev. Lett. **37**, 1662 (1976).
5. M. T. Grisaru, Phys. Lett. **66B**, 75 (1977).
6. S. Deser, J. Kay, and K. Stelle, Phys. Rev. Lett. **38**, 527 (1977).
7. S. Ferrara and P. van Nieuwenhuizen, Phys. Lett. **78B**, 573 (1978).
8. S. Deser and J. Kay, Phys. Lett. **76B**, 400 (1978).
9. For a review, see M. T. Grisaru and P. van Nieuwenhuizen in "Deeper Pathways in High-Energy Physics," in Proceedings of the 1977 Coral Gables Conference.
10. M. Fischler, Phys. Rev. **D20**, 396 (1979).
11. S. Ferrara and P. van Nieuwenhuizen, Phys. Lett. **74B**, 333 (1978), and K. Stelle and P. C. West, Phys. Lett. **74B**, 331 (1978).

12. E. Fradkin and M. A. Vasiliev, Phys. Lett. **72B,** 70 (1977); R. Kallosh, Nucl. Phys. **B141,** 141 (1978); G. Sterman, P. K. Townsend, and P. van Nieuwenhuizen, Phys. Rev. **D17,** 1501 (1978).

13. M. T. Grisaru, H. Pendleton, and P. van Nieuwenhuizen, Phys. Rev. **D15,** 996 (1977). In this reference the spin (2, 3/2) multiplet is derived.

14. P. Breitenlohner, unpublished.

15. S. Ferrara and P. van Nieuwenhuizen, Phys. Rev. Lett. **37,** 1669 (1976).

16. D. Z. Freedman and A. Das, Nucl. Phys. **B120,** 221 (1977); E. Fradkin and M. Vasiliev, Lebedev Institute preprint.

17. E. Cremmer, B. Julia, J. Scherk, S. Ferrara, L. Giarardello, and P. van Nieuwenhuizen, Nucl. Phys. **B147,** 105 (1979).

18. J. Scherk and J. Schwarz, Phys. Lett. **82B,** 60 (1970), and Nucl. Phys. **B153,** 61 (1979).

19. E. Cremmer and B. Julia, Nucl. Phys. **B** (forthcoming), Phys. Lett. **80B,** 48 (1976).

20. M. Gell-Mann, unpublished.

21. S. Ferrara, M. T. Grisaru, and P. van Nieuwenhuizen, Nucl. Phys. **B138,** 430 (1978).

22. J. Fang and C. Fronsdal, Phys. Rev. **D18,** 3630 (1978).

23. F. A. Berends, B. DeWitt, J. W. van Holten, and P. van Nieuwenhuizen, Nucl. Phys. **B154,** 261 (1979).

24. J. Schwinger, *Particles, Sources, and Fields,* Vol. 1 (Addison-Wesley, Advanced Book Program, Reading, Mass., 1970).

25. Berends et al., Phys. Lett. **83B,** 188 (1979).

26. Ibid., J. Phys. **A13,** 1643 (1980).

27. S. Ferrara and P. van Nieuwenhuizen, Phys. Lett. **76B,** 404 (1978); K. Stelle and P. C. West, Phys. Lett. **77B,** 376 (1978).

28. Ibid.

29. P. K. Townsend and P. van Nieuwenhuizen, Phys. Rev. **D19,** 3592 (1979).

30. For a superspace approach, see the work of Arnowitt and Nath, Wess and Zumino, Brink and Gell-Mann, and Ramond and Schwarz, Siegel, Gates, Taylor, and others. Space-time forbids details.

31. Townsend and van Nieuwenhuizen, op. cit. in n. 29.

32. S. M. Christensen and M. Duff, Nucl. Phys. **B154,** 301 (1979). Phys. Lett. **83B,** 88 (1979), and Nucl. Phys. **B154,** 261 (1979), and J. Phys. **A** (forthcoming).

33. J. A. M. Vermaseren and P. van Nieuwenhuizen, Phys. Lett. **65B,** 263 (1976).

34. S. Ferrara, J. Scherk, and B. Zumino, Nucl. Phys. **B121,** 393 (1977).

Open Discussion

Following Papers by Y. NE'EMAN and P. VAN NIEUWENHUIZEN

Chairman, V. F. WEISSKOPF

R. Penrose (Oxford U.): I'd like to make one remark and pose one question that could be directed to either of the speakers. The *remark* is in connection with what Yuval Ne'eman said; it is that I would really like to register my disagreement in connection with the question of spin being an intrusion from quantum mechanics. I would like to say that if by spin you mean spinors, then they fit in very, very beautifully with the classical theory. In fact I do not think of them as a quantum mechanical thing at all, for they fit in remarkably with classical general relativity. The *question,* which is partly a comment as well, has to do with the background and fundamental problems posed by quantum gravity. There has been a great deal of work done in connection with quantum gravity from all sorts of different angles that has gone on for many years, and there are lots of very fundamental questions that have been raised. I just wonder how, in fact, supergravity comes to terms with them. If I may just mention one of them: In quantum field theory, one of the basic requirements is that field operators should commute at spacelike separations. But one does not have a concept of what a spacelike separation is, because it depends upon the metric, and the metric itself is something that is supposed to be a quantum operator. How does one come to terms with something of this sort in supergravity theory?

Weisskopf: Do you want to answer this Yuval, or maybe both of you should answer this.

Ne'eman: To the first part, I agree that, let us say, the twistor formalism has taken features into account and one can regard some features of spin as classical. On the other hand the only reason I know for the two-valuedness of spin (that is, the half integer representations of the Poincaré group) *being at all relevant to physics* is that quantum mechanics tells us that the theory deals with *amplitudes,* but we measure *probabilities;* that is, the phase does not count.

Now to a second point. Those theories that we know how to renormalize are gauge theories. Then we assume that in order to renormalize we have to be able to represent gravitation in such a picture. In the original Einstein example in which you absorb a change, for instance in the elevator or when there is a being in outer space that suddenly pulls the chest and you have an acceleration, you have a space-time dependent Lorentz transformation that is a gauge transformation for the Poincaré group in that case. In general relativity in the spirit of the theory, everything is absorbed in general coordinate transformations. However, when you have spin, *the intrinsic spin transformations cannot be reabsorbed as far as the coordinates are concerned.* Now, it is true that geometrically, because of representation theory, you can have other manifolds (you leave space-time) and other representations, like working in twistor space. There, you can have manifolds where spin transformations can be reabsorbed by a coordinate change. On the other hand, orthodox general relativity — I mean the one that Einstein gave us — could only fit spin in as an appendix, the way it was done in 1928. In gauge gravity and supergravity it is the group picture that regulates everything, and the general coordinate transformation is really only a secondary feature. We do have automatically the additional dimensionality of the gauge bundle, and there is room for absorption of the spin changes in the coordinates of the intrinsic variable. This is even a physical question: there is a physical distinction between using first-order formalism or second-order formalism as Hermann Weyl showed. Even though it may be a small difference, it exists. The first-order formalism comes into being only in the modern gauge picture — it then becomes the more elegant one and the more fundamental one. The connection is no longer identified with the Christoffel symbols. The additional degrees of freedom of spin feed the connection. It thus has its own spin-gauge contribution, beyond the Christoffel symbols. So, this is the second point I am making about the spin's quantum nature and its being foreign to orthodox general relativity. Now to your second question. It is true that fundamental questions arise with respect to what happens to causality, to causal commutators that relate to the foundations of quantum field theory. The answers that we seem to get produce puzzles. With respect to what precisely happens, I am not sure of my answer. But what seems to be happening, I think, is that the gauge theory is taking care of itself, in the sense that it will compensate and correct the path so that the commutator will be physical.

van Nieuwenhuizen: Yes, this problem you mentioned of course has always been mentioned at every conference and is always true. We do not know how to do canonical quantization. It is true that if you have the metric field classically and you choose first two points to be timelike, and after you quantize, the quantized metric could very well make these points spacelike so we would be in a problem. My point of view is the following: We start not with canonical quantization but with what we technically call covariant quantization, which is a Feynman path integral approach. It means only postponing the problem of showing that these two are equivalent, which in the end we hope to do. We have not arrived there, and it is one of the main problems. I have no solution to offer. However, looking at these results and doing them daily at your desk, you see all these striking regularities happen in these calculations, which is really a marvelous thing to see. And I cannot help feeling that these regularities point to something. Maybe it is not the ultimate quantization, but there is something.

C. N. Yang (State U. of New York, Stony Brook): I would like to comment on Professor Penrose's remark. It is certainly true that the mathematicians have developed very beautiful and powerful views and methods about things related to general relativity. Yet, I think it is important to remember that physics is not mathematics, just as mathematics is not physics. Somehow nature chooses only a subset out of the very beautiful and complex and intricate mathematics that mathematicians develop, and that precise subset is what the theoretical physicist is trying to look for. I think that to say that everything about spin is understood because the mathematics of it is, is from a physical viewpoint, not a correct thesis. In fact, I would think that that is an area where the most fruitful interaction between physicists and mathematicians is likely to come about in the next ten years.

Ne'eman: Two more remarks. One with respect to spin and how foreign it is. The way it appears, spin exists in the tangent. And general relativity teaches you on the contrary not to regard the tangent as something special. Its philosophy is to treat curved space as such, and not to look for the particular role of the tangent, but rather to live in the curved space. Spin does, however, pick the tangent. In the last two years I happen to have found the infinite spinors of the general coordinate transformation group, but they are not the ones that fit the electron, for example. They might fit the hadrons and they exist even in the curved part. Still, ordinary spinors require the tangent. Now, going further to the translations: in Sciama–Kibble gauging you are preserving the structure of the Lorentz group. In "classical" general relativity you say that there can be no role for the Poincaré group! In gauging, the whole structure of the manifold that I described is a preservation of that group. So this is why it stands a chance of being more complete. In fact it was shown, for instance, that the positivity of energy comes about because the two supersymmetric generators continue to give a "modified" translation even in the curved situation. So you have an additional set of features that comes because of the particular role that the group holds in the gauge procedure.

Weisskopf: Is there anything else?

J. Ehlers (U. Munich): Just with respect to the question of spin, one additional remark. I think everybody can speak only for himself if talking about understanding. I have understood spin essentially from Weyl and Pauli. The essential feature of spin, in Pauli's words, is a nonclassical two-valuedness, and the structural basis for that is that the homogeneous Lorentz group has a twofold covering group. If one wants to incorporate that into general relativity, then it seems to me the conceptually essential step is to say when I want to describe fields of a boson character, then it suffices to take as the seat of these fields either space-time itself or its linear frame bundle or its bundle of orthonormal frames. However, if one wants to deal with spinors, one has to construct a twofold covering of the latter bundle; I do not see why you bring in Hilbert space at that stage. The Hilbert space comes about, as far as I understand, if one constructs a state space or wants to put a Hilbert space structure on the space of sections of bundles associated to that spinor bundle or orthonormal bundle via representations of the group. So, I am on Professor Penrose's side more or less in this dispute.

I would like to ask a question related to the last remarks that have been made by Professor Ne'eman, which concerns the so-called gauging of the translation part of the Poincaré group on the one hand and the rotation part on the other. There is a point I never understood, particularly when I read your papers, Professor Ne'eman. Namely, the homogeneous part of the Lorentz group has an obvious place in general relativity. It simply is the group that operates on the orthonormal frames at one point; it is a Lie group acting on a principal fiber bundle. Now, if one wants to incorporate translations, one says, one "gauges the translations." One first writes down a translation in flat space $x' = x + \zeta$ where ζ is just a set of four numbers, and then one says one gauges that by making these ζ's position-dependent. From the point of view of geometry, it seems that the only meaning one can attach to a "position-dependent translation" is to say, well, they are diffeomorphisms. And, in fact, your "softened translations" do not form the basis of a Lie algebra, nor do they generate a group action on some bundle, in contrast to the Lorentz generators. Instead, the corresponding commutation relations form a part of the Cartan structural relations, which does not describe a Lie group structure. Is "gauging translations" just a new phrase for describing this situation? There is an essential difference between translations and rotations.

Ne'eman: It is as you describe it except for one further complication. If you do it only as a general coordinate transformation, as a diffeomorphism, the index would be a holonomic index. This index is not rotated by the Lorentz generators. Replace it by a vierbein index by multiplying the covariant derivative by an inverse vierbein. This corresponds to ξ^ν being replaced by $\xi^r e_\nu{}^a$ so that the transformation is also field-dependent. Then you preserve the commutation relations between "rotations" and "translations." This is related to what has been done years ago by Bergmann et al. and to the "shifts" and "lapses" of the Wheeler school.

P. Bergmann (Syracuse U.): I think I have to apologize for being the singular speaker in the discussion who has no desire to say anything about spin. I would like to first make a short historical footnote remark. Yuval mentioned that Rosenfeld and Møller insisted that there was no logical need for quantizing general relativity, which is true of course. I would like to add to this that the first person who attempted quantization was the late Leon Rosenfeld, whose early papers go back to 1930 in the *Annalen der Physik* and the *Annals de l'institute Poincaré*. Just so that Rosenfeld's contribution is not totally buried, I learned about it very late and I have been apologizing for my ignorance ever since. Now, I would like to make two remarks, both of which will sound critical, but which I will qualify. The first is that the real difference between a Yang–Mills type of gauge theory and general relativity is not to my mind to say that we are pushing the group diffeomorphisms into the structure group or into the gauge group or anything like this. But that, aside from coordinatization, the underlying structure of general relativity is a purely topological one. Only the notion of neighborhood is given a priori and any other fields, including the metric field, go into the dynamics. That, I think, is the basic difference, and that is what makes a perturbation expansion away from Minkowski suspect. You destroy the basic absence of metric structure of an a priori metric structure to begin with. That was one remark. The other is that I have my suspicions about the Grassmann algebra. The Grassmann algebra is only one possible hypercomplex algebra and, in my opinion, suffers from the serious shortcoming for our purposes, that it is homomorphic, in a very trivial sense, to the algebra of ordinary complex numbers. The specific implication of this remark is that in writing down the algebra, the graded algebra of the generators of the graded group that you essentially copied vertatim from the Wess–Zumino paper, the translation that is the result of the anticommutation of the super gauge transformation is not a translation, it is a "generalized translation" because the Δx that appears, or the p, is a nilpotent commuting number. Now, this has some serious implications. It has the implication that even in the ultimate formulation of the theory, we will always be able to pull out the nonsuper part invariantly. I think I know two possible formal ways out of this fix. One is to go over, and that has been mentioned many times but I do not think it has been carried out successfully, to go over from a Grassmann to a Clifford algebra, in which this homomorphism does not exist — where you can get back to ordinary numbers as a product of two anticommuting Grassmann elements. The other possibility is to go to a nonassociative algebra in which you have only anticommuting fermions — no Clifford — but in which $(ab)c$ is not a $a(bc)$. Whether either of these modifications would lead to a removal of this formal difficulty, I don't know. But I would like to go back, and that is why I say I will qualify these apparently critical remarks, to the point that Peter van Nieuwenhuizen made. There are so many beautiful, intriguing things in the new approaches that I think even if we discover a number of imperfections in the present state of the theory, that we should not drop it but, on the contrary, pursue it, being aware of the defects and hope that in the next few years we will see more clearly.

Yang: Since we have in the audience a person who taught us all what is the spin, I wonder what he wants to say about the gravitational theory of the spin? Where is Professor Dirac?

Weisskopf: One more remark. Eugene Wigner.

E. Wigner: I have a very simple question that shows essentially my ignorance. Most of the theories that I know use a complex Hilbert space in which two vectors, the components of which differ by the same factor, are identical. Has it been investigated what happens if one uses other types of Hilbert spaces? The real Hilbert spaces even I would know, but one could think of quarternion Hilbert spaces.

[Several people say, "Professor Gursey!"]

Ne'eman: If Professor Gursey is here, I think he could give the answer because he has done it. Is Professor Gursey here?

F. Gursey (Yale U.): The quarternion Hilbert spaces were first considered by von Neumann actually — so it's very old. And since then they have been reinvestigated by Yao and his collaborators and Y. Finkelstein and others. The quarternion Hilbert spaces arise in a rather natural way. If you look at the ordinary Hilbert space, it is essentially a projective space. If you take a point with homogeneous coordinates, then the homogeneous coordinates correspond to a ket. And when the ket is complex, then what corresponds to it is a projective space — in complex space. You identify a ket — you say that it represents the same physical state if you multiply it by an arbitrary complex number. And that is the same thing for identifying a point with the homogeneous coordinates. You multiply the homogeneous coordinates by an arbitrary complex number — it is still the same point. And you can extend it to quarternions. You can consider the quarternionic projective space. That'll be a model of quarternionic Hilbert space. You take quarternionic points with $n + 1$ quarternionic coordinates and you identify the quarternionic point with another one that is a quarternionic multiple of this quarternionic ket. Now, you can choose whether to multiply this ket by a quarternion on the left or on the right. So you have two possibilities. So if you multiply it on the right, then you have a phase ambiguity corresponding to $SU(2)$. And you can get the nonhomogeneous coordinates by dividing, say, by one of the coordinates so that corresponds to n inhomogeneous coordinates. You start form $n + 1$ homogeneous coordinates. And this process can be carried out so that you have quarternionic Hilbert space and the geometric meaning is that of a quarternionic projective space. The only extension of that is with octonions,. But then you have to do it only with three components and you cannot go to higher.

Wigner: I am very grateful for your answer. I am not sufficiently familiar with the subject and I do not even know whether the representations of the Poincaré group in these spaces have been established. [Someone says, "Yes."] Yes, thank you very much. Who did it? [Someone answers, "Ernst."] Thank you very much. Now I'll have an answer to my question. In both quarternion. . . [Someone speaks, "Yao. He was a student of Yao."] and in octonion space? [Someone speaks, "No, he did it in quarternion space."]

X. Einstein and the Unity of Theoretical Physics

29. THIRTY YEARS OF KNOWING EINSTEIN

Eugene P. Wigner

Recollection of Early Days

Among those assembled here, I am surely one of the oldest and knew Einstein also in his earlier years. Permit me, though, to express my regrets that Cornelius Lanczos can no longer be with us. He not only knew Einstein much before I knew him, he also continued to maintain an active interest not only in him personally, but also in all his writings. Lanczos's book, *The Einstein Decade,* gives a truly marvelous account of Einstein's thinking, of his contributions to more fields of physics than most of us can recall.[1] I can recommend it wholeheartedly. The recent book by Hoffmann and Miss Dukas is also very informative though less on the scientific and more on the human side.

I, myself, first saw Einstein at the physics colloquia held at the University of Berlin each Thursday afternoon. I attended them from 1920 on. Together with other notables — M. Planck, W. Nernst, M. von Laue, Richard Becker, to mention only a few — Einstein sat in the first row and listened to the reviews of the papers chosen for this purpose by von Laue, three or four papers every Thursday. This was very good; people could maintain interest not only in the subject of their own work but in physics in general. If the review of the paper presented a clear picture, no one in the first row made any comment; most of the questions and comments came anyway from the rest of the audience. However, if the article's meaning did not become clear, there were questions from the first row, principally from Einstein. The answers to these questions, and the questions themselves, contributed greatly to the clarification of the new information contained in the paper discussed. And this was good even in those old days. I remember particularly the discussion of the Stern–Gerlach experiment that astounded all of us, including also Einstein. The results of the Bothe-Geiger, and of the Compton–Simon experiments pleased all of us, but, I must admit, Einstein took them for granted — he firmly believed in the microscopic validity of the energy and momentum conservation principles. And this played a great role also in his later years.

The physics colloquia acquainted us with the clarity of Einstein's thinking, with his simplicity and modesty, and also with his skill of exposition. However, few in the audience knew each other personally — about sixty people attended these colloquia. Personal acquaintance with Einstein came to most of us in the seminar on statistical mechanics that he organized. Actually, Einstein's

Harry Woolf (ed.), Some Strangeness in the Proportion: A Centennial Symposium to Celebrate the Achievements of Albert Einstein
ISBN 0-201-09924-1

appointment in Berlin came from the Prussian Academy of Sciences even though he also had the title of professor at the university and also that of director of a not truly existing physics research institute (Kaiser Wilhelm). He had no teaching obligations. He organized the seminar, I feel, because he wanted to establish contact with his young colleagues, because he wanted to know about their ideas and attitudes. He chose statistical mechanics as a subject because he liked it, because he admired its initiators. He was also too modest to suggest the relativity theories as subjects because the fact that he was the initiator of these would have given him a special status, would have elevated him above the other participants. It would have been difficult, if not impossible, to create the feeling of equality, the feeling that every participant has learned and not created the subject, had this been relativity theory. All this did not mean that Einstein did not supplement or even improve on our presentations — statistical mechanics uses intricate methods, and Einstein had more skill to clarify these than any of us — but he did that in a most simple and cooperative way so that the speakers (with one exception) did not feel embarrassed. Incidentally, the exception later received the Nobel Prize. But we all felt that Einstein acted as a friend and not as a supervisor. In fact, it is tempting here to quote Lanczos, who wrote in the book that was mentioned before, *The Einstein Decade,* "The magic of his personality imposed itself on almost everybody who came in touch with him and the very name Einstein has assumed a charismatic sound" (p. ix). This was true also in Princeton and we all not only admired but also liked him.

Unfortunately, the relation to his colleagues became much less close when he came to Princeton. Part of the reason for the change was that, even though he could speak it, he never felt at home with the English language. A second reason was his deep concern with the political situation in Europe, with the danger apparent at the time he came to Princeton, the danger that Hitler would defeat the democracies. This occupied much of his thoughts and attention. A third reason was that his interest, deeply devoted to a modification of the theory of general relativity so as to form a common basis for all physics and perhaps even for all science, was very different from the prime interest of most of his colleagues and even the students in Princeton. Most physicists were, at that time, most interested in the application of quantum mechanics to a variety of phenomena, including the theory of atoms and molecules, the properties of solids, especially of metals, and also to the basic principles of chemistry. This work also contributed to the unification of science, particularly that of chemistry and physics, but not to the unification of the fundamental principles of all physics which was Einstein's interest. Some of the mathematicians, including L. P. Eisenhart, were greatly interested in Riemannian geometry, the basis of general relativity, but their interest was centered on the rigorous mathematics thereof, not on an extension of relativity theory to encompass electromagnetic and perhaps other phenomena. They were averse to speculations. For all these reasons, Einstein's true friendship and interest was confined to the small circle of his collaborators and perhaps a few friends of his earlier days. But these truly enjoyed his friendship.

Perhaps I should mention the reason I believe Einstein did not share the interests of most of us, why he restricted his attention in his later years to macroscopic physics.[2] He did not like the statistical nature of quantum mechanics and he knew that all physical theories are temporary. He "played dice" hoping that quantum mechanical principles would soon be superseded by a better and deterministic theory.

Let me now proceed to the true subject of my address, only superficially touched so far: Einstein's interest in the unity of theoretical physics, in fact the unity of all science. I will introduce this subject, though, with a few remarks about the generality of Einstein's interest in the various branches of physics.

The Extension of Einstein's Interest in Physics

The generality of Einstein's interest in all problems of physics is perhaps best demonstrated by the variety of papers he published. During the period between the discovery of the special and the general theories of relativity — about ten years — he published more than sixty papers. Many of these were, naturally, on the theory of relativity. But there were even more on other subjects: statistical mechanics, including the theory of the Brownian motion (proposed also by Smoluchowski); quantum theory of radiation, including his simplified derivation of Planck's law, now the basis of our description of the functioning of lasers; solid state physics, including his explanation of the decrease of the specific heat with decreasing temperature; opalescence; straight electrodynamics. He also discovered the law of the photoelectric effect, for which he received the Nobel Prize. It is in fact clear that he believed in the reality of the quantum phenomenon more firmly than its original proponent, M. Planck. He felt that the quanta of radiation are as definite and indivisible as the electric charges of electrons. In his Berlin days he had made as yet no direct attempt toward the development of an all-embracing basis of theoretical physics, but his interest in practically all areas of physics was evident. This manifested itself also in the colloquia; he was ready to comment, explain, or even argue, about all papers that did not appear to be very clear.

Einstein's interest in new ideas, about most of which he heard in the colloquia mentioned before, manifested itself frequently. Perhaps the most remarkable such manifestation was his immediate acceptance, and in fact enthusiasm, for Bose's modification of the classical theory of statistical mechanics. He reviewed it at the colloquium himself and was convinced of its validity. On the other hand, when the Bohr–Kramers–Slater theory was put forward, that the energy and momentum conservation laws are valid only in the statistical sense, he expressed great skepticism. He was convinced of the universal validity of these conservation laws and took the outcomes of the Bothe–Geiger and of the Compton–Simon experiments for granted. Of course, he was interested in the experiments just the same and was pleased by their success, but it was for him a success of experimental skill.

His firm belief in the conservation laws manifested itself in another way also. He was very early well aware of the wave-particle duality of the behavior of light (and also of particles); in their propagation they show a wave character and show, in particular, interference effects. Their emission and absorption are instantaneous, they behave at these events like particles. In order to explain this duality of their behavior, Einstein proposed the idea of a "guiding field" (*Führungsfeld*). This field obeys the field equation for light, that is Maxwell's equation. However, the field only serves to *guide* the light quanta or particles, they move into the regions where the intensity of the field is high. This picture has a great similarity with the present picture of quantum mechanics and has, obviously, many attractive features. Yet Einstein, though in a way he was fond of it, never published it. He realized that it is in conflict with the conservation principles: at a collision of a light quantum and an electron for instance, both would follow a guiding field. But these guiding fields give only the probabilities of the directions in which the two components, the light quantum and the electron, will proceed. Since they follow their directions independently, it may happen that in one collision the light quantum is strongly deflected, the electron very little. In another collision, it may be the other way around. Hence the momentum and the energy conservation laws would be obeyed only statistically — that is, on the average. This Einstein could not accept and hence never took his idea of the guiding field quite seriously. The problem was solved, as we know, by Schrödinger's theory in which the guiding field progresses in configuration space so that the joint configuration of the collliding particles is guided, not the two separately and independently. In Einstein's view, Schrödinger's great accomplishment was the idea of a guiding field in configuration space — surely much less picturesque than separate guiding fields in our ordinary space for the separate particles, but enabling the formulation of a theory in which the conservation laws hold not only on the average but for each process, in particular each collision, individually.

This story was told to illustrate the firmness of Einstein's conviction in the validity of some fundamental principles that played such a decisive role in his efforts to provide a basic principle that would contain all laws of nature. The conservation laws for energy, momentum, and angular momentum belonged in this category and, after Schrödinger's work, these did not cause further problems. But some other convictions, such as the universal validity of causality, or the total microscopic validity of the space-time concepts, did.

Perhaps I should also mention that, when attending the Berlin colloquia, and even somewhat later, I often had the impression that many of the participants, particularly those in the first row, feared that man may not be wise and clever enough to create an attractive basic theory for microscopic atomic physics. This was only an impression of mine, not supported at that time by any statement of Einstein, for instance, but it was my impression. I have not mentioned this before but, a short time ago, Dr. Dirac told me of a similar impression he had, and I now dare to put it forward. It shows not only the significance and unexpected nature of the discovery of quantum mechanics but also explains the natural doubts concerning its final nature that were present in many of the mature minds.

Einstein's Interest in "The Unity of Theoretical Physics"

Let me begin by repeating what we mean by unity of theoretical physics. It is the establishment of some basic principles from which all known laws of physics would follow or, rather, of which these laws appear as valid in limiting situations, situations in which some, or most, effects can be neglected. Thus Maxwell's equations give a unity to the physics of electromagnetism: the laws governing the behavior of magnets can be drived from them by considering the situations in which the electric charges are infinitely small everywhere; the older laws governing the behavior of electric charges can be derived from them by assuming that the motions of these are infinitely slow as compared with the velocity of light. Also, the laws governing the propagation of light follow for situations in which no electric charges or magnetic poles and, in fact, no material bodies, are present, that is, for vacuum. Needless to say, the general law representing the unity of all theoretical physics would have to be even more encompassing, very much more encompassing than is Maxwell's electromagnetic theory. But Einstein always hoped that such a theory would eventually be established, at least for physical phenomena, that it *will* be created even if not by him. Whether this theory, a theory of theoretical physics, would also describe the regularities of the phenomenon of life, is a question I never discussed with him and, until I came to our centennial symposium, I had the impression that he did not expect theoretical physics to extend that far. But here, my attention was called to his address at the celebration of M. Planck's sixtieth birthday. He said there (my translation): "They [the laws of theoretical physics] should form the basis from which a picture of all processes of nature can be derived by thoughtful deduction — and these include also the process of life." But, on the other hand, he also said that "It would be a true miracle if man could discover a common basis for all sciences — physics, biology, psychology, sociology and others. We are striving toward such a goal but we also can give arguments against its attainability." In other words, he felt that the human intelligence also has limitations, as has that of other animals. I must confess that the latter statement is much closer to my own thinking than the former. But, of course, all of us hope for some progress in every direction.

Having attributed the immense magnitude of the task implied by "unity of theoretical physics" (even if we confine our attention to inanimate systems), let me also admit that Einstein and his collaborators did not strive directly after the whole of this distant ideal. One reason for this was Einstein's innate modesty, his realization that no one can achieve everything. But it must also be admitted that his familiarity with the totality of theoretical physics had decreased by the time he came to Princeton, by the time this subject came to occupy his chief interest. This had several reasons. One of these was that there was, in Princeton, no "colloquium" in the sense of the Berlin colloquia at each of which three or four papers were reviewed, the papers having been selected (by von Laue) because of their apparent interest. Only one paper is presented at our colloquia and this is one written, or to be written, by the speaker. As a result, many more papers of importance fail to be presented at our colloquia than was the case in Berlin. To duplicate the Berlin type of colloquia is unfortunately impossible at present, or it seems so. Another reason for the contraction of Einstein's interest was the fact that the number of reasonably separated areas of theoretical physics had greatly increased. This, and the resulting specialization of most of us theoretical physicists, was mentioned before. A third reason for the decrease of Einstein's earlier encompassing interest in virtually all areas of theoretical physics was that he had much less close contact with his colleagues here than he had before, and virtually no contact with the students, graduate or undergraduate. Finally, it must be admitted that, as he became older, his ability to absorb ideas had decreased and also that his interest turned increasingly toward philosophical problems, more removed from the immediate problems of physics. Lévy-Leblond pointed out the generality of this phenomenon. But I think the main reason for the limited scope of his and his collaborators' efforts toward the unification of physics was their realization that even a modest step into that direction would be a significant accomplishment, well worth a thorough and concentrated effort. And with this we can fully agree.

Which parts of theoretical physics did Einstein and his collaborators strive to unify? Naturally, general relativity was one of these and, perhaps equally naturally, electromagnetic theory the other. Both are field theories; that is, they describe the physical states by functions of space-time. This is true as long as quantum effects are disregarded, and, somewhat to our surprise, they were: the union was attempted essentially in the macroscopic domain; it did not apply as much to the actual field strengths as to their averages over a space-time region exceeding the quantum restrictions. One sign for this was that their fields were functions of space-time; they did not use any substitute for the configuration space. This, of course, surprised those of us who were familiar with Einstein's early interest in, and support of, the quantum hypothesis, but it appeared to make the effort toward uniting the two theories much less difficult. Another, more technical circumstance that at least appeared to help was the nature of the fields of the two theories. Einstein's gravitational equations are formulated in terms of the metric tensor, which is a symmetric 4-by-4 tensor. The electromagnetic field is a similar but antisymmetric tensor, so that the union of the two is easily formed; it is a general 4-by-4 tensor, in general neither symmetric nor antisymmetric.

The attempt at unification disregarded, of course, in addition to the existence of quantum effects, the so-called strong, that is nuclear, interaction, and also the weak interaction, responsible for the β-decay. But these were less close to Einstein's interest than the gravitational and electromagnetic interactions. However, in spite of these simplifications, quite drastic ones, and in spite of the fortunate circumstance mentioned before, the effort at unification, in which several of his collaborators, including Peter Bergmann, Banesh Hoffmann, and Leopold Infeld participated, was not truly successful. At least Einstein was not satisfied with the results. And *The Einstein Decade* also says: The "unified field theory" on which Einstein worked for the last thirty years of his life, remained an unfulfilled dream (p. 30).

It would be a mistake, of course, to consider the work of Einstein's last thirty years totally ineffective. The result that he, Hoffmann, and Infeld obtained, and that is considered most important, is the realization that the equations of the general theory of relativity describe not only the time behavior of the gravitational (that is, metric) field but also the motion of the objects creating this field. In this way, these equations do unify the equations determining the development of the field with the equations determining the motion of the objects creating this field. Unfortunately, I must confess that I am not totally convinced of the validity of the underlying argument. But Einstein did write several other very interesting articles, and his influence on his collaborators, and also some of his other colleagues in physics, was even more important.

The question naturally arises why Einstein's and his collaborators' effort to unify the gravitational and electromagnetic theories remained unsuccessful, and a few remarks on this question will be added.

The Problem of a Unified Theoretical Physics

It is, of course, possible that the laws of nature cannot be given a joint, simple, and mathematically beautiful formulation, or at least that man's mind is not powerful enough to discover such a formulation. The community of theoretical physicists is, however, committed toward striving for such a formulation and the possibility of the success of such an effort will therefore be assumed in most of the following discussion. The aforementioned question is therefore natural to ask: Why were Einstein's and his collaborators' efforts in that direction not more successful?

An answer to this question was proposed by Peter Bergmann:[3] the effort was premature; it was undertaken at a time when no full theory of the other interactions, strong and weak, was available. (It is not impossible, of course, that even further types of interactions will be discovered.) Bergmann implies that only to all interactions together can be given a satisfactory joint basis, not however to just an arbitrarily chosen pair of them. This view seems to be contradicted by S. Weinberg's and A. Salam's unification of the electromagnetic and weak interaction theories, but, of course, the fact that two of the presently known interactions can be given a joint basis does not show that any arbitrarily chosen pair from the four can be so treated.

I agree with the spirit of Bergmann's remark but would go much further. He believes that even if we exclude life and consciousness from the areas of physics, and even if a physics of the limiting situation in which life and consciousness play no role is possible, physics is as yet very far from perfection, and some of Einstein's assumptions, and those of present-day physics, may have to be revised. The very fundamental problems raised by quantum theory should be given more weight also in philosophical discussions.

The first such question is the separability of space-time points. Present-day physics is in agreement with Einstein's attribution of definite properties to all space-time points. It assumes that the electric and magnetic field strengths, for instance, can be measured pointwise. Quantum mechanics does not attribute definite field strengths to these points as Einstein's and his collaborators' unification attempts do, but does assume that the field operators corresponding to space-time points are meaningful. This also implies that the field strengths at such definite points are measurable. Yet the arguments put forward to support this are anything but convincing, and, as far as particles are concerned, we know that their position measurments' possibility is highly questionable. If a world line could be established for them, a relativistically invariant position operator would exist, and we know that this is not the case. But even a nonrelativistic position operator's meaningfulness is questionable as has been demonstrated particularly by Fleming and Hegerfeldt: in present quantum mechanics there is no wave function that would restrict the position of the particle to a finite space region in such a way that that particle would not appear outside the light cone of that region at a later time. This indicates that, in addition to Einstein's macroscopic revision of the space-time concept, a microscopic revision may be necessary, that only average positions or certain probability distributions of positions are meaningful in the microscopic domain, in the area of quantum phenomena.

The second problem concerns the principle of causality or determinism. We know how firmly Einstein believed in this concept — "God does not play dice" is one of his famous sayings. And indeed all equations of present-day physics are deterministic. The probabilistic outcome of quantum mechanical measurements is blamed on the measuring process the probabilistic nature of which, in spite of many attempts in this direction, cannot be reconciled with the equations of quantum mechanics. This has been discussed a great deal, also by me, for I am convinced that, if one wishes to be entirely consistent, one can reconcile the probabilistic nature of the measurement or observation process with quantum mechanics only by admitting that the equations of the latter do not describe the behavior of living beings, of beings with consciousness. As is well known, some of our colleagues go further and assume implicity, though rarely explicitly, that those equations do not apply to the measuring apparatus "because it is macroscopic."

What may be worthwhile to point out in this connection is that an observation of the German physicist Zeh seems to justify this and in fact leads to much more drastic conclusions. Zeh points out that the energy levels of even a moderately macroscopic system, such as a cubic centimeter of gas, lie so close together that even a single electron, many yards away, causes transitions between its states. It follows that macroscopic objects cannot form isolated systems, and since the equations of quantum mechanics — in fact all deterministic equations of physics — apply only to isolated systems, they are not strictly valid for macroscopic bodies. It can be claimed, in fact, that since the electrons — and protons and other nuclei — in the so-called empty space all interact with each other, that no system remains "isolated" for a macroscopic length of time, and if the equations of quantum mechanics were valid for such long intervals, they could be valid only for the wave function, or the state vector, of the whole universe. This is evidently meaningless: such a state vector would be surely unascertainable; in particular it would also embrace the "observer," that is, myself. What I believe follows from this is that perhpas God does not play dice but that, as far as we are concerned, the notion of the deterministic theory, deterministic for macroscopic objects and for long time intervals, is meaningless. It is not easy to accept this conclusion, but it is not easy to avoid it either. Perhaps I may add that I am hoping to deal with this problem as some of our colleagues, including Belinfante, already have. Naturally, we know that the microscopic changes of a macroscopic object, induced by the distant electron or positive charge, do not, or virtually never do, change its macroscopic properties, but they do induce significant changes of its state vector in Hilbert space. It would be good to have all this demonstrated mathematically, and this is being planned.

The preceding arguments seem to show that, accepting the state of our world, the validity of quantum mechanics is just as clearly restricted to nonmacroscopic situations as are our macroscopic theories, including the general theory of relativity, are restricted to nonmicroscopic systems. In addition, it appears that determinism may have to be given up. All this appears to be true even if we restrict our physics to inanimate nature. Naturally, the description of the phenomena of life and consciousness would necessitate further changes in our picture of the world. It follows that one can well agree with Peter Bergmann's conclusion that the present state of physics cannot permit a full unification of its basic theories — no theory's validity can be assumed to be universal. Nevertheless, I think that we do not need to admit that the time chosen by Einstein to unify at least two theories was premature, even though these theories are only partially valid.

NOTES

1. *The Einstein Decade* (Academic Press, 1974).
2. A system will be called microscopic if it consists of only a few atoms — or only one. Ordinary objects, of everyday life or even larger ones, such as stars or planets, that consist of billions of billions of atoms — will be called macroscopic.
3. G. J. Whitrow: "Einstein, the Man and His Achievement," pages 72-73.

Open Discussion

Following Paper by E. P. WIGNER

Chairman, V. F. WEISSKOPF

Weisskopf: Wigner has proved his last sentence. He has given us pleasure and satisfaction. But I am sure the satisfaction is not complete and there will be some comments and questions.

B. Kursunoglu (U. of Miami): May I have your permission to read a short letter from Erwin Schroedinger received in 1951 by me? Erwin Schroedinger writes to me:

> Supposing all your computations are correct. Your suggested field equations are decidedly a possibility. It would be interesting to know whether the right-hand side of equations 15 would yield in first approximation the conventional Einstein tensor as sources of the gravitational field. I suppose you know that neither in Einstein's nor in my version of the non-symmetric theory it does. With all that, I must confess that I have no great confidence in your version. The way in which the symmetric parts of both the covariant and contravariant g^{ik} intervene seems to me rather artificial, as can be seen particularly well from the Lagrangian L you mention at the end of your paper. To modify a frequently quoted remark of Pauli's — if you do decide to join together what God has separated, namely a symmetric and anti-symmetric tensor, you should at least not have to divorce them again in order to get the fundamental equations.
>
> Yours truly, Erwin Schrödinger

The letter is dated June 1951. I would like to mention that I have quite a few letters here both from him and from Albert Einstein on this subject matter when I was a student in Cambridge, and I somehow, despite all the dangers in it, did not give it up. And recently I solved the equations just to see what happens. Of course I know Professor Wigner is rarely wrong — it may be that this time he can afford to be wrong once. I solved these equations, and I have the following rather interesting result. That is, it is expected that to just unify electromagnetism and gravitation should not really produce exciting results in the light of so many other things. The solution in question is that an elementary particle appears to consist of stratified layers with alternating signs of magnetic charge with decreasing amounts as one recedes from the origin; the total amount of magnetic charge is zero. No monopoles. The total gravitational binding of such a structure I have computed to be of the order of 10^{21} MeV. And since you have this stratified structure, the resultant field is obviously a short-range field, and according to this result, gravitation causes its binding. Its structure has a spin, and therefore spin and gravitational force are perhaps related to the origin of spin and the theory is finite everywhere. The electric field is zero at the origin, but at asymptotic distances it is a Coulomb force. So therefore there are only three fundamental interactions: magnetic of the type I have described, electromagnetic, and gravitational. Other forces must be derivable from these three, including weak interactions and molecular and atomic. So this is one thing I thought I would mention without — well, hoping that this does justice to the concept of unified field theory in some way.

Weisskopf: Do you want to answer?

Wigner: No. I hope you are right, and that these unifications of the field theories will succeed. The reasons that I gave for my doubts in present quantum mechanics are many, and you did not mention the strong interaction that seems to exist. But I don't feel that that answers the fundamental questions of the possibility of localization and of macroscopic bodies and the impossibility of answering it for microscopic bodies. I would also like to mention (It's in my paper) that according to a recent idea of Dr. Dirac, the ratio of gravitational and electromagnetic forces constantly decreases. If this is true and I think there is good reason to believe it is true — it is not contained in the equations that you mentioned. I don't know whether this has any . . .

Weisskopf: I think this is probably enough for today and I hope the discussion will be continued privately. Yuval wanted to make a remark.

Y. Ne'eman (Tel-Aviv U.): Two nonphysics comments and one in physics. One nonphysics comment: I would suggest a note with respect to Einstein's role in these last thirty years that comes to one's mind in comparing him with the other great physicist of this century — that is, Heisenberg. Both really were interested in unified field theory in their last years and over a long period. Both had the right to become interested in even a very extremely difficult problem of that nature because they had done so many other things that they could risk failure even over thirty years. What I would say in favor of Einstein is an impression that one has — since I never knew him while I knew Heisenberg a little bit, though I certainly was not close to him. Einstein did not try to convince the whole community of physics that his was the way and his was the answer. He knew it was a risk and he was sort of ready to take the risk himself. Whoever wanted to could attach himself to his ideas and work with or follow him, but the rest of physics could do what it was doing and try other ways. In contradistinction, for many years Heisenberg felt so sure about his being the only right way, that he even tried to convince others to stop theirs. For example, at the time he fought the construction of new accelerators. I recall his telling me that there would be no need for the machines, since his theory contained all the answers. He changed afterward and realized that the situation was different. Maybe I have the wrong impression; maybe Einstein was doing things that I don't know about. I feel a man can take any risk upon himself, provided he does not involve the whole community in this venture.

The other nonphysics point that I make is: English is the language of physics now and to most of us who do not have it as their first language, the distinction between microscopic and macroscopic is a near impossibility. I often go to Texas and there the term that means small is pronounced *macroscopic.* And so, that's even worse. May I suggest that we try to boycott these terms and use phrases such as large-scale, small-scale.

The physics comment concerns the same program of unifying electromagnetism and gravitation. I just thought it might be interesting to the audience — to those people who have not followed the work in supergravity — Professor van Nieuwenhuizen mentioned the $n = 2$ case. This case is exactly the one in which one unifies gravity and the electromagnetic field, and it needs a charged spin 3/2 field as a mediator between them. And then it works! Now, when I say it works — as a classical theory — it works. Einstein was not worried beyond that. However, if one wants also to have a local gauge invariance for the charge gauge or phase, then there are difficulties, and the difficulties are connected with the appearance of a cosmological term that is extremely large in that case. I think that that small exercise in supergravity already has gone beyond the results of Einstein's thirty years with respect to that particular problem! It is not a solution but an interesting side phenomenon.

Weisskopf: Eugene, do you want to answer to this?

Wigner: No. I am sorry to admit that I agree entirely with what Dr. Ne'eman said, so I don't have anything to add. I hope I didn't say the opposite in my speech.

Dirac: I would like to comment on a fact that has been mentioned by Wigner and by other people at this meeting. Namely, that Einstein received the Nobel Prize for his work on the photoelectric effect and not for his work on relativity. Why is that so? Now the deliberations of the Nobel committee are kept strictly secret; but one knows the principles according to which they work. It was the wish of Nobel that the prizes should be awarded for discoveries in experimental physics. That meant, in the first place, that the prizes should go to experimenters. But that was extended somewhat so as to allow theoretical workers to receive the prize if their work was of direct benefit to the experimental development of physics. In the case of Einstein, his work on the photoelectric effect was very definitely connected with experiment and has great importance. His work on relativity, of course, was really more fundamental, but its fundamental character was not realized at the time, and people were uncertain about how it would influence the development of physics, and so the committee felt on safer ground in awarding the prize just for the photoelectric effect. One sees another example of this later on in the case of the award of the Nobel Prize to Heisenberg. One of the early successes of the Heisenberg theory was in predicting two kinds of hydrogen molecules, those with even and those with odd angular momenta. These two kinds can be separated to some extent and they are fairly stable: they last a few hours; one can do experiments with them; and one can see that they have different specific heats. The experiment was done, confirmed Heisenberg's theory, and that provided a very definite basis for a Nobel Prize to Heisenberg. It was mentioned in the citation, but they were a bit vague about his influence on physics in general.

Wigner: That may be right, but you see the special theory of relativity also had many experimental consequences; in particular, the acceleration of heavy particles close to light velocity. And many other things — for example, there is the nonexistence of objective aberration; the light emitted by two stars that move around each other arrives at the same time. The star that moves toward us does not emit light with greater velocity than the star that moves away from us. So I think the prize could have been given for his creation of the special theory of relativity, but, as you say, it was easier to admit the significance of his work relating to the photoelectric effect than to admit the validity of the theory of relativity, particularly since the experimental evidence was not yet as unique as it is today.

P. A. M. Dirac (Florida State U.): No. I'm afraid that the reasons may have been more mundane, and the opposition to relativity theory was so great that the Nobel committee wanted to be objective.

Wigner: Yes, that is what I wanted to say.

I. I. Rabi (Columbia U.): I asked Einstein at one time how it is that he did not discover the de Broglie relations, because he had h/p, and in a certain sense it was almost trivial, at least from hindsight. He told me he had thought of that, but there was no experimental evidence and he did not publish. I bring this point up because we have been listening to some wonderful mathematical theory all morning — that's perfectly marvelous — and then we heard Wigner, and my respect for him knows no bounds. But he is a little far from experiment. Now, as an experimenter — former experimenter — I would like to ask him just what place does actual experiment (also, the other mathematical physicists around) what place does experiment or actuality have in their thinking, because you always have the problem: What shall I do when I go into the laboratory? what is interesting? what shall I measure? Of course, I have to assemble equipment, and it is very complicated equipment as Professor Wigner has just pointed out. It is influenced by the most distant stellar phenomena, and, of course, I have got at least to do something that I can describe to some committee that will fund me. So I have a feeling that this dialogue, which is fascinating, is not new, because this kind of theoretical discussion goes back certainly to the beginning of this century and — if you want to pursue it — further, much further than that, and a lot of it is accumulating dust in the libraries, and a lot of it, being one of the older people here, I studied as a student. They were full of — I don't want to criticize them, though, but just describe it — but consider how much of it there was and how little connected it was with experiment. So I am concerned about this. I think Hilbert was once asked, in a certain mathematical colloquium, what he thought of a paper. And he said, "Kreide" ("chalk"). And this is just a slight reminder that there is a real world!

Wigner: It looks that way. I confirm one of the points that Dr. Rabi made; namely, that Einstein knew the connection between the so-called de Broglie wavelengths and the velocity of the particle. And, as I mentioned, he did not publish a paper on it because he could not formulate it in such a way that it satisfied the conservation laws of energy and momentum. Of course, the experiment — now, who was it who carried out that experiment at General Electric? [Someone interjects, "Davison and Germer at Bell Labs."] Yes, Davison and Germer, the Bell Lab experiment. Einstein could have proposed that experiment, but he did not.

XI. Working with Einstein

WORKING WITH EINSTEIN

Chairman, HANS BETHE

Panel, BANESH HOFFMANN, *Moderator*
VALENTINE BARGMANN
PETER G. BERGMANN
ERNST G. STRAUS

Bethe: Of all the sessions we have had, this one perhaps, comes closest to Einstein. Its title is "Working with Einstein" and there will be a panel of people who have actually worked with Einstein. I am very happy to introduce them. They are: Banesh Hoffmann who will be the moderator, Valentine Bargmann, Peter Bergmann, and Ernst Straus. Professor Uhlenbeck unfortunately was not able to come here, so there will be four participants in this panel.

Hoffmann: Before we begin, I wondered if I might make a remark that may be of use to some people. You were all given, on registration, a copy of a book, *Albert Einstein, The Human Side.* Many people, it seems, have not realized that the book that you received was a special printing and that it has your name on the first of the two title pages. Now, to get down to the business of this meeting. I would like to introduce the panelists again because I can at least point to them as I mention the names. I am going to introduce first Peter Bergmann of Syracuse University; he worked with Einstein from 1936 to 1941. Next, Valentine Bargmann of Princeton University, who worked with Einstein from 1938 to 1943. Then, Ernst Straus of the University of California at Los Angeles; he worked with Einstein from 1944 to 1948. Myself, Banesh Hoffmann, Queen's College; I worked with Einstein from 1936 to 1937. You will note from that that I am, in a certain sense, the senior person, and this is probably why I was asked to be the moderator. I held a meeting with my colleagues and it did not go at all the way I had hoped. I did not want to be the first speaker — nobody wants to be the first speaker. So, I suggeted that the moderator, out of politeness, ought to speak last. Unfortunately, I was absolutely outvoted. I have never been on a committee so unanimous — even against me, and they voted that we should be in order of seniority starting with the most senior, and I couldn't budge them, so I, therefore, have to start the proceedings and I do have a sort of theory about it. You know, as you get older, your memory begins to fade, and they

Harry Woolf (ed.), Some Strangeness in the Proportion: A Centennial Symposium to Celebrate the Achievements of Albert Einstein ISBN 0-201-09924-1

must have thought that I was at the stage where every minute counted. Well, no one of us knows at all what any of the other panelists is going to say. So, I am going to be as curious as all of you as to what transpires.

In a more serious vein, I myself feel, and I think the other panelists also, that we are in a rather frightening situation because, unlike what it would be if we were just merely giving a paper on the history of relativity or something of that sort, every single word we say will be taken down as gospel and will be repeated verbatim at the bicentennial and forever after. So we have to watch our language and feel that perhaps we will be excused for any lapses in memory.

The main thing that occurred to me as I worked with Einstein was that what I had been taught as to the nature of science was all wrong. I have the strong impression that science, as done by an Einstein at least, is one of the creative arts. I do not know whether the other panelists feel the same way; but you could see that Einstein was motivated not by logic in the narrow sense of the word, but by a sense of beauty. He was always looking for beauty in his work. Equally, he was moved by a profound religious sense fulfilled in finding wonderful laws, simple laws in the universe. It was really a religious experience for him, of the most profound sort, even though he did not believe in a personal god, as you know. I should mention — someone said to me, "Yes, mention it even though everyone knows it" — I asked him once about a theory and he said, "When I am evaluating a theory, I ask myself, if I were God, would I have made the universe in that way." If the theory did not have the sort of simple beauty that would be demanded of a God, then the theory was at best only provisional.

Now, the second thing that comes to mind sharply about working with Einstein is that he never gave up. He persisted when Infeld and I were falling into despair. I will come back to that particular thing in a moment. Another thing is that when we (I should say that I did not work alone with Einstein; I worked in collaboration with Leopold Infeld) were working with him, he treated us as equals. There was never any attempt by him to browbeat us; quite the contrary, he made us feel extremely comfortable intellectually and emotionally, and I will illustrate this as I proceed. We went to Einstein and asked him whether we could work with him and he said yes, and he suggested two topics. We chose one of them and it actually worked out in the end. Einstein had already created the basic architecture of this work. And the architecture involved a new method of approximation. He had the main ideas. We treated matter as singularities, and we were looking for the equations of motion of these singularities and seeing whether we could not get them out of the field equations of gravitation. Infeld and I wanted immediately to say, "Well, take the singularity, move it a little bit, delta, epsilon or something like that," and Einstein said, "No, you mustn't do that, because if you move an infinity a tiny distance, you've made a big change in the field." Infeld and I, at a loss to know what to do, sort of wanted to continue to move these singularities and then Einstein used the word — he said, "You must not sin." I recall that one of the speakers in an earlier session on a previous day also told about Einstein saying, "You mustn't sin."

There were often times when we were up against a blank wall, and when that happened, Infeld and I were apt to despair, maybe because we were the ones doing the dirty work of the calculations, because Einstein did not do the calculations, but without doing them, he still managed to stay absolutely *au courant* with everything that was going on. That in itself is rather remarkable because I find, if I do not do the calculations, I do not know what the person is talking about. Einstein did not need to do the calculations. But, there were times when we were very despondent; on one occasion, I recall Einstein using essentially these words, "The world has waited this long, another few months won't make much difference.'' And so he was ready to go ahead and sort of ''waste'' some more months on this particular kind of thing.

I want to mention a couple of incidents that stick very strongly in my mind. The first one has to do with about the only significant contribution I made to the collaboration. We were making approximations in a certain way and from the method of approximation we were taking something on the left-hand side, and putting on the right-hand side surface integrals that had certain values. As we proceeded from one stage of approximation to the other, we added more of these messy surface integrals on the right. And we kept doing this for a long time. And all of a sudden, it occurred to me — as if to say the ''Emperor has no clothes'' — that as soon as we put something on the right, we no longer had the field equations, and this was a very sad thing because we wanted to solve the field equations. I remember when I said that that Einstein looked at me with the most extraordinary combination of admiration and a sort of surprise. It was the nicest thing that ever happened to me in that respect. And he was very happy indeed and marvelled that I could have noticed something, which essentially was a simple thing because I am not the sort of person who can think of other than simple things. Of course, Infeld and I then became very despondent — it looked as if all our work was not only bad already, but at each approximation, was going to get worse. But Einstein did not give up, and he twisted this around into an advantage reminiscent of the way in which he took the principle of equivalence, which appeared as evidence against the relativity of acceleration, and twisted it to show that it was for the relativity of acceleration. Einstein took this awful situation of ours and said, Now I know what to do; we'll take this mess that we have accumulated and we will set it equal to zero. Then, at once, you have the field equations, for ''this mess put equal to zero'' was indeed the equations of motion that we were seeking. So, we had then a victory snatched from defeat by Einstein.

I now want to tell of the second incident. The number of equations of motion that we were going to have per gravitating object was four. You could see four coming up because there are four dimensions. Now, however, all we needed was three, and four was too much. Again, Infeld and I were despondent. But not Einstein. And, I think, Bram Pais spoke of something like this — I didn't know it at the time; but Einstein said, ''Oh, this is very good.'' We said, ''What?!'' And he said, ''We will have an overdetermined set of equations and we may therefore get quantum conditions analogous to the allowed Bohr orbits.'' You can see what a marvelous situation that would be, that, out of general relativity, you get unallowed and allowed orbits. Unfortunately, it did not work out that way. Infeld, in a very ingenious maneuver, showed that at each stage, by renormalizing the time coordinate, we could get one of the equations to amount to zero equals zero. So, we were back with just three — we didn't get the quantum effect — nevertheless, Einstein beamed gratefully at Infeld for having found this particular thing.

I will close by telling of what happened when we were really baffled; some people know this story already — I apologize to them. We did the discussions in English for my benefit; my German was not as good as Einstein's or Infeld's, of course. When we got into a situation where we did not quite know what to do next, without realizing it, Einstein would lapse into German. Infeld immediately switched to German, and I did my best to do the same. We went on like that, and very often just doing the German sufficed and we found the solution to the difficulty. But there were occasions when switching to German did not work. And then, and this happened at least three times, Einstein would say — I mean we would be looking at each other in blank despair — and Einstein would say, ''I will a little think;'' that was his best way of saying, ''I will think a little.'' And then he would twirl his hair like this and he would walk up and down or stand still and his face had no sign of strain at all. He seemed as if he was in another part of the universe, only his body being with us, and Infeld and I kept absolutely silent. I simply do not know how long this went on. There was Einstein thinking like this and after a while he would suddently relax, come back to earth, look at us, smile at us, and say, Yes, we should do such and such. Of course, it worked, and that was how we got out of those very deep difficulties. Now, Infeld and I had hoped that by working with Einstein, we would learn how it is done, to some extent. Here we were, looking at the thing being done in front of us and there was not the least clue how it was. I must say that I have tried this [twirling of the hair]. Thank you. I will now call on the next in seniority and that is Peter Bergmann.

Bergmann: I guess my chances for doing any very useful work are rapidly diminishing. I had noted down here a few points, which I shall have to modify, because Banesh has anticipated a couple, which is something we anticipated. [Hoffmann: "It's your fault that I spoke first."] Absolutely not, so I will enlarge on the others. I would like to comment first on the difference in the situation in which young scientists find themselves today versus in the thirties. We all have been — that is, those of us who are professors — very much concerned with how to help our young graduating students to find appropriate postdoctoral positions. There was nothing to worry about in the thirties. There just were not such positions. And, for somebody like myself, who was German and Jewish and had gotten his Ph.D. in an out-of-the-way place like Prague, the question of how to advance my career was a question that you simply did not pose because there was no answer.

To decide what to do, I was guided essentially by the question: What sort of physics would I like to do? And with all the mature experience born of twenty-one years of life, I decided that, if it was at all possible, I would like to work with Einstein. I think I can honestly say that I was in no way — or at least only in a very minor way — intrigued with Einstein's fame and reputation. I had read a number of his papers and I thought this was the kind of physics that fascinated me, that I would like to do. And so, I wrote to Einstein a letter saying I was about to get my degree in Prague, could I come to Princeton and work with him. No answer, for several weeks. I wrote a second letter saying I was now *really* getting my degree and I sent him a copy of my thesis and repeated my question. And then I got a response saying, well, he was associated with an institution in Princeton — not a university, but the Institute — and, if I should decide to go to the Institute, there might be an occasion for scientific contact.

What I did not know in my naiveté was that even a person as generous as Einstein would like to know a little bit more than there is somebody in Prague who is interested in working with him. The reason that I did not get a response was that he wrote not to me but to Philipp Frank, the director of our Institute, and apparently got a very generous response in support of myself from Philipp Frank. Then he answered me. I suspect this holds for the other people who have been associated with Einstein as well as for me — that there was at the beginning no hope to get any salary and no hope to climb on the back of a famous man who clearly was a refugee himself in the country in which he was living now, but simply enthusiasm for the kind of work that Einstein did. When I came to Princeton, actually a few months later, Einstein no longer had two topics to dispose of, because, in the meantime, Banesh Hoffmann and Leopold Infeld had picked one. Einstein suggested that I work on a classical model of an electron, one that would not be spherically symmetric — in other words, not the Reissner-Nördström situation, but what he visualized at that time (what we would now call a Kerr model), for which nobody knew of a solution. Eventually we drifted into several different kinds of unitary field theories — a field that I think has lain fallow roughly from the time of Einstein's death until a few years ago, when it was reactivated by the interst in Einstein–Cartan theory, in complexification of space-time and supersymmetry and supergravity.

One point on which I can be very brief, because Banesh has beautifully anticipated it, is what a collaboration with Einstein was like. It was a very intense affair. Obviously Einstein was then working with two different groups — although I had the privilege of sitting in on yours [B. Hoffmann's] occasionally, and I'm sure you sat in on mine — but, essentially, we met every weekday for several hours in the morning and then separated ourselves for lunch and usually worked by ourselves in the afternoon. That did not preclude the frequent use of the telephone in the afternoon, and during the first year, when we lived in a furnished place that had no telephone at all, Einstein would not be adverse to coming over if he had made a discovery that seemed to be relevant to our work. Banesh has indicated, and several other speakers have indicated, Einstein's tremendous persistence, that he was willing to spend decades, a whole professional life, on considering the questions that he perceived to be important. Banesh has also indicated [Einstein's extraordinary creativity]; not only in that very large canvas, but also in the small, Einstein was, in my experience at least, by far the most creative person I have ever met. Perhaps I should add that I think what made Einstein extraordinary (a common characteristic is that many creative persons are very uncritical toward their own ideas; they may be very critical toward the ideas of others) was that he could work up tremendous enthusiasm for a new unitary field theory — and during the five years I was in Princeton certainly there was a large number of new unitary field theories — but there would always come the moment of truth. Einstein would discover a fatal flaw in what he himself had initiated and ruthlessly cut off that attempt, only to take on a new idea for work usually within a few days.

Let me comment on one other thing: it has often been said, and I think to some extent it is true, that Einstein did not create, at least not during the time that he was in Princeton, what you might call a school. There are so many people here who usually, by being associated with a university, have had students, not to say disciples, who then in turn had students and disciples, so that you have a large number of people who consider themselves the spiritual heirs of one individual. I think at least during his later years when I knew him, the intensity of his collaboration with a few young people went hand in hand with a relative isolation, for instance, from the university in town and even from the more purely mathematical work that went on at the Institute. At that time there were six professors in the school of mathematics, of whom he was the only theoretical physicist — the other five were all pure mathematicians — although Johnny von Neumann obviously had maintained some degree of interest in physics too.

However, I think that it was posthumously that Einstein, perhaps with Wolfgang Pauli as midwife, created a school of sorts. During the last few months of Einstein's life, Pauli organized in Bern a fiftieth anniversary meeting on relativity, which took place in the summer of 1955. Pauli had consulted with Einstein on the scientific program in considerable detail; for me it was the first international meeting in which I ever took part, and it was saddened only by Einstein's death a few months earlier. But this Bern conference has developed into a series of conferences that take place every three years and that are now conveniently numbered as GR-1, -2, -3, -4, and so forth, Bern itself being GR-0. We are now between GR-8 and GR-9 and, in connection with these conferences, a fairly loose international organization has developed — the International Society for General Relativity and Gravitation — whose principal activities are the sponsorship of these conferences and a journal. In trying to keep the bureaucratic machinery to an absolute minimum, I believe we are acting in the spirit of Einstein. But there is now a loose conglomeration of several hundred people who are actively at work in general relativity and its marginal areas, thus continuing at least one of Einstein's principal interests.

Hoffmann: Thank you, Peter. We did decide that all four of us would speak first and then we would interact and then leave time for questions from the audience. So I'd like to call on Valentine Bargmann to speak next.

Bargmann: Of course, much that I could or would have said has already been said by my two predecessors. Maybe it will be best to say a few words about how I came to work with Einstein. In the summer of 1937, a year after I had obtained my Ph.D. at the University of Zurich, I came as a refugee to this country. At the start, I did not have any definite plans, but on the advice of several American physicists whom I met, I decided to try to get to the Institute for Advanced Study — where, as we all know, Einstein was a professor in those days. There Hermann Weyl, who happened to know me, soon introduced me to Einstein. Hermann Weyl was always very much interested in the refugees who came to the Institute and looked after them. So I met Einstein, and he asked me to come again for a somewhat longer and more extensive conversation. This I did, and it was my real introduction to Einstein.

Clearly, when I entered Einstein's office, I was quite conscious of the fact that I stood in front of one of the greatest scientists of all time. It would not have been strange if I had been entirely tongue-tied. But this is not what happened. Einstein included me in the conversation that had been going on when I entered his office, and after five to ten minutes, I had forgotten my difficulties and felt entirely at ease.

In those days, Einstein, as you have already heard, worked with Peter Bergmann and Leopold Infeld. Now Einstein was extremely kind and asked me to come whenever I felt like it, and this became a very pleasant habit, for I visited his office about once a week for a while. Mainly the discussion was on the unified field theory, which was then being worked out by Einstein and Peter Bergmann; in particular the discussion concerned the mathematical questions that arose. So I got acquainted with the theory and I could participate in the discussion, and after a while — it was by now 1938 — Einstein asked me whether I would like to collaborate more regularly as a kind of unofficial assistant. I gladly agreed, and this was then the start of our collaboration.

Peter Bergmann and I were Einstein's assistants from 1938 to 1941, and we worked on that unified field theory about which you have heard. There are many things that one can say about Einstein, but much of it would be a repetition of what you have heard already. Let me talk about something more specific. I think you all know that Einstein liked to formulate his ideas about various things — scientific and nonscientific — as epigrams or aphorisms. One I remember was this. One of Einstein's famous quotations, which you can even see in the old Fine Hall above the mantelpiece, reads in German, "Raffiniert ist der Herr Gott aber boshaft ist Er nicht," of which an English translation may read as follows: "God is subtle, but He isn't malicious." And the interpretation usually was: "It might be difficult to find the laws of nature, but it is not impossible." There is also a somewhat different interpretation, because once Einstein said to us, "I have had second thoughts. Maybe God is malicious after all." But what he meant was something very specific. It was that God makes us believe that we understand something, when in reality we are very far from it. And Einstein was very much concerned that one should not be uncritical enough to be misled in this way.

Hoffmann: That will last to the bicentennial!

Bargmann: Another aphorism that I want to mention has to do with something that was discussed this morning. Namely, in his various attempts at creating a unified field theory, Einstein always talked in fact only about gravitation and electromagnetism. What about other forces? What about other elementary particles? During the time Peter and I were working with Einstein, new particles were being discovered, and yet it seemed that Einstein did not show any very great interest in them. And once he said to us, "You know, it would be sufficient to really understand the electron." He explained to us more clearly what he meant. The electron was, of course, entirely alien to any previous theories, starting with the theory of electrons. Consequently, in order to understand the electron properly, one would have to have a deeply modified new theory. And, if you were to succeed in this, then the chances are you would also understand the other elementary particles. Now, of course, it would have been quite wrong to take this rigidly as Einstein's final assessment of the state of affairs. Clearly, he could and would change his mind if further thoughts or further experience would suggest it.

Hoffmann: Well, Ernst, it's your turn at last.

Straus: I see, in contrast to the others, I was wise in not even making myself little notes, because by now I would have had to disregard them all anyway. By virtue of my position, I feel impelled to tell a little Einstein story that, in contrast to the others that you have heard, has absolutely no cosmic significance, but that I read and hope remember roughly right. It was told by one of the many sculptors who made a bust of Einstein — by the way, Professor Woolf liked what I said in recalling Einstein's comment about them — Einstein said that when in the future they dig out these sculptures, they will not believe that there was such a person. They will say that it is a representation of the Hair God that was being worshipped in the middle of the twentieth century. In any case, the sculptor — I think it was Davidson — was sitting in Einstein's office, and Einstein was talking animatedly and in German to a somewhat glum looking young man who was sitting next to him, and every now and then, the young man would put in a discouraging word, also in German, like saying, "No," or, "That won't work either," or, "That's hopeless." After half an hour, Einstein ran out of conversation and the young man left. Einstein then turned to the sculptor and said, "That was my assistant, Valentine Bargmann. He's wonderful. I wouldn't know what to do without him."

Since we have all gone into a bit of autobiography, I must say I made really no effort at all to get to meet Einstein. In fact, I thought it rather foolish of a mathematician who had met me to have proposed my name to Einstein as a replacement of the very capable assistant whose name ends in "mann" — whom he had lost. In fact, I did not know it was possible to be an assistant to Einstein unless your name ended in mann!

When I first met him at his friend Bucky's home in New York, he was lying upstairs and resting, and as I came in, he said, as I think was his custom in such things (in German they have a saying for it: "Fall into the house with the door" — that means do not make any preliminaries, but get to the point), so, he said to me, "You know, I have recently lost confidence in the principle of no action at a distance and therefore have tried to base physics on the following ideas.'' And then he talked for about an hour explaining — and I must say my courage was greatly raised by this, because I could see that there were mathematical questions there that I could at least understand. Maybe I asked some intelligent questions, for at the end of this exchange, he said, ''How much money would you need as an assistant? We don't pay very much at the Institute.'' And I told him that I was getting married and he said, ''Oh, then you can get $400 more.'' So then, when he reached for the phone, the enormity of the thing struck me, and I said, ''I must tell you, I know absolutely no relativity theory.'' He said, ''That doesn't matter. I know relativity theory. You can read up on the necessary part.'' And so, that is how I came to him.

I discovered some missing links in this story only last summer when we visited Margot [Einstein] and Miss Helen Dukas, because Miss Dukas told me that he came down and asked Bucky [Einstein's friend and physician], ''What do you know about this young man?'' And he said, ''I really don't know anything about him. I have heard that his father was a well-known Zionist in Munich.'' And so, Miss Dukas said, ''Oh, he's that Straus. I know him. I was at his circumcision.'' So, in contrast to the remarkable caution and circumspection that he showed relative to hiring Peter Bergmann, he really took very little evidence in employing me. I asked him a year or two later why, in getting himself an assistant, he went about it in such a cavalier way. He said, ''Oh, I was under a completely different impression, I thought that you had moved heaven and earth to get to meet me.''

I would like to talk a little bit about the kind of ideas, as I saw them, that he was working on. Of course, there is the conventional description, I think, of Einstein's work after 1926, that he went into this blind alley of unified field theories that people do not have much confidence in, maybe because of an unreasoning opposition to quantum theory, or because his physical inventiveness had declined. I cannot judge this but I think I can judge that he seemed to be following the same path that he had been following all along. I know the description of the two things that, in my opinion, were governing his work and they both occur in Goethe's Faust where Faust describes what he is trying to do. First he describes his goal: ''um zu verstehen was die Welt im Innersten zusammen hält'' — to understand what holds the world together at the core. That is really what he was looking for and I think he often told me, sometimes in criticism of others, that he felt that scientific greatness, in the end, was not a matter of intelligence. It was a matter of character. He said, ''You have to find the central question and then you have to pursue it through everything — no matter what the difficulties seem to be. Especially, you must never permit yourself to be seduced by any problem no matter how difficult it is.''

The second point, which he emphasized especially in his later years, was, again to quote from Faust when he criticizes his assistant, who is a kind of experimental physicist (if you'll pardon the expression), in which he says, ''und was sie deinem Geist nicht offenbart, das zwingst du ihr nicht ab mit Hebel und Schrauben'' (''And what she [Nature] won't reveal to your spirit, you won't force from her with levers and screws''). Now, Einstein formulated this in various ways, saying that there is no bridge between experience and concept. And he believed and expressed it in many ways, that the concept is the primary thing. That you have to look for the right concept and that the rightness of the concept will reveal itself in the beauty, the correctness — he called it logical simplicity — of the theory to which it leads. Then you proceed to find the consequences and test them against experience to see whether in fact you were guided correctly or whether the somewhat malicious God had given you the image of success when you had actually failed. Of course Einstein knew both from experience and conviction that this happened very often.

Again I realize, and several people have said it at this meeting, that this is at least a seeming change of view from the young Einstein, who, at least in his writing, seemed to emphasize that you must build on the experiment. He told me and he believed it at the time, that this was carelessness of the way he had expressed his conviction in his youth. Now that he was getting older, he was more careful in how to state what had been his conviction all along. Those of you who think they know better will probably still know better. Because, as he himself often said, "You don't know the twenty-five-year-old man when you are sixty-five." And that is also true.

The inventiveness that he brought to the theories that we tried to think about, was essentially, then, an inventiveness of mathematical structures; and I must say it was extraordinarily rich. I think Professor Chern mentioned some of those things. Very often it was much richer than his mathematical power permitted him to pursue. Professor Pais quoted some of Einstein's remarks that he considered words of despair. I must say I considered them almost the exact opposite. Usually he would speak that way after he had indulged in a grandiose vision of various possible models for the correct theories for which, however, he felt that his mathematical powers were not great enough, and time did not permit him to indulge in pursuing things for which more mathematics would have to be learned that he felt he could still learn. This could be described as despair, but I do not think it was. On the contrary, I would think it was some kind of exultation — a feeling that, even if I don't know enough, if my mathematics isn't strong enough, is not rich enough to pursue these things, all possibilities are there. He especially emphasized the ideas of topological characterizations of the meaningful universe, although he called it "analysis situs" — which made my life difficult with the mathematicians. I remember I could never be at a party where Lefschetz was present without some discomfort. Because every time Einstein met him, he would ask him: "What is new with analysis situs?" Lefschetz would store and build up the rage to let it out on me afterward and say, "How can you stay with a man who after fifty years still calls topology 'analysis situs'?"

Another idea of Einstein's I think, which in various ways has been dwelled upon by others here, was the idea of the lack of freedom that God has, which I think many people mentioned in connection with Einstein's disapproval of singularities. Not only did he disapprove of singularities because they are parts where the laws of physics do not hold, but he disapproved of them, I think, even more because there are parts where free parameters could be introduced, free boundaries — parts where the universe was not compact. His conviction was that, not only are singularities to be deplored, but any kind of boundary conditions that allow free parameters are really not right. He usually phrased it in a way in which mathematicians would probably not phrase it, by saying, you need the kind of overdetermination that will only allow discrete parameters. What he had in mind was a kind of eigenvalue theory for nonlinear differential and integral operators.

Maybe I can take a minute or two to discuss the theories, because at least one of them, as far as I can tell has not been published before. None of my three neighbors has apparently discussed it with him and we never published anything on it. One of the ideas was only touched on briefly by Professor Chern.

The first thing that he described to me when I saw him was this idea. He said he had observed that if you take the exponential of i times the Euclidean or Minkowski distance as a kernel of an integral transform, then the inverse of that transform has the conjugate complex expression as kernel. In other words, he was saying that this is a Hermitian operator — to use mathematical slang — into which I must say I was translating it for myself. And he was wondering what is the set of metrics that this property characterizes; that is to say, you want to look at the totality of all possible metrics so that if you take e to the i times the distance as the kernel of an integral transform, this kernel is Hermitian. The way I killed his interest in the project was by finding too large a manifold of solutions. He felt that physics could not permit that many solutions and if you now put in additional conditions they again assume a differential nature so at least the principle on which he wanted to build this attack, namely, that maybe there is action at a distance, would then again be violated. This was a half year's work; I was going to lunch, as I usually did in those days, with Pauli and told him about the collapse of the theory. By the way, Pauli liked it very much and urged me to go on looking into it because he said he did not agree with Einstein that the manifold of solution that I had found was sufficiently discouraging. But, he said, "You know there is one thing I have never been able to understand about Einstein. If I work for half a year on a theory and it collapses under me, I'm depressed. I don't do anything for a month. But Einstein has no downs." And sure enough, the very next morning, when I picked him up, on the way to the Institute, he said, "You know Mr. Straus, I have a new idea." And it was an entirely different idea and there was absolutely no mourning for the one that we had just abandoned. The entirely new idea — I won't enlarge on it because, as I say, Professor Chern mentioned it, was to build the foundations of physics on the Cayley–Menger characterization, which is purely algebraic, of the Euclidean or Minkowski distance. Cayley characterizes the property of four points to be embeddable in the Euclidean plane by giving you one very simple algebraic relation among the six distances that they determine. And Menger has developed this into a whole theory. There is a rather nice book by Blumenthal on metric geometries based on such algebraic conditions. The idea was to look at minimal weakenings of the condition of flatness of space and see what they would lead to. I think this again failed for the same reason that perhaps one should have expected a priori. Algebraic conditions are just not sufficiently restrictive to get that kind of purely discrete solution manifold that Einstein was hoping for.

The next theory is more familiar, and various people have spoken of it. The idea was to base the universe on a four-dimensional complex manifold, that is to say, to employ eight real dimensions and to make the fundamental tensor Hermitian. I was most taken by that theory because I guess it is attractive to a mathematician to complexify things; but Einstein was beset by two things — first, by Pauli who was absolutely scathing in his denunciations, and second, by his own conviction that of these four complex dimensions, the observable part is the real part of the space coordinates and the pure imaginary part of the time coordinate. In other words, you have to determine something that, to Einstein's physicist mind, was really eight-dimensional, so much so, that it is really very much determined even in the four-dimensional cross-section. And he said, "This is too difficult." That was the only time I think I argued with him on the principle, because I did not think of myself as an architect but only as a hod-carrier in this work. Nevertheless I think that time I felt that he was abandoning something that had inherent attractiveness. He made the point that I (Strauss), may well be right, but that he did not feel that he had enough command of mathematics to be able to bring such a theory to fruition, and that his lifetime had to be allocated as fruitfully as possible. What he could do was to develop the theory that appeared first, I think, in a joint paper on keeping the asymmetry of the fundamental tensor that had come out of this complex attempt, but abandoning the complexity of the universe — I mean the mathematical complexity, not the physical complexity.

Since I see I am running into our discussion time, I would like to make only two quick points. One, I think, was mentioned by Professor Dyson yesterday: the idea that an ultimate theory, which Einstein obviously very strongly had in mind, would be something almost as unattractive as the idea of converting all mathematicians into some kind of glorified computer using Hilbert's algorithm, which Gödel had effectively killed. Now, since, like most people, I had the attitude, that scientific theories get more and more perfect and keep changing, and that there is no ultimate theory, I kept probing Einstein's conviction on that point, but it was absolutely firm. And while I do not have an aphoristic quotation on the point, I have the impression that the way he responded to me was this — he said, "Well, if the ultimate theory is established, then physics will be in the situation in which mathematics is now. That is to say, then you start proving theorems. At the moment what the great physicists do is look for axioms. Mathematicians don't do that. They have axioms and they look for theorems. And that is what the physicists will be doing then." So he did not consider that as a kind of a dead-end, as several people here considered it.

Finally, I think I can combine the two points. No story of Einstein in Princeton would be complete without mentioning his really warm and very close friendship with Kurt Gödel. They were very, very dissimilar people, but for some reason they understood each other well and appreciated each other enormously. Einstein often mentioned that he felt that he should not become a mathematician because the wealth of interesting and attractive problems was so great that you could get lost in it without ever coming up with anything of genuine importance. In physics, he could see what the important problems were and could, by strength of character and stubbornness, pursue them. But he told me once, "Now that I've met Gödel, I know that the same thing does exist in mathematics." Of course, Gödel had an interesting axiom by which he looked at the world; namely, that nothing that happens in it is due to accident or stupidity. If you really take that axiom seriously all the strange theories that Gödel believed in become absolutely necessary. I tried several times to challenge him, but there was no out. I mean, from Gödel's axioms they all followed. Einstein did not really mind it, in fact thought it quite amusing. Except the last time we saw him in 1953, he said, "You know, Gödel has really gone completely crazy." And so I said, "Well, what worse could he have done?" And Einstein said, "He voted for Eisenhower."

Hoffmann: Thank you, Ernst. I wonder whether any of the panelists is ready to ask any other panelists a question or make a remark. Let us entertain questions from the audience then.

I. I. Rabi (Columbia U.): This is a marvelous opportunity for me because I have always wondered what the opinion of those who knew Einstein well would be. Once here in Princeton Einstein remarked — it was during one of his fruitless discussions with Niels Bohr — that when he was a young man he thought all that a physicist had to know of mathematics was addition and multiplication.

[Someone says: "Trigonometry"]

Rabi: Trigonometry. He didn't mention it at that time but still he wouldn't object. But he went on to say, "I've since changed my mind." Now my question is this. When you think of Einstein's career from 1903 or 1902 on to 1917, it was an extraordinarily rich career, very inventive, very close to physics, very tremendous insights; and then, during the period in which he had to learn mathematics, particularly differential geometry in various forms, he changed. He changed his mind. That great originality for physics was altered, well I won't say gone; but it was a different Einstein after 1917 than the fifteen or sixteen years before that. Whether, perhaps, he had lost interest in physics and had an unfortunate love of mathematics or something else happened — I suppose that may happen to all of us after a certain period — this original creativity that one gets when one is first introduced to the subject appeared to be absent. I would certainly enjoy and be enlightened by comments from the panel — who are all mathematicians!

Straus: Actually, since you started out by referring to one of the fruitless discussions with Niels Bohr, I sat in on two or three hours of such discussions and I am not sure that I understood; I mean, I'm quite sure that I didn't understand all the physics. Of course, those of you who knew Niels Bohr, knew that acoustically you at best understood one tenth. Nonetheless, I do have a clear feeling of what went on. Namely, that Niels Bohr said, just as Einstein had made physics richer and more important by introducing the idea of non-Euclidean geometry in an organic way into physics, so he, Bohr, had made physics richer and more important by introducing the idea of non-cummutative rings of operators into physics. Both of them were arguing mathematics. That is to say, they felt that their inspiration was mathematical. Now, as I said, I can only take Einstien's own assessment of himself. I have no right and no intention to make an assessment of whether mathematical invention is poorer than physical invention; but mathematical invention was rich and was there.

Bargmann: I would like to make one remark that I think is quite important. Namely, that Einstein in a certain way underestimated the mathematics that he knew. What I have in mind is this: Take Einstein's early papers on statistical mechanics. From a very sophisticated mathematical point of view you might say the mathematics is not particularly difficult; it isn't. But conceptually, I think, in 1902 or in 1903, among run-of-the-mill physicists, 20 billion dimensional spaces over which you integrate — they are not so trivial. And Einstein handled all the analysis that you need in statistical mechanics as a master. And, also, from our point of view, the tensor calculus that he had to learn is not so difficult after all. It's a bit of linear mathematics which everybody can learn. So I don't think he had to learn so much new mathematics. If one looks at Einstein's work, one finds he always knew enough mathematics or could create enough new mathematics in order to solve the problems he wanted to solve.

P. A. Schilpp (Southern Illinois U. at Carbondale): I am neither a physicist nor a scientist, but only an unusually ignorant philosopher. I would like to add a footnote to what Professor Bergmann related about Professor Einstein's unusual, almost unbelievable ruthless self-criticism. After he had sent in his reply to my critics for my Einstein volume, and it had gone to the publisher, I received two and one-quarter handwritten pages from him in which he said, "I have now the solution to the unified field theory and here it is." Naturally I was thrilled to have finally in hand what he had been searching for all these years. So I immediately sent it off to the publisher. Within three days I got a telegram, "Send those two and ½ pages back to me — it's all false."

J. Stachel (Boston U.): I feel too humble after hearing these very beautiful and affecting remarks to comment on what the panelists have said, but I would like to comment on what Professor Rabi said — on the basis of historical information that goes back to a period beyond even the most senior to the coworkers of Einstein present today. It would not be correct to think that, up until 1917, Einstein had the physics community with him and that they went along with what he did. If you look at what Einstein was doing in general relativity, Planck was dead set against it all along.

He even took the occasion of Einstein's 1914 *Eintrettsrede* to his speech, when he joined the Prussian Academy, to make a reply in which he criticized Einstein's work on general relativity. Einstein was disturbed enough by that to write a letter to Planck about the question. Abraham and Mie, who were working on rival theories of gravitation, thought they had slain Einstein in their polemics against him. What Einstein was doing was looked upon as an aberration at that time. Now, suppose Einstein had failed, that he had not come up with general relativity. What would one say then? Then one might say, "Oh, up until 1912 Einstein was a great physicist and then something went wrong." [Someone says: "You're wrong."] May I finish? I defer, obviously, to Professor Rabi.

Rabi: You are a better historian from documents than orally, because I did not talk about Einstein's popularity of his ideas. I spoke of the richness of invention at that time.

Stachel: Well, that was the point I was just about to get to. Thank you. What was it that enabled Einstein to succeed in formulating general relativity? It was precisely that his physical intuition led him some time in that miraculous autumn of 1912 to the realization that it was what we would now call the metric tensor. I do not know whether he even knew that term at that time, but let us say a four-dimensional analog of the Gaussian two-dimensional metric tensor, which was needed to correctly describe mathematically the gravitational field. He realized that was what he needed, and he realized he was out of his mathematical depth at that point. When he returned to Zürich he turned to Marcel Grossmann and said, "What is the mathematics I need to pursue this conceptual insight I have into the correct way to represent the gravitational field?" And it was then, with the help of Grossmann, that he discovered that great tradition from the nineteenth and early twentieth century that Professor Chern talked about — Riemann, Christoffel, Ricci, Levi-Civita, and so forth. And it was on the base of the mastery of that tradition, which was not common knowledge among physicists at that time, that Einstein was able to succeed in his creative endeavor. If he had not been willing to learn the mathematics or to have help in learning the mathematics, he would have been blocked at that point and we might still be looking for the general theory of relativity. It was of course because he succeeded that he then took the physics community along with him. You have picked the date 1917 as your turning point. But I would say, again on the basis of the documents, that after 1917 when people turned to him with their problems (and his correspondence is full of examples of this) there is no sign in Einstein's comments of any flagging of his creative ability; or the ability to understand physical concepts of a more conventional nature.

Einstein was still quite capable, to the extent that he was interested, in mastering other people's problems and seeing how to help them. I think the comments we heard about the Berlin Colloquium, for example, shows that. The point was that he was pursuing what he saw as the central core of his interests, which was inner-directed. This inner-directed core led him quite naturally, and largely on his own, from special relativity to general relativity and then to unified field theory. If the rest of the physics community's interests diverged from that, well and good. Einstein was going to pursue his interests to the end; and that is what he did — like it or not!

Straus: I think you misunderstood Professor Rabi's demurrer. All he was trying to tell you is that his judgement of physics now is better than Planck's was then.

A. G. Shenstone (Princeton U.): I would like to ask Professor Bargmann if he would make some remarks about his relationship with Einstein on the subject of music. And I would like to tell one little story about Einstein and music at the same time. My wife knew him very well and was a pianist. And when Einstein had given up the violin, he and my wife were talking, Einstein said, "I've given up the violin. It's too hard. I'm going to take up something easier — the piano."

Bargmann: Since 1937, the year I came to Princeton and had met Einstein, we regularly played together about once a week for several years. And, I have to say a number of things. First of all, there are, I think, somewhat odd ideas abroad about what kind of musician Einstein was. Now, Einstein, of course, was not a professional musician, but he had a deep love and a deep understanding of music. It was extremely gratifying to make music with him. Some of you might be interested in his taste. At least in the years I played with him, his taste was very classical. He did not like the romantics of the nineteenth century. His favorites, probably, as far as I can make out, were Mozart, Bach, Vivaldi, and possibly Schubert. Schubert he liked very much. Now, we mostly played sonatas and also, very rarely, other chamber music when a few other people were around. I know that he stopped playing the violin, but I must make here the following remark: I visited Switzerland two weeks ago and met an excellent violinist with whom I had played in the past, and she told me that for a violinist at a certain age, around sixty years or so, it becomes very hard to play the violin. Nothing like that happens to a pianist. Playing the piano is never particularly hard, but playing the violin seems to be extremely hard for the body of the violinist. And I wonder whether this hasn't something to do with Einstein's giving up the violin. It was at just about that age — sixty-five or so.

A. Hermann (U. of Stuttgart): I want to ask a question that has something to do with the remark of Professor Rabi. In his Berlin years, it was a special ability of Einstein to follow in the physics colloquia and to give interesting remarks on the new ideas presented, and I wonder about the Princeton years. Was there something similar here?

Bergmann: I must admit that during the years that I was in Princeton, if such affairs took place, I think I did not go to them very regularly, at least not in the first two or three years, simply because my command of English was not adequate. Later on I went quite regularly to the colloquium at the university, but that was not where Einstein went.

Bargmann: There was something in Princeton in those early years that was similar to the Berlin Colloquia. This was called the Journal Club. It met once a week, I think it was Monday night, eight o'clock or so, and there were reports on current literature. It was not quite on the level of the Berlin Colloquium, but it was quite interesting, and new ideas were being discussed. Einstein, as far as I know, did not come to the Journal Club. At the Institute, Einstein was the only professor of theoretical physics.

Straus: I guess in later years Oppenheimer did introduce such a colloquium, but of course, that was much later, and as far as I can tell, Einstein hardly ever attended.

E. Wigner: I wanted to say again that Einstein was not at home in English. And if you listen to a speech in a language in which you are not at home, it is more difficult to absorb it, and it is even more difficult to make comments. Einstein always said, instead of E square, "E Quadrat." Of course, that's just as good, but it isn't English. And this was a great difficulty also. His relation to his colleagues was much less close than in Berlin, much less close, as Professor Bargmann said.

XII. Einstein and the Physics of the Future

EINSTEIN AND THE PHYSICS OF THE FUTURE

Chairman, HANS BETHE

Panel, MARVIN L. GOLDBERGER, *Moderator*
STEPHEN L. ADLER
STEVEN WEINBERG
FREEMAN DYSON
C. N. YANG

Bethe: We have had a glimpse of the past and now we shall have, of course, an equally definite preview of the future. The moderator of this panel is Professor Goldberger of the California Institute of Technology.

Goldberger: Thank you Hans. There has been an enormous upsurge in activity in recent years in Einstein's theory of gravitation or what is popularly referred to as the general theory of relativity. There are at least three reasons for this. First, a very small number of influential physicists who, in addition to their intellectual assets, were rather charismatic characters, took an interest in the theory about twenty-five years ago and produced a generation of brilliant disciples. Among the more prominent people of the twenty-five-year period were John Wheeler and Dennis Sciama. The second reason, and in my view probably the most important, is that there have been a number of astronomical discoveries in recent years that necessarily require for their complete understanding the full machinery of the theory of gravitation. I think the absence of contact with experiment for so many years dampened the interests of many physicists and made the theory of gravitation the realm of mathematicians or highly mathematically oriented physicists, and that was not the best thing for the subject. Finally, there have been great technical advances in quantum field theory that have led to renewed efforts to construct a quantum theory of gravity. A solution to this fundamental problem is still lacking. Nevertheless, there is much action in this area. Furthermore, the development of quantum theory in the presence of strong background gravitational fields has led to the remarkable discovery of the so-called Hawking radiation, and this too has stimulated much theoretical work. I now leave it to the panel to discuss the extent to which these developments and I am sure many others related to the eventual unification of strong, elec-

Harry Woolf (ed.), Some Strangeness in the Proportion: A Centennial Symposium to Celebrate the Achievements of Albert Einstein
ISBN 0-201-09924-1

tromagnetic, weak, and, I hope, gravitational interactions, will figure in the future. The gentlemen who are so bold or perhaps foolish enough to make such predictions are, in the order in which they will speak: Stephen Adler, Professor in the School of Natural Sciences at the Institute for Advanced Study; Steven Weinberg, Higgins Professor of Theoretical Physics at Harvard University; Freeman Dyson, Professor in the School of Natural Sciences at the Institute for Advanced Study; and Frank Yang, Einstein Professor of Physics at the State University of New York at Stony Brook.

Adler: I am going to comment on one specific aspect of the influence of Einstein on the physics of the future, the possible impact of analogies with general relativity on the development of our understanding of gauge theories and particularly on progress in quantum chromodynamics. Let me begin by making some brief remarks about general relativity. To a good first approximation, most analytic calculations in relativity fall into two categories. Category A comprises perturbation theory calculations, where one is working around a nearly flat metric. In this category one would put things like parameterized post-Newtonian studies of general relativity in solar system experiments, gravitational radiation theory, and renormalizability studies. In a second category, category B, I would put strong coupling phenomena, which involve things such as horizons that are not treatable as small perturbations about a flat metric. In this category, I would put cosmology and black-hole physics. The salient feature of category B is that it is dominated by the study of certain special exact solutions of the field equations, for example, the Robertson–Walker metric in cosmology and the Kerr family of solutions in black-hole physics. Somewhat bridging the gap between these two categories and permitting the study of less symmetric geometries are numerical methods that have become prominent in the last few years. As emphasized in the papers of Chern and Regge, relativity and gauge theories are close kin with many formal similarities in structure. Hence, one might expect useful analogies between the methods that are useful for studying physically relevant problems in the two cases.

Let me now turn to gauge theories. Again, in gauge theories, most analytic work falls into two categories which are analogs of those that I just described in the case of general relativity: category A — perturbation theory — and here I might mention first the tree-Lagrangian work in the Weinberg–Salam model, which yields all the neutral current predictions, and then the various radiative correction calculations to it, which are finite because the theory is renormalizable. A second example would be asymptotic freedom calculations for deep inelastic processes, and recent work on jet phenomena, in quantum chromodynamics. In category B — strong coupling phenomena — I put first instanton physics and secondly monopole solutions and their relatives.

Let me attempt to tie together these comments about general relativity and gauge theories. Again, as a first approximation, the following analogies are valid: Analogy 1) — the cosmological solutions in general relativity correspond to the instanton solutions in gauge theories. The cosmological solutions are time-dependent metrics describing universes of various types. The instanton solutions are time-dependent solutions describing tunneling of Yang–Mills fields from one classical minimum to another. A concrete example of this analogy is the fact that the Euclidean continuation of the de Sitter metric is isomorphic to a 4-sphere of constant curvature, and by an appropriate mapping that in turn is isomorphic to the single instanton solution discovered by Polyakov and collaborators. Analogy 2) — black-hole solutions in general relativity correspond to monopole solutions in gauge theories. The black-hole solutions are time-independent solutions of the source-free field equations, and when the electric charge of the hole is zero, they are Euclidean solutions as well as Minkowski solutions of the field equations. The monopole solutions can be regarded, and I think this the most useful point of view for treating them, as time-independent solutions of the source-free static Euclidean Yang–Mills equations. The word Euclidean is important here, because for technical reasons that may turn out to have fundamental physical consequences, Euclidean continuation is essential for the correct quantization of a quantum field theory. This is a fact that has been important in recent work in both relativity and gauge theories. As a concrete example of analogy (2), I mention that Louis Witten recently showed that all of the axially symmetric stationary metrics of relativity can be transformed into axially symmetric static self-dual Yang–Mills solutions. Unfortunately, they do not give physically interesting Yang–Mills solutions — the relativity solutions map into topologically trivial (topological quantum number $n = 0$) singular solutions of the gauge equations, whereas the physically interesting solutions are the topologically nontrivial but nonsingular solutions.

A further remark about both the instanton and the monopole solutions: The instanton solutions — that is, the gauge solutions that are analogous to cosmological solutions in relativity — have been completely classified as a result of the work of Atiyah and collaborators. The monopole solutions — the solutions that in some sense are analogs to the black-hole solutions — have not been completely analyzed. Much less is known about them. The problem of studying them is not compactifiable, and the classification problem for these solutions is still open.

Based now on what I have said, let me attempt to peek into the future. I wil begin by saying that I, and I think most of the physicists I know, are very stingy when it comes to gambling with money, and that is because we are big gamblers in our time. What I have been working on for the past year, and I expect it will probably take six months to a year more to get final answers, is the hunch that the static Euclidean Yang–Mills solutions (that is, the analogs of the black hole solutions in relativity) may be the dominant feature in the dynamics of heavy quark systems. The general formalism has a lot of properties that look right, and I have some evidence that there is a much richer structure of solutions than has hitherto been thought. Future work on this, to try and see whether this hope is realized, will probably heavily involve numerical methods of the type used in numerical relativity. It appears to be well within the state of the art to get definite answers.

Weinberg: Well, ten minutes is really very little time to talk about weak, strong, electromagnetic, and gravitational interactions, so I will restrict my comments to their unification. That is, what are the prospects for a future theory that is sufficiently simple and elegant to attract our allegiance and that unifies all the forces of nature? This morning, Professor Wigner mentioned the unification of the weak and electromagnetic interactions, and Steve Adler has just talked about quantum chromodynamics, the other ingredient in our present understanding of elementary particle interactions. Quantum chromodynamics has a crucial property that opens up I think for the first time a meaningful unification of at least the weak and electromagnetic interactions on one hand and the strong interactions on the other, and that is the property of asymptotic freedom, that the strong interaction, although strong at low energies, giving rise to all the richness of hadron physics, becomes weaker at high energies. At the energies that we are reasonably familiar with, the strong interactions look very different from the weak and electromagnetic interactions, but because of the property of asymptotic freedom, the strong interaction coupling constant gradually decreases as we go to high energy, so that there is some chance of an extremely high energy at which weak, strong, and electromagnetic interactions all come together.

What does it mean for them all to come together? Well, clearly it will not do just to say that the coupling constants become equal, because that definition would then depend on how precisely you normalize the currents. What particular thing do you mean by the $SU(2)$ generators — do you mean the Pauli matrices or half the Pauli matrices? There is, however, a natural way of defining what we mean by the coupling constants of weak, electromagnetic, and strong interactions all coming together. If there is an overall gauge group of which the $SU(2) \otimes U(1)$ of the weak and electromagnetic interactions and the $SU(3)$ of the strong interactions are subgroups, and if that overall great gauge group is of the type that mathematicians call simple, then such a theory will impose a condition of equality among the coupling constants — equality in the very well-defined sense that if you normalize the currents so that the traces of the squares of the generators in any particular representation are equal, then the coupling constants are equal. So if we find some set of particles that we feel forms a more-or-less complete representation of the overall grand unified gauge group, then we know what the trace of the squares of these generators are, and we can give a precise meaning to the coupling constants being equal.

Well, we know about lots and lots of fermions, we know about a great many quarks and leptons and they seem to fall into neat generations. We have now about three generations; there may be more. If one makes the not unreasonable assumption that all or almost all the fermions that exist fall into this familiar pattern of generations of quarks and leptons, then we have a handle on what we mean by equality of couplings, and one can say precisely what one means by equal couplings: with the conventional definitions, the g of the weak and electromagnetic interactions is equal to the strong coupling and the g' is the square root of three-fifths of it. This result is true for a very large class of theories; it is true for any theories in which the fermions form just essentially the kinds of families that we know about. So it is worth pursuing this to see what its consequences are.

One of its consequences is that the one parameter that was at the beginning unknown in the unified theory of weak and electromagnetic interactions, the Z-photon mixing parameter $\sin^2\theta$, is predicted. It is predicted to be about 0.2 (roughly from 0.19 to 0.21). When this calculation was first done, the experimental result was 0.35, but since then the reported value of this quantity $\sin^2\theta$ has decreased. The best present value from neutrino experiments at CERN is 0.24 ± 0.02, and the best present value from the SLAC–Yale experiment on deep inelastic electron scattering is 0.224 ± 0.02. So it is quite striking that one gets from theoretical considerations a number that turns out to be not unlike the experimentally determined number.

One can, by this analysis, calculate one other interesting number: the energy at which the couplings all become equal in the sense I have described. It is a very simple calculation and gives a very simple result: the natural logarithm of the ratio of mass scales — that is, the ratio of the superlarge mass scale where the couplings become equal to the mass scale of quantum chromodynamics — is close to $137\pi/11$. Of course since Landau, and perhaps before that, people have talked about very large numbers that arise when you ask at what energy the radiative corrections in quantum electrodynamics build up, so that electromagnetism becomes strong. Here I am asking at what energy the strong interactions become of electromagnetic strength. Well, if you have one of these little pocket calculators, you can figure out that the energy $\exp(137\pi/11)m_p$ is about 10^{16} GeV, a number that on the surface seems ridiculous, vastly beyond anything we are studying today experimentally, and yet on reflection not entirely inappropriate for us to begin to think about now.

There is no one theory that we have that unifies weak, electromagnetic, and strong interactions and that is so attractive that we feel we must take all its consequences very seriously, but at least we can ask: What are the consequences of this general point of view? What are the general consequences of the idea that there is some very simple theory characterized by energies of the order of 10^{16} GeV in which weak and electromagnetic and strong interactions all have the same couplings? Well, one consequence is the prediction of $\sin^2\theta = 0.2$, which I have already mentioned. It will be very interesting to see as the experiments become more precise whether that prediction is borne out.

We can also look at the question more generally and ask: If it is really true that there is a simple theory at those very high energies, is there any chance of seeing some other sign of it at ordinary energies? Or to put it another way: What would physics at ordinary energies look like from the point of view of this grand unified theory? Well, if you wonder what you would be able to detect at ordinary energies if the fundamental scale of physics were of the order 10^{16} GeV, one thing you would be able to detect are those particles that for various reasons happen to have masses very small compared to 10^{16} GeV. Those, presumably, are the quarks and leptons and guage bosons that we are all familiar with, and the interactions that you would be able to detect of these particles, the interactions that are not suppressed by negative powers of 10^{16} GeV, would be their renormalizable interactions. So you would expect to find a renormalizable theory of quarks, leptons, and gauge and scalar bosons, which is what we do find in today's physics.

However, there might be some additional very weak effects, which although enormously suppressed by powers of 10^{16} GeV in the denominator, nevertheless have peculiar properties that allow them to be detected despite their incredible weakness. In particular, if there happened to be, as a survivor from this grand unified theory, a mass zero spin-2 particle, its couplings would be precisely those of the quantum of gravitational radiation in Einstein's theory — the graviton — and furthermore, to an enormously high degree of accuracy, its interactions would be precisely described by the field equations that Einstein wrote down. This is simply because higher terms in the Lagrangian, terms quadratic in the curvature for example, would be suppressed by two or more powers of 10^{16} GeV. And that is what we see, except that the coupling constant of the gravitational intereaction is not $(10^{+16}\text{ GeV})^{-2}$; it is $(10^{+19}\text{ GeV})^{-2}$.

There would be one other class of phenomena that we might expect to see. These theories, because they put quarks and leptons in the same representations, naturally have as a consequence the nonconservation of baryon number. Several groups have analyzed the ways in which such theories might conserve baryon number, and I think it is fair to say, the conclusion is that you can make them conserve baryon number by standing on your head, but it is much more reasonable for them not to conserve it. Consequently, it becomes quite reasonable to expect that there will be a very low, but nonzero, rate of baryon decay. If you take the kind of mass scale that I mentioned, 10^{16} GeV, and make reasonable estimates, the lifetime of the proton (this has been worked on by a number of people now) comes out to be of the order 10^{30} to 10^{32} years. The present experimental lower bound is 10^{29} years. Thus the time is ripe for an assault on the next few orders of magnitude in the proton lifetime. The discovery of proton decay could be a clue to something that otherwise is experimentally inaccessible at present: the incredibly high energy range, where according to the best guess we have the four forces of nature all come together.

Dyson: I will talk mostly about the past, because it is easier to talk about the past than about the future. There are two kinds of mathematics. School mathematics is a tool that physicists and carpenters may easily master and apply every day to the practical needs of their work. A physicist uses school mathematics to design apparatus, to reduce data, to compare experimental results with theory. But real mathematics is something different. Real mathematics is not a tool but a form of art. "A mathematician," said my teacher G. H. Hardy, "like a painter or a poet, is a maker of patterns. If his patterns are more permanent than theirs, it is because they are made with ideas." Real mathematics is the structure of intellectual patterns built up over the centuries by the combined efforts of a succession of great artists. Real mathematics in the nineteenth century was Riemann surfaces, Cantor sets, and Lie groups. Real mathematics in the twentieth century is cohomology, sheaves, and categories. My discussion of the future of physics begins with a question: Do we have any reason to expect that the concepts created by contemporary mathematicians out of their own imagination should be of any help to us in understanding the workings of nature?

The answer to this question is not obvious. In the nineteen-fifties, when I was an arrogant young physicist, I would have answered the question unhesitatingly in the negative. In the nineteen-fifties, theoretical physics was jumping from triumph to triumph. We had in rapid succession the Yang–Lee theory of parity violation, the Bardeen–Cooper–Schrieffer theory of superconductivity, the Gell-Mann–Nishijima theory of strange particles. It was a great time for calculations. We went ahead, calculating everything and getting quantitative agreement with experiment, using only school mathematics and not real mathematics. We needed no help from the mathematicians. We thought we were very smart and could do better on our own. We listened with a certain contempt to our mathematical colleagues talking of fiber bundles and topological obstructions. It seemed that real mathematics had become irrelevant. The distinguished moderator of this panel [Marvin L. Goldberger] is alleged to have asked, when a mathematically inclined young man was recommended for promotion, "Has that guy ever calculated a cross-section?" You do not need topology for calculating cross-sections.

Looking further back into the past, we can see clearly that real mathematics has been of decisive importance in many of the great developments of physics. Einstein's general relativity had its roots in the differential geometry of Gauss and Riemann. Schroedinger's wave mechanics had its roots in the deep mathematical analysis of classical mechanics by Hamilton and Jacobi. Finally, the modern view of particle theory, with the subnuclear world a playground of interlocking broken and unbroken symmetries, had its roots in Felix Klein's *Erlanger Programm* of 1872, in which Klein for the first time described the various branches of geometry and algebra and analytical dynamics existing in his day as special cases of a unifying general theory of symmetric spaces with continuous transformation-groups. In this connection it is interesting to note that when Klein's collected works were published forty-nine years later, Klein added a historical note describing how Sophus Lie happened to be in Erlangen during the weeks in which the *Erlanger Programm* was written. "Lie reacted with the greatest enthusiasm to my idea of a group-theoretical classification of geometries." Klein was then a full professor at the age of twenty-three; Lie was thirty and had been recently released from a French jail. During the Franco-Prussian war he had the bad luck to be picked up by some French vigilantes who thought he was a Prussian spy. So Lie provides us with a direct link between the *Erlanger Programm* of 1872 and the Lie groups and Lie algebras that dominate the thinking of particle-physicists one hundred years later. Every physicist who learns to classify particles in terms of multiplets and broken symmetry is, whether he knows it or not, talking the language of Felix Klein.

If we take a long view, the history of every branch of science follows a pattern that was first described by Res Jost. According to Jost, each branch of science goes through three stages of development. The first stage is qualitative; facts are collected and described and general principles of classification and explanation established. The second stage is quantitative; magnitudes are measured, and laws are expressed by equations. The third stage is again qualitative; the concepts of a quantitative theory are understood in depth, and the resulting regularities of natural phenomena are described by theorems rather than by equations. Now comes the paradox. School mathematics, the mathematics of elementary computation, is important in the second, quantitative, stage of a science. Real mathematics, the mathematics of ideas and proofs, is important in the qualitative stages one and three, but not in stage two.

If we interpret the history of physics according to the three-stage model, then it is easy to understand why the young physicists of the nineteen-fifties regarded real mathematics as irrelevant. The decade of the nineteen-fifties was an anomalous period, because it happened that at that time all the fashionable branches of physics were simultaneously in stage two. Yang and Lee, Bardeen, Cooper and Schrieffer, Gell-Mann and Nishijima did not need any real mathematics to make their discoveries. Only gradually, as we moved into the sixties and the seventies, the bandwagon of stage-two physics began to slow down and the historically normal balance between the three stages was slowly reestablished. Now, at the end of the seventies, there are clear signs of a renaissance in several branches of physics that belong to stages one and three.

Let me take a minute or two to run through various areas of physics as they exist at present. I begin with areas that are now in stage two. These are areas that have been the mainstream of physics for the last thirty years but are now showing signs of diminishing excitement. Nuclear physics, solid state, quantum optics, and phenomenological particle physics. Of course these areas are very far from being exhausted. They are still vigorously alive, but they are not as full of basic unsolved problems as they were in the nineteen-fifties.

Areas of physics that are now in stage three are classical mechanics, classical general relativity, partial differential equations, statistical mechanics, and quantum field theory. Each of these areas has experienced an upswing during the last decade. Each of them is more lively in 1979 than it was in 1969. Classical general relativity has been rejuvenated by the global point of view introduced by Penrose and Hawking. The singularity theorems of Penrose and Hawking gave us a new qualitative understanding of the meaning of Einstein's equations. For the first time it has been possible to make precise statements about all solutions of Einstein's equations instead of trying to construct solutions one at a time. A similar rejuvenation in the study of other partial differential equations was produced by the Atiyah–Singer index theorem.

Finally, there are a few areas of physics that are still in stage one. These are in many ways the most exciting of all, the areas in which qualitative exploration has not yet passed through a quantitative phase. They are, quantum gravity, supergravity, and gauge-field models of particle physics. The reinaissance of quantum gravity began with the recent discovery by Hawking of a deep-lying unification of general relativity, statistical mechanics, and quantum theory, in which the area of an event horizon is identified with its entropy. The equality of area and entropy is a revolutionary idea. It is as surprising, and it will probably be as fruitful, as Einstein's identification of mass with energy. Supergravity is in my opinion the only extention of Einstein's 1915 theory that enhances, rather then diminishes, the beauty and symmetry of the theory. Of all the "unified field theories" that have attempted to link gravity with other physical fields, supergravity is the first that deserves, on aesthetic grounds alone, to be true. Whether nature will concur with this aesthetic judgment remains to be seen. Lastly, the gauge-field models of particle physics, pioneered by Yang and Mills, are the latest triumph in the long tradition of group-theoretical marriage between geometry and dynamics that began with Felix Klein.

So far as the future is concerned, one fact is clear. Real mathematics will be required in large doses if we are to make substantial progress in understanding either supergravity or gauge-fields. Supergravity is closely linked with the mathematical theory of graded Lie algebras. The behavior of gauge-fields is in many circumstances dominated by configurations that the physicists call instantons, and the classification of instantons is in essence a topological problem in which the Atiyah–Singer index theorem turns out to be relevant. So it is not only the old stage-three branches of physics that will be forcing physicists to learn functional analysis. Even the newest, brightest, most fashionable stage-one branches are running into problems that can only be handled with sophisticated mathematics.

It is always unwise to extrapolate from the past into the future. Unexpected surprises will drive the future in directions that we cannot foresee. Nevertheless I will disregard my own warning and make an extrapolation. The purpose of an extrapolation is not to be believed but only to provide a basis for discussion.

If I extrapolate from twenty-five years back, through the present, to twenty-five years ahead, what do I see happening to the development of physics? I see the connections between real mathematics and physics continuing to grow closer and firmer. In mathematics itself, the last twenty-five years have been a period of extensive unification, with the separate disciplines of algebra, geometry, analysis, and topology becoming internally united into a tightly linked structure. Extrapolation suggests that for the next twenty-five years the internal unification of mathematics will continue and will be carried over into those areas of physics that real mathematics will penetrate during the same period. As the end result of this process of unification and penetration of physics by mathematics, I predict that in the next twenty-five years we shall see the emergence of unified physical theories in which general relativity, group theory, and field theory are tied together with bonds of rigorous mathematics. When that day comes, the sign that Plato placed over the door of his academy in Athens, "Let Nobody Ignorant of Geometry Enter Here," will have to be put up over the door of The Institute for Advanced Study.

Yang: I have been told that the difference between a mathematician and a physicist is that the physicist uses a viewgraph and the mathematician does not.

Every scientist is born in and nurtured by the traditions of his field before him. And if he is successful, he would mold, develop, and perhaps dramatically change these traditions. Einstein in his career had molded two specific traditions of physics in a very deep way: statistical mechanics and field theory. It is perhaps particularly interesting today, in this centennial celebration of his birth, to observe that field theory is again moving center stage in fundamental physics. If one reads what Einstein said in his autobiographical notes near the end of his life, one cannot help but be impressed by his repeated emphasis that field theory is a fundamental problem of physics — perhaps *the* fundamental problem of physics. And I suspect that this fact will stay with us for many years to come.

Einstein also left behind a very important new tradition which, I think it is correct to say, was started by him, and that is the principle that *symmetry dictates interactions*. Before Einstein, interactions were experimentally observed first. We all know the four Maxwell equations as mathematical expressions of experimental laws. Einstein reversed this. He created general relativity at least partly, out of the principle that it is the symmetry that dictates the interaction. After that, in the sixty years or so after general relativity, there developed a number of other ideas that are a continuation of this principle. These are gauge fields and supersymmetry and suppergravity. The topics that have been discussed at this conference testify to the vigor of this line of development.

Together with the specific ideas of using symmetry to generate interactions, there goes the interpretation of these symmetries, and in all these cases, in some sense, the symmetry is related to some *geometry*. There has been lots of discussion about the relation between mathematics and physics in various directions throughout this conference, including the remarks of Freeman Dyson. Let me add a few more remarks. The first comment I would like to make derives from a statement made and emphasized repeatedly by Einstein, that beautiful mathematics is the language of fundamental physics. This is not an empty statement. Let us take the case of supergravity. There are of course only two possibilities. Either supergravity turns out eventually not to be related to physics, or it is deeply related to it. If supergravity turns out eventually not to be related to physics, then it is an interesting and not necessarily a beautiful type of mathematics. But if it should turn out to be a part of physics, then it is obviously going to be a very fundamental part of physics because it gives rise to a basic understanding of a unification of matter and field. If that is the case and we believe in Einstein's remark, then we are bound to accept the statement that the mathematics of supergravity must be very beautiful, which I believe it is not at this moment. Of course there are algebras that could generate supergravity or are deeply related to the mathematics of supergravity. Maybe it is my prejudice — maybe it is my ignorance — but I do not believe that any of these graded Lie algebras has the intrinsic and fundamental beauty of Lie algebras and Lie groups, not as yet!

However, we must not go overboard; therefore my second remark. Not all beautiful mathematics is useful to physics. If we do not understand this point we are likely to be led astray. In the twentieth century we have observed tremendous revolutionary changes in physics that brought in very beautiful and profound mathematics. Examples worth mentioning are Hilbert space and Riemannian geometry and now fiber bundle geometry and of course continuous groups. But they are, in fact, only a small part of the very exuberant and profound developments of the mathematics of the twentieth century. I believe it was Professor Marshall Stone of Chicago who made a very remarkable statement. He said that the greatest achievement of twentieth-century mathematics is that it had finally liberated itself from the shackles of physics. How much of the profound mathematics that has been developed is going to enter into physics? There is no way in which one can quantify this, but I would say that order-of-magnitude-wise, it is a few percent. If you believe this statement, you would believe as I do that if one starts by scanning all beautiful mathematics in order to find that part of it that fits physics, then one is most likely going to fail.

Another point I would like to make concerns an extremely interesting statement, I believe first made by Professor Dirac in 1931 in his famous article on the Dirac monopole. He argued that there are two routes to the mathematical concepts that are useful to fundamental physics. One route is to try to invent concepts that would help to organize experimental information into physical laws. The other route is to start from beautiful mathematics and then use that to relate to experimental reality. Perhaps I can illustrate this thinking by a diagram (Fig. XII.1). The first route starts from experiments and you try to formulate theory out of the experiments. Of course there are always feedbacks, but the main part of the conceptual information flows in the direction of the solid arrow in the figure. And the theory is strongly coupled to mathematics and to mathematical beauty. The second route is the reverse; you start from the theory in the main and go to experiments. It was very interesting that Dirac remarked in that paper that as time goes on, because of the increasing complexity of experiments, perhaps the first route is going to prove progressively less important compared to the second. I suspect there is great truth in this conjecture. Later in 1963, Dirac expanded on this theme and speculated on the possibility that theories may be developed, the physical meanings of which are not understood until later. Again there is great truth in this statement. Quantum electrodynamics and renormalization is one example. Black holes another. Non-Abelian gauge fields still another.

Goldberger: We'll now give the panel an opportunity to make remarks to or at each other if they choose to do so.

Dyson: I would just like to ask Frank [Yang]: How did you discover the Yang–Mills theory?

Yang: Well, it is a rather straightforward generalization of Maxwell's equation. [Laughter] It is.

Dyson: We all know what an amazing piece of creation that was, but the question is: Did it come purely from considerations of formal beauty or did it come from physics?

Yang: When I was a graduate student in Chicago, I was deeply impressed by the gauge invariance of electromagnetism that we all learned from Pauli's article in the *Reviews of Modern Physics.* So, I thought one should generalize them. But there were repeated failures, and it was only in 1954 in collaboration with Bob Mills that I succeeded in writing down one possible theory. But there was another motivation for it. You probably remember that around that time mesons were being discovered, and there were quite a few of them, and we were all busily trying to write down all kinds of interactions. I thought that we must have a principle to guide us to write down interactions. It was obvious at that time that there was one principle (I'm not talking about general relativity) that did tell us how to write down interactions and that was electromagnetism. Therefore we said that we have got to generalize it.

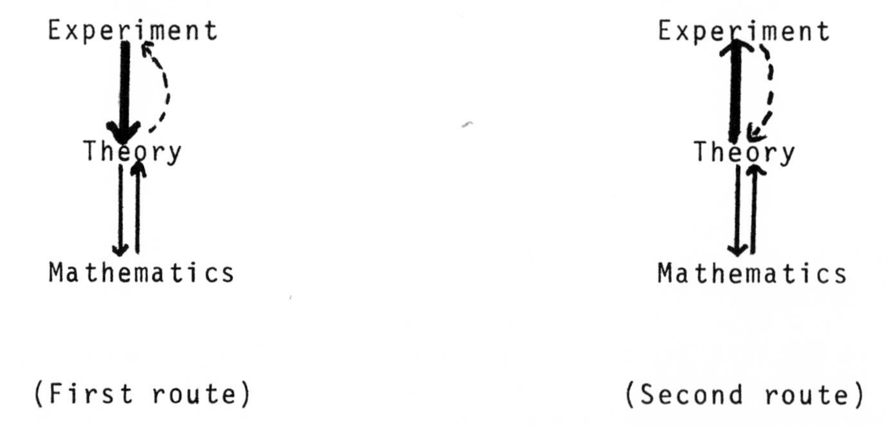

Fig. XII.1 Dirac's two routes to the mathematical concepts that are useful to fundamental physics.

Goldberger: I have some faint recollections of this because, if I remember correctly, you had taken a year off to go to Brookhaven and we shared an office the summer after you had begun thinking about these things. And I remember being impressed with the elegance of it, but was too stupid to realize how profound it was. Any other comments? Steve?

Adler: Well, I thought I would make a comment on the remarks that mathematics will be of increasing importance in doing theoretical physics. I would agree with that. I would like to make a distinction between mathematics at the level of technical mastery and mathematics at the level of mathematical awareness. I think professional mathematicians and those physicists who do mathematical physics — who really prove things rigorously — have to have mathematics at the level of technical mastery at their command. I think many of us operate in a rather different style. I feel that I play the game of quantum field theory by trying to guess the right answer and then worrying later about whether I can derive or prove it. But in order to do that you still need mathematics — but it is mathematics, let us say, at the level of mathematical awareness. You want to make sure that your guesses are ones that will not readily be disproved by somebody who has technical mastery. You have to have enough feeling for the intuitive basis of the theorems so that you don't fall into the more obvious pitfalls. But you do not have to have enough mastery so you can prove everything that you make as a guess.

Weinberg: Perhaps I can comment on that. It seems to me that very often theoretical physicists do get inspired by the beauty of a conception and then they pursue it, very often for a long time, without any support from experiment, and eventually they hope that they can make contact with experimental reality. Sometimes, though, I have heard discussions of this that make me nervous, because they talk about the beauty of the mathematics inspiring the physics. Well, that may be true for some physicists, but it seems to me that there is a slight confusion there. What sometimes inspires the theorist is the beauty of a theoretical principle. The theoretical principles underlying thermodynamics are beautiful. I don't think the particular mathematical formulation in terms of $dS = dQ/T$ inspires any affection.

General relativity is a good example. The idea of building a theory of gravity based on the principle of the equivalence of inertia and gravitation was a beautiful idea and 90 percent of general relativity follows from that idea. I would say that even if the precession of the perihelion of Mercury were not known at that time — at the time of Einstein's formulation of general relativity in 1915–16 — it would be entirely reasonable for a theorist to say, this must be the theory of gravity. But not because the equations were beautiful or because of any analogy with a well-developed branch of mathematics that had been developed one hundred years before, but simply because the physical idea was so grand.

There was another part of Einstein's theory, the restriction of the field equations to second-order differential equations, which you could very well describe as a demand of elegance in the equations. However, this restriction has never seemed to me to be very persuasive, except as an example of what we all do in theoretical physics. That is, when you do not know what else to do, assume the simplest thing possible, so that you have some chance of solving the equations in order to see what happens. And this, as we know, worked magnificently and the tests of general relativity all succeeded. Although this may be a feature of elegance in the theory, I would say that it is precisely that kind of elegance that should leave us discontented and looking for something else.

Yang: I think one point about what Steve said I would beg to differ.

Goldberger: Which Steve?

Yang: Steve Weinberg. It is about the question of whether the mathematics of statistical mechanics is beautiful. The little book by Gibbs is a beautiful book. It is truly remarkable because he knew that what he was doing was contradictory to experimental evidence. He said so in the beginning — I think it is in the introduction to the book. And yet he developed it, and I would say it is both a profound physical insight and a beautiful mathematical development. I would also add another remark. We have heard at this conference repeatedly that thermodynamics was one of the important concerns of Einstein. Let me freely admit that I have not understood what has been repeatedly spoken of here about the Hawking radiation. I believe it is a profound thing but I must admit that I have not really gotten the hang of it. I agree with, I think it is Freeman, that it is a very profound thing and that it is another example of the importance of a deep understanding of statistical mechanics and thermodynamics in the development of physics.

Goldberger: Let me now ask for questions from the floor. Professor Segrè?

E. Segrè (U. California, Berkeley): I have the deepest admiration and respect for the illustrious panel but I would like to point out that five of you are distinguished theoretical physicists. Now, I do not claim to represent experimental physicists, but as an experimentalist I would like to introduce a word of caution. Unless you make some recourse to experiment once in a while, you will make beautiful and more beautiful theories and who knows where they will lead and who knows what contact, if any, they will have with reality. After all, physics purports to a certain extent, to be a science connected with the world as it is, not the way it should be according to Lie groups.

Goldberger: Emilio, as I told you before this began, I think everyone in this panel would agree that physics is an empirical science, and again to recall Frank's search for the theory of gauge theories. He was thinking very much about physics at the time — I mean because that much I do remember.

I. I. Rabi (Columbia U.): I'll follow along on Emilio with a little anecdote that took place in Göttingen where I was a fresh-caught Ph.D. brought to visit the great theorist Max Born. As you understand, this was a time of very great excitement in the world of theoretical physics. And Max Born very calmly told us, Otto Stern and myself, that physics will be over in six months. This was not just a statement taken out of thin air. Our colleague, Professor Dirac, had just done his celebrated theory of the electron, which was a very, very heady achievement, one of the greatest in the history of theoretical physics, and there it was: we had the electron. And he said in six months we will have the proton. And he said by that time physics as we have known it will be over. There will be a lot to do of course, filling in many details, but physics will be over. And along with this piece of reminiscence I recommend a certain modesty.

Segrè: In that connection and about that time Otto Stern explained to a seminar the work he was doing in measuring the magnetic moment of the proton. He had not measured it yet. After his presentation, he passed around a sheet of paper, and he asked those present to write the value that he would find. It was *obvious* to everybody it had to be one nuclear magneton, and Pauli said to Stern: "Why do you waste your time?" Pauli was simply off by a factor of about 3 . . .

Goldberger: I find this attempt to drive a wedge between theorists and experimentalists very unpleasant, and I think we should abandon this particular tack in the discussion. Professor Ne'eman?

Ne'eman: Two short comments on Professor Yang's remarks and then a response to Professor Weinberg's. First to Professor Yang. It is said that beauty is in the eye of the beholder. And as somebody who has spent many years in Lie groups, I must say that I am rather enamored with super groups or the graded Lie algebras. As far as their abstract beauty is concerned, in my eyes they have nothing to be ashamed of, as compared to the Lie groups! In fact they are in the same family. The question of usefulness in physics is really the key question,and there I have some hopes and I am basing much of my present work on super groups. I hope that it will be productive, but the outcome is not really with us, it is with Mother Nature. Now another comment: just an historical one that perhaps Professor Yang is not aware of. I happened to come across the proceedings of a conference — the Kamerlingh Onnes Conference in 1953 — the proceedings are in *Physica,* and there Abraham Pais presented an idea about how to explain what we know now as strangeness. He suggested it was orbital isospin. Pauli, in the audience then asked him if he had even thought of trying to turn isospin into a local gauge principle. The physical importance of such an approach was thus apparently realized by Pauli, but he of course did not try to work it out. The principle allowing generalized phases to be functions of position was invented, of course, by Yang and Mills around that time.

The other remark is to Professor Weinberg. We spent this morning in a long discussion about the question of Einstein's involvement with a unified field theory in his last years, and whether or not it was, as Professor Rabi suspects, the love of mathematics that weakened or diverted him. Some of us thought that it was perhaps not mathematics but the fact that he had managed once before to really do something fantastic of just that nature — the unification he embodied in general relativity. He had worked eight years and it came out and it was a very great thing. He therefore decided to try again a similar long-range effort, but it turned out that this time it was time misspent. Heisenberg did the same, and I have an impression that our colleagues who are jumping from the successful unification of the weak and electromagnetic interactions directly to the grand unification are doing it again. However, there is a Talmudic expression that may apply here: "Whoever still makes prophecies after the destruction of the temple is naive." And so, I have an impression that there is still so much structure to develop between the present level of energies and the 10^{16} or 10^{19} GeV of the Planck mass, and so much additional structure that may come out of the experiments in terms of new types of fundamental fields to appear. For example, who knew that there should be more quarks than four? (Some people thought so, but for wrong reasons, it turns out.) Thus, to jump to the grand unification may turn out to be premature. Physics itself will save us from getting immediately to the final theory, by inventing many things along the way. They will make life more interesting for many years to come.

A. Pais: I would not have brought this up but since Yuval spoke about it, I should tell you the following. I did some work on these questions that you have mentioned immediately and I got a reaction from Pauli. And Pauli said this must be localized. This was the question of some kind of an internal space. And Pauli did write down the formalism and at first thought that he was not quite in agreement with you, Frank, and then I remember a letter I got in which he said, yes, it is exactly the same and the letter ends with a German quotation, "Now we fall in each other's arms." This is written down in letters that are, among other places, in the CERN collection of letters by Pauli. That was totally independent, and it was essentially the same idea. Pauli never published this, but there was one man who once published some lectures by Pauli on it. But it is in the correspondence.

Ne'eman: Frank had already done it too.

Pais: Absolutely.

Goldberger: Professor Weinberg wanted to respond in part to Professor Ne'eman's comment.

Weinberg: I think Yuval's comment that we may have great richness between the scale at which we now work of a few hundred GeV up to the enormous energy that is given to us by the inverse square root of Newton's constant, the Planck mass 10^{19} GeV (or the similar mass that comes from estimates of grand unification — 10^{16} GeV). That's a perfectly reasonable point of view. It may very well be true, and I think anyone who now tries to bridge this gap in one jump is taking a great leap — to make a tautology out of it. But I do not think that this leap is utterly unreasonable either. After all, we know of one huge scale of mass in physics; it is the one provided by Newton's constant; it is 1.22×10^{19} GeV. We know that couplings that are of electromagnetic strength at that energy will become strong at an energy that is smaller than that by a factor like e to the power 137 times some numerical constant that we think is something like $\pi/11$. Thus, the existence of these two grossly different scales does not seem to be absurd or unreasonable. It is something that would naturally happen in a very wide variety of theories.

But maybe it is much more complicated. Maybe there is a whole ladder of physics, and we'll have an endless series of Rochester conferences about it. Maybe to assume simplicity here is an example of what Professor Rabi said about the confidence of the theoretical physicists; I don't think anyone is saying now that physics is almost over, but even to think that we are on the right track may be overconfident.

In a sense, I think the theoretical physicist is like the drunk in the story who has lost a quarter. He has no idea of where he lost it, but he's looking under a lamp post because that is where the light is good. However, I always sympathize with the drunk. Because it is true. He doesn't really know where he lost the quarter, but if he looks for it anywhere else but where the light is good, he is sure not going to find it.

Goldberger: On that profound note I think we'll bring this session to a close. Thank you.

XIII. Closing Ceremony: The State and the Nation

CLOSING CEREMONY: THE STATE AND THE NATION

H. Woolf: The closing evening of the symposium had a uniqueness of its own, graced by the participation of the governor of New Jersey, the Honorable Brendan Byrne and the science advisor to the president, Frank Press. Each brought an important message to the assembled participants and guests, with Dr. Press also acting as President Carter's personal courier in bringing his greetings and warm good wishes to the celebration. So as to preserve the flavor of the centennial's final ceremony, the text that follows reflects its actual unfolding pattern.

M. L. Goldberger: To have lived through a very small portion of the era where the influence of Albert Einstein on science has been so profound has been a great privilege. Personally, I am extremely pleased to be here and I would like to draw attention to the extraordinary job that Harry Woolf has done to bring together this unusual program, this unusual group of people to pay tribute to this one great man of our time. Thank you Harry very much for letting us participate.

H. Woolf: Thank you Dr. Goldberger, very much indeed. I would like to acknowledge the presence of the governor of New Jersey, *Brendon Byrne,* and ask him to step up and bring us the greetings of his office and the state.

Thank you, Harry Woolf and distinguished guests and ladies and gentlemen.

Even so great a mind as Professor Einstein's could not have figured a formula that would have predicted that I would be here in 1979 — as the governor of the state. As most people are celebrating the centennial of Einstein's birth, we reckon back, and those of us who even saw Albert Einstein value and treasure the memory. And I remember as an undergraduate seeing Albert Einstein twice: once crossing the campus on a rainy evening, completely oblivious to the rain; the second time I saw him was at the bicentennial commencement of Princeton University. And I think it is significant that we do conjure up these recollections of ways to identify with this great mind. And so, as governor of a state in which he worked and lived we try to rationalize why he chose New Jersey. Perhaps by coincidence, perhaps because he recognized what a great state this was, the unique opportunity which is found only here for the friction of great minds, the

Harry Woolf (ed.), Some Strangeness in the Proportion: A Centennial Symposium to Celebrate the Achievements of Albert Einstein ISBN 0-201-09924-1

freedom of expression which I find excessive at times but which most people who are admirers of Albert Einstein would find refreshing.

In any event and for whatever reason in the coincidences of history, he is identified with New Jersey and we are proud of that fact. And so proud that the legislature at our recommendation has just voted, and I will sign next week, a bill creating an Einstein Chair at three locations in New Jersey. The first, here at the Institute, the second at the State University, and the third at the New Jersey Institute of Technology. I look forward to signing that legislation on the fourteenth of March, just as I looked forward to being here tonight and enjoying a portion of the evening with so many distinguished world leaders. Thank you.

Woolf: Thank you, Governor Byrne. We are grateful for your thoughts — and your deeds!

We have a man who came here this evening with a double mission — as a courier for the president of the United States and as a distinguished spokesman in his own right. The first of his missions was to bring this message from *President Jimmy Carter:*

March 14, 1979 marks the one-hundredth anniversary of the birth of Albert Einstein — a man who profoundly influenced the shape of science and the course of history.

Albert Einstein set the tone for nearly a century of physics. He took a science that could no longer explain phenomena through the concepts of Newton and greatly expanded its viewpoint. His insights form the basis of much of our twentieth-century comprehension and control of matter and energy. We are still following the path he outlined, and his genius remains a powerful stimulus and guide for future scientific discovery.

But Albert Einstein left his mark on humanity by more than just his brilliant scientific achievements. He will be remembered by all of us for the simplicity of his life, the humility and willingness with which he shared his talents, and the dedication with which he pursued the greatest good of all mankind.

He believed that it was the nature of man to inquire thoroughly and endlessly. In a lecture at Oxford he stated: "The deeper we search, the more we find there is to know, and as long as human life exists, I believe it will always be so." And he believed such search was worthwhile. "The most incomprehensible thing about the world is that it is comprehensible," he said.

Einstein abhorred oppression. He fled from it in his native land and found freedom and friendship in the United States. His gentle nature led him towards pacificism, but he was at the same time deeply committed to the defense of freedom and rights of free people everywhere.

He sought and found order, understanding, and beauty in the universe. He gave his findings freely to all the world. Our tallest tribute to him in this centennial year of his birth is to reaffirm our commitment to build vigorously on his enduring legacy of scientific discovery and social progress.

Woolf: Our special guest and speaker this evening is President Carter's advisor on science and technology, Dr. Frank Press. A graduate in geophysics from Columbia University and later a faculty member at the California Institute of Technology, he has specialized in seismic studies. In 1965, he became chairman of the department of earth and planetary sciences at the Massachusetts Institute of Technology, a post from which he took leave of absence to join the Carter administration on June 1, 1977. Frank Press's government service includes membership on the president's science advisory committee during the Kennedy and Johnson administrations, as well as on committees that were precursors to the Office of Science and Technology during the Ford administration. He also served as a consultant to the Defense and Interior departments, NASA, the Arms Control and Disarmament agency and the Agency for International Development and was a member of the United States delegation to the nuclear test ban negotiations in Geneva and Moscow in the early 1960s.

Among other achievements he is credited with having helped to secure more funds in this year's federal budget for basic research — a matter of considerable direct interest to most of this audience, and with having encouraged a major administrative study of the disturbing decline in American technological innovation and productivity. We are honored by his presence here tonight at the Institute for Advanced Study on this final evening of the Einstein Centennial Symposium. Ladies and gentlemen, Dr. Frank Press.

It is a privilege for me to participate with this distinguished international group in the celebration of the Einstein Centennial. Somewhere Albert Einstein may be surveying this scene, smiling at what he would perceive as the pretense of this occasion. But the one-hundredth anniversary of the birth of such an individual *is* an occasion to celebrate. And no place could be more appropriate to do so than at the Institute for Advanced Study, a place where Einstein enjoyed so many years of peace, scientific work, and friendship.

Perhaps as much as a time to celebrate, it is a time to reflect on the legacy of this great scientist, and to contemplate on where that legacy may be leading us.

I will not attempt to review Einstein's scientific accomplishments. This has been done in your outstanding symposium. I doubt if a worthier assemblage of Einstein's colleagues and disciples has ever been convened to pay homage to the man, and to discuss his influence on science.

I would like to speak to the broader spirit of Einstein's ideas, to how they pervade our thinking related to science today, and to the role of science and technology in our society.

In reviewing those ideas one thing impresses me most: Einstein's inherent optimism. He was convinced that there was order in a seemingly chaotic universe. And he believed us capable of finding it. He felt compelled to pursue it personally. This was central to his life. But he also knew it was a search central to all human existence and to human progress. In short, as far as humanity was concerned, research was an act of life, and life, he believed, was better for the truths and knowledge it brought.

Einstein also believed we could bring order and harmony to the human community — to the family of nations. And he devoted much of his life to this cause. He saw science and scientists playing key roles in this. He stressed their internationalism, and he particularly felt that this country — his adopted homeland — was a stronghold of that internationalism, and thus a place where we would strive to abolish war, promote world peace, and develop our science and technology to serve people everywhere.

How well are we doing this? What steps are we taking toward world peace? What are we doing to advance our science and technology? And how are we using them to advance mankind? Is the United States, the America that Einstein placed such hope in, pursuing the best course in its science and technology policy?

Let me attempt some answers to these important questions.

Perhaps I should begin with a few general comments about how this country feels about science. I think this is important, because our science policy is public policy, ultimately based on the understanding and support of our people. It is therefore gratifying to know that science is held in high esteem by the public. Several recent surveys have indicated that people still consider science as one of the most important forces in our society. They credit science for past success. They look to science as a principal factor for our future progress. This feeling, I believe, is not confined to the United States. It exists throughout many parts of the world.

I take heart in this, because in this period of much general skepticism and even cynicism, confidence in science is all the more important. It will allow scientists and engineers more of the support necessary to confront some of our most pressing problems. And in doing this it may help to restore some of the self-confidence, inner strength, and faith in the future our society seems to need so badly today.

But people not only have continued confidence in science, they are increasingly intrigued by its intellectual challenge, by the search itself. They are fascinated with the new areas it is probing, and with the probing questions it is raising. Many of these grip the emotions as well as the mind. This would have delighted Einstein, who once said, "Science ennobles anyone who is engaged in it, whether a scholar or merely a student." He would have also been pleased that we still ardently support fundamental scientific research — not only in hope that its results might someday provide practical applications, but because it could reveal some universal truth. Science fills a basic human need.

That scientific research is central to our survival today, is becoming increasingly evident. We are a knowledge-intensive society. Everything in our lives calls for, and depends on, an increase in our understanding of nature, of life, of ourselves, and of our society. We could no more turn off or reduce our pursuit of such understanding than we could stop breathing. Together, those of us celebrating this centennial could list a thousand reasons why this is so. But let me just cite a few that are on many minds today.

Perhaps the foremost is the urgent necessity to find ways to sustain and provide a decent life for the more than six billion people who will inhabit the earth by the time we enter the twenty-first century. This can no longer be done by exploiting the resources of the earth in the ways we have. We must exploit human intelligence to raise substantially our level of efficiency in dealing with such resources. We have barely begun this process. Nature holds an untold number of secrets as to how this can be done. And once having discovered them, further research and engineering skills must be developed to apply what we have learned.

A similar expansion of knowledge is needed to understand more fully life itself, its bases and processes. As we probe deeper into the molecular structure of the cell and work with recombinant DNA, some feel we are at the threshold of a revolution in our biological knowledge. And it could be one that would radically alter our approaches to human health, to the production of our food, and to our relationship with our environment.

Surely we must pursue these new paths. Currently we are plagued by many environmental uncertainties. And in a world where science-based activities have begun to have a pervasive and often unanticipated impact on ourselves and our world, more and better science is essential.

A third broad argument for the expansion of scientific knowledge lies in the fact that we are forging a global community that must rest on new economic and human relationships. New technologies, new industries, new products and services, new opportunities, and potentially new means of making all this available to more people, have raised the expectations and aspirations of people everywhere. Left unfilled, particularly during the next two decades when more than 1 billion young people arrive at working age and reach for opportunities for a decent life, these hopes become fuel for incredible political and social unrest and upheaval.

To avoid this it becomes imperative to make our world work. And this must be based on an underpinning of new knowledge.

A final broad concern for science and scientists — one very close to Albert Einstein — relates to the matter of arms control and disarmament. World peace depends on this. Our civilization can afford neither the risk nor the cost of an arms race out of control. Restraints, and eventually a reversal of arming, must succeed — starting with a limitation on strategic nuclear arms. This is a most vital concern of our times.

Einstein, were he with us today, would have been appalled that the world now spends more than $350 billion a year on arms and more than $30 billion on their research and development. These military expenditures equal the annual income of almost 2 billion people in the 36 poorest nations of the world. If there is anything that the science community can do to honor the memory of Albert Einstein today, it is to support efforts towards arms control; and specifically at this time, the approval and implementation of the SALT II agreements. Any journey for world peace must take this important step.

Having touched so broadly on these major concerns of science, let me sharpen my focus. What are some of the things this country — and this administration — are doing to apply their advances to our national and global concerns?

First let me turn to the matter of the nation's support of its science and technology. Earlier I mentioned the attitude of the public related to this. Its confidence in and hopes for science and technology are reflected, perhaps magnified, by those of the president. We are fortunate, I believe, to have a president who recognizes the vital roles that science and technology play in our society, and particularly who views basic research as an important investment in the nation's future.

This was not the case in previous years. As a result, our federal support of research eroded seriously during the period of 1967 through 1976. We have now reversed that declining support. And I am pleased to say that over the two years of the Carter administration, the average annual increase in the support of basic research will amount to more than 12 percent.

What influenced the president, in a time when he is placing such emphasis on fiscal restraint, to increase the budget for basic research? What other conditions and considerations were on his mind? It would seem that in a budget that was "lean and austere," and in a time when emphasis is on immediate economic improvements, there would be little or no room for increased support for basic research. Some might argue that the support of such research, with its long-term pay-off, could easily be sacrificed to meet our more immediate social, economic, and political needs.

But that was not the president's thinking. It is not the thinking of the leader of a great nation, acting responsibly and wisely in the best interests of the nation. Nor is it the thinking of a statesman who would consider the interests of other peoples and future generations.

Let me articulate a few of the factors considered by the president in viewing the research and development budget as an investment in the future.

First, research and development are essential to the solution of many of the problems that beset our society. Without advances in research and development:

> We cannot gain the scientific and technical information to set intelligent health, safety, and environmental regulations.
>
> We cannot produce goods in a safe and acceptable way.
>
> We cannot gain the knowledge of diseases essential to their prevention or cure, nor broadly improve health care.

We cannot make the transition to the new energy sources we will need.

We cannot provide more efficient and safer modes of transportation.

We cannot conserve our natural resources, or find new sources, or substitutes.

We cannot understand, or prepare ourselves to live with the consequences of, changing climate and weather and the hazards of natural disaster.

To deal effectively with all or any of these things requires vast extensions of our scientific knowledge and technological capabilities, and that requires continued research and development.

We also must face the fact that we deal with all these things within the framework of our economic system, which is challenged in many ways. It is essential that we in the United States increase our economic growth, that we improve our productivity, control inflation, and raise the level of our industrial innovation. All this is necessary for us to maintain our economic competitiveness and leadership in the world. Our economic strength is essential to that of the rest of the world — to global economic growth and stability.

The president's economic advisers believe very strongly that over the long run research and development and technological innovation will be major factors in providing that economic strength. And I share that view. We and the other members of the administration — the department and agency heads with science and technology-related programs — will be working very closely with the Congress to assure that the nation's research and development programs receive the necessary support.

Returning to the matter of research — more specifically, basic research — we are more aware than ever of the predominant role our universities and colleges play in this vital enterprise. We are also aware of the problems they face in relation to declining enrollments, increasing costs, aging facilities and equipment, and the responsibilities of reconciling the need for opportunities for young faculty and researchers with the rights and privileges of older ones. On top of this there has been an increasing burden associated with the universities' responsibilities to government — one that grew from the cost and complexities of accountability for public funds.

These are difficult problems, and much effort has been going into new approaches to resolve them. A National Commission on Research, composed of members of the university and research communities, has been established to explore such approaches. We look forward to their recommendations and will work closely with them.

Cost-sharing research facilities are being established. New efforts to improve the management of research proposals are being tried, ways to cut paperwork, reduce review and approval time, and generally improve administrative procedures. And we are working with increased research budgets and other ideas to provide more opportunities for your scientists, and thus keep American science young and vital.

There is another matter that is receiving growing attention. It is the need to establish a better linkage — an increased flow of ideas and closer cooperation — between our universities and industry. These two sources of America's strength need each other. And the country needs the additional knowledge, know-how and output that could result by their working together better. We believe this can be done without compromising the independence and quality of university research, or industry's needs to remain competitive, seek essential profits, and carry out its future plans. We think the government can play a constructive role as catalyst in bringing these two forces together, without impinging on the virtues of either. We are looking into new ways to do this.

I mentioned before the importance of industrial innovation. It is a subject of major concern these days. And we know that while research and development are important contributors to it, innovation is a complex process involving various economic incentives, the availability of capital, industry structure, patent policy, regulations, and many other factors that encourage or discourage the inventor, entrepreneur, and business leader. Last year the president initiated a major cabinet-level study of this complex subject. The results of that study, involving some thirty federal agencies and scores of advisory groups and other contributors from the private and public sectors, will be on the president's desk by late spring. We believe it will contain important recommendations for presidential actions and legislation to stimulate innovation and help strengthen the economy.

In any discussion of science and technology we invariably must look beyond our own borders. For we know the universality of science and the global need for technology. Scientific knowledge has long been viewed as a common heritage of humanity. There have even been cases when warring nations allowed scientists to move freely across their borders, and scientific expeditions were given safe passage across the seas.

Today we recognize science and technology as international enterprises for several reasons. International cooperation is essential to conduct the global experiments necessary to understand the earth's climate, its atmosphere and oceans, its geologic structure and forces. We need to understand their past to project their future. We need to understand how these global forces may impact on us, and how our activities bear on them. Therefore we are involved in such large international programs as the Global Atmospheric Research Program, and the International Decade of Ocean Exploration.

In addition, we place great importance on our bilateral and cooperative science and technology agreements with nations around the world. The most recent of these are the newly signed agreements with the Peoples Republic of China, which cover such areas as energy, agriculture, health, and space. We have also just recently renewed our joint working agreements in science and technology with the USSR.

A major incentive for cooperative research with other nations is the large and growing cost of some research. The sharing of such costs among two or more nations is quite helpful and could mean quicker research results in some important fields. Last year we concluded such agreements in energy research involving nuclear fusion and coal liquifaction with Japan and Germany and other countries. And we plan similar cooperative programs on other important research projects, such as experimental drilling on the offshore continental margins.

Perhaps the most important reason for international cooperation in science and technology is the need to foster the development of the poorer countries of the world. This will be the subject of a major United Nations conference this summer. I will not cite all the humanitarian reasons for development. You know them only too well. And they constitute major reasons why this administration is so ardently pursuing programs to deal with the issues of world health and world hunger. But the matter of development today is more than a matter of half of humanity reaching out to help the other half — to heal the sick, feed the hungry, and provide some hope for the future. Development has become a matter of survival for all. The One World of the yesterday's idealists is the One World of today's realists. Economic, social, and political forces bind us together as never before. And there is growing agreement that the future of the developed world is inevitably tied to the future of the developing world. For many decades to come the largest part of humanity will reside in the poorest nations. The rest of us must work with these people to help them sustain themselves and eventually become productive members of the human family. Science and technology can make major contributions toward this end. Used cooperatively and wisely it can make the poor productive — which is the principal aim of development. And as this happens, as these people gain the ability to feed themselves, improve their health, and attain the education and skills to raise their standards of living, their growth will shift from being a drain on the global economy to a source of new development, new markets, and new progress for all. If there is to be a new economic order, as many are calling for, that is what it should be.

But the path to all this is still long and arduous, and close cooperation in science and technology will be essential.

One of the United States' proposals for improving such cooperation lies in the idea of an institute for technological cooperation. This is an organization first proposed by the president in a speech to the Venezuelan Congress last spring. Since then comprehensive plans have been made to develop such an institute as a major part of a revitalized U.S. foreign assistance program.

The primary purpose of such an institute would be to improve the availability and application of technology to solve the problems that impede development, and to expand the knowledge and skills of developing peoples. The institute would work with all developing countries, including the middle-income countries, in collaborative programs. It would address problems of mutual concern, including the pressing needs of nutrition, health, and education, and the global problems of energy, resource development, and environmental protection.

Legislation to establish this institute for technological cooperation has been introduced in the Congress. We hope it will receive the full support of the science community and the general public.

No comments at an Einstein Centennial would be complete without some thought as to how Einstein might have viewed our ventures into space. His intellectual penetration of that frontier greatly preceded and influenced today's thrust into the universe.

I believe he would have been pleased that through our space science programs we are embarked on a great exploration of the planets and the cosmos.

> We have explored Mars and Venus.
>
> We have initiated the complex and difficult Galileo mission to probe deep into the atmosphere of Jupiter.
>
> We will be placing in orbit in a few years a Space Telescope — a permanent orbital astronomical research facility — that should allow us to study celestial phenomena that range back in time and out in space to the beginning and the edge of the universe.
>
> We will be initiating the International Solar Polar Mission — a mission that will provide new knowledge of the sun's structure and activities, and a new understanding of how they shape earth's space environment and affect its biosphere.

Many other scientific investigations in space have been initiated or planned, all to extend our reach out into the universe, physically and intellectually.

I believe that Albert Einstein would also have been pleased that we are using space to help fulfill the needs of humanity here on earth. Space has become one of our greatest cooperative ventures for extending our knowledge of the earth and man's activities and influences on it. From our platforms in space, and through ever more sophisticated instrumentation, we are gaining the capability to study the earth's resources, its geologic features and environment, and to investigate and evaluate our activities such as agriculture and forestry, and the development and impact of our industrial growth. Space satellites also provide us remarkable means for advancing our communications, navigation, educational systems, and other means of aiding and uniting mankind.

We are going to pursue all of these great benefits of science and technology. We are going to advance our intellectual achievement and capacity, to extend our physical capabilities, and to extend our ability to improve the lives of people everywhere.

To do this we need not only continue in the light of Einstein's scientific legacy, but heed his words concerning truth and courage. And I would like to close with a few of those words:

> Times such as ours have always bred defeatism and despair. But there remains, nonetheless, some few among us who believe man has within him the capacity to meet and overcome even the greatest challenges of his time. If we want to avoid defeat, we must wish to know the truth and be courageous enough to act upon it. If we get to know the truth and have courage, we need not despair.

Woolf: Thank you, Dr. Press, for the wide-ranging issues you have placed before us and the thoughtfulness of your comments. We are comforted by the knowledge that sensitivity and intelligence such as yours are present in high places in this government. May I, before closing, also acknowledge the welcome presence of the counsul general from Israel, Paul Cador, and the presence of our friends and neighbors, to represent Princeton University, the institution that gave us hospitality when we first came here almost fifty years ago, President and Mrs. William Bowen. Thank you. Goodnight.

XIV. Epilogue

30. REMINISCENSES OF EINSTEIN

Ilse Rosenthal-Schneider

Whoever listened to Einstein lecturing, speaking in a discussion, or only briefly answering a question will have realized the deep and lasting impact of his words. Not only did you learn to see a problem and its possible or impossible solutions in a completely new light; you felt that your own thinking was strongly influenced by his methodical approach, causing you to scrutinize fundamental concepts and deep-rooted assumptions that were perhaps not permissible at all, whereas you had taken it for granted that they were correct, acceptable, and applicable to the purpose for which you had used them.

Quite apart from all the essential features that made Einstein's explanations and especially his exemplifying and illustrating analogies so memorable, it was that indefinable quality, his sense of humor, that gave them their unique flavor.

This sense of humor could be observed whenever you listened to him. The most consequential, intricate ideas were often introduced in such a way as to help even the dullest student to understand at least the gist of what was meant.

Still more impressive than in university lectures or public addresses was Einstein's manner of expressing his thoughts when he discussed anything with you personally. Each time an unforgettable experience! And it was my good fortune that I had this experience twice a week when Einstein was lecturing at the university in Berlin. When he accepted the call to the Akademie Der Wissenschaften in Berlin he was not obliged to give any lectures or to have anything to do with examinations. But fortunately he did lecture to students on special relativity theory as well as on the general theory. After the lectures students were allowed to ask questions; and this was how I came to know him. Soon after he had started lecturing in Berlin I found him sitting in the tram by which I usually went to and from the university. He indicated to me to sit down next to him. So it happened that I had quite regularly long tram rides with him to and from the university and often long periods of waiting for the tram at the tram stop — to me they never seemed long!

You may like to know what we talked about. About problems of theoretical physics — talking shop, so to speak — mainly, but by no means exclusively. We discussed many epistemological problems too and discussed some general ideas on cultural values that we were interested in. It was well known that Einstein was easily approachable. But I was amazed that he, with his creative, immensely superior and powerful mind, would talk to me about his ideas and conjectures. And more

Harry Woolf (ed.), Some Strangeness in the Proportion: A Centennial Symposium to Celebrate the Achievements of Albert Einstein
ISBN 0-201-09924-1

than this: that he could be at all interested in knowing what I was working on at the time of our tram trips. This was mainly the relation of his theories to philosophy and, in particular, to the philosophy of Kant. At many universities such discussions became controversial themes of the 1920s. Often these current discussions became more than controversial. It was frequently claimed that Einstein's theories had "refuted" Newton's and that therefore Kant's ideas of space and time were refuted too, proved to be "wrong" because they are based on Newton's physics.

Such superficial judgement may still be found in publications by self-styled philosophers or scientists. Einstein himself was always an unbiased judge of his predecessors, whose achievements he would not only acknowledge but admire. In his unpretentiousness he used to speak of "those who made my work possible." General relativity was created when the space-time continuum had received curvature. As far as Newton is concerned, a few lines may be quoted from Einstein's letter to Sommerfeld of November 28, 1915: "now the marvelous thing which I experienced was the fact that not only Newton's theory results as first approximation but also the perihelion motion of Mercury as second approximation. . ."

Einstein was also most interested in the epistemological implications of his theories that had started the then current heated controversy with all its misinterpretations.

Quite unexpectedly I received a postcard from him to call on him if I liked to and had the time to read or peruse with him a paper on the controversial subject. Of course, I accepted this invitation. This was a very interesting and enjoyable discussion not only for the brilliant ideas he uttered in his simple and natural manner, which enabled me to answer his questions and to express my own view, but because of the humorous way in which he presented these ideas.

He seemed to create his thoughts while he was talking, then giving them a clear and precise formulation, and finally adding the humorous flavor with a little playfully mischievous smile.

This smile was characteristic of him when he intended to tease me, as for instance with the remark: "Isn't the whole of philosophy like writing in honey? It looks wonderful, at first sight, but if you look again it is all gone, only the plane surface, the pulp, is left." He himself, however, had not only studied the writings of the philosophers, but he liked to discuss such topics as Spinoza's metaphysics in the Ethices as well as the epistemology of Hume or Kant — and I loved that too.

When we had discussed some intricate problem concerning Kant's views on the general universal laws of nature in their relation to geometry — which, by the way, are very similar to Einstein's — the characteristic smile made me expect another beautiful parallel and there it was: "Kant is a sort of highway with lots and lots of milestones. Then all the little dogs turn up and each deposits its contribution at the milestones." Pretending to feel indignation I said: "But, what a comparison!" With his loud, boyish laughter he remarked: "But what will you have? Your Kant is the highway after all, and that is there to stay."

Einstein's interest in Kant's treatment of the space-time problem was obvious to me when I received his first card with the invitation to read the paper on relativity and epistemology with him. Later in 1920 I received a note reading: "When will your relativized Kant come out? I am looking forward to it." I said that it may have turned out to be a Kantianized Einstein.

In the many wonderful letters I received from Princeton, which I treasure as most precious possessions, there is so much important information that I cannot even start enumerating the most important points. Many of them will be published very soon, I hope, in a book *Reality and Scientific Truth — Discussions with Einstein, von Laue, and Planck*. This book contains besides many of Einstein's letters and some of Planck's also many from Einstein's close friend Max von Laue.

They all deal with problems about which I had ventured to put forward questions and an opinion. And I am most grateful for having received a wealth of information through their letters.

To mention a very special occasion: Einstein was discussing some problems with me in his study when he suddenly interrupted his explanation and handed me a cable from the windowsill with the words, "This may interest you." It was the news from Eddington confirming the deviation of light rays near the sun that had been observed during the eclipse. I exclaimed enthusiastically, "How wonderful, this is almost what you calculated." He was quite unperturbed. "I knew that the theory was correct. Did you doubt it?" When I said, "Of course not, but what would you have said if there had not been such a confirmation?" He retorted, "Then I would have to be sorry for dear God. The theory is correct."

The many decades over which my recollections of personal contact with Einstein extend have not dimmed my memory. All my reminiscences are as vivid as if the happenings occurred only yesterday.

31. SOME REMINISCENCES ABOUT EINSTEIN'S VISITS TO LEIDEN

George E. Uhlenbeck

In the nineteen twenties Einstein came regularly for a month or so during the fall to the University of Leiden in Holland as a kind of visiting professor. It is well known that Einstein and Ehrenfest were very close friends, and also that Einstein had a great admiration for Lorentz, who was retired but still came to Leiden every Monday for his famous lecture. So perhaps these were the reasons why Einstein liked to come, although I suspect that he also did not mind to get away from Berlin for a while! Anyway, in Leiden Einstein seemed quite at home. He worked in the Institute for Theoretical Physics (at that time called the Leeskamer Bosscha), talking mainly with Ehrenfest and/or some visitors. He may have given once in a while a lecture at the famous Wednesday colloquium, but I do not remember them. But he was always present, and he often took part in the lively discussions. I also remember vividly the way he walked around in the Institute in his woolen socks (he always put off his shoes) and woolen sweater and always with his pipe! The pipe was an *enormous* privilege. Ehrenfest was fanatically against smoking in the Institute, and he made only an exception for Einstein. (Lorentz, who also had an office at the Institute did not smoke, thank God, so that Ehrenfest was not put into quandary!)

I got to know Einstein for the first time in the fall of 1925. The reason why I had not met him earlier was because of my job in Rome, where I tutored the son of the Dutch ambassador during the academic year. This ended in June 1925, and then began the wonderful summer when the late Sam Goudsmit and I worked together and which led to the electron spin hypothesis. I have described this period in an article in *Physics Today*[1] so let me just try to recall the attitude of Einstein in the fall when I talked with him about our idea. It was quite different from the reaction of Ehrenfest and Lorentz. Ehrenfest was doubtful but encouraging, Lorentz was interested but skeptical, while Einstein's attitude I can only characterize by the words, "Why not?" He was clearly not deeply interested in the details of the theory of atomic spectra. Still he gave me the essential hint how to understand and calculate the spin-orbit coupling, a problem with which Sam and I had struggled after having received a mysterious letter of Heisenberg. I have described all this in my article, and as I said there, I believe that this argument of Einstein really convinced Bohr of our idea.

The next year — 1926 — was of course the year of the wave mechanics, and in the fall Ehrenfest and I worked very hard to understand how to reconcile the so-called new statistical

Harry Woolf (ed.), Some Strangeness in the Proportion: A Centennial Symposium to Celebrate the Achievements of Albert Einstein ISBN 0-201-09924-1

methods of Bose and Fermi with the classical theory of Boltzmann and Gibbs. I am almost sure that in this time Einstein was again in Leiden and that we talked with him about these questions. However, I have no clear memory about it, so I may be wrong. Still, I believe that as a result of these discussions I started to study in the beginning of 1927 the paper of Einstein in which he claims that an ideal gas like helium would show a condensation phenomena as a consequence of the Bose statistics.[2] This seemed very paradoxical, and to my surprise I came to the conclusion that Einstein was wrong! In my opinion his error was to replace the partition function over the discrete energy levels of the particles by an integral that was not allowed near the condensation point. The exact formula did not show any kind of singularity, so that, for instance, the equation of state would be quite smooth. I was of course quite excited, especially because Ehrenfest was convinced that I was right when I told him my reasoning. I know that he lectured about it at a few places and also that he wrote a letter to Einstein in which at the end he quoted some phrases of Schiller, I think, which I do not quite remember except that the sense was: If the masters fail, then the young people must do the work! I hope that Martin Klein can dig up the letter.[3]

In retrospect I think one should say that, although I was technically right, Einstein had somehow intuitively understood that a phase transition like the Bose condensation can occur only for large systems and is a kind of limit property. This should justify the replacement of sums by integrals. Of course Einstein did not prove this! In fact I know from a later conversation with him, in the thirties, that he agreed with my criticism. The insight that only in the so-called bulk or thermodynamic limit the partition function of a system can show singularities like phase transitions became clear and generally accepted in the late thirties. My late student Boris Kahn played an important role in this clarification process.

As a result Einstein's suggestion of the condensation of an ideal Bose gas became again in fashion, especially through the work of F. London. Einstein had mentioned in the same paper that perhaps the phenomenon had something to do with liquid helium. Again he somehow was at least partially right. Of course, the Bose condensation has nothing to do with the ordinary liquefaction of helium, but the remarkable low-temperature, liquid-phase He-II has surely something to do with the Bose condensation, although there are many unsolved problems left. Einstein could of course not have known about He-II (it was discovered by Keesom in 1928), but somehow he smelled that there might be a connection!

NOTES

1. Physics Today **29,** 43 (1976).
2. Berl. Ber. p. 3 (1925).
3. Martin Klein has found the letter of Ehrenfest. It is in the form of a manuscript to the *Z. f. Physik* and it is very witty, although difficult to read. I have tried to translate the beginning to give some idea of the style. The title is:

   Does the Bose–Einstein Statistics Lead for Ideal Gases to
   a Condensation in the Degenerate State?
   G. E. Uhlenbeck and P. Ehrenfest (Leiden)

   Motto: When kings are building, then the sewer cleaners
   have work to do.

   Summary: No!

   In this way it goes on. There is also an answer by Einstein, who appreciated the jokes, but who was at this point not yet convinced.

32. ON PLAYING WITH SCIENTISTS: REMARKS AT THE EINSTEIN CENTENNIAL CELEBRATION CONCERT BY THE JUILLIARD QUARTET

Robert Mann

The first theoretical physicist with a passion for chamber music to interrupt my comfortable isolation as a profession-oriented music student was not Albert Einstein but a young Columbia University professor, Hy Goldsmith. One morning in 1939, while slowly saturating an ulcer with a container of milk in the Juilliard School student lounge, I struck up a conversation with this slightly older man, obviously not a music student. He inquired whether I enjoyed playing string quartets. I responded with a sweeping critique of what was wrong with most current chamber music performances. He invited me to join some musicians at his apartment for an informal session of quartet readings, adding that I certainly would have ample opportunity to demonstrate the proper approach to such music. At the appointed time I knocked at his door; said door was opened by Isaac Stern. Moments later that same door produced Gregor Piatigorsky, and the finishing touch so exquisitely wrought by physicist Hy Goldsmith was no less a violist than William Primrose. As we were about to play, Hy whispered in a gently, wry, playful tone, "Now Robert, show me what chamber music is really about." From that moment my concept of a physicist's view of chamber music was traumatically resolved into the equation: $E = CM$ to the fourth order. That is, emotion equals chamber music acted upon by four participants: a string quartet.

A new element joined that equation when, as I passed through Chicago on an army furlough in 1944, Hy invited Leo Szilard to one of his chamber music evenings. Dr. Szilard read a Sunday edition of *The New York Times* from cover to cover in a somewhat noisy accompaniment to Haydn, Mozart, Beethoven, and Schubert. Aware of my reaction to his newsgathering during the proceedings, Dr. Szilard insisted that he heard and digested every note. Thus my new physicist–chamber music equation now appeared as: $E = CM$ to the fourth order $\div$ the well-known universal axiom, VLNYTR (for those of you unaware of this axiom I proffer the acronym velocity of light *New York Times* reading.)

None of these brief encounters prepared me for the incandescent perihelion of my chamber music–physicist orbit: The Juilliard Quartet's evening in 1952 spent with Dr. Albert Einstein here in Princeton following a Sunday afternoon concert at the university. Aware of his lifelong devotion to chamber music both as a listener and participant, we offered to play for him at his home.

Harry Woolf (ed.), Some Strangeness in the Proportion: A Centennial Symposium to Celebrate the Achievements of Albert Einstein ISBN 0-201-09924-1

He graciously accepted. A fervent hope inspired a devious plan that necessitated our bringing along an extra viola and the music of Mozart's two viola quintets. He received us in a genial mood and an extremely comfortable attire. As we played both Beethoven and Bartok, he listened a room removed (not wishing for a visual distraction), and, while he did not seem to share the general aversion to twentieth-century music that seemed to infect his fellow physicists, it was evident that Dr. Einstein's heart belong to the earlier music.

After the Bartok we launched our surprise attack. Producing the Mozart score and the extra viola, we said, "It would give us great joy to make music with you." He protested that he had not played for years due to a hand injury. With the insensitive exuberance of youth we persisted, and he, the true scientist recognizing an irresistable force versus immovable object contretemps gently acceded to the inevitable. Our second violinist handed over his violin and picked up the extra viola. Dr. Einstein without hesitation chose the great, brooding G minor quintet.

Supersensitive to the great man's remark that he had not played in seven years, we immediately from the very first sounds regrouped around Dr. Einstein in complete rapport with his deliberate but purposeful momentum; basking in the warm glow of making Mozart's music with this disarmingly unassuming human being. Slow, slower, and slowest crept each successive movement. In direct reaction to the slow tempi did the intensity of our manic happiness grow. Dr. Einstein hardly referred to the notes on the musical score (notes in his mind were never to be forgotten), and, while his out-of-practice hands were fragile, his coordination, sense of pitch, and concentration were awesome. The mood of the Juilliard Quartet as we finished the Mozart was beatific. Time was late; so we gathered instruments and reluctantly bid goodbye.

At the door he, obviously satisfied that we had coerced him to join us, seemed to want to say more than just a thank you for the music. Finally he remarked, "I love America and American musicians. They are wonderful. My only complaint is that like the tempo of life here in this country, American musicians tend to play their music too fast." I replied, "You know, Dr. Einstein, I am sorry to report that the Juilliard Quartet is also accused of this indiscretion." He was silent — one could sense his thoughts touching on the evening's Mozart, so unhurried, so lovingly respectful of his own tempo desires. "You are criticized for playing too fast?" He shook his head and said, "I don't understand why."

Needless to add, the Juilliard Quartet with our friend John Graham can think of no better way to honor this man and remember that long ago wonderful evening than by playing the same Mozart quintet for you.

We hope, Albert, that you will approve our tempi.

INDEX*

INDEX*

*Index by person and institution.
Authors are not indexed.
Footnotes are indicated by parentheses.
Einstein and his theories are not indexed.